AF581885

Postharvest Management of Fruits and Vegetables

Postharvest Management of Fruits and Vegetables

Editors

Eleni Tsantili
Jinhe Bai

MDPI • Basel • Beijing • Wuhan • Barcelona • Belgrade • Manchester • Tokyo • Cluj • Tianjin

Editors
Eleni Tsantili
Agricultural University of Athens (AUA)
Greece

Jinhe Bai
Horticultural Research Laboratory
USA

Editorial Office
MDPI
St. Alban-Anlage 66
4052 Basel, Switzerland

This is a reprint of articles from the Special Issue published online in the open access journal *Foods* (ISSN 2304-8158) (available at: https://www.mdpi.com/journal/foods/special_issues/Postharvest_Management_of_Fruits_and_Vegetables).

For citation purposes, cite each article independently as indicated on the article page online and as indicated below:

LastName, A.A.; LastName, B.B.; LastName, C.C. Article Title. *Journal Name* **Year**, *Volume Number*, Page Range.

ISBN 978-3-0365-3535-7 (Hbk)
ISBN 978-3-0365-3536-4 (PDF)

Contents

About the Editors

Eleni Tsantili is Professor of Pomology at the Agricultural University of Athens (AUA), Greece. She was awarded a diploma of Agriculture from AUA in 1981 and PhD degree from Wye College, University of London, UK, in 1988, for postharvest research. She conducted a sabbatical for research purposes and visited the Postharvest Laboratory, Department of Horticulture at Cornell University, USA. Since 2015, she has served as Full Professor of Pomology and Director of the Postharvest Laboratory. She teaches theoretical and lab courses of General Pomology, Temperate and Subtropical Trees, and Postharvest Fruit Biology and Technology to graduate and postgraduate students. She specializes in the physiology, biochemistry and technology of fruits during their postharvest life. She has participated in projects in England, USA and Greece. Her current interests include the development of defense responses to abiotic conditions, growth regulators, pre- and postharvest treatments, fruit storage in controlled or modified atmosphere, fruit quality and nutritional value.

Jinhe Bai, PhD, is a research chemist in the United States Department of Agriculture, Agricultural Research Service, Horticultural Research Laboratory at Fort Pierce, Florida, USA. He completed his BSc and MSc in Horticulture at Shanxi Agriculture University and Northwest A & F University (both in China) and PhD at Osaka Prefecture University (Japan). He has worked in the horticultural and food industries, research institutes, and universities before his present position. His main research interest concerns preservation of the postharvest quality of fruits and vegetables. Specific areas of expertise include controlled-atmosphere storage, modified atmosphere packaging, edible coating technologies, and the discovery of how internal and environmental factors influence metabolism and further impact the flavor and nutritional quality of horticultural crops. His research efforts are supported by well-equipped laboratory equipment and state-of-the-art technologies, such as GC–MS and LC–MS/MS, used to identify and quantify volatile and nonvolatile metabolites.

MDPI

Article

Edible Coatings from *Opuntia ficus-indica* Cladodes Alongside Chitosan on Quality and Antioxidants in Cherries during Storage

Miltiadis V. Christopoulos [1,*], Dimitrios Gkatzos [1], Mina Kafkaletou [2], Jinhe Bai [3], Dimitrios Fanourakis [4], Giorgos Tsaniklidis [5] and Eleni Tsantili [2,*]

1 Institute of Technology of Agricultural Product, Hellenic Agricultural Organization-DEMETER, S. Venizelou 1 Str., Lycovrissi, 14123 Attica, Greece; d.gkatzos@gmail.com
2 Laboratory of Pomology, Department of Crop Science, Agricultural University of Athens, Iera Odos 75, 11855 Athens, Greece; mkafkaletou@gmail.com
3 Horticultural Research Laboratory, USDA-ARS, 2001 S. Rock Rd., Ft Pierce, FL 34945, USA; jinhe.bai@usda.gov
4 Laboratory of Quality and Safety of Agricultural Products, Landscape and Environment, Department of Agriculture, School of Agricultural Sciences, Hellenic Mediterranean University, Estavromenos, 71004 Heraklion, Greece; dimitrios.fanourakis82@gmail.com
5 Institute of Olive Tree, Subtropical Plants and Viticulture, Hellenic Agricultural Organization, ELGO-Dimitra, P.O. Box 2228, 71003 Heraklion, Greece; giorgos.tsaniklidis@gmail.com
* Correspondence: miltchrist@yahoo.gr (M.V.C.); etsantili@aua.gr (E.T.); Tel.: +30-694-697-4146 (M.V.C.); +30-693-422-6533 (E.T.)

Abstract: The aim of this work was to investigate the effect of edible coatings (ECs) prepared from extracts of *Opuntia ficus-indica* (OFI) cladodes in comparison with a commercial chitosan formulation on the quality of 'Regina' cherries packaged in macro-perforated bags and stored for up to 28 d (1 °C, 90% RH). The coating concentrations were 25% and 50% aqueous OFI extract (approximately 0.59 and 1.18% dry matter, respectively), 1% OFI alcohol insoluble polysaccharide and 1% chitosan. The variables evaluated included weight loss (WL), respiration rates (RR), peel color, firmness, microbial decay, total antioxidants (phenolics, flavonoids, anthocyanins, antioxidant capacity), individual phenolic compounds (anthocyanins, hydroxycinnamic acids, flavan-3-*O*-ols), and pedicel removal force. The main results show that all coatings reduced WL and RR similarly, enhanced firmness throughout storage and antioxidants after 28 d of storage compared to the controls. Among treatments, chitosan resulted in much higher peel glossiness and firmness in comparison to OFI extracts. On day 28, all ECs resulted in higher antioxidants than controls, OFI extracts resulted in higher cyaniding-3-*O*-rutinoside than chitosan, while 50% OFI treatment resulted in the highest catechin concentration. Therefore, OFI extracts are promising ECs for cherry storage since they exhibited no negative effect, improved quality and extended storage life by one week compared to the controls.

Keywords: *Prunus avium*; edible coatings; *Opuntia ficus-indica* extracts; chitosan; storage; quality; anthocyanins; phenolic compounds; total antioxidant capacity

Citation: Christopoulos, M.V.; Gkatzos, D.; Kafkaletou, M.; Bai, J.; Fanourakis, D.; Tsaniklidis, G.; Tsantili, E. Edible Coatings from *Opuntia ficus-indica* Cladodes Alongside Chitosan on Quality and Antioxidants in Cherries during Storage. *Foods* **2022**, *11*, 699. https://doi.org/10.3390/foods11050699

Academic Editor: Evandro Leite de Souza

Received: 29 January 2022
Accepted: 24 February 2022
Published: 26 February 2022

1. Introduction

Cherries are early summer fruits and are highly appreciated by consumers due to their size, color and flavor. They are also a rich source of nutritive compounds, containing sugars, acids, potassium, melatonin, dietary fiber, vitamins C, A, E and B, phenolic acids and anthocyanins with only low caloric content [1]. Because of their composition, they contribute to health promotion and prevent diseases, such as cardiovascular disorders and types of cancer. Most of their beneficial effects are attributed to phenolic compounds [2].

Fresh cherries are exposed to the market for a very short period of time. They are very susceptible to physiological disorders and microbial decay, rendering the fresh produce sensitive to transport and storage. Moreover, they are harvested with pedicels, indicating the fruit's freshness as long as they remain green and attached to the fruit [3]. Low-temperature management remains crucial for quality retention throughout the whole fresh cherry chain. Postharvest handling, such as ethanol treatment [4], modified clamshells with reduced water loss [5], or lowered oxygen concentration and elevated carbon dioxide in modified atmosphere packaging (MAP) in combination with low temperature maintain cherries' sensory and quality characteristics for up to 6–7 weeks [6–8]. However, prolonging the shelf life of small and very sensitive fruit further still remains a challenge [9].

An alternative method, based on MAP, is the implementation of edible coatings (ECs). They form semipermeable barriers on the surface of each single fruit to moisture, solute and gas (O_2, CO_2 and other volatiles) transport, regulating their exchange between fruit and the surrounding atmosphere, provided that the fruit does not induce anaerobic respiration [10]. Therefore, the reduced water vapor pressure (WVP) caused by the coating results in reduced WL, O_2 uptake and CO_2 evolution and consequently reduced RR and in ripening delay. In stone fruit during storage, the effects of ECs composed of polysaccharides are the most studied, exhibiting reduced weight loss (WL) and ripening delay. They are applied to fruit as a liquid solution by immersion, spraying and dripping/brushing. Most studies use one of the two types of polysaccharide coating: chitosan or natural gel (mucilage) extracted from plant sources [10–12]. Chitosan is the most common edible coating applied to fruit, such as strawberries [13], sliced mangoes [14], and cherries [15–17], with promising effects on quality characteristics and storability [12] and inhibiting microbial decay [10,18].

Plant extracts applied to cherries, such as *Aloe vera* [19], guar gum with ginseng extract [20] and Arabic or almond gum [21], or to other fruit, such as *Aloe vera* to tomato [22] comprise a few examples exhibiting promising results during storage by delaying ripening processes. Recently, increasing interest has been focused on the novel ECs with material derived from wild plants that are rich in polysaccharides, such as the *Opuntia ficus-indica* (OFI) or *Opuntia cactus* (Cactaceae), commonly called prickly pear or cactus pear. It is a xerotrophyte plant, cultivated in Central and South America, Asia and South Europe. The mucilage of the cladodes has been studied in citrus [23], kiwi slices [24], strawberries [25], figs [26] and a few other cases, improving their shelf life, but not yet in cherries.

This work aimed to investigate some extracts of OFI cladodes (mucilage and gel solutions) alongside chitosan as ECs on cherry quality during low-temperature air storage. The variables evaluated were the fruit WL, respiration rates (RR), peel color, firmness, total phenolics (TP), total flavonoids (TF), total anthocyanins (TAN), total antioxidant capacity (TAC), individual phenolic compounds, pedicel removal force (PRF) and fruit microbial decay.

2. Materials and Methods

2.1. Plant Materials

Cherry (*P. avium* L.) fruits of cv. 'Regina' were harvested from trees grafted on 'Gisela 6′ rootstock and grown in a commercial orchard located at Tegea, Arkadia County, Greece (37°26′ 20″ N, 22°24′47″ E). The fruit were harvested (16 July 2019) soon after the commercial maturity stage and transported to the laboratory within 2.5 h. Upon arrival at the laboratory, the fruits were sorted and only healthy fruits of uniform maturity, macroscopically free of disorders and diseases, were selected for the experiment. Cactus pear (*Opuntia ficus-indica*) cladodes were collected from vegetatively propagated/cladode rooting trees of a red/dark pink fruit variety grown in a commercial orchard located in Gerakas, Attica County, Greece (38°00′ 34″ N, 23°52′28″ E). One- and two-year old cladodes were collected on 1 July 2019, transported to the laboratory within 2 h and stored (4 °C, 85% RH) until processing (7–11 July 2019).

2.2. Mucilage Extraction from Opuntia ficus-indica Cladodes and Preparation of the Edible Coatings

The mucilage from cladodes of cactus pear was extracted in two forms for the preparation of different ECs (Figure 1). The first mucilage material, aqueous polysaccharide extract, was prepared according to Del Valle et al. [25] with some modifications. Cladodes were peeled and sliced (1 cm width) (Step 1), and homogenized in a blender (Model CB15, Waring, Torrington, CT, USA) (Step 2). The slurry was incubated in a water bath at 40 °C for 90 min (Step 3) and then centrifuged (Model 4239R, Alc International, Srl, Cologno Monzese Srl, Cologno Monzese, Italy) at 10,000× *g* for 10 min (Step 4). The collected supernatant was thermally pasteurized in a water bath at 77 °C for 1 min and stored at 4 °C until use. The final material was in a gel (Code: Gel or G) form with specifications: 2.36% (*w*/*v*) dry matter; 0.69% (*w*/*v*) ash; 2.0 °brix; 4.47 pH. The second mucilage material was prepared according to Allegra et al. [24] with some modifications consisting of the alcohol-insoluble polysaccharide content of cladodes. This procedure was similar to the described aqueous-based polysaccharide extract, following the same steps 1–4, as a first part of the processing. The supernatant from step 4 was boiled at 100 °C until reaching a concentration of 50–60% of the initial volume (step 5), and then centrifuged at 3600× *g* for 5 min (Step 6). In the collected supernatant, an equal volume of ethanol (96%) was added (1:1) for precipitation of the polysaccharides (step 7) and the mixture was incubated at 1 °C for 24 h (step 8). After incubation, the mixture was centrifuged at 3600× *g* for 5 min (Step 9) and the pellet was collected containing the alcohol insoluble polysaccharides. Finally, the collected solid material was dried in a vacuum (−0.7 bar) oven at 50 °C for 12 h (Step 10), and the dried solid was ground and sieved (18 mesh) (Step 11) for use as the final cactus pear polysaccharide material (Code: Polys).

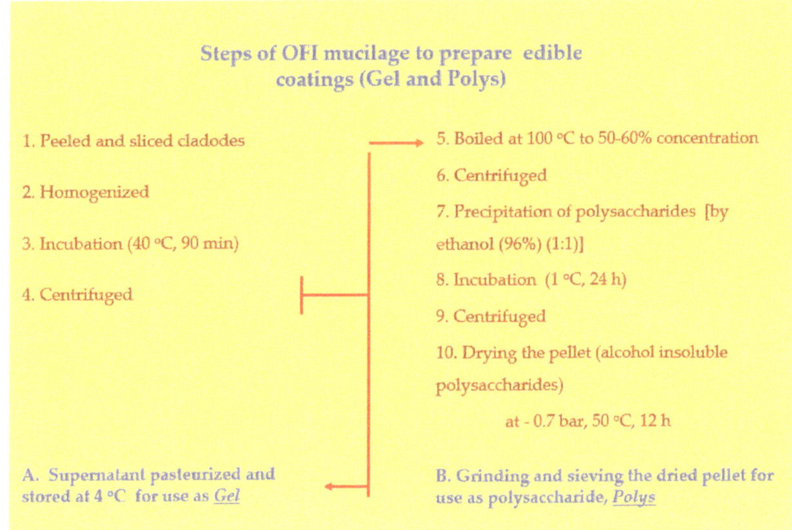

Figure 1. Diagram of the steps during preparation of edible coatings (ECs) from cladodes of *Opuntia ficus-indica* (OFI). A, the alcohol insoluble polysaccharides content, called Gel; B, the alcohol insoluble polysaccharides content, called Polys.

2.3. Edible Coating Treatments and Storage of Cherry Fruit

Preliminary experiments were conducted for the determination in the final edible coating application solution of the (i) concentration of the Gel material (tested 0, 20%, 25%, 33%, 50% and 100% Gel (*v*/*v*)); (ii) concentration of the Polys material (tested 0, 1%, 5% and 10% Polys (*w*/*v*)); (iii) concentration of plasticizer (tested 0%, 1%, 5% and 10% glycerol (*v*/*v*)); and (iv) an extra step for gelation (tested 0%, 1% and 5% $CaCl_2$ (*w*/*v*)). Edible coating solutions were prepared (Gel ± glycerol, Polys ± glycerol) and applied by dipping treatments on cherry fruit followed or not by a successive gelation step (±$CaCl_2$).

Increases in glycerol, $CaCl_2$ and pH increase the viscosity but decrease the values of wettability and adhesion coefficients [12]. The selection criteria for the desired coating solution specifications and application procedure were the (i) uniform coverage of the whole fruit after drying of the coating; (ii) the presence of abnormal fruit appearance and (iii) fruit weight loss after 5 days at 1 °C, 85–90% RH. Based on the results of the preliminary experiments, 25% and 50% Gel and 1% Polys concentrations were selected, all with the addition of 5% glycerol as a plasticizer and no requirements for an extra gelation step.

For the final experiment, the tested coating solutions in deionized (DI) water were:

(i) a liter of 25% Gel (*v/v*) containing 5% glycerol (code: G25),
(ii) a liter of 50% Gel (*v/v*) containing 5% glycerol (code: G50),
(iii) a liter of 1% Polys (*w/v*) containing 5% glycerol (code: Polys),
(iv) a liter of 1% (*w/v*) chitosan in acetic acid (0.5% *v/v*) containing 5% glycerol,
(v) a liter DI water containing 5% glycerol used as a control for 7, 14, 21 and 28 d of storage.

The chitosan solution was prepared by dissolving 5 g of chitosan (Chitosan from shrimp shells; degree of deacetylation ≥0.75; color, white to beige; Aldrich Chemistry; Product) in 1 L of DI water containing 0.5% (*v/v*) acetic acid under stirring at 40 °C for 24 h.

Batches of about 300 g of cherries (a total of 6 batches per treatment) were dipped for 1 min in each coating solution, the excess coating was drained, and the coated fruits were dried under forced air at 20 °C for 60 min. Coated cherries were packaged in polypropylene (PP) macro perforated packages with 10 × 20 cm dimensions and 3 holes cm^{-1} perforation of 500 μm diameter (15 fruits per package, averaged 152.35 ± 10.98 g of fruit per package, 12 packages per treatment), sealed, and stored in a cold chamber at 1 °C and 90% RH for up to 28 days. Each package served as a biological replicate. Quality and chemical attributes were analyzed, just before coating treatments (day 0) and at 7, 14, 21 and 28 d after storage, on 3 replicates per treatment at each sampling day. Weight loss was measured immediately after removal from the store and packages, whereas the remaining variables were evaluated after removal from packages and temperature equilibration at 20 °C for 15 h. During each sampling, soon after the quality parameters' evaluation on fresh fruit per package (10 per 15 fruit selected randomly), fruits were frozen (−20 °C) until the extraction of phytochemicals.

2.4. Total Soluble Solids, Titratable Acidity, pH, Peel Color, Weight Loss, Moisture/Dry Matter and Respiration

The total soluble solids (TSS) of the flesh was estimated in each fruit separately by an Atago 8469 (Atago Co., Ltd., Tokyo, Japan) hand refractometer. Titratable acidity (TA) was measured by the titration of 10 g of fruit sap to pH 8.2 with 0.1 M NaOH. pH was measured by a pH-meter (Jenway 3310; Jenway Ltd., Dunmow, UK).

The color grade was evaluated according to the color program developed by the Centre Technique Interprofessionnel des Fruit et Legumes (CTIFL, Paris, France), in which 1 = light pink and 7 = dark mahogany. The present cherries at harvest were evaluated as of color grade 6, indicating the advanced maturity stage.

Accurate peel color determinations were carried out on 10 fruits per replicate on the opposite sides of each fruit with a Minolta chromatometer (CR-300; Minolta, Ahrensburg, Germany) according to Tsantili et al. [27]. The measurements are expressed as chroma (intensity of color), hue angle (actual color, or redness), and L^* value (lightness ranging from 0 = black to 100 = white). In particular, the recorded values of a^* and b^* were converted into hue angle ($h\circ$) and C^* according to the following equations:

$$h\circ = \tan^{-1} (b^*/a^*) \text{ when } a^* > 0 \text{ and } b^* > 0$$

$$h\circ = 180^o + \tan^{-1} (b^*/a^*) \text{ when } a^* < 0$$

$$h\circ = 360^o + \tan^{-1} (b^*/a^*) \text{ when } a^* > 0 \text{ and } b^* < 0$$

$$C^* = (a^* + b^*)^{1/2}$$

Fruit weight loss was measured immediately after removal of the packages from storage and expressed as the percentage difference between the fruit weight (15 cherries) immediately after drying at day 0 and the weight at sampling (%, *w/w*).

Fruit moisture/dry matter was determined according to AOAC method 934.06 on each sampling day by the difference in weight of ~5 g pulp from 10 fruit before and after drying at 105 °C until constant weight.

Fruit respiration rates (RR) assessed as CO_2 production were measured using a closed portable infrared gas analyzer (LI-6400; LI-COR, Lincoln, NE, USA) connected to a 750 mL airtight jar at a flow rate of 900 μmol s^{-1} [28]. On each sampling day and for each coating treatment, the RRs were measured on 10 randomly selected fruits per replicate after temperature equilibration at 20 °C. The CO_2 production rates were expressed in nmol kg^{-1} h^{-1}.

2.5. Fruit Firmness and Microbial Decay

The texture analysis was performed using an HD-Plus texture analyzer (Stable Micro Pedicels Ltd., Godalming, UK) and the Texture Expert Exceed Software for the data analysis. The determination of the textural characteristics of whole fruits was conducted with a cylindrical probe of 2 mm diameter and movement speeds of 1 mm/s during the test, 5 mm/s for the pre-test and 10 mm/s for the post-test. The compression depth was set at 5 mm, the measurement was conducted at the equatorial zone in each fruit and the results were expressed as the maximum recorded force in N.

Each fruit was visually assessed for decay incidence (molds and other infections), and the presence (=1) or absence (=0) of any decay symptom was expressed as the mean of 10 fruit per replicate.

2.6. Resistance to Pedicel Removal

Resistance to pedicel removal (PRF) was measured with a HD-Plus texture analyzer equipped with a hook probe. Each fruit was immobilized in the moving probe, the tip of the pedicel was connected to the stable base of the instrument, and the probe was moving upwards in a perpendicular direction to the horizontal plane until pedicel removal. The movement speeds were 10 mm/s during the test, 10 mm/s for the pre-test, and 10 mm/s for the post-test, and the results were expressed as the maximum recorded force in N.

2.7. Extraction of Phytochemicals

The extraction procedure of phytochemicals was carried out according to Blackhall et al. [29] after some modifications. Frozen cherries (three replicates of 10 cherries each) were de-stoned and homogenized in a blender (Model 38BL40, Waring commercial, New Hartford, CT, USA) for 15 s. Approximately 2 g of cherry pulp and 20 mL of methanol containing 0.1% 10 N HCl were homogenized using an Ultra-Turrax (Model T25, Ika Labortechnik, Germany) for 1 min at 9500 rpm min^{-1}. The homogenate was incubated in a supersonic bath for 60 min at 37 °C, centrifuged at 4000 rpm for 6 min and the supernatant was recovered and used for the analyses.

2.8. Determinations of Total Phenolics, Flavonoids, Anthocyanins and Antioxidant Capacity

Total phenolics (TP) was measured by the Folin–Ciocalteu method according to Tsantili et al. [30], recording the absorbance at 750 nm versus a blank using a spectrophotometer (Model Cary 50, Varian Inc., Walnut Creek, CA, USA). Total flavonoids (TF) were measured by a colorimetric method using a 0.3 mL cherry extract for reactions and absorbance recording at 510 nm [31], as described by Tsantili et al. [30]. Total anthocyanins (TAN) were measured according to Meyers et al. [31], as described by Tsantili et al. [30], recording absorbance at 510 and 700 nm in buffers at pH 1.0 and 4.5, and converted to cyanidin 3-rutinoside (keracyanin) equivalents (c-3-rut) using a molar extinction coefficient of 28,840 L mol^{-1} cm^{-1}. Total antioxidant capacity (TAC) was evaluated using both ferric reducing antioxidant power (FRAP) [30] and radical scavenging capacity (2,2-diphenyl-1-picrylhydrazyl, DPPH) [32] assays according to Christopoulos and Tsantili [33]. For all determinations, triplicate reactions per replicate were performed, and the results of TP, TF, TAN and TAC were expressed as equivalents of gallic acid (GAE), catechin (CAE), c-3-rut and Trolox acid (TAE), respectively, all on a DW basis.

2.9. Determinations of Individual Phenolic and Anthocyanin Compounds

Individual phenolic and anthocyanin compounds were determined according to Durst and Wrolstad [34] by an HPLC system equipped with a pump Nexera X2 (LC-30 AD), an autosampler system (SIL-30AC), a diode array detector (SPDM20A) (Shimadzu, Kyoto, Japan) and a Macherey–Nagel HPLC column C18 (250 × 4.6; 5 μm, Nucleodur PolarTec at 30 °C. An aliquot of 5 mL of the extract of phytochemicals (point 2.7) was evaporated under N_2 stream at 37 °C and the residue was dissolved in 1 mL MeOH (HPLC grade). The extract was filtered through a Chromafil AO-45/25 polyamide filter (0.45 μm pore size), 20 μL was injected and the flow rate was set at 1 mL min^{-1}. The elution solvents were (A) 100% acetonitrile and (B) aqueous formic acid 1%. The separation of the compounds was achieved according to the gradient: 0–15 min, 35% A; 15–30 min, 10% A; 30–80 min, 15% A; 80–100 min, 50% A; and finally washing and reconditioning of the column (equilibration time), 100–105 min 5% A. Identification of compounds was carried out by comparing retention times and their UV–Vis spectra from 200 to 700 nm. Each compound was quantified in comparison with a multipoint calibration curve obtained from the corresponding authentic standard (Extrasynthese, Genay, France) and expressed as mg g^{-1} DW. Chlorogenic and neochlorogenic acid were monitored at 320 nm, flavan-3-ols at 280 nm and anthocyanins at 510 nm. The data analyses were carried out using LabSolutions LC/GC 5.82 (SkyCom, Tokyo, Japan).

2.10. Statistical Analyses

The significance of the treatment effects (Ecs), storage days and their interaction on the determined variables was estimated by two-way ANOVA. In controls, one-way ANOVA was also performed. During the last two sampling dates and also when denoted, partial analysis of data was performed in addition to two-way ANOVA of all data from day 7. Mean comparisons were performed using the Tukey-HSD multiple range test ($\alpha = 0.05$) with standard error (SE) values calculated from the residual variances. Data of weight loss (WL), pedicel removal force (PRF), and the analyses of respiration rates (RR) of controls, hue angle of controls and pedicel removal force of controls were transformed to log10, while c-3-glc and analyses of controls of c-3-rut, catechin, epicatechin, chlorogenic and neochlorogenic acids were transformed to the square root. The data presented were back transformed. Decay data were analyzed without transformation after checking the residuals for normality, according to the Shapiro–Wilk test, and the plot of the residuals for homoscedasticity. All statistical analyses were performed using STATGRAPHICS Plus (Statgraphics Technologies Inc., The Plains, VA, USA).

3. Results

3.1. Total Soluble Solids, pH and Titratable Acidity

At harvest, the total soluble solids (TSS) was 18.33 ± 0.06 (%), pH 4.09 ± 0.06 and titratable acidity (TA) 0.49 ± 0.02% (*w/w*) malate (Table 1).

Table 1. Total soluble solids (TSS), pH values and titratable acidity (TA) of juice in cherries at harvest.

	Harvest	
TSS [a] (%)	**pH**	**TA (%, *w/w*)**
8.33 ± 0.06	4.09 ± 0.06	4.29 ± 0.02

[a] Numbers are means of three replicates of 10 cherries each ± standard deviations.

3.2. Fruit Color, Weight Loss and Respiration Rates

The color parameter L^* was 29.9 at harvest (Figure 2a). Changes in controls during storage were significant, exhibiting the highest value on day 14 and the lowest on day 28. Analysis of all data from 7 d showed that L^* remained almost stable in controls, G25, G50 and Polys, whereas in chitosan-treated fruit, it increased substantially, reaching the highest value on day 14 and then decreased gradually, being significantly lower at the end of storage, but similar to the remaining treated fruit. The treatment effect and storage days were significant, but not their interaction.

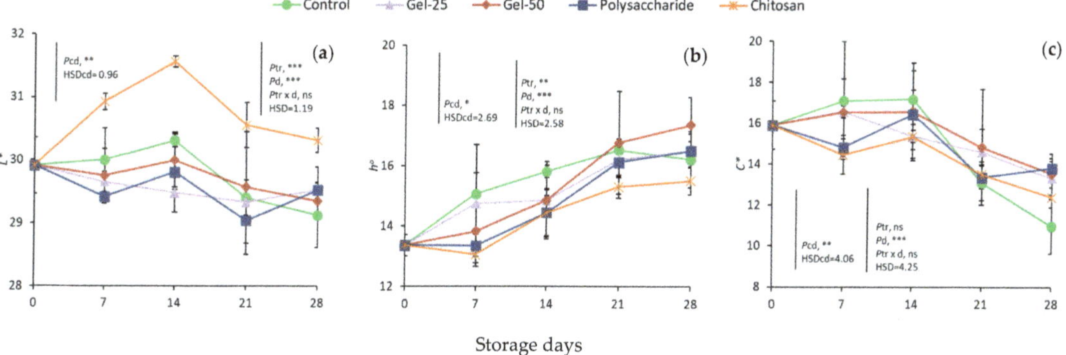

Figure 2. Effect of edible coatings on changes in peel color parameters, L^* (**a**), hue angle (**b**) and C^* (**c**), in cherries during storage. Points are means of three replicates of 10 cherries each; bars on the points, ± standard deviations. NS, non significant; * significant at $p < 0.05$; ** significant at $p < 0.01$; *** significant at $p < 0.001$. *Pcd*, probability of storage days in controls (one-way ANOVA from day 0); HSDcd, honest significant difference calculated from ANOVA. *Ptr*, probability of edible coating treatment; *Pd*, probability of storage days; *Ptr* × *d*, probability of interaction (two-way ANOVA from day 7); HSD, honest significant difference calculated from ANOVA.

Values of hue angle increased gradually during storage in controls from 13.4 on day 0 to 16.5 on day 21 and remained stable thereafter (Figure 2b). From day 7, increases in hue angle were observed in all samples, with chitosan-treated fruit exhibiting the lowest increases. The treatment effect and storage days were significant, but not their interaction.

In controls, the parameter C^* was 15.9 at harvest and remained almost stable up to 14 d and then decreased until 28 d (Figure 2c). Factorial analysis showed the significant effect of days, but no effect of treatment or their interaction.

Weight loss (WL) averaged 1.61% on day 7 in all samples and increased progressively during storage up to day 28 (Figure 3a). WL in controls, being higher than in the remaining treatments from day 14 up to the end of storage, reached 6.48% after 28 d. WL increased in the other treatments, but at a slower and similar rate, averaging 4.86% on day 28. Indeed, the effect of treatments and days in storage were both highly significant, but not their interaction. The factor treatments, as the main effect, showed significant differences between controls and treatments, but no difference among treated fruit.

At harvest, CO_2 production rates were approximately 311 nmol kg^{-1} s^{-1}, but thereafter elevated sharply in controls up to 21 d, reaching 578 nmol CO_2 kg^{-1} s^{-1}, and then decreased (Figure 3b) ($Pcd < 0.001$). In all treated fruit, increases were consistent after 7 d, but much lower than controls throughout storage. The effect of treatment and storage days were both highly significant, in contrast to their non-significant interaction.

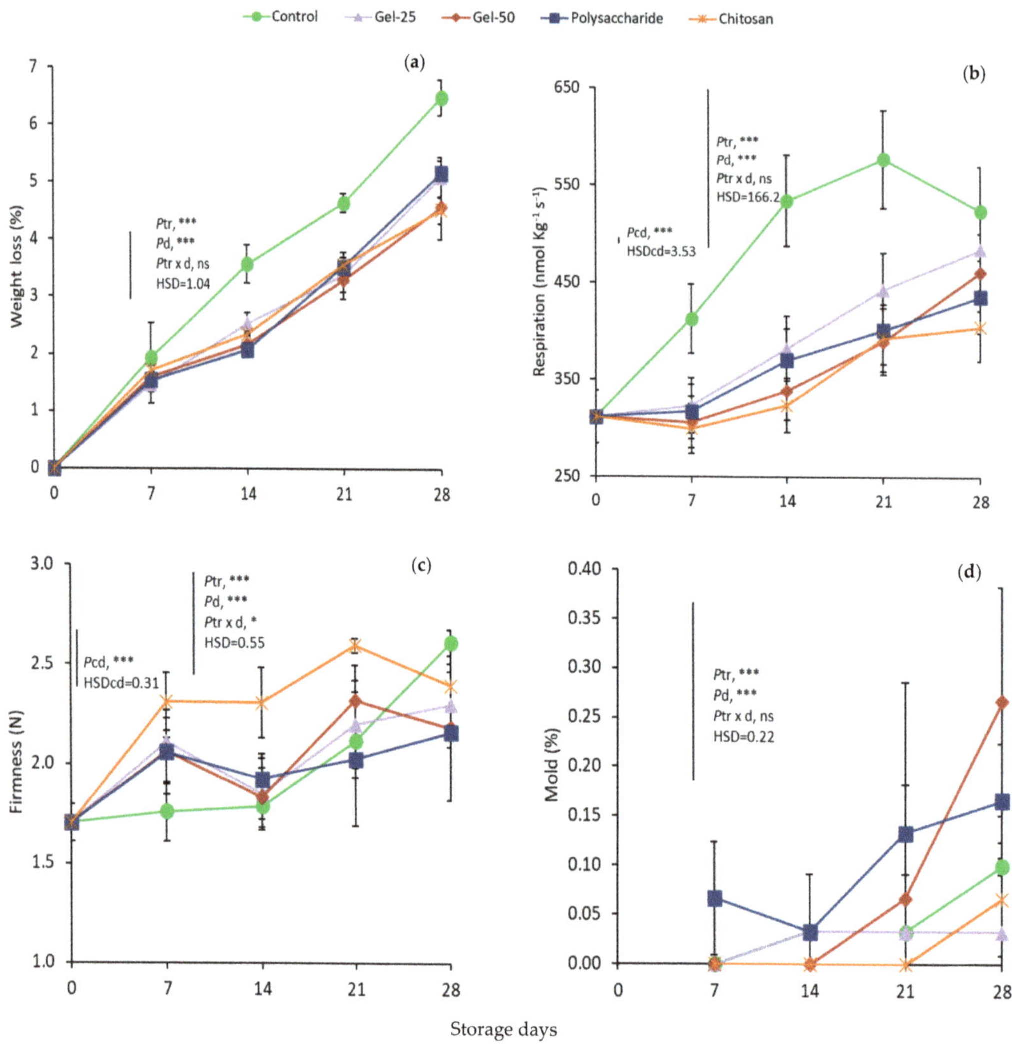

Figure 3. Effect of edible coatings on changes in weight loss (WL) (**a**), respiration (**b**), firmness (**c**) and microbial decay (**d**) in cherries during storage. Points are means of three replicates of 10 cherries each; bars on the points, ± standard deviations. NS, non significant; * significant at $p < 0.05$; *** significant at $p < 0.001$. *Pcd*, probability of storage days in controls (one-way ANOVA from day 0); HSDcd, honest significant difference calculated from ANOVA. *Ptr*, probability of edible coating treatment; *Pd*, probability of storage days; *Ptr* × *d*, probability of interaction (two-way ANOVA from day 7); HSD, honest significant difference calculated from ANOVA.

3.3. Fruit Firmness and Decay, and Pedicel Removal Force

The initial firmness value was 1.71 N, and after 14 d it increased gradually in controls, reaching the level of 2.61 N on day 28 (Figure 3c). However, chitosan-treated cherries showed the highest average levels in store (2.4 N).

In particular, chitosan exhibited a burst in firmness on day 7 (2.31 N), remained at almost stable levels up to day 14, peaked at 2.59 N on day 21 and declined to 2.39 N at the end of storage. Firmness in G25 and G50 also showed increased levels similar to the respective days in chitosan on days 7 and 21, while averaging 2.22 N between 7 and 28 d. Polys showed the lowest increases on average. The effect of treatments and days in storage were both highly significant, but not their interaction.

The mold development (%) of fruit increased during storage, with an overall mean of 0.05% and averaging 0.02% in both controls and chitosan and 0.026% in G25 during storage (Figure 3d). In Polys and G50, the decay averaged 0.084% and 0.1%, respectively. Additionally, on day 28, the decay increased in all treatments, being significantly higher in G50 than in G25.

At harvest, pedicel removal force (PRF) in controls was 7.18 N and then fluctuated between 6.68 and 7.67 N (Figure 4). In all treatments, changes in PRF were not consistent during storage. The effects of treatments, storage days and their interaction were all insignificant.

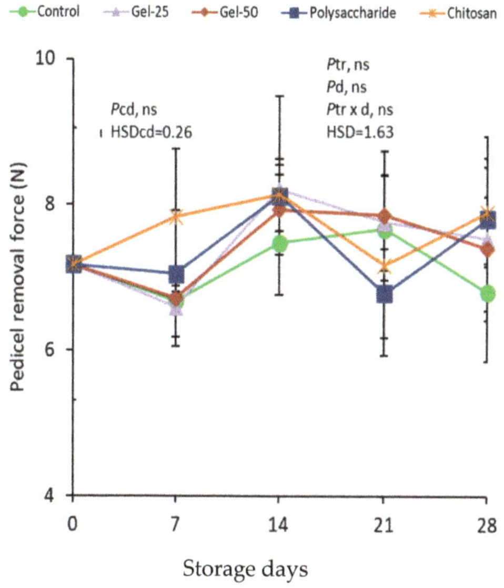

Figure 4. Effect of edible coatings on changes in pedicel removal force in cherries during storage. Points are means of three replicates of 10 cherries each; bars on the points, ± standard deviations. NS, non significant. *Pcd*, probability of storage days in controls (one-way ANOVA from day 0); HSDcd, honest significant difference calculated from ANOVA. *Ptr*, probability of edible coating treatment; *Pd*, probability of storage days; *Ptr* × *d*, probability of interaction (two-way ANOVA from day 7); HSD, honest significant difference calculated from ANOVA.

3.4. Total Phenolics, Total Flavonoids, Total Anthocyanins and Total Antioxidant Capacity Determined with DPPH and FRAP Methods

At harvest, the values of TP, TF and TAN were 9.4 mg GAE g^{-1} DW, 3.3 mg CE g^{-1} DW and 4.5 mg c-3-rutg^{-1} DW (Figure 5a,b; Table 2). In controls, the values of all three variables decreased gradually and significantly up to day 21 and then increased to levels lower than at harvest. Treatments and controls showed a very similar pattern of changes during storage in each determined variable. The two-way analyses showed that treatments had no effect on changes in TP, TF and TAN in all three variables, but storage days were significant for TP and TF and the interaction of treatments with days was significant only for TAN. The lowest values of TP, TF and TAN were observed on day 21 in controls and reached approximately 4.6 mg GAE g^{-1} DW, 1.0 mg CE g^{-1} DW and 2.0 mg c-3-rut g^{-1} DW, respectively. In G25, the TP and TF values decreased on day 14 before the subsequent increases (partial analysis of data on day 14). Increases were observed in TP and TF variables in all treatments and controls on day 28, averaging 7.34 mg GAE g^{-1} DW, and 2.79 mg CE g^{-1} DW, respectively, compared to day 21, averaging 6.18 mg GAE g^{-1} DW, and 2.45 mg CE g^{-1} DW, respectively (partial analysis of data on days 21 and 28). The corresponding TAN values were 2.99 mg c-3-rut g^{-1} DW and 3.18 mg c-3-rut g^{-1} DW on days 21 and 28, respectively (partial analysis of data on days 21 and 28).

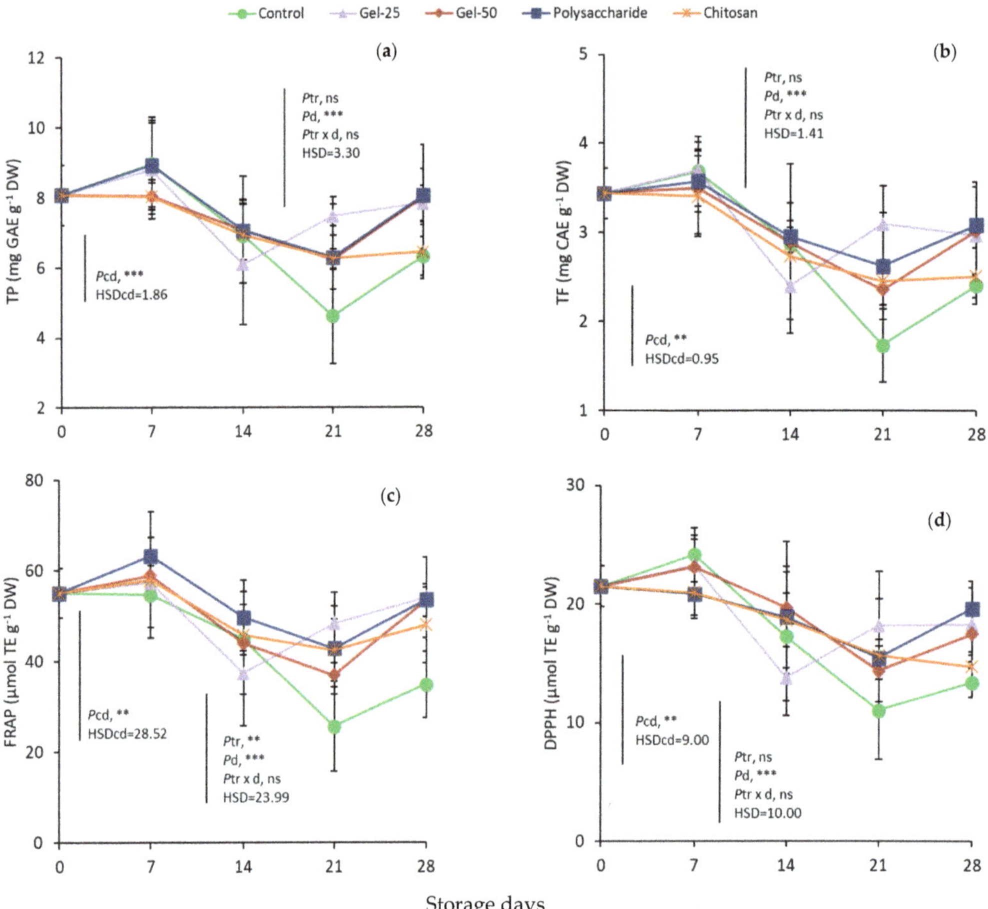

Figure 5. Effect of edible coatings on changes in total phenolic compounds (**a**), total flavonoids (**b**), total antioxidant capacity (TAC) determined by FRAP assay (**c**) and DPPH assay (**d**) in cherries during storage. Points are means of three replicates of 10 cherries each; bars on the points, ± standard deviations. NS, non significant; ** significant at $p < 0.01$; *** significant at $p < 0.001$. *Pcd*, probability of storage days in controls (one-way ANOVA from day 0); HSDcd, honest significant difference calculated from ANOVA. *Ptr*, probability of edible coating treatment; *Pd*, probability of storage days; *Ptr* × *d*, probability of interaction (two-way ANOVA from day 7); HSD, honest significant difference calculated from ANOVA.

Table 2. Effect of edible coatings on changes in total antioxidant anthocyanin content, on cyaniding-3-*O*-rutinoside, cyaniding-3-*O*-glucoside, peonidin-3-*O*-glucoside, malvidin-3-*O*-glucoside, chlorogenic acid, neochlorogenic acid, catechin and epicatechin in cherries during storage.

	Treatments	Storage Days 0	7	14	21	28	*Pcd*	HSDcd	*Ptr*	*Pst*	*Ptr* × *st*	HSD
	Control	4.56 ± 1.13	3.56 ± 0.83 [a]	3.02 ± 0.86	2.04 ± 0.83	2.90 ± 0.19	ns [b]	2.72 [c]				
	Gel-25	4.56 ± 1.13	3.30 ± 0.38	2.25 ± 0.90	3.45 ± 0.74	3.56 ± 0.53						
TAN (mg g^{-1} DW)	Gel-50	4.56 ± 1.13	3.53 ± 0.51	2.34 ± 0.82	2.51 ± 0.30	3.64 ± 0.54			ns	ns	**	1.93 [d]
	Polysaccharide	4.56 ± 1.13	4.21 ± 1.12	2.83 ± 0.94	2.70 ± 0.30	4.00 ± 0.16						
	Chitosan	4.56 ± 1.13	3.10 ± 0.73	2.41 ± 0.43	2.50 ± 0.52	2.95 ± 0.26						
						Anthocyanins						
	Control	2.27 ± 0.53	1.75 ± 0.41	1.28 ± 0.34	1.13 ± 0.38	1.51 ± 0.15	*	1.35				
	Gel-25	2.27 ± 0.53	1.80 ± 0.20	1.15 ± 0.31	1.76 ± 0.18	2.82 ± 0.41						
Cyanidin 3-*O*-rutinoside (mg g^{-1} DW)	Gel-50	2.27 ± 0.53	2.37 ± 0.22	1.83 ± 0.60	1.73 ± 0.21	3.07 ± 0.41			***	***	ns	1.37
	Polysaccharide	2.27 ± 0.53	2.27 ± 0.26	1.95 ± 0.87	1.65 ± 0.19	2.95 ± 0.69						
	Chitosan	2.27 ± 0.53	2.26 ± 0.08	1.35 ± 0.23	1.43 ± 0.24	1.86 ± 0.14						
	Control	0.056 ± 0.006	0.025 ± 0.022	0.005 ± 0.003	0.016 ± 0.014	0.023 ± 0.005	**	0.01				
	Gel-25	0.056 ± 0.006	0.023 ± 0.011	0.004 ± 0.004	0.020 ± 0.004	0.051 ± 0.016						
Cyanidin 3-*O*-glucoside (mg g^{-1} DW)	Gel-50	0.056 ± 0.006	0.024 ± 0.003	0.024 ± 0.008	0.019 ± 0.002	0.052 ± 0.016			*	***	ns	0.001
	Polysaccharide	0.056 ± 0.006	0.031 ± 0.008	0.022 ± 0.020	0.019 ± 0.004	0.046 ± 0.020						
	Chitosan	0.056 ± 0.006	0.028 ± 0.007	0.007 ± 0.006	0.011 ± 0.010	0.024 ± 0.009						
	Control	0.104 ± 0.017	0.062 ± 0.031	0.063 ± 0.022	0.020 ± 0.017	0.082 ±0.007	**	0.05				
	Gel-25	0.104 ± 0.017	0.135 ± 0.037	0.072 ± 0.052	0.076 ± 0.040	0.192 ± 0.069						
Peonidin 3-*O*-glucoside (mg g^{-1} DW)	Gel-50	0.104 ± 0.017	0.158 ± 0.023	0.133 ± 0.013	0.097 ± 0.020	0.208 ± 0.017			***	***	ns	0.11
	Polysaccharide	0.104 ± 0.017	0.185 ± 0.007	0.106 ± 0.021	0.078 ± 0.026	0.201 ± 0.025						
	Chitosan	0.104 ± 0.017	0.167 ± 0.035	0.110 ± 0.011	0.085 ± 0.034	0.169 ± 0.025						
	Control	0.037 ± 0.009	0.035 ± 0.014	0.025 ± 0.004	0.029 ± 0.016	0.030 ± 0.004	*	0.01				
	Gel-25	0.037 ± 0.009	0.017 ± 0.001	0.026 ± 0.011	0.029 ± 0.007	0.042 ± 0.026						
Malvidin 3-*O*-glucoside (mg g^{-1} DW)	Gel-50	0.037 ± 0.009	0.051 ± 0.007	0.051 ± 0.006	0.044 ± 0.009	0.067 ± 0.006			***	**	ns	0.03
	Polysaccharide	0.037 ± 0.009	0.072 ± 0.008	0.049 ± 0.008	0.041 ± 0.006	0.065 ± 0.009						
	Chitosan	0.037 ± 0.009	0.050 ± 0.011	0.048 ± 0.012	0.039 ± 0.014	0.054 ± 0.003						
						Phenolic acids						
	Control	0.023 ± 0.010	0.043 ± 0.018	0.022 ± 0.004	0.012 ± 0.002	0.019 ± 0.001	*	0.001				
	Gel-25	0.023 ± 0.010	0.029 ± 0.001	0.019 ± 0.005	0.021 ± 0.003	0.033 ± 0.003						
Chlorogenic acid (mg g^{-1} DW)	Gel-50	0.023 ± 0.010	0.041 ± 0.005	0.020 ± 0.016	0.024 ± 0.002	0.036 ± 0.004			ns	***	*	0.01
	Polysaccharide	0.023 ± 0.010	0.039 ± 0.003	0.033 ± 0.006	0.022 ± 0.004	0.036 ± 0.002						
	Chitosan	0.023 ± 0.010	0.035 ± 0.005	0.032 ± 0.004	0.028 ± 0.006	0.031 ± 0.002						
	Control	0.62 ± 0.03	0.43 ± 0.06	0.37 ± 0.09	0.18 ± 0.04	0.35 ± 0.01	***	0.01				
	Gel-25	0.62 ± 0.03	0.49 ± 0.03	0.29 ± 0.12	0.37 ± 0.06	0.58 ± 0.09						
Neochlorogenic acid (mg g^{-1} DW)	Gel-50	0.62 ± 0.03	0.74 ± 0.12	0.49 ± 0.19	0.37 ± 0.03	0.63 ± 0.04			ns	***	ns	0.61
	Polysaccharide	0.62 ± 0.03	0.70 ± 0.06	0.50 ± 0.13	0.32 ± 0.04	0.60 ± 0.06						
	Chitosan	0.62 ± 0.03	0.61 ± 0.06	0.46 ± 0.05	0.44 ± 0.09	0.53 ± 0.04						
						Flavan-3-ols						
	Control	1.77 ± 0.16	1.07 ± 0.05	1.08 ± 0.23	0.61 ± 0.08	0.88 ± 0.07	**	0.01				
	Gel-25	1.77 ± 0.16	1.27 ± 0.17	1.03 ± 0.20	1.13 ± 0.09	1.41 ± 0.13						
Catechin (mg g^{-1} DW)	Gel-50	1.77 ± 0.16	1.55 ± 0.17	1.98 ± 0.71	1.54 ± 0.37	2.15 ± 0.27			***	**	ns	1.02
	Polysaccharide	1.77 ± 0.16	1.45 ± 0.09	2.02 ± 0.67	1.00 ± 0.06	1.85 ± 0.42						
	Chitosan	1.77 ± 0.16	1.66 ± 0.49	1.46 ± 0.19	1.21 ± 0.17	1.20 ± 0.11						
	Control	0.22 ± 0.06	0.20 ± 0.14	0.06 ± 0.01	0.04 ± 0.01	0.07 ± 0.01	***	0.01				
	Gel-25	0.22 ± 0.06	0.11 ± 0.03	0.07 ± 0.04	0.06 ± 0.04	0.19 ± 0.04						
Epicatechin (mg g^{-1} DW)	Gel-50	0.22 ± 0.06	0.15 ± 0.03	0.14 ± 0.02	0.11 ± 0.01	0.23 ± 0.02			ns	***	**	0.05
	Polysaccharide	0.22 ± 0.06	0.18 ± 0.01	0.11 ± 0.03	0.09 ± 0.03	0.20 ± 0.03						
	Chitosan	0.22 ± 0.06	0.18 ± 0.02	0.12 ± 0.01	0.10 ± 0.03	0.17 ± 0.02						

[a] Numbers are means of three replicates of 10 cherries each ± standard deviations. [b] NS, non significant; * significant at $p < 0.05$; ** significant at $p < 0.01$; *** significant at $p < 0.001$. [c] *Pcd*, probability of storage days in controls (one-way ANOVA from day 0); HSDcd, honest significant difference calculated from ANOVA. [d] *Ptr*, probability of edible coating treatment; *Pd*, probability of storage days; *Ptr* × *d*, probability of interaction (two-way ANOVA from day 7); HSD, honest significant difference calculated from ANOVA.

The patterns of changes in TAC values, determined with both methods, were similar to each other and close to the patterns observed in TP, TF and TAN. Initial samples showed 66.4 µmol TE g^{-1} DW with FRAP and 27.8 µmol TE g^{-1} DW with DPPH (Figure 5c,d). The lowest value in the controls was 25.7 µmol TE g^{-1} DW and 11 µmol TE g^{-1} DW with FRAP and DPPH, respectively, and found on day 21 (*Pcd* < 0.01 for FRAP and DPPH). When all data from day 7 were analyzed, both treatment and storage days were significant for FRAP, but not their interaction (Figure 5c). In FRAP, controls averaged 40.06 µmol TE g^{-1} DW during storage, being the lowest average value among treatments. During storage, FRAP averaged 52.39, 49.33, 48.58, and 48.35 µmol TE g^{-1} DW in Polys, G25, chitosan and G50, respectively, with all these values being similar, but significantly higher than the respective values in controls. FRAP increased significantly to 48.82 µmol TE g^{-1} DW on day 28 on average, compared to 39.28 µmol TE g^{-1} DW on day 21. Additionally, FRAP decreased up to 14 d in G25 and increased thereafter, as in total antioxidants (TP, TF, TAN) and in contrast to the remaining treatments, which showed decreases up to day 21 and increases afterwards. On day 28, treated fruit exhibited similar FRAP values, but significantly higher than controls. However, DPPH values were not affected by treatment significantly, but only by days and the interaction of treatment with days, while changes followed those of FRAP (Figure 5d).

3.5. Anthocyanins

The major anthocyanin, cyanidin-3-*O*-rutinoside (c-3-rut), was 2.27 mg g^{-1} DW at harvest (Table 2). In the controls, it decreased gradually but significantly during storage, reaching the levels of approximately 1.13 mg g^{-1} DW on day 21, and remained almost stable thereafter. The two-way analysis from day 7 showed the significant effect of treatment and days, but not their interaction, with controls exhibiting the lowest values. Values of G25-treated fruit decreased from 7 d up to 14 d, reaching 1.15 mg c-3-rut g^{-1} DW, and then increased to 1.76 mg c-3-rut g^{-1} DW on day 21 and to 2.82 mg c-3-rut g^{-1} DW at the end of storage. C-3-rut in chitosan-treated fruit showed decreases after day 7, while it increased afterwards up to 1.86 mg c-3-rut g^{-1} DW on day 28. Partial analysis (data on 21 and 28 d) showed that cherries coated with G25, G50 and Polys averaged 2.95 mg g^{-1} DW, being 1.3-fold higher than the initial value, in contrast to the chitosan-coated and control fruits, being 0.82- and 0.66-fold lower, respectively.

Cyanidin-3-*O*-glucoside (c-3-glc) was found at 0.06 mg g^{-1} DW at harvest (Table 2), being at the highest level of the whole experiment, whereas the lowest level during storage was 0.01 mg g^{-1} DW on day 14 in the controls. In all treatments, c-3-glc decreased in the middle of storage, being lower in controls, G25 and chitosan than Polys and G50 on day 14, and then increased, averaging 0.2 mg g^{-1} DW in controls and chitosan and 0.5 mg g^{-1} DW in the remaining treatments on day 28. The effect of days was significant in controls. The two-way analysis showed that the effect of treatment and days were significant, but not their interaction.

Peonodin-3-*O*-glucoside (p-3-glc) was 0.2 mg g^{-1} DW at harvest (Table 2). In the controls, it decreased significantly to 0.02 mg g^{-1} DW on day 21 and showed an insignificant increase on day 28. When all data were analyzed, the treatment effect and storage days were also both significant, but not their interaction. During storage, c-3-glc averaged 0.06 mg g^{-1} DW in controls compared to the significantly higher values of the remaining treatments, averaging 0.13 mg g^{-1} DW, being approximately two-fold higher than the controls. On day 28, c-3-glc in all treatments also showed an increase, averaging 0.17 mg g^{-1} DW in comparison to the respective value of 0.07 mg g^{-1} DW on day 21. On day 28, the values in the controls were significantly lower than those in G25, G50 and Polys, which reached the levels at harvest, whereas those in chitosan were intermediate and close to those at harvest.

Malvidin-3-*O*-gluoside (m-3-glc) was 0.037 mg g^{-1} DW at harvest and remained almost stable in controls during storage. In the two-way analysis, the treatment effect and storage days were significant, but not their interaction. Values fluctuated from 0.02 to 0.072 mgg^{-1} DW, with the average values in the controls and G25 being 0.029 mg g^{-1} DW compared to 0.052 g^{-1} DW in the other treatments. On day 28, in treated cherries, m-3-glc averaged 0.05 mg g^{-1} DW compared to 0.03 on day 21 mg g^{-1} DW, with controls and G25 being significantly lower than the remaining treatments on day 21.

3.6. Phenolic Acids and Flavan-3-ols

Chlorogenic acid was 0.02 mg g^{-1} DW initially and ranged from 0.01 to 0.05 mg g^{-1} DW, with the highest value in the controls observed on day 7 and the lowest on day 21 (Table 2), and the effect of storage days was significant. In the two-way analysis, the effect of storage days was significant, the effect of treatment was non-significant and that of the interaction was significant. In particular, chlorogenic acid averaged 0.19, 0.15, 0.14 and 0.17 mg g^{-1} DW on days 7, 14, 21 and 28, respectively, during storage.

Neochlorogenic acid, from 1.9 mg g^{-1} DW at harvest, decreased progressively during storage up to day 14 in G25, while up to day 21 in the remaining treatments (Table 2). The lowest value of neochlorogenic acid was observed in the controls on day 21, being 0.18 mg g^{-1} DW, but increased significantly thereafter ($Pcd < 0.001$). From the two-way analysis, the effect of storage days was significant, with the treatment effect and the interaction being non-significant. The averaged treatment values were 1.43, 1.02, 0.82 and 1.24 mg g^{-1} DW, on days 7, 14, 21 and 28, respectively.

In controls, catechin was 1.77 mg g^{-1} DW, decreased gradually during storage, reaching 0.61 mg g^{-1} DW on day 21 (Table 2), which was the lowest value in the whole experiment, and increased to 0.88 mg g^{-1} DW on day 28. Catechin in Polys- and G50-treated cherries exhibited significant changes during storage, but not consistent, averaging 2.37 and 2.69 mg g^{-1} DW, respectively, resulting in 1.85 and 2.15 mg catechin g^{-1} DW, respectively, on day 28. In chitosan and G25, it decreased progressively during storage, reaching 1.2 and 1.41 mg g^{-1} DW at the end of storage. The effects of treatment, storage days and their interaction were all significant.

Epicatechin values were 0.22 mg g^{-1} DW at harvest (Table 2).The controls decreased until day 21 and increased at the end of storage significantly but slightly. When all data were analyzed, the treatments showed similar values which decreased up to day 21 and increased thereafter. The effect of treatment was not significant, but those of storage days and their interaction were significant. In all treated cherries, values averaged 0.20, 0.12, 0.09 and 0.21 mg g^{-1} DW on days 7, 14, 21 and 28, respectively.

Partial analysis of 21 and 28 d data (two-way ANOVA) showed the significant effect of treatment on chlorogenic, neochlorogenic, catechin and epicatechin, with the control values being the lowest in all cases, and all treatments equally effective in increasing the acids, whereas G50 was the most effective in increasing catechin.

4. Discussion

4.1. Peel Color

The present TSS, pH and TA values (Table 1) are in general agreement with other 'Regina' studies [35–37], and the present color values agree with those of the work of Harb et al. [6]. 'Regina' is appreciated by consumers who prefer the mahogany cherries, while at advanced maturity stage cherries obtain a good eating quality [38] with high antioxidant levels [39]. In this experiment, the control fruit deteriorated after 21 d of storage, in contrast to the treated ones that were marketable even after 28 d.

Chitosan treatment resulted in considerably higher L^* values than all other cherries throughout storage, but in lower hue angle increases up to 21 d. Therefore, chitosan-treated cherries showed the best shininess with a mahogany color, exhibiting an appearance improvement [10]. Increases in hue angle during low-temperature storage were also found in 'Regina' controls harvested at a relatively high TSS and low hue angle values by others [6] and in alginate-treated cherries [40]. The increases in hue angle could be attributed to a loss of anthocyanins and/or to a lower rate of their synthesis. In other cherry studies, the hue angle decreased during storage [40], and the differences in hue angle changes could be attributed to the maturity stage at harvest. However, the decreases in hue angle were also reduced in treated ones with aloe [19], alginate [41], almond or Arabic gum [21] and chitosan [17,42]. In this work, because of the mahogany peel color, no browning could be observed.

4.2. Weight Loss and Respiration Rates

Here, the WL of controls, averaged 4.15% during storage, was comparable to other cherry studies [17,19,43,44]. Bai et al. [4] found that WL in macro-perforated packages was <1% after 6 weeks of storage at low temperature. Indeed, the perforated material and the number, area and frequency of perforation were different between the studies. In addition, the weight of fruit per package was very low in the present work, justifying the increased WL. Additionally, in this experiment there was no plan to calculate the contribution of the macro-perforated packages to the reduction in WL.

In 'Regina', the WL of the treated cherries was consistently and similarly reduced by all ECs in comparison to the controls from day 7 to the end of storage. Indicatively, the averaged WL of treated cherries was 0.74- and 0.75-fold lower than the controls on days 21 and 28, respectively. In cherries, it was shown that chitosan lowered WL [15,17], with the reduction being increased by increasing the chitosan concentration [15]. Similarly, reduced WL in cherries was achieved with other ECs [19–21] or in other species treated with OFI mucilage [24–26]. In another work on cherries, the increased hydrophobicity increased the reduction in WL and firmness [45]. However, WL results in lower fruit volume in cherries due to the lack of peel flexibility [46]. The incorporation of a plasticizer, such as glycerol, reduces the rigidity of the coating, by increasing its strength of elongation, although it also increases the WVP and WL. However, some cracks or flakes due to WL or mechanical damage are eliminated after the plasticizer's addition [12]. Therefore, there is no recommendation for the hydrophobic/hydrophilic ratio. The properties of the coating depend on many factors, such the particular coating composition, conditions of storage and properties of the peel.

Cherries belong to non-climacteric fruit [43,47], although this classification is considered oversimplified [48]. Increases in ethylene production, although limited, and in RR were observed in ripening cherries [21,28], but these increases do not strictly comply with the non-climacteric behavior. Since the RR in cherries is considerable [7,43], and along with glycolysis reflects energy status, reduced RR is required to avoiding consuming high energy levels [6]. Reduced RR increases were observed in cherries treated with guar gum [20], almond or Arabic gum [21], chitosan [15], as well as in strawberries with OFI mucilage [25]. On the contrary, in several cherry studies, the RR decreased during storage, but ECs again exhibited their beneficial effect on increased reductions [17]. The difference in the direction of respiration changes is attributed to the maturity stage at harvest and conditions during and after storage.

Here, the rates of respiration are comparable to other cultivars [20,28], while all ECs exhibited a similarly positive effect on reduced RR. On day 21, the RR of G25-, G50-, Polys- and chitosan-treated fruit was lower than controls by 0.77-, 0.66- 0.7- and 0.68-fold, respectively, while on day 28 it was lower by 0.92-, 088-, 0.83- and 0.77-fold.

4.3. Firmness

The final firmness values comprise the result of non-equivalent rates of softening and of increases in firmness due to WL [28]. A similar trend of firmness increases in cherries during storage was found in three out of six cultivars [49] and in alginate-treated ones [40]. This discrepancy between the increased and decreased firmness is associated with the cultivar [49], maturity stage and the conditions during their shelf life.

In this work, the contribution of all ECs to firmness increases seems to be higher than that of WL, as compared to the firmness and WL of controls (Figure 3a,c). These results are in line with chitosan or other ECs applied to cherries [19–21,41,50] or to other commodities [12], and with OFI mucilage on strawberries [25] and cut kiwi fruit [24].

However, here, the best effect on firmer fruit was observed by chitosan throughout storage, by comparison.

The present beneficial effect of treatments on firmness could be ascribed, at least partially, to the lower enzyme activities related to the firmness, and has been demonstrated by the enzyme activities and their reaction products. The chitosan effect on reduced softening in Chinese cherries was explained by the reduced gene expression of pectin methylesterase (PME) genes, PME activity, the lower content of sodium carbonate soluble pectins (SCSP), the lower rate of pectin demethylation and the loss of main and side chain neutral sugars of rhamnogalacturonan I (primary structure of SCSP in cherries) [50]. Additionally, in kiwi slices, the higher firmness of the tissue treated with OFI mucilage was attributed to the higher total pectin and protopectin concentrations during storage, implying the lower respective enzyme activities in comparison to controls [24]. Enzyme activities connected with firmness loss require O_2 and ethylene. Here, the levels of O_2 were reduced by ECs, but it was not known whether chitosan that resulted in higher firmness suppressed the O_2 levels more than the other ECs used. Ethylene synthesis is low in cherries, but there is no ethylene limit that inactivates ethylene action even under low temperatures. Moreover, increases in ethylene along with firmness loss and enhancement of SP content have been observed in cherries during low-temperature air storage [28].

4.4. Microbial Decay, Pedicel Removal

Ripe fruits are very vulnerable to microbial decay. Chitosan is known to prevent the fungal decay of fresh produce. Its positive charge of amino groups reacts with the negatively charged microbial membrane, inhibiting DNA replication, but also it binds to metals, inhibiting microbial growth [18]. Additionally, it triggers fruit defense responses by increasing the activities of chitinase, β-1,3 glucanase (directly preventing the microbial growth) and phenylalanine ammonia lyase (PAL) (inducing the synthesis of phenolic compounds) [18]. Here, although G50 and Polys exhibited higher decay than the other samples on day 28, the decay still remained at very low levels. Moreover, decay was also limited in controls during storage. Although glycerol has antimicrobial activity [51], it is suggested that decay prevention was the main outcome from the small fruit groups. Allegra et al. [24] found that OFI mucilage increased the growth of yeast slightly in kiwi slices, whereas it resulted in the low development of *Enterobacteriacea* in figs [26]. Further research is needed to investigate the effect of these ECs on the decay of cherries packaged in larger groups.

Pedicel removal force (PRF) is an indicator of the adhesion and retention of pedicels during postharvest life, which is essential to reduce WL and microbial contamination. Additionally, the appearance of the attached fresh green pedicels plays a crucial role in market value. Here, the PRF was not affected by treatment, while it exhibited inconsistent changes during storage, which might be the interaction of ECs, and WL of both fruit and pedicels. In other studies of uncoated cherries, PRF decreased [35,43] in contrast to this work.

4.5. Total Antioxidants and Total Antioxidant Capacity

Mahogany cherries exhibit higher antioxidant concentrations than those with a bright red color [38]. The present results comply well with other studies on 'Regina' [36,52] and are rather close to the highest values among cherry cultivars [35,53].

Here, it is of interest that the patterns of changes in TP, TF and TAN during storage had very close similarity, exhibiting a decline up to day 21 d. It is known that cherry antioxidants increase during ripening [39], and it seems that maxima concentrations of TP, TF and TAN in the ripe 'Regina' were observed at harvest. However, an increase was observed in total antioxidants on day 28 at levels up to the initial ones. All ECs and controls resulted in similar decreases and increases during storage.

There has been an argument about the effect of low temperature storage on cherry anthocyanin changes. For example, the elevation of anthocyanin concentration was found by Gonçalvez et al. [54,55], in contrast to decreases presented by others [17]. A further study on cherries with regard to this issue found reduced transcription of genes coding for the enzymes anthocyanidin synthase (ANS) and flavonol 3-*O*-glucosyltransferase (UFGT) (crucial enzymes for anthocyanin synthesis) and of PAL (the initial enzyme of the phenylpropanoid pathway) and limited TAN increase, but stable TP levels in cherries at low non-chilling temperature compared to those at harvest, indicating a complex regulation of phenolic compounds [56]. Here, it is suggested that ripe cherries had probably almost exhausted their ability to synthesize more phenolic compounds, at least at low temperature, whereas an effort to recover was observed towards the end of storage. The decreases might be the result of the deceleration phenolic synthesis under low temperature [56] and their depletion to defend the reactive oxygen species (ROS) that were inevitably produced after harvest [57]. Indeed, a pattern of PAL activity shown by Dang et al. [15] during air storage complies well with the present pattern of TP, TF and TAN. In addition, in this work TAN decreases are in general agreement with hue angle increases.

Chitosan treatment in other cherry studies resulted in lower decreases in antioxidants than controls [17]. Enhanced activities of PAL and antioxidant enzymes and prevention of PPO and lipoxygenases (LOX) due to chitosan treatment contributed to the extension of cherry storage life [15,16]. PAL increases at late periods of storage were also attributed, at least partially, to chitosan effect [18], and all these comply with most antioxidant changes here. Nevertheless, the effectiveness of any EC on fruit depends on the cultivar and/or its composition/properties as well as the method of preparation and application. For example, increasing the alginate concentrations above 1%, as an edible coating in one study [40], but also decreasing to 1% in another study [41], had an impact on extending storage life, maintaining antioxidants at elevated levels in stored cherries. Additionally, guar gum applied to cherries maintained a higher concentration of ascorbic acid and increased TP levels, but suppressed the TAN levels during cold storage compared to controls [20], indicating the complexity of phenolic compound synthesis as well. OFI mucilage on figs in the store did not have any effect on TP, but a positive one on total carotenoids [26].

Generally, TAC renders the cherries as a source of natural antioxidants [2]. Anthocyanins comprise a large part of the TP amount, with c-3-rut primarily contributing to TAN [52]. C-3-rut possesses a high antioxidant capacity [58]. 'Regina', being a dark-colored cultivar, exhibited a relatively high TAC determined either by the FRAP or DPPH assays, similar to another 'Regina' study [37]. The different results obtained between the FRAP and DPPH assays here were expected and agree with another cherry study [43], since they are based on different methods. It is important that all EC-treated cherries possessed similarly higher FRAP levels (Figure 5c) than controls after 21 and 28 d of storage.

Considering that controls were non-marketable after 28 d of storage due to their appearance (attributed to high WL), the present results confirm the beneficial effect of coatings not only on extending the cherry storage, but also on maintaining the antioxidant concentrations close to the initial levels.

4.6. Phenolic Compounds

Phenolic compounds also play an important role in cherry quality since they influence the appearance, taste and nutritional value of fruit [1]. The most representative classes of cherry phenolic compounds are anthocyanins, hydroxycinnamic acid derivatives and flavan-3-ols, while flavanols (such as rutin) have also been determined [1,54,59].

C-3-rut is the major anthocyanin in cherries, comprising approximately up to 95% of TAN [1]. The present values of c-3-rut ranging between 1.13 and 3.07 mg g^{-1} DW, are in agreement with other cherry studies [36,52]. It is of interest that the averaged c-3-rut values in controls were the lowest throughout storage, whereas those in fruit coated with OFI extracts were the highest after 28 d of storage, and indeed were higher than the initial values. P-3-glc showed a similar trend of changes, but after 28 d of storage the levels of all treated cherries with ECS were close to the initials, being 2.4-fold higher than controls. C-3-glc and m-3-glc were minor anthocyanins in 'Regina'. The present c-3-glc values are very similar to other 'Regina' studies [36,52], while p-3-glc and m-3-glc are in accordance with those found in other cultivars by Gonçalvez et al. [54] and Martini et al. [59], respectively. Regarding the acids chlorogenic and neochlorogenic, as well as the flavan-3-*O*-ols catechin and epicatechin, their concentrations are comparable with other 'Regina' studies [36,52], while their changes followed the pattern of anthocyanins. At the end of storage, G50 exhibited the highest catechin concentration.

Therefore, at the end of storage, the OFI extracts were the most effective in containing higher c-3-rut levels even than the initial ones, while G50 in the highest catechin concentration.

5. Conclusions

The results here show that all ECs reduced WL and RR similarly and increased firmness values compared to the controls, resulting in extending the cherry storage by one week. Among treatments, chitosan was superior in enhancing peel glossiness and firmness when compared to OFI extracts. OFI extracts exhibited elevated c-3-rut and G50 catechin levels after 28 d of storage and no negative effect among the determined variables.

The use of raw materials from plant residues is desirable, while OFI cladodes are rich in dietary fibers, calcium, potassium, carbohydrates and polyphenols. Nevertheless, *Opuntia* genotype, cladode age, cultivation area and pruning season affect their composition. Cladodes are abundant after pruning in cultivation areas of cactus pear, and comprise an alternative for ECs preparation at low cost, while in a dried form would facilitate their storage and transport [60]. In OFI cladode extracts, incorporation of other materials or compounds along with plasticizers could improve the properties and functionalities of the coating. However, more experiments are needed to improve the effectiveness of OFI extracts on cherry storage. After all, *Opuntia* cladode extracts have potential as ECs to extend the storage life and improve the quality of cherries.

Author Contributions: Conceptualization, M.V.C., J.B., M.K., E.T.; methodology, M.V.C., M.K., J.B., G.T., D.F., E.T.; investigation, M.V.C., D.G., M.K.; writing—original draft preparation, M.V.C., M.K., E.T.; writing—review and editing, M.V.C., D.G., M.K., J.B., D.F., G.T., E.T. All authors have read and agreed to the published version of the manuscript.

Funding: This research received no external funding.

Data Availability Statement: Data is contained within the article.

Acknowledgments: The authors thank G. Papadopoulos for statistical advice.

Conflicts of Interest: The authors declare no conflict of interest.

References

1. Serradilla, M.J.; Hernández, A.; López-Corrales, M.; Ruiz-Moyano, S.; de Guía Córdoba, M.; Martín, A. Composition of the Cherry (*Prunus avium* L. and *Prunus cerasus* L.; *Rosaceae*). In *Nutritional Composition of Fruit Cultivars*; Simmonds, M.S.J., Preedy, V.R., Eds.; Academic Press: San Diego, CA, USA, 2016; pp. 127–147. [CrossRef]
2. Kent, K.; Hölzel, N.; Swarts, N. Polyphenolic Compounds in Sweet Cherries: A Focus on Anthocyanins. In *Polyphenols: Mechanisms of Action in Human Health and Disease*, 2nd ed.; Watson, R.R., Preedy, V.R., Zibadi, S., Eds.; Academic Press: San Diego, CA, USA, 2018; pp. 103–118. [CrossRef]
3. Linke, M.; Herppich, W.B.; Geyer, M. Green peduncles may indicate postharvest freshness of sweet cherries. *Postharvest Biol. Technol.* **2010**, *58*, 135–141. [CrossRef]
4. Bai, J.; Plotto, A.; Spotts, R.; Rattanapanone, N. Ethanol vapor and saprophytic yeast treatments reduce decay and maintain quality of intact and fresh-cut sweet cherries. *Postharvest Biol. Technol.* **2011**, *62*, 204–212. [CrossRef]
5. Bai, J.; Baldwin, E.; Tsantili, E.; Plotto, A.; Sun, X.; Wang, L.; Kafkaletou, M.; Wang, Z.; Narciso, J.; Zhao, W.; et al. Modified humidity clamshells to reduce moisture loss and extend storage life of small fruits. *Food Packag. Shelf Life* **2019**, *22*, 100376. [CrossRef]
6. Harb, J.; Saquet, A.; Bisharat, R.; Streif, J. Quality and biochemical changes of sweet cherries cv. Regina stored in modified atmosphere packaging. *J. Appl. Bot. Food Qual.* **2006**, *80*, 145–149.
7. Wang, Y.; Long, L.E. Respiration and quality responses of sweet cherry to different atmospheres during cold storage and shipping. *Postharvest Biol. Technol.* **2014**, *92*, 62–69. [CrossRef]

8. Wang, Y.; Bai, J.; Long, L.E. Quality and physiological responses of two late-season sweet cherry cultivars 'Lapins' and 'Skeena' to modified atmosphere packaging (MAP) during simulated long distance ocean shipping. *Postharvest Biol. Technol.* **2015**, *110*, 1–8. [CrossRef]
9. Wani, A.A.; Singh, P.; Gul, K.; Wani, M.H.; Langowski, H.C. Sweet cherry (*Prunus avium*): Critical factors affecting the composition and shelf life. *Food Packag. Shelf Life* **2014**, *1*, 86–99. [CrossRef]
10. Bai, J.; Plotto, A. Coatings for fresh fruits and vegetables. In *Edible Coatings and Films to Improved Food Quality*, 2nd ed.; Taylor & Francis Group: Boca Raton, FL, USA, 2012; pp. 185–242.
11. Park, H.J. Development of advanced edible coatings for fruits. *Trends Food Sci. Technol.* **1999**, *10*, 254–260. [CrossRef]
12. Riva, S.C.; Opara, U.O.; Fawole, O.A. Recent developments on postharvest application of edible coatings on stone fruit: A review. *Sci. Hortic.* **2020**, *262*, 109074. [CrossRef]
13. Han, C.; Lederer, C.; McDaniel, M.; Zhao, Y. Sensory Evaluation of Fresh Strawberries (*Fragaria ananassa*) Coated with Chitosan-based Edible Coatings. *J. Food Sci.* **2005**, *70*, S172–S178. [CrossRef]
14. Chien, P.-J.; Sheu, F.; Yang, F.-H. Effects of edible chitosan coating on quality and shelf life of sliced mango fruit. *J. Food Eng.* **2007**, *78*, 225–229. [CrossRef]
15. Dang, Q.F.; Yan, J.Q.; Li, Y.; Cheng, X.J.; Liu, C.S.; Chen, X.G. Chitosan Acetate as an Active Coating Material and Its Effects on the Storing of *Prunus avium* L. *J. Food Sci.* **2010**, *75*, S125–S131. [CrossRef]
16. Pasquariello, M.S.; Di Patre, D.; Mastrobuoni, F.; Zampella, L.; Scortichini, M.; Petriccione, M. Influence of postharvest chitosan treatment on enzymatic browning and antioxidant enzyme activity in sweet cherry fruit. *Postharvest Biol. Technol.* **2015**, *109*, 45–56. [CrossRef]
17. Petriccione, M.; De Sanctis, F.; Pasquariello, M.S.; Mastrobuoni, F.; Rega, P.; Scortichini, M.; Mencarelli, F. The Effect of Chitosan Coating on the Quality and Nutraceutical Traits of Sweet Cherry During Postharvest Life. *Food Bioprocess Technol.* **2015**, *8*, 394–408. [CrossRef]
18. Romanazzi, G.; Feliziani, E.; Sivakumar, D. Chitosan, a Biopolymer With Triple Action on Postharvest Decay of Fruit and Vegetables: Eliciting, Antimicrobial and Film-Forming Properties. *Front. Microbiol.* **2018**, *9*, 2745. [CrossRef]
19. Martínez-Romero, D.; Alburquerque, N.; Valverde, J.M.; Guillén, F.; Castillo, S.; Valero, D.; Serrano, M. Postharvest sweet cherry quality and safety maintenance by Aloe vera treatment: A new edible coating. *Postharvest Biol. Technol.* **2006**, *39*, 93–100. [CrossRef]
20. Dong, F.; Wang, X. Guar gum and ginseng extract coatings maintain the quality of sweet cherry. *LWT* **2018**, *89*, 117–122. [CrossRef]
21. Mahfoudhi, N.; Hamdi, S. Use of Almond Gum and Gum Arabic as Novel Edible Coating to Delay Postharvest Ripening and to Maintain Sweet Cherry (*Prunus avium*) Quality during Storage. *J. Food Process. Preserv.* **2015**, *39*, 1499–1508. [CrossRef]
22. Athmaselvi, K.A.; Sumitha, P.; Revathy, B. Development of Aloe vera based edible coating for tomato. *Int. Agrophys.* **2013**, *27*, 369–375. [CrossRef]
23. Riaz, S.; Sultan, M.T.; Sibt-e-Abass, M.; Irman, M.; Ahmad, R.S.; Hussain, M.B.; Shariati, M.A.; Kosenko, I.S.; Kleymenova, N.L.; Egorova, G.N. Extraction of polysaccharides from opuntia cactus for its potential application in edible coating to improve the shelf life of citrus (Kinnow mandarin) fruit. *J. Microbiol. Biotechnol. Food Sci.* **2018**, *8*, 745–750. [CrossRef]
24. Allegra, A.; Inglese, P.; Sortino, G.; Settanni, L.; Todaro, A.; Liguori, G. The influence of Opuntia ficus-indica mucilage edible coating on the quality of 'Hayward' kiwifruit slices. *Postharvest Biol. Technol.* **2016**, *120*, 45–51. [CrossRef]
25. Del-Valle, V.; Hernández-Muñoz, P.; Guarda, A.; Galotto, M.J. Development of a cactus-mucilage edible coating (Opuntia ficus indica) and its application to extend strawberry (*Fragaria ananassa*) shelf-life. *Food Chem.* **2005**, *91*, 751–756. [CrossRef]
26. Allegra, A.; Sortino, G.; Inglese, P.; Settanni, L.; Todaro, A.; Gallotta, A. The effectiveness of Opuntia ficus-indica mucilage edible coating on post-harvest maintenance of 'Dottato' fig (*Ficus carica* L.) fruit. *Food Packag. Shelf Life* **2017**, *12*, 135–141. [CrossRef]
27. Tsantili, E.; Rouskas, D.; Christopoulos, M.V.; Stanidis, V.; Akrivos, J.; Papanikolaou, D. Effects of two pre-harvest calcium treatments on physiological and quality parameters in 'Vogue' cherries during storage. *J. Hortic. Sc. Biotechnol.* **2007**, *82*, 657–663. [CrossRef]
28. Christopoulos, M.V.; Tsantili, E. Storage of fresh walnuts (*Juglans regia* L.)—Low temperature and phenolic compounds. *Postharvest Biol. Technol.* **2012**, *73*, 80–88. [CrossRef]
29. Blackhall, M.L.; Berry, R.; Davies, N.W.; Walls, J.T. Optimized extraction of anthocyanins from Reid Fruits 'Prunus avium' 'Lapins' cherries. *Food Chem.* **2018**, *256*, 280–285. [CrossRef]
30. Tsantili, E.; Shin, Y.; Nock, J.F.; Watkins, C.B. Antioxidant concentrations during chilling injury development in peaches. *Postharvest Biol. Technol.* **2010**, *57*, 27–34. [CrossRef]
31. Meyers, K.J.; Watkins, C.B.; Pritts, M.P.; Liu, R.H. Antioxidant and Antiproliferative Activities of Strawberries. *J. Agric. Food Chem.* **2003**, *51*, 6887–6892. [CrossRef]
32. Benzie, I.F.F.; Strain, J.J. Ferric reducing/antioxidant power assay: Direct measure of total antioxidant activity of biological fluids and modified version for simultaneous measurement of total antioxidant power and ascorbic acid concentration. In *Methods in Enzymology*; Lester, P., Ed.; Academic Press: Cambridge, MA, USA, 1999; Volume 299, pp. 15–27.
33. Brand-Williams, W.; Cuvelier, M.E.; Berset, C. Use of a free radical method to evaluate antioxidant activity. *LWT Food Sci. Technol.* **1995**, *28*, 25–30. [CrossRef]
34. Durst, R.W.; Wrolstad, R.E. Separation and Characterization of Anthocyanins by HPLC. *Curr. Protoc. Food Anal. Chem.* **2001**, F1.3.1–F1.3.13. [CrossRef]

35. Karagiannis, E.; Michailidis, M.; Karamanoli, K.; Lazaridou, A.; Minas, I.S.; Molassiotis, A. Postharvest responses of sweet cherry fruit and stem tissues revealed by metabolomic profiling. *Plant Physiol. Biochem.* **2018**, *127*, 478–484. [CrossRef]
36. Milinović, B.; Dragović-Uzelac, V.; Kazija, D.H.; Jelačić, T.; Vujević, P.; Čiček, D.; Biško, A.; Čmelik, Z. Influence of four different dwarfing rootstocks on phenolic acids and anthocyanin composition of sweet cherry (*Prunus avium* L.) cvs 'Kordia' and 'Regina'. *J. Appl. Bot. Food Qual.* **2016**, *89*.
37. Commisso, M.; Bianconi, M.; Di Carlo, F.; Poletti, S.; Bulgarini, A.; Munari, F.; Negri, S.; Stocchero, M.; Ceoldo, S.; Avesani, L. Multi-approach metabolomics analysis and artificial simplified phytocomplexes reveal cultivar-dependent synergy between polyphenols and ascorbic acid in fruits of the sweet cherry (*Prunus avium* L.). *PLoS ONE* **2017**, *12*, e0180889.
38. Serrano, M.; Diaz-Mula, H.M.; Zapata, P.J.; Castillo, S.; Guillén, F.; Martinez-Romero, D.; Valverde, J.M.; Valero, D. Maturity stage at harvest determines the fruit quality and antioxidant potential after storage of sweet cherry cultivars. *J. Agric. Food Chem.* **2009**, *57*, 3240–3246. [CrossRef]
39. Serrano, M.; Guillén, F.; Martínez-Romero, D.; Castillo, S.; Valero, D. Chemical constituents and antioxidant activity of sweet cherry at different ripening stages. *J. Agric. Food Chem.* **2005**, *53*, 2741–2745. [CrossRef]
40. Chiabrando, V.; Giacalone, G. Effects of alginate edible coating on quality and antioxidant properties in sweet cherry during postharvest storage. *Ital. J. Food Sci.* **2015**, *27*, 173–180.
41. Díaz-Mula, H.M.; Serrano, M.; Valero, D. Alginate Coatings Preserve Fruit Quality and Bioactive Compounds during Storage of Sweet Cherry Fruit. *Food Bioprocess Technol.* **2012**, *5*, 2990–2997. [CrossRef]
42. Zhang, Y.-L.; Cui, Q.-L.; Wang, Y.; Shi, F.; Liu, Y.-P.; Liu, J.-L.; Nie, G.-W. Effect of carboxymethyl chitosan-gelatin-based edible coatings on the quality and antioxidant properties of sweet cherry during postharvest storage. *Sci. Hortic.* **2021**, *289*, 110462. [CrossRef]
43. Kafkaletou, M.; Christopoulos, M.V.; Ktistaki, M.-E.; Sotiropoulos, T.; Tsantili, E. The influence of rain cover on respiration, quality attributes and storage of cherries (*Prunus avium* L.). *J. Appl. Bot. Food Qual.* **2015**, *88*, 87–96. [CrossRef]
44. Toivonen, P.M.; Kappel, F.; Stan, S.; McKenzie, D.-L.; Hocking, R. Firmness, respiration, and weight loss of Bing 'Lapins' and Sweetheart' cherries in relation to fruit maturity and susceptibility to surface pitting. *HortScience* **2004**, *39*, 1066–1069. [CrossRef]
45. Rojas-Argudo, C.; Pérez-Gago, M.B.; del Río, M.A. Postharvest Quality of Coated Cherries cv. 'Burlat' as Affected by Coating Composition and Solids Content. *Food Sci. Technol. Int.* **2005**, *11*, 417–424. [CrossRef]
46. Kritzinger, I.; Theron, K.I.; Lötze, G.F.A.; Lötze, E. Peel water vapour permeance of Japanese plums as indicator of susceptibility to postharvest shriveling. *Sci. Hortic.* **2018**, *242*, 188–194. [CrossRef]
47. Biale, J.; Young, R. Respiration and ripening in fruits. In *Recent Advances in the Biochemistry of Fruits and Vegetables*; Friend, J., Rhodes, M.J., Eds.; Academic Press: London, UK, 1984; pp. 1–39.
48. Abdi, N.; McGlasson, W.B.; Holford, P.; Williams, M.; Mizrahi, Y. Responses of climacteric and suppressed-climacteric plums to treatment with propylene and 1-methylcyclopropene. *Postharvest Biol. Technol.* **1998**, *14*, 29–39. [CrossRef]
49. Toivonen, P.; Kappel, F. Effect of cold storage duration on shelf quality attributes of 'Bing'and five sweet cherry cultivars developed in the Summerland breeding program. In Proceedings of the XXVIII International Horticultural Congress Science and Horticulture for People (IHC2010): International Symposium on 934, Lisbon, Portugal, 22–27 August 2012; pp. 1011–1016.
50. Xin, Y.; Jin, Z.; Chen, F.; Lai, S.; Yang, H. Effect of chitosan coatings on the evolution of sodium carbonate-soluble pectin during sweet cherry softening under non-isothermal conditions. *Int. J. Biol. Macromol.* **2020**, *154*, 267–275. [CrossRef]
51. Chiu, P.-E.; Lai, L.-S. Antimicrobial activities of tapioca starch/decolorized hsian-tsao leaf gum coatings containing green tea extracts in fruit-based salads, romaine hearts and pork slices. *Int. J. Food Microbiol.* **2010**, *139*, 23–30. [CrossRef]
52. Kim, D.-O.; Heo, H.J.; Kim, Y.J.; Yang, H.S.; Lee, C.Y. Sweet and Sour Cherry Phenolics and Their Protective Effects on Neuronal Cells. *J. Agric. Food Chem.* **2005**, *53*, 9921–9927. [CrossRef]
53. Chockchaisawasdee, S.; Golding, J.B.; Vuong, Q.V.; Papoutsis, K.; Stathopoulos, C.E. Sweet cherry: Composition, postharvest preservation, processing and trends for its future use. *Trends Food Sci. Technol.* **2016**, *55*, 72–83. [CrossRef]
54. Gonçalves, B.; Landbo, A.-K.; Knudsen, D.; Silva, A.P.; Moutinho-Pereira, J.; Rosa, E.; Meyer, A.S. Effect of ripeness and postharvest storage on the phenolic profiles of cherries (*Prunus avium* L.). *J. Agric. Food Chem.* **2004**, *52*, 523–530. [CrossRef]
55. Gonçalves, B.; Silva, A.P.; Moutinho-Pereira, J.; Bacelar, E.; Rosa, E.; Meyer, A.S. Effect of ripeness and postharvest storage on the evolution of colour and anthocyanins in cherries (*Prunus avium* L.). *Food Chem.* **2007**, *103*, 976–984. [CrossRef]
56. Tsaniklidis, G.; Kafkaletou, M.; Delis, C.; Tsantili, E. The effect of postharvest storage temperature on sweet cherry (*Prunus avium* L.) phenolic metabolism and colour development. *Sci. Hortic.* **2017**, *225*, 751–756. [CrossRef]
57. Toivonen, P.M.; Hodges, D.M. Abiotic Stress in Harvested Fruits and Vegetables. In *Abiotic Stress in Plants-Mechanisms and Adaptations*; Shanker, A., Ed.; InTech: Changzhou, China, 2011; pp. 39–58. Available online: http://www.intechopen.com/books/abiotic-stress-in-plants-mechanisms-andadaptations/abiotic-stress-in-harvested-fruits-and-vegetables (accessed on 10 December 2021).
58. Rice-Evans, C.A.; Miller, N.J.; Paganga, G. Structure-antioxidant activity relationships of flavonoids and phenolic acids. *Free Radical Biol. Med.* **1996**, *20*, 933–956. [CrossRef]
59. Martini, S.; Conte, A.; Tagliazucchi, D. Phenolic compounds profile and antioxidant properties of six sweet cherry (*Prunus avium*) cultivars. *Food Res. Int.* **2017**, *97*, 15–26. [CrossRef] [PubMed]
60. Di Bella, G.; Vecchio, G.L.; Albergamo, A.; Nava, V.; Bartolomeo, G.; Macrì, A.; Bacchetta, L.; Turco, V.L.; Potortì, A.G. Chemical characterization of Sicilian dried nopal [*Opuntia ficus-indica* (L.) Mill.]. *J. Food Compos. Anal.* **2022**, *106*, 104307. [CrossRef]

MDPI

Article

Functional Characteristics of Aldehyde Dehydrogenase and Its Involvement in Aromatic Volatile Biosynthesis in Postharvest Banana Ripening

Yoshinori Ueda [1,†], Wei Zhao [2], Hideshi Ihara [3], Yoshihiro Imahori [4], Eleni Tsantili [5], Sumithra K. Wendakoon [6], Alan Chambers [7] and Jinhe Bai [2,*]

1 Center for Research and Development of Bioresources, Osaka Prefecture University, 1-1 Gakuen-cho, Nakaku, Sakai 599-8531, Osaka, Japan
2 U.S. Horticultural Research Laboratory, Agricultural Research Service, U.S. Department of Agriculture, 2001 S. Rock Rd., Ft Pierce, FL 34945, USA; Wei.zhao@usda.gov
3 Department of Biological Science, Graduate School of Science, Osaka Prefecture University, 1-1 Gakuen-cho, Nakaku, Sakai 599-8531, Osaka, Japan; ihara@b.s.osakafu-u.ac.jp
4 Graduate School of Life and Environmental Science, Osaka Prefecture University, 1-1 Gakuen-cho, Nakaku, Sakai 599-8531, Osaka, Japan; imahori@plant.osakafu-u.ac.jp
5 Laboratory of Pomology, Department of Crop Science, Agricultural University of Athens, Iera Odos 75, 118 55 Athens, Greece; etsantili@aua.gr
6 Department of Bioresource Science, Faculty of Agriculture, Ryukoku University, 1-5 Yokotani, Seta Oe-cho, Otsu 520-2194, Shiga, Japan; swendakoon@agr.ryukoku.ac.jp
7 Tropical Research and Education Center, University of Florida, 18905 SW 280th St., Homestead, FL 33031, USA; ac@ufl.edu
* Correspondence: jinhe.bai@usda.gov; Tel.: +1-772-462-5880; Fax: +1-772-462-5986
† Deceased.

Abstract: Butanol vapor feeding to ripe banana pulp slices produced abundant butyl butanoate, indicating that a portion of butanol molecules was converted to butanoate/butanoyl-CoA via butanal, and further biosynthesized to ester. A similar phenomenon was observed when feeding propanol and pentanol, but was less pronounced when feeding hexanol, 2-methylpropanol and 3-methylbutanol. Enzymes which catalyze the cascade reactions, such as alcohol dehydrogenase (ADH), acetyl-CoA synthetase, and alcohol acetyl transferase, have been well documented. Aldehyde dehydrogenase (ALDH), which is presumed to play a key role in the pathway to convert aldehydes to carboxylic acids, has not been reported yet. The conversion is an oxygen-independent metabolic pathway and is enzyme-catalyzed with nicotinamide adenine dinucleotide (NAD^+) as the cofactor. Crude ALDH was extracted from ripe banana pulps, and the interference from ADH was removed by two procedures: (1) washing off elutable proteins which contain 95% of ADH, but only about 40% of ALDH activity, with the remaining ALDH extracted from the pellet residues at the crude ALDH extraction stage; (2) adding an ADH inhibitor in the reaction mixture. The optimum pH of the ALDH was 8.8, and optimum phosphate buffer concentration was higher than 100 mM. High affinity of the enzyme was a straight chain of lower aldehydes except ethanal, while poor affinity was branched chain aldehydes.

Keywords: *Musa* AAA; ALDH; aroma volatile; ester; enzyme characteristics

Citation: Ueda, Y.; Zhao, W.; Ihara, H.; Imahori, Y.; Tsantili, E.; Wendakoon, S.K.; Chambers, A.; Bai, J. Functional Characteristics of Aldehyde Dehydrogenase and Its Involvement in Aromatic Volatile Biosynthesis in Postharvest Banana Ripening. *Foods* **2022**, *11*, 347. https://doi.org/10.3390/foods11030347

Academic Editor: Arun K. Bhunia

Received: 24 December 2021
Accepted: 23 January 2022
Published: 26 January 2022

1. Introduction

Aldehyde dehydrogenases (ALDH, aldehyde:$NAD(P)^+$ oxidoreductases, EC 1.2.1) use nicotinamide adenine dinucleotide (NAD^+) or nicotinamide adenine dinucleotide phosphate ($NADP^+$) as a cofactor to convert aldehydes to their corresponding carboxylic acids plus NADH or NADPH. ALDH in plants are currently receiving considerable attention because they are involved in processing many aldehydes that serve as biogenic intermediates in a wide range of metabolic pathways [1]. They often function as an 'aldehyde scavenger', thus removing reactive aldehydes generated during the oxidative degradation, especially under environmental stress, such as exposure to salinity, drought, cold, and heat [1].

The highly abundant volatiles in fresh bananas are aldehydes, ketones, alcohols, carboxylic acids, and esters. It has been recognized that the production of straight chain alcohols, aldehydes, ketones, and acids in fruit is largely derived from α-oxidation, β-oxidation, or the lipoxygenase pathway [2,3]. Branched chain volatiles are derived from branched chain amino acids [4,5]. Recently, Sugimoto et al. [6,7] proposed that some branched and straight chain alcohols and acids come from the citramalate pathway.

Many enzymes regarding the conversions between volatile alcohols, aldehydes/ketones, acids, and esters have been studied extensively in fruits and plants, and/or adapted from microorganism studies [8–10]. In banana fruit ripening, gene expression and enzyme activity of alcohol dehydrogenase (ADH, short and medium chains), which facilitates the interconversion between alcohols to aldehydes [11,12], acyl-CoA synthetase (ACS), which activates carboxylic acids to acyl-CoAs [13] thus can be used to biosynthesize esters, and alcohol acetyl transferase (AAT), which catalyzes ester biosynthesis, have been intensively studied (Figure 1) [14–17]. ALDH dehydrogenizes aldehydes to carboxylic acids in different plant tissues [1,18]. However, to our knowledge, there is no report on the role of ALDH in volatile flavor metabolism in fruits.

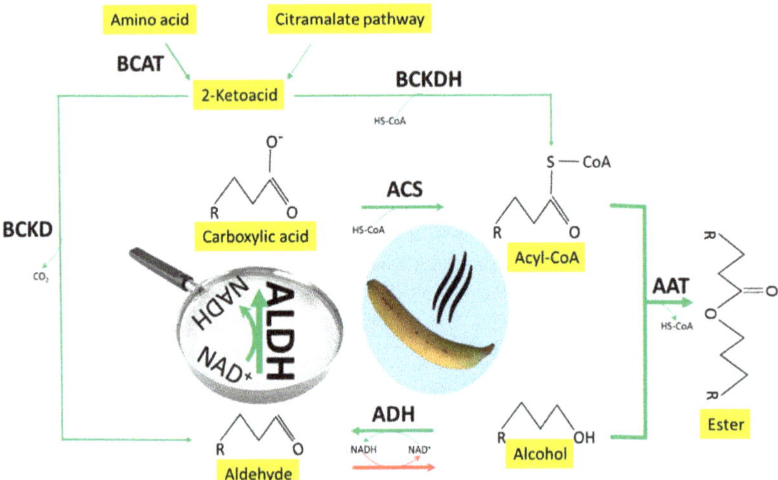

Figure 1. Role of aldehyde dehydrogenase (ALDH) in the dehydrogenation from aldehydes to carboxylic acids and the entire ester production pathways. ADH, alcohol dehydrogenase; ACS, acyl-CoA synthetase; AAT, alcohol acyl-CoA transferase; BCAT, branched-chain amino transferase; BCKD, branched-chain α-ketoacid decarboxylase; BCKDH, branched-chain α-ketoacid dehydrogenase.

Beekwilder, Alvarez-Huerta et al. [16] showed that when incubating petunia leaves with 3-methylbutanol vapor for 24 h, 3-methylbutyl 3-methylbutanoate was the dominant volatile, in addition to 3-methylbutanol in the headspace, indicating a strong dehydrogenization of 3-methylbutanol to 3-methylbutanal, further to 3-methylbutanoic acid, then passing through 3-methylbutanoyl-CoA, and finally synthesizing 3-methylbutyl 3-methylbutanoate, catalyzed by ADH, ALDH, ACS, and AAT, respectively [13–16].

The purpose of the current study was to determine the enzyme which converts volatile straight and branched aldehydes to carboxylic acids, to identify their involvement in volatile metabolisms in banana fruit (Figure 1), and to reveal the functional characteristics of ALDH. Due to both ADH and ALDH activities being determined by the change in NADH concentration, efforts were made to minimize the effect of ADH [19].

2. Materials and Methods

2.1. Chemicals and Reagents

Folin and Ciocalteu's phenol reagent and 2-keto-4-methylpentanoic acid were obtained from Sigma-Aldrich (Tokyo, Japan). Polyclar-VT was purchased from Gokyo (Osaka, Japan). Bovine serum albumin, dithiothreitol (DTT), nicotinamide adenine dinucleotide (NAD^+), nicotinamide adenine dinucleotide phosphate ($NADP^+$), and the reduced forms NADH and NADPH, tris(hydroxymethyl)aminomethane (Tris), 2-[4-(2,4,4-trimethylpentan-2-yl)phenoxy]ethanol (Triton X100), 4-methyl pyrazole, sodium hydroxide (NaOH), phosphoric acid, C2–C6 branched and straight chain alcohols, aldehydes, and carboxylic acids were purchased from Wako (Osaka, Japan).

2.2. Plant Materials

Banana fingers (*Musa* spp. AAA group, Cavendish subgroup) ripened to "yellow with green tips", "yellow", or "yellow, flecked with brown", unless otherwise stated [20], were produced in the Philippines and purchased from local grocery stores, and experiments were conducted in the Osaka Prefecture University lab. For ripening stage influence experiments, unripe fruit were purchased and ripening was triggered by 100 $\mu L\ L^{-1}$ of ethylene at 20 °C for 24 h. For all samples, each replicate contained five fingers, and each treatment had three replicates. For fruit pulp slice samples, cylindrical pulp discs were taken from the central section of a banana finger and further divided into 4 wedges. Pulps were homogenized in some experiments.

2.3. Substrate Feeding Experiments

2.3.1. Feeding of Alcohols and 2-Keto-4-Methylpentanoic Acid

The feeding experiments were conducted by incubating 10 g of banana pulp slices with different exogenous substrates: 50 µmol of C3–C6 straight chain or branched alcohols, or 2-keto-4-methylpentanoic acid, in a 150-mL conical flask for 90 min at 35 °C. The substrates were spotted onto a 40 × 5 mm filter paper strip prior to incubation to enhance evaporation (Figure 2a). After incubation, 1 mL of headspace gas was taken with a glass syringe, and analyzed by gas chromatography (GC, Shimadzu GC-15A, Kyoto, Japan) to target alcohols, aldehydes, or esters. A flame ionization detector was equipped with a packed glass column (2 m × 3 mm) with 20% of tween 20 or 5% of polyethylene glycol 6000. Injector and detector temperatures were 190 °C, and oven temperature was 60–120 °C, depending on the molecular weight of the target esters. Flow rate of carrier nitrogen gas was 20 mL min^{-1} and flaming hydrogen gas was 30 mL min^{-1}.

2.3.2. Feeding of Aldehydes and Trapping of Carboxylic Acids

When feeding experiments were conducted by incubating 10 g of banana pulp slices with 50 µmol of exogenous C2–C6 straight chain or branched aldehydes, the incubation conditions were similar to Section 2.3.1 (Figure 2a,b). Due to the targeted products being carboxylic acids, specific procedures were applied as Figure 2b. Filter paper strips treated with NaOH (1 M) were dried before spotting exogenous substrate aldehydes to remove any possible acid impurities in the substrate aldehydes. A recovery flask (250-mL) was used for the incubation. After incubation, the pulp slices were well soaked with 1.5 mL of phosphoric acid (0.4 M) in the flask. Then, the flask was attached to a rotary evaporator in which a cotton ball was inserted at back-flow prevention glass wear. The cotton ball for the acid trap was previously treated with NaOH (0.1 M) and dried. After rotary evaporation (40 °C for 60 min), the cotton ball was put into a 17 mL vial and then mixed well with 5 mL of phosphoric acid (0.4 M). For ethanoic acid measurement, 5 mL of acidified ether was used instead of the phosphoric acid solution. For the phosphoric acid solution with carboxylic acids, 1 µL was injected into GC to determine the targeted carboxylic acids (Figure 2b). The GC conditions were: a glass column (3 m × 3 mm) of SP1200 (10%) + phosphoric acid (1%), and oven temperature 60–120 °C depending on molecular weight of the target acid. The recovery rate of this method was 53.3% ± 3.2 for ethanoic acid, 81.4% ± 4.7 for propanoic acid, 92.6 ± 1.6 for 2-methylpropanoic acid, 95.6 ± 6.1 for butanoic acid, 97.2% ± 7.4 for 3-methylbutanoic acid, 94.9% ± 7.2 for pentanoic acid, and 97.6% ± 1.3 for hexanoic acid (n = 3).

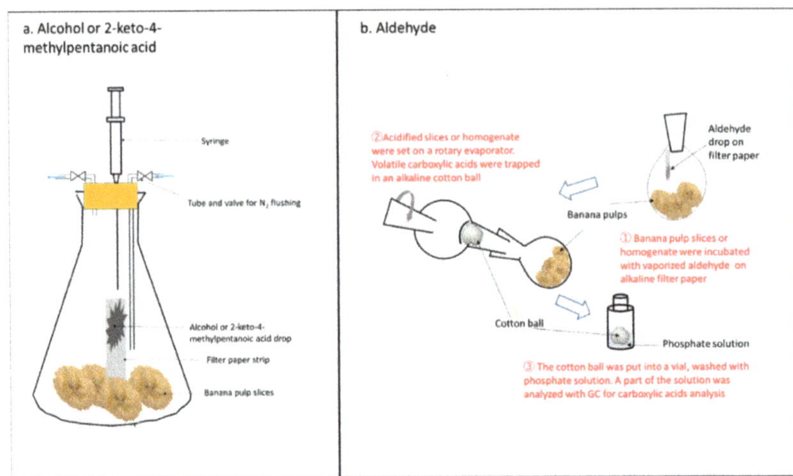

Figure 2. Feeding experiments procedures. (**a**) Feeding alcohol or 2-keto-4-methylpentanoic acid feeding to banana pulps to produce alcohol/aldehyde/esters in air or nitrogen gas. (**b**) Incubating aldehyde vapor with banana pulps to produce corresponding carboxylic acid, and separation of the carboxylic acid from pulps.

The same method as mentioned above was used for the feeding of aldehydes and trapping of acids when pulp homogenate was used instead of pulp slices (Figure 2b). Ten grams of pulp slices were macerated together with 10 mL of phosphate buffer (0.2 M, pH 8.8), 0.8 g of Polyclar-VT, 1.0 mg of DTT, and 3 g of quartz sand with a mortar and pestle in an ice bath. The pH of the homogenate was adjusted to pH 8.8 with 0.1 M NaOH just after macerating.

2.3.3. Setting and Monitoring of O_2 and CO_2 in Incubation

To determine the effect of oxygen on the volatile metabolism, anaerobic conditions were created by flushing the conical flasks with pure nitrogen gas (Figure 2a). The headspace gas sample, 1 mL, was tested to confirm the conditions by using GC-thermal conductivity detector (TCD). A stainless column packed with active charcoal (4 m × 3 mm) was used for oxygen and a stainless column with active aluminum (1 m × 1 mm) was used for carbon dioxide analysis. The flow rate of carrier nitrogen was 10 mL min^{-1}, and oven temperature was 50 °C.

Headspace gas samples during incubation were tested periodically to confirm that there were no severe anaerobic conditions, $O_2 > 15$ kPa and $CO_2 < 5$ kPa when using air as the incubation gas.

2.4. Crude ALDH Extraction from Banana Pulps, and Functional Characteristics of the Enzyme

2.4.1. Crude Protein Extraction

Ten grams of banana pulps were macerated together with 10 mL of Tris buffer (0.2 M, pH 8.8), 0.8 g of Polyclar-VT, 1.0 mg of DTT, and 3 g of quartz sand with a mortar and pestle in an ice bath for 5 min. The final homogenates were adjusted to pH 8.5. From this point, the procedures were conducted under 4 °C in an ice bath. The homogenates were centrifuged at 8300× *g* for 15 min, and supernatants were collected. The pellets were carefully resuspended into 30 mL of Tris buffer (0.1 M, pH 8.8), and centrifuged again in the same condition. The first and second supernatants were merged for ALDH activity test (protein-S). The pellets were resuspended into 30 mL of Tris buffer (0.1 M, pH 8.8, containing 1.0 mg of DTT) with the assistance of ultrasonic wave for 10 min. Then, centrifugation (8300× *g* for 15 min) was applied. The supernatants were used for pellet ALDH activity test (protein-P).

For each of the above supernatants (protein-S and protein-P), solid ammonium sulfate was added to the supernatants (80% saturation: 5.61 g 10 mL^{-1}). After stirring for 2 h, the precipitations were separated by centrifugation at 25,000× *g* for 15 min. The precipitations were solubilized by 10 mL of Tris buffer (0.01 M, pH 8.8) with DTT (1 mg) and dialyzed by using a total of 2 L of Tris buffer (0.01 M, pH 8.8) for 3 h where the buffer was renewed twice. After dialysis, the insoluble pellets were removed by centrifugation at 25,000× *g* for 15 min, and the supernatants were collected as crude protein-S and protein-P, respectively.

A crude protein-PT sample was also prepared to test whether Triton X-100 nonionic surfactant improves protein extraction and enzyme activity of crude protein-P. The procedures were similar to the above for protein-P, and the only change was to add 0.1% Triton X-100 to the Tris buffer during the ultrasonic treatment.

The crude proteins were stored at −20 °C until enzyme characteristics analysis. The activity was stable for at least one month under DTT condition.

2.4.2. Functional Characteristics of ALDH

Substrate specificity: reaction mixture (2 mL) included 1660 μL of phosphate buffer (0.2 M, pH 8.8), 100 μL of low C2–C4 aldehydes (50 mM), 100 μL of crude protein extraction (protein-S, -P or -PT), 40 μL of NAD^+ (50 mM), with or without 100 μL of 4-methyl pyrazole (100 mM). For C5–C6 aldehydes, 500 μL (10 mM) was added with reduced volume of buffer (1260 μL). ALDH activity was determined by monitoring the increase in NADH at 340 nm at 35 °C by a spectrophotometer with a 10 mm path length UV cuvette. Km value of NAD^+ was measured with butanal as the substrate. An extinction coefficient of 6.2 mM^{-1} cm^{-1} was used for the activity calculation.

Optimum pH and buffer concentration: ALDH activity was determined by monitoring the increase in NADH at 340 nm at 35 °C by a spectrophotometer, as described above. Butanal was used as the substrate. For the optimum pH searching, bicine buffer (0.1 M) was used at a pH range of 6.5–10.0. For the optimum buffer concentration searching, phosphate buffer (pH 8.8) was used at a range of 0–200 mM.

Protein contents in the crude ALDH extraction: Lowry method [21,22] was adapted to determine protein content with Folin and Ciocalteu's phenol reagent and bovine serum albumin was used as the standard.

2.4.3. Activity of ADH in Crude Protein Extractions and Inhibition by 4-Methyl Pyrazole

To confirm the purity of ALDH, and exclude the potential influence from other enzymes, especially ADH, the ADH activity in the crude protein extraction listed in Sections 2.4.1 and 2.4.2 was determined by monitoring the decline in NADH at 340 nm at 35 °C by a spectrophotometer in a mixture comprised of 1660 μL of phosphate buffer (0.2 M, pH 8.8), 100 μL of butanol (50 mM), 100 μL of crude protein extraction, and 40 μL of NAD^+ (50 mM) with or without 100 μL of 4-methyl pyrazole (100 mM), where the total volume was 2 mL (balanced with water to make the volume). An extinction coefficient of 6.2 mM^{-1} cm^{-1} was used for the activity calculation [11,19].

2.5. Statistical Analysis

Data were present as the means and standard deviations of three replicates. Statistical analysis was performed with JMP software (version 11.2.2; SAS Institute, Cary, NC, USA). Differences were tested using Tukey's honestly significant difference (HSD) with the significance level at 0.05.

3. Results

3.1. Feeding Alcohols to Banana Pulps to Produce Esters

When butanol vapor was fed to banana pulp slices (at the "yellow with green tips" stage), butyl ethanoate as well as butyl butanoate were accumulated (Figure 3a). Under anaerobic conditions, those esters were barely accumulated (Figure 3a). Green banana (immediately after ripening was triggered by ethylene) pulp was not able to biosynthesize such esters, but obtained the capacity within one day (Figure 3b). Butyl butanoate production increased continually until the "yellow with green tips" stage (day 3 after ethylene treatment), then gradually declined toward senescence (Figure 3b). Similarly, feeding butanol to fruit pulp of ripe muskmelon, pineapple, pawpaw, strawberry, European pear, and apple, also produced butyl butanoate (data not shown). Figure 3a,b results were confirmed when incubating a banana finger in a 1 L jar with butanol vapor, with a much lower yield (about 20%) of butyl butanoate in comparison to that of pulp slices (data not shown). When feeding other C3–C6 branched or straight chain alcohols to banana pulp instead of butanol, propyl propionate and pentyl pentanoate were produced in comparable amounts to butyl butanoate, but productions of hexyl hexanoate, 3-methylbutyl 3-methylbutanoate, and 2-methylpropyl 2-methylpropanoate were much lower (Figure 3c).

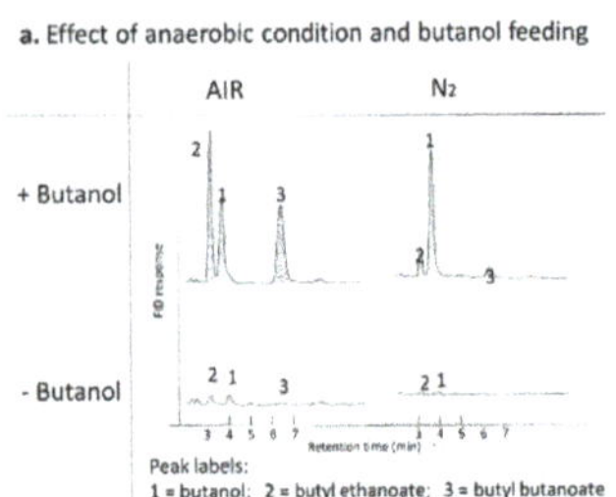

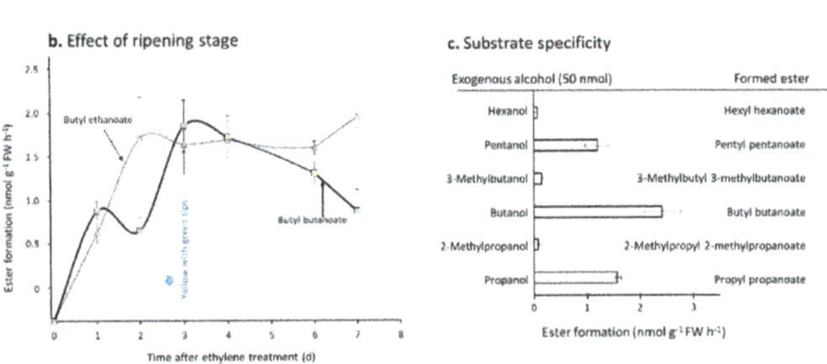

Figure 3. Alcohol feeding experiments and production of corresponding esters in banana pulps. Exogenous alcohol (50 µmol) on filter paper was fed into 10 g of banana pulp tissues. (**a**) Effect of anaerobic conditions and butanol feeding to pulps at the "yellow with green tips" stage; (**b**) effect of ripening stage; (**c**) substrate specificity of C3–C6 branched and straight chain alcohols at the "yellow with green tips" stage. Vertical/horizontal line at each marker/column shows average ± SD (n = 3).

3.2. Feeding Aldehydes to Banana Pulps to Produce the Corresponding Acids

When feeding C2–C6 branched and straight chain aldehydes to the banana pulp slices (at the "yellow with green tips" stage), the acids with the highest production were hexanoic acid, followed by butanoic acid, pentanoic acid, 2-methylpropanoic acid, propanoic acid, and 3-methylbutanoic acid. The acid that was produced in the lowest quantities was ethanoic acid (Figure 4a).

When feeding butanal to banana pulp homogenate samples from fruit from different ripening stages, butanoic acid produced more in over ripe fruit—"yellow with brown flecks" and "brown" stages—and less in the "yellow" stage fruit. Butanoic acid production in fruit at the "yellow with green tips" stage was even lower (Figure 4b).

To confirm the ALDH activity in pulp homogenates, fruit at the "yellow with brown flecks" stage was combined with butanal feeding experiments. Little butanoic acid was detected when the banana pulp homogenate was heated with a 600 W microwave oven for 1 min (Figure 4c). However, anaerobic conditions did not affect the butanoic acid production (Figure 4c). When adding 50 µM NAD^+ to the reaction system, butanoic acid production increased by two-fold in comparison to the addition of 10 µM NAD^+, which was no different from non-NAD^+ addition control (Figure 4c).

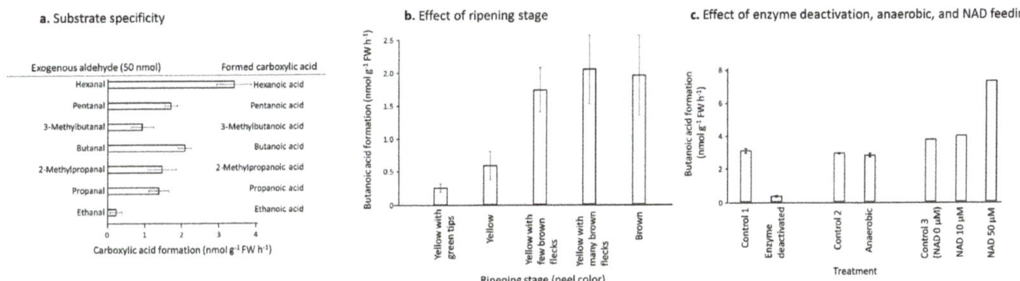

Figure 4. Aldehyde feeding experiments and production of corresponding carbolic acids in banana pulps. Exogenous aldehyde (50 μmol) on filter paper was fed into 10 g of banana pulp tissues. (**a**) Substrate specificity of C2–C6 branched and straight chain aldehydes at the "yellow with green tips" stage (banana pulp slices); (**b**) Effect of ripening stage (pulp homogenate); (**c**) Effects of enzyme deactivation by heat, anaerobic incubation, and feeding of NAD+ at different levels (pulp homogenate). Vertical/horizontal line at each column shows average ± SD (n = 3).

3.3. Extraction and Functional Characteristics of ALDH

As shown in materials and methods, crude proteins were extracted in three ways: from supernatant (protein-S), or from pellets by sonication-assisted extraction with or without Triton X-100 (protein-PT and protein-P, respectively). The protein-S possessed more than 95% of ADH activity, while both protein-P and protein-PT had very little ADH activity (Figure 5a). On the other hand, ALDH activity was evidenced in all protein extractions (Figure 5b,c).

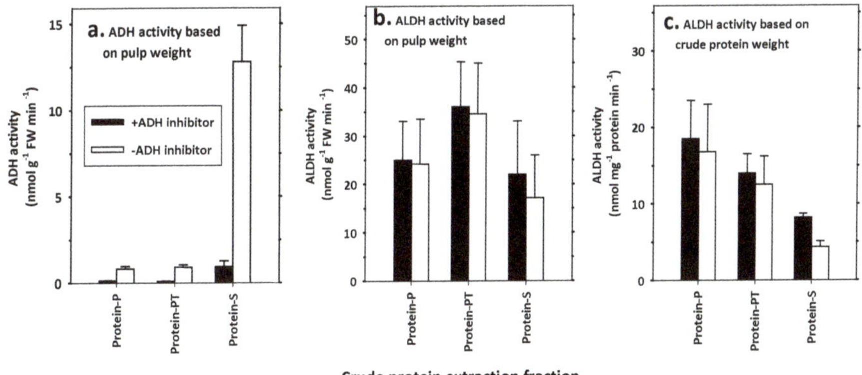

Figure 5. ADH and ALDH activity in crude proteins extracted from different fractions and effect of Triton X-100 (protein extraction enhancer) and 4-methyl pyrazole (ADH inhibitor). Proteins were extracted by ammonium sulfate precipitation from Tris buffer extraction of banana pulp tissues at the "yellow with brown flecks" stage. The activity was presented based on fresh sample mass (**a**,**b**) or protein mass (**c**). Protein-S: extraction from supernatant merged from 2 × Tris buffer extractions; protein-P: extraction from pellets was washed 2 × Tris buffer and solubilized by sonication; protein-PT: extraction from pellets after 2 × Tris buffer washing and solubilized by sonication with addition of Triton X-100. Vertical line at each column shows average ± SD (n = 3).

When based on weight of pulp tissue, calculated ALDH activity distribution for protein-P was slightly more than for protein-S; however, the distribution at pellet extraction was increased when Triton X-100 was used (protein-PT) (Figure 5b). When 4-methyl pyrazole was added into the reaction mixture, ALDH activity distribution at protein-PT was about 60%, and at protein-S, it was about 40% (Figure 5b). However, when calculation was based on protein weight, the highest ALDH activity was found in protein-P, followed by protein-PT and then protein-S. This was partially because Triton X-100 caused more extraction in non-ALDH proteins (protein-PT), and protein-S contained more non-ALDH proteins (Figure 5b,c). ALDH-PT contained more yellow pigments as contaminants. Protein-S had the lowest ALDH activity at protein basis (Figure 5). Thus, protein-P was chosen as crude extract of ALDH in the following experiments.

ALDH activity was very low when the phosphate buffer concentration was less than 20 mM (Figure 6a). The activity increased slightly until the buffer concentration reached saturation at 100 mM (Figure 6a). Optimum pH of the ALDH was 8.8 as shown in Figure 6b, and at pH 8.5–9.2, ALDH activities were kept > 80% capacity (Figure 6b). Km values of butanal and NAD^+ of the ALDH were about 250 μM and 25 μM, respectively, from the Lineweaver–Burk plot (data not shown). When replacing NAD^+ by $NADP^+$, the ALDH activity reduced to about one third in comparison to NAD^+ (data not shown).

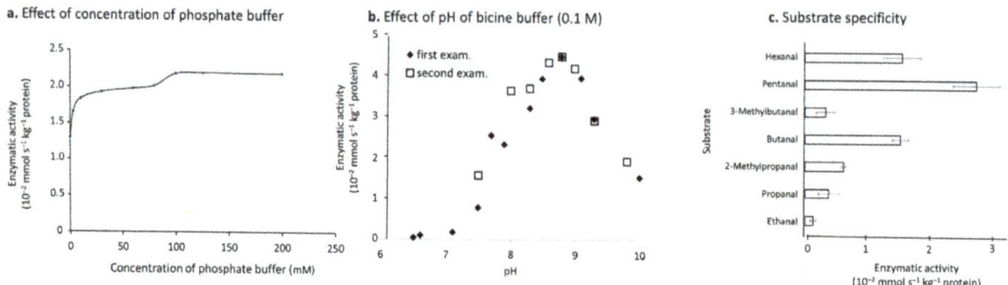

Figure 6. Activity of crude ALDH. (**a**) Optimum concentration of phosphate buffer; (**b**) Optimum pH; and (**c**) Substrate specificity to C2–C6 branched and straight chain aldehydes. Horizontal line at each column shows average ± SD (n = 3).

Substrate specificity of the ALDH was shown in Figure 6c. Lower molecular weight straight chain aldehydes, except ethanal, had high affinity to ALDH, while poor affinity was detected for branched chain aldehydes (Figure 6c).

3.4. *Feeding 2-Keto-4-Methylpentanoic Acid to Banana Pulps to Produce Branched Alcohol, Aldehyde and Ester*

To test the production of branched chain alcohol, aldehyde and ester from keto acid, 2-keto-4-methylpentanoic acid was fed to banana pulps. As expected, 3-methylbutanol and 3-methylbutal were detected in the headspace (Figure 2a). However, there was no 3-methylbutyl 3-methybutanoate in the headspace, although 3-methylbutyl ethanoate was detected (Figure 7).

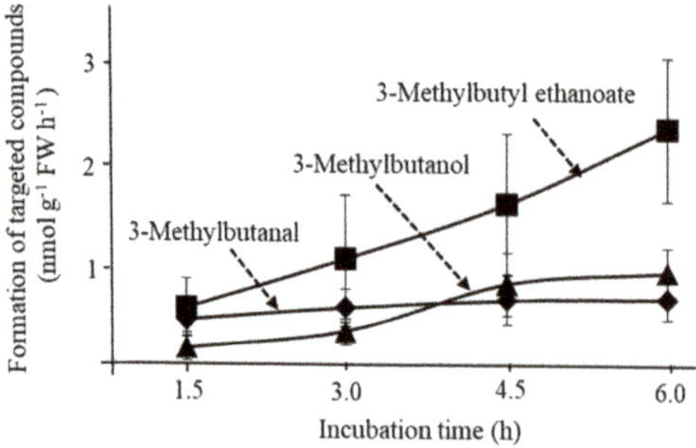

Figure 7. Formation of 3-methylbutanol, 3-methylbutanal and 3-methylbutyl ethanoate from exogenous 2-keto-4-methylpentanoic acid (50 nmol) after incubation with pulp slices (10 g) of bananas at the "yellow with green tips" stage. Vertical line at each marker shows average ± standard deviation (n = 3).

4. Discussion

Under anaerobic conditions, exogenous butanol fed to banana pulp was not effectively converted to butyl butanoate (Figure 3a), indicating that oxygen was required in the pathway from alcohols to esters (Figure 1). Due to ADH, ACS, and AAT all not directly requiring oxygen, aldehyde oxide (ALO) was suspected as the enzyme which catalyzes butanal to butanoic acid, although only several specific substrates were catalyzed by ALO in plants, some of which are precursors of important plant hormones (indole-3-acetate (IAA) and abscisic acid (ABA)) [23] (Figure 1). However, the butanal feeding experiment showed that oxygen was not directly required; under anaerobic conditions, banana pulp converted exogenous butanal to butanoic acid, meanwhile, NAD^+ was the cofactor (Figure 4c), indicating that ALDH, not ALO, is the enzyme. Another side-by-side experiment in Figure 4c showed that enzyme-deactivated pulp tissue by microwave heating lost the ability to convert butanal to butanoic acid. Nevertheless, there is not a clear answer why under anaerobic conditions, feeding with butanol did not produce butyl butanoate (Figure 3a). It seems that a deficit of NAD^+ may correlate to the cascade conversions from alcohols to carboxylate, and dehydrogenation of NADH to NAD^+ was blocked under anaerobic conditions (Figure 1).

In vitro experiments showed that the optimum phosphate buffer concentration to ALDH was >100 mM (Figure 6a). This might be due to phosphate ion accelerating the activity of ALDH, which was observed in germinating peanut cotyledon [24]. NAD^+ was a better coenzyme to ALDH and the efficiency of $NADP^+$ was only one third in comparison to NAD^+ (data not shown), although NAD^+ and $NADP^+$ perform similar redox functions within the cell. The latter is also more confined to biosynthetic pathways and redox protective roles in general [25].

During ripening, ALDH activity in the homogenate increased continually until the "yellow with many brown flecks" stage, and then it remained unchanged or slightly decreased in the butanal feeding experiment (Figure 4b). However, when butanol was fed into banana slices at different ripening stages, the peak of butyl butanoate production appeared the "yellow with green tips" stage (Figure 3b), indicating that experiments using pulp slices and homogenate may lead to different results in enzyme activities: ALDH could have high activity at younger ripening stage, i.e., at the "yellow with green tips" stage (Figure 3b), but at that stage, polyphenol and tannin contents were much higher than at the "yellow with many brown flecks" stage, which were mixed with enzymes during homogenization, thus inhibiting the enzyme activity of ALDH (Figure 4b) [26].

In the substrate specificity experiments feeding C2–C6 branched or straight chain aldehydes, the ones with the highest affinity to ALDH were even and straight chain C6 and C4 aldehydes, and that with the lowest was C2 ethanal (Figure 4a). This means that ethanal formed during fermentation or senescence cannot leak to ethanoic acid, which usually is a key function in detoxification of exogenously and endogenously generated aldehydes in mammals [27]. Branched chain aldehydes were generally lower in affinity to ALDH in comparison to the straight chain aldehydes with the same carbon number (Figure 4a). The trends are similar to the affinity of the acyl-CoA/carboxylate to AAT [14]: the even and straight chain substrates had high affinity, but branched chain substrates had low affinity (Figure 4a). A conflicting observation was that even though the affinity of ALDH for hexanal was high (Figure 4a), very low conversion occurred from exogenous hexanol to hexyl hexanoate (Figure 3c). One of the possible reasons is due to slow partition of hexyl hexanoate in headspace, it did not build up enough vapor when direct headspace sampling was used, even though hexyl hexanoate was abundant in the pulps or solutions. For example, hexyl hexanoate was one of the most abundant esters in Gala apples when Tenax GC trap or solid-phase microextraction (SPME) trap sampling methods were used [28]. However, when direct headspace sampling method was used, hexyl hexanoate was not detectable [29].

TCA cycle and β-oxidation of fatty acids in fruit pulps continually provide ethanoyl-CoA (acetyl-CoA)/ethanoate to the background in the pulp. One of the consequences is that when feeding ethanol, numerous ethyl ethanoates are produced, and it is difficult to differentiate whether the ethanoyl-CoA/ethanoate are from the fed ethanol or the background substrates. Thus, in Figure 3c, ethanol feeding was not shown. Thus, in bananas, there are rich sources of ethanoate/ethanoyl-CoA via TCA and β-oxidation, and ALDH is not the major way to produce ethanoate (Figure 4a).

It is well known that butanoate esters are the second most produced esters after ethanoate esters in banana fruits [30], and the high affinity of butanal to ALDH ensured sufficient butanoate (Figure 4a). However, ester profiles changed with fruit senescence [30,31], controlled/modified atmosphere [32–34], or other treatments which may extend shelf life with sacrifice of flavor quality [29]. Due to ester production being more dependent on alcohol and carboxylic acid substrate availability than substrate specificity of AAT [14,15,35], ALDH plays a key role in the ester profile and flavor quality of fresh fruit.

Feeding of 2-keto-4-methylpentanoic acid to banana pulps produced 3-methylbutanol, and 3-methylbutyl ethanoate, but no 3-methylbutanoate and the esters (Figure 7). The results confirmed that the affinity of ALDH for branched chain aldehydes is low. There are reports that some fruits such as apples and melons produce a relatively high amount of esters consisting of branched chain carboxylic acid [36,37]. It is worth continuing research on ALDH in branched chain carboxylic acid rich fruits.

The most critical question for this feeding model research was how to obtain high purity ALDH and minimize the effect of other enzymes, especially ADH, which is also determined by monitoring the change in NADH concentration. Furthermore, ADH facilitates the interconversion between alcohols and aldehydes with the redox between NAD^+ and NADH. In this study, two actions were taken to eliminate the potential effect of ADH. The first was to remove proteins that had high ADH activity but much less ALDH activity. We removed elutable proteins (protein-S) which contained 95% of ADH, but only about 40% of ALDH activity, and obtained proteins which had higher binding force to pellets—they were extracted under sonicator-assistance (Figure 5). In such pellet protein, there was little ADH activity (Figure 5). The second action involved using 4-methyl pyrazole, an ADH inhibitor, to block any potential ADH activity in the reaction mixture for ALDH activity determination (Figure 5).

5. Conclusions

ALDH, which converts aldehydes to carboxylic acids, was found in banana pulps and may play a key role in the conversion between alcohols, aldehydes, carboxylic acids and esters, and the formation of fruit aromas. Crude ALDH tests showed that the enzyme required NAD+ as a cofactor, and the optimum pH was 8.8. Lower molecular weight straight chain aldehydes, except ethanal, had high affinity to ALDH, while poor affinity was detected to branched chain aldehydes. Further research is needed to confirm whether ALDH is an enzyme in the routine pathway for volatile production associated with fruit ripening or just a consequence of aldehyde scavenging.

Author Contributions: Conceptualization, Y.U. and J.B.; methodology, Y.U., W.Z., H.I., Y.I., E.T., S.K.W., A.C. and J.B.; statistical analysis and the analysis of the results, Y.U. and J.B.; writing, Y.U., J.B., W.Z., H.I., Y.I., E.T., S.K.W. and A.C. All authors have read and agreed to the published version of the manuscript.

Funding: This research received no external funding.

Data Availability Statement: Data is contained within the article.

Acknowledgments: This manuscript is dedicated to Yoshinori Ueda, in memory of his devotion in enzymology of fruit volatiles and postharvest plant physiology of horticultural crops. He will be greatly missed by those who were privileged to know him. We would thank Akira Wadano for his technical advice, and Hannah Clarke and Alice Bai for language improvement.

Conflicts of Interest: The authors declare no conflict of interest.

References

1. Brocker, C.; Vasiliou, M.; Carpenter, S.; Carpenter, C.; Zhang, Y.; Wang, X.; Kotsoni, S.O.; Wood, A.J.; Kirch, H.-H.; Kopečný, D.; et al. Aldehyde dehydrogenase (ALDH) superfamily in plants: Gene nomenclature and comparative genomics. *Planta* **2013**, *237*, 189–210. [CrossRef] [PubMed]
2. Schwab, W.; Davidovich-Rikanati, R.; Lewinsohn, E. Biosynthesis of plant-derived flavor compounds. *Plant J.* **2008**, *54*, 712–732. [CrossRef] [PubMed]
3. Dudareva, N.; Klempien, A.; Muhlemann, J.K.; Kaplan, I. Biosynthesis, function and metabolic engineering of plant volatile organic compounds. *New Phytol.* **2013**, *198*, 16–32. [CrossRef]
4. Wyllie, S.G.; Fellman, J.K. Formation of volatile branched chain esters in bananas (*Musa sapientum* L.). *J. Agric. Food Chem.* **2000**, *48*, 3493–3496. [CrossRef] [PubMed]
5. Tressl, R.; Drawert, F. Biogenesis of banana volatiles. *J. Agric. Food Chem.* **1973**, *21*, 560–565. [CrossRef]
6. Sugimoto, N.; Forsline, P.; Beaudry, R. Volatile profiles of members of the USDA Geneva Malus Core Collection: Utility in evaluation of a hypothesized biosynthetic pathway for esters derived from 2-methylbutanoate and 2-methylbutan-1-ol. *J. Agric. Food Chem.* **2015**, *63*, 2106–2116. [CrossRef]
7. Sugimoto, N.; Jones, A.D.; Beaudry, R. Changes in free amino acid content in 'Jonagold'apple fruit as related to branched-chain ester production, ripening, and senescence. *J. Am. Soc. Hortic. Sci.* **2011**, *136*, 429–440. [CrossRef]
8. Tavaria, F.K.; Dahl, S.; Carballo, F.J.; Malcata, F.X. Amino acid catabolism and generation of volatiles by lactic acid bacteria. *J. Dairy Sci.* **2002**, *85*, 2462–2470. [CrossRef]
9. Beck, H.C.; Hansen, A.M.; Lauritsen, F.R. Metabolite production and kinetics of branched-chain aldehyde oxidation in Staphylococcus xylosus. *Enzym. Microb. Technol.* **2002**, *31*, 94–101. [CrossRef]
10. Dickinson, J.R.; Harrison, S.J.; Dickinson, J.A.; Hewlins, M.J. An investigation of the metabolism of isoleucine to active amyl alcohol in Saccharomyces cerevisiae. *J. Biol. Chem.* **2000**, *275*, 10937–10942. [CrossRef]
11. Bai, J.; Baldwin, E.A.; Imahori, Y.; Kostenyuk, I.; Burns, J.; Brecht, J.K. Chilling and heating may regulate C6 volatile aroma production by different mechanisms in tomato (*Solanum lycopersicum*) fruit. *Postharvest Biol. Technol.* **2011**, *60*, 111–120. [CrossRef]
12. Yang, X.; Song, J.; Fillmore, S.; Pang, X.; Zhang, Z. Effect of high temperature on color, chlorophyll fluorescence and volatile biosynthesis in green-ripe banana fruit. *Postharvest Biol. Technol.* **2011**, *62*, 246–257. [CrossRef]
13. Wendakoon, S.K.; Ueda, Y.; Imaahori, Y.; Tomimasu, Y. Presence of acetyl-CoA synthetase in ripened bananas. *Food Preserv. Sci.* **2005**, *31*, 75–80. [CrossRef]
14. Ueda, Y.; Tsuda, A.; Bai, J.-H.; Fujishita, N.; Chachin, K. Characteristic pattern of aroma ester formation from banana, melon and strawberry with reference to the substrate specificity of ester synthetase and alcohol contents in pulp. *J. Jpn. Food Sci. Technol.* **1992**, *39*, 183–187. [CrossRef]
15. Jayanty, S.; Song, J.; Rubinstein, N.M.; Chong, A.; Beaudry, R.M. Temporal relationship between ester biosynthesis and ripening events in bananas. *J. Am. Soc. Hortic. Sci.* **2002**, *127*, 998–1005. [CrossRef]
16. Beekwilder, J.; Alvarez-Huerta, M.; Neef, E.; Verstappen, F.W.; Bouwmeester, H.J.; Aharoni, A. Functional Characterization of Enzymes Forming Volatile Esters from Strawberry and Banana. *Plant Physiol.* **2004**, *135*, 1865–1878. [CrossRef]
17. Ueda, Y.; Fujishita, N.; Chachin, K. Presence of alcohol acetyltransferase in melons (*Cucumis melo* L.). *Postharvest Biol. Technol.* **1997**, *10*, 121–126. [CrossRef]
18. Dong, Y.; Liu, H.; Zhang, Y.; Hu, J.; Feng, J.; Li, C.; Li, C.; Chen, J.; Zhu, S. Comparative genomic study of ALDH gene superfamily in *Gossypium*: A focus on *Gossypium hirsutum* under salt stress. *PLoS ONE* **2017**, *12*, e0176733. [CrossRef]
19. Modig, T.; Lidén, G.; Taherzadeh, M.J. Inhibition effects of furfural on alcohol dehydrogenase, aldehyde dehydrogenase and pyruvate dehydrogenase. *Biochem. J.* **2002**, *363*, 769–776. [CrossRef]
20. USDA-AMS Bananas, Market Inspection Instructions. 2004; p. 16. Available online: https://www.ams.usda.gov/sites/default/files/media/Bananas_Inspection_Instructions%5B1%5D.pdf (accessed on 11 November 2021).
21. Lowry, O.H.; Rosebrough, N.J.; Farr, A.L.; Randall, R.J. Protein measurement with the Folin phenol reagent. *J. Biol. Chem.* **1951**, *193*, 265–275. [CrossRef]
22. Peterson, G.L. Review of the Folin phenol protein quantitation method of Lowry, Rosebrough, Farr and Randall. *Anal. Biochem.* **1979**, *100*, 201–220. [CrossRef]
23. Omarov, R.; Dräger, D.; Tischner, R.; Lips, S.H. Aldehyde oxidase isoforms and subunit composition in roots of barley as affected by ammonium and nitrate. *Physiol. Plant.* **2003**, *117*, 337–342. [CrossRef] [PubMed]
24. Oppenheim, A.; Castelfranco, P.A. An acetaldehyde dehydrogenase from germinating seeds. *Plant Physiol.* **1967**, *42*, 125–132. [CrossRef] [PubMed]
25. Ying, W. NAD+/NADH and NADP+/NADPH in cellular functions and cell death: Regulation and biological consequences. *Antioxid. Redox Sign.* **2008**, *10*, 179–206. [CrossRef] [PubMed]
26. Mura, K.; Tanimura, W. Change of polyphenol compounds in banana pulp during ripening. *Food Preserv. Sci.* **2003**, *29*, 347–351. [CrossRef]
27. Crabb, D.W.; Matsumoto, M.; Chang, D.; You, M. Overview of the role of alcohol dehydrogenase and aldehyde dehydrogenase and their variants in the genesis of alcohol-related pathology. *Proc. Nutr. Soc.* **2004**, *63*, 49–63. [CrossRef] [PubMed]
28. Mattheis, J.; Buchanan, D.; Fellman, J.K. Volatile compounds emitted by 'Gala 'apples following dynamic atmosphere storage. *J. Am. Soc. Hortic. Sci.* **1998**, *123*, 426–432. [CrossRef]

29. Bai, J.; Baldwin, E.A.; Mattheis, J.P. Effect of 1-MCP pretreatment, CA storage and subsequent marketing temperature on volatile profile of 'Gala' apple. In Proceedings of the Florida State Horticultural Society, Sheraton World Resort, Daytona Beach, FL, USA, 8–10 June 2003; Volume 116, pp. 400–404.
30. Macku, C.; Jennings, W.G. Production of volatiles by ripening bananas. *J. Agric. Food Chem.* **1987**, *35*, 845–848. [CrossRef]
31. Bai, J.; Hagenmaier, R.D.; Baldwin, E.A. Volatile response of four apple varieties with different coatings during marketing at room temperature. *J. Agric. Food Chem.* **2002**, *50*, 7660–7668. [CrossRef] [PubMed]
32. Imahori, Y.; Yamamoto, K.; Tanaka, H.; Bai, J. Residual effects of low oxygen storage of mature green fruit on ripening processes and ester biosynthesis during ripening in bananas. *Postharvest Biol. Technol.* **2013**, *77*, 19–27. [CrossRef]
33. Ueda, Y.; Bai, J.; Yoshioka, H. Effects of polyethylene packaging on flavor retention and volatile production of 'Starking Delicious' apple. *J. Jpn. Soc. Hortic. Sci.* **1993**, *62*, 207–213. [CrossRef]
34. Ueda, Y.; Bai, J. Effect of short term exposure of elevated CO_2 on flesh firmness and ester production of strawberry. *J. Jpn. Soc. Hortic. Sci.* **1993**, *62*, 457–464. [CrossRef]
35. Lara, I.; Miró, R.M.; Fuentes, T.; Sayez, G.; Graell, J.; López, M.L. Biosynthesis of volatile aroma compounds in pear fruit stored under long-term controlled-atmosphere conditions. *Postharvest Biol. Technol.* **2003**, *29*, 29–39. [CrossRef]
36. Bauchot, A.D.; Mottram, D.S.; Dodson, A.T.; John, P. Effect of aminocyclopropane-1-carboxylic acid oxidase antisense gene on the formation of volatile esters in cantaloupe Charentais melon (cv. Vedrandais). *J. Agric. Food Chem.* **1998**, *46*, 4787–4792. [CrossRef]
37. Mir, N.A.; Perez, R.; Schwallier, P.; Beaudry, R. Relationship between ethylene response manipulation and volatile production in Jonagold variety apples. *J. Agric. Food Chem.* **1999**, *47*, 2653–2659. [CrossRef]

Article

Antioxidative Responses to Pre-Storage Hot Water Treatment of Red Sweet Pepper (*Capsicum annuum* L.) Fruit during Cold Storage

Jirarat Kantakhoo and Yoshihiro Imahori *

Laboratory of Postharvest Physiology, Graduate School of Life and Environmental Sciences, Osaka Prefecture University, 1-1 Gakuen-cho, Nakaku, Sakai, Osaka 599-8531, Japan; jirarattim@gmail.com
* Correspondence: imahori@plant.osakafu-u.ac.jp; Tel.: +81-72-254-9418

Abstract: The effects of hot water treatments on antioxidant responses in red sweet pepper (*Capsicum annuum* L.) fruit during cold storage were investigated. Red sweet pepper fruits were treated with hot water at 55 °C for 1 (HWT-1 min), 3 (HWT-3 min), and 5 min (HWT-5 min) and stored at 10 °C for 4 weeks. The results indicated that HWT-1 min fruit showed less development of chilling injury (CI), electrolyte leakage, and weight loss. Excessive hot water treatment (3 and 5 min) caused cellular damage. Moreover, HWT-1 min slowed the production of hydrogen peroxide and malondialdehyde and promoted the ascorbate and glutathione contents for the duration of cold storage as compared to HWT-3 min, HWT-5 min, and control. HWT-1 min enhanced the ascorbate-glutathione cycle associated with ascorbate peroxidase, monodehydroascorbate reductase, dehydroascorbate reductase, and glutathione reductase, but it was less effective in simulating catalase activity. Thus, HWT-1 min could induce CI tolerance in red sweet pepper fruit by activating the ascorbate-glutathione cycle via the increased activity of related enzymes and the enhanced antioxidant level.

Keywords: *Capsicum annuum* L.; hot water treatment; chilling injury; ascorbate-glutathione cycle

Citation: Kantakhoo, J.; Imahori, Y. Antioxidative Responses to Pre-Storage Hot Water Treatment of Red Sweet Pepper (*Capsicum annuum* L.) Fruit during Cold Storage. *Foods* **2021**, *10*, 3031. https://doi.org/10.3390/foods10123031

Academic Editors: Eleni Tsantili and Jinhe Bai

Received: 8 November 2021
Accepted: 29 November 2021
Published: 6 December 2021

1. Introduction

Red sweet pepper (*Capsicum annuum* L.) is an important crop of the dominant horticultural products in terms of the global economy. It is grown worldwide in tropics and subtropics. More than 65% of the production area is in Asia; China is the largest producer, followed by Mexico, Turkey, Indonesia, Spain, and the USA [1]. Red sweet pepper fruit contains nutrition value (a good source of vitamin A and C) and bioactive materials, such as phenolic compounds and carotenoids [2,3]. However, red sweet pepper fruit are a perishable commodity due to mechanical damage and microbial deterioration related to softening along with fruit ripening when stored at ambient temperature, resulting in a short shelf-life after harvest. Thus, a suitable storage method is necessary to maintain the quality of red sweet pepper fruit.

Low-temperature storage maintains the quality of horticultural products, and is required to extend the storage duration for horticultural products, particularly for long-distance transport or export quarantine treatments that take 21–24 days before marketing [4,5]. The shelf life of peppers in storage is limited by decay, loss of water during storage, and sensitivity to cold stress. It is reported that the optimum storage temperature for peppers is indicated to be 12 °C [6]. However, at storage temperatures below the optimum temperature, peppers often occur chilling injury (CI) according to their chilling sensitivity. CI symptoms indicates surface pitting, calyx discoloration, water-soaked areas, and shriveling related with moisture loss [7]. The development of CI symptoms relates to several factors. The severity level of CI is subject to the storage time and storage temperature interaction. Generally, after exposure to lower chilling temperatures and longer duration, horticultural products suffer more CI [8].

Several stress in plant is caused by change in the environment conditions [9]. Environmental stress excessively induces reactive oxygen species (ROS) production and provokes oxidative stress to the plant. Oxidative stress is exhibited when the ROS produces over the capacity that maintains homeostasis in plant cellular redox. Imbalances between the generation and detoxification for ROS are CI responses in horticultural crops [10,11]. Lipid peroxidation in a plant is the first evidence of the illustration of CI. Malondialdehyde (MDA) is generated by the non-enzymatic oxidation and enzymatic destruction of polyunsaturated fatty acids in plant cell membranes. The level of MDA is usually a marker of plant oxidative stress, and its level is used to evaluate lipid peroxidation [9,12]. Plants react to oxidative stress by non-enzymatic and enzymatic systems that sustain the equilibrium between ROS generating and scavenging. This system includes both antioxidants of ascorbate (AsA) and glutathione (GSH), and scavenging enzymes that consist of catalase (CAT) and ascorbate-glutathione (AsA-GSH) cycle-related enzymes composed of ascorbate peroxidase (APX), monodehydroascorbate reductase (MDHAR), dehydroascorbate reductase (DHAR), and glutathione reductase (GR). The AsA-GSH cycle, which implicates antioxidants and antioxidative enzymes, improves the chilling resistance of plants [13,14].

Hot water treatment can be utilized to inhibit fruit ripening or to increase CI tolerance. The alleviative effect of CI by hot water treatment has been revealed in many horticultural products [15], such as cucumber [16], zucchini [17], mume [18], banana [19], orange [20], and many other crops [21–23]. The mechanism to mitigate CI in fresh fruits and vegetables by postharvest heat treatment could be attributed to (1) improving the membrane integrity by the increase of unsaturated fatty acids and saturated fatty acids ratio; (2) enhancing heat shock protein gene expression and accumulation; (3) enhancing the antioxidant system; (4) enhancing the arginine pathways, which result in the accumulation of signaling molecules that play critical roles in improving chilling tolerance; (5) altering phenylalanine ammonialyase and polyphenol oxidase enzyme activities; and (6) enhancing sugar metabolism [21].

Hot water immersion is a nonchemical and safe postharvest treatment [13]. Prestorage with moderate heat treatment of fresh commodities not only improves their heat stress resistance but also promotes tolerance to other stresses [22]. In scanning electron microscopy analysis of hot water-treated sweet pepper fruit, it was indicated that hot water immersion markedly firmed the surface and that cracks in the epidermis were sealed due to recrystallization or melting of the wax layer, thus maintaining the fruit quality during cold storage [24]. Exposing to sub-lethal temperatures induces thermos-tolerance, which can protect horticultural products from subsequent introduction into lethal temperatures [15]. However, the responses of commodities to hot water treatments differ depending on the treatment temperature and time of exposure to hot water. It is important that the suitable conditions of hot water immersion are established for low-temperature storage.

Thus, the object in our study was to analyze the influences of different immersion durations (1, 3, and 5 min) of hot water immersion at 55 °C on antioxidant responses by monitoring the oxidative stress level and the AsA-GSH cycle in red sweet pepper fruit during low-temperature storage at 10 °C, and to determine the suitable conditions of hot water treatment.

2. Materials and Methods

2.1. Plant Materials and Treatments

Red sweet pepper (*Capsicum annuum* L.) cv. Habataki fruits were obtained at the red ripening stage from an agricultural field in Kochi Prefecture, Japan. Fruits of uniform size that were free of defects and had been delivered immediately to the laboratory were selected for this study. In total, 144 fruits of sweet pepper fruits were cleaned with water, air-dried at 20 °C for 15 min, and divided into four groups (36 fruits each). According to a preliminary experiment based on cold storage in fruit treated by hot water at different temperatures, suitable temperatures for hot water treatment were selected. The results showed that hot water treatment at 55 °C had better effects on alleviating CI. One group (control) was kept without being treated. The others were immersed in hot water at 55 °C for 1 (HWT-1 min), 3 (HWT-3 min), and 5 min (HWT-5 min), respectively, and then quick cooled until room temperature. After air drying, 3–4 fruits each were packed in a polyethylene film bag (38 cm × 26 cm with 0.03 mm thickness and four 5 mm holes on each side). In all treatments, three bags of fruits represented three replications. The pepper fruits were stored in a laboratory refrigerator at 10 °C and relative humidity (RH) of 95% for 4 weeks. Samples were taken every 7 days during storage to measure the quality attributes (weight loss; L^*, a^*, and b^*; and overall quality); CI index; and CI incidence; and to determine electrolyte leakage. Fruit pericarp tissues were cut into roughly 1 cm pieces, immediately immersed in liquid nitrogen, and then preserved at −80 °C until further analysis.

2.2. Chilling Injury

The chilling injury symptom was evaluated on the level of surface pitting in accordance with the methods described by Wang et al. [25]. CI was measured according to a scale of 6 levels; 0, no pitting; 1, trace (0% < injury ≤ 10%); 2, slight (10% < injury ≤ 20%); 3, regular (20% < injury ≤ 30%); 4, moderate (30% < injury ≤ 50%); 5, severe (injury ≥ 50%). CI index and CI incidence were determined by the following formula: CI index = (sum (CI scale × number of fruit at the CI scale))/total number of evaluated fruit; CI incidence (%) = (the number of CI fruit/the total number of evaluated fruits) × 100.

2.3. Quality Attributes

Red sweet peppers were weighed at the beginning and the end of each storage period. The variance between the two weigh values was calculated to be the total weight reduction and was indicated as a percent on a fresh weight basis.

Fruit color was indicated by measuring L^* (lightness), a^* (redness), and b^* (yellowness) using a color difference meter (Nippon Denshoku, ZE 6000, Tokyo, Japan). Each fruit was measured with three replications.

The overall quality of the fruit was assessed by the method of Özden and Bayindirli [6]. The quality was evaluated subjectively for their flavor and appearance according to the following scale: 5 = excellent, 3 = marketable level, and 1 = unusable.

2.4. Electrolyte Leakage

Electrolyte leakage was determined according to the method of Endo et al. [13]. Twenty disks (10 mm × 2 mm) of fruit pericarp tissue from each fruit were cut using a cork borer and then rinsed with distilled water before immersion in 30 mL of double-distilled water in glass vials. After incubation for 2 h at 25 °C, the initial electrolyte conductivity was evaluated with a digital electrolyte conductivity meter (DKK-TOA, MM-41DP, Japan). The disks were boiled for 30 min to complete the electrolyte leakage, and the final electrolyte conductivity was determined. The percent of electrolyte leakage was expressed as a relative between the initial and final electrolyte conductivity.

2.5. Malondialdehyde Content

The level of lipid peroxidation was expressed by the MDA level based on the thiobarbituric acid (TBA) reaction following the method of Dipierro and De Leonardis [26]. Two grams of frozen sample were extracted in 10 mL of cold 0.1% trichloroacetic acid (TCA). The extract was passed through filter paper and centrifuged at 20,000× g for 10 min at 4 °C. Then, 1 mL supernatant of extract was mixed with 4 mL of 0.5% TBA in 20% TCA and then was heated at 95 °C for 30 min and then quick cooled to room temperature. The absorbance was determined spectrophotometrically at 532 and 600 nm (Jasco V-530, Jasco, Tokyo, Japan), after being centrifuged at 20,000× g for 10 min at 4 °C. The MDA content was calculated using the extinction coefficient, 155 $mM^{-1}cm^{-1}$.

2.6. Hydrogen Peroxide Content

The content of hydrogen peroxide (H_2O_2) was determined by the chromogenic peroxidase-coupling method following the procedures of Veljovic-Jovanovic et al. [27]. Three grams of frozen sample were extracted in 12 mL of cold 1 M $HClO_4$, then passed through filter paper. After centrifugation at 12,000× *g* for 10 min at 4 °C, the supernatant of extract was neutralized to pH 5.6 by 5 M K_2CO_3 and then centrifuged at 12,000× *g* for 10 min at 4 °C to eliminate insoluble $KClO_4$. To oxidize ascorbate before incubation, the supernatant was incubated with 1-unit ascorbate oxidase for 10 min. Then, 1 mL of neutralized supernatant was reacted with the reaction mixture solution of 0.1 M phosphate buffer (pH 6.5) containing 16.5 mM 3-(dimethylamino) benzoic acid, 0.35 mM 3-methyl-2-benzothiazoline hydrazine, and 250 ng horseradish peroxidase. The absorbance was measured spectrophotometrically at 590 nm (Jasco V-530, Jasco, Tokyo, Japan) and monitored at 25 °C. The content of hydrogen peroxide was determined from a 25–100 μM H_2O_2 standard curve.

2.7. Antioxidant Content

2.7.1. Ascorbate and Dehydroascorbate

The contents of ascorbate and dehydroascorbate (DHA) were determined following the procedures of Stevens et al. [28]. Three grams of frozen sample were homogenized in 12 mL of cold 6% TCA. The extract was passed through two layers of Miracloth (Calbiochem, San Diego, CA, USA). Two assays were performed on each sample to analyze the total ascorbate and reduced ascorbate. The first assay measured the total ascorbate (incubation with 5 mM dithiothreitol (DTT) for reduction of the oxidized ascorbate), and another assay measured the reduced ascorbate (without DTT). The content of DHA was determined from the difference between the total and the reduced ascorbate content. In each sample, after the filtrate (1 mL) was mixed with 1 mL of 0.4 M phosphate buffer (pH 7.4) with 5 mM DTT or 0.4 M phosphate buffer (pH 7.4), it was incubated at 37 °C for 20 min. After incubation, the reaction solution was added with 5 mL of 0.5% *N*-ethyl malemide for the total ascorbate assay, or 5 mL of 0.4 M phosphate buffer (pH 7.4) for the reduced ascorbate assay, and left for 1 min at room temperature, and finally was added with 4 mL of color reagent. The absorbance was determined spectrophotometrically at 550 nm (Jasco V-530, Jasco, Tokyo, Japan) after 40 min incubation at 37 °C. The AsA content was calculated from a standard curve. The color reagent was as follows: solution A comprised of 31% orthophosphoric acid, 4.6% TCA, and 0.6% iron chloride; and solution B, which included 4% 2,2-dipyridyl (made up of 70% ethanol). Solutions A and B were mixed 2.75 parts (A) to 1 part (B). The content of AsA was estimated by a standard curve.

2.7.2. Reduced Glutathione and Oxidized Glutathione

The contents of GSH and GSSG contents were determined using the 5,5′-dithiobis-(2-nitrobenzoic acid) (DTNB) and glutathione reductase procedure as detailed by Griffith [29], with certain modifications. Frozen fruit tissue (5 g) was homogenized with 10 mL of cold 5% sulphosalicylic acid and passed through two layers of Miracloth (Calbiochem, San Diego, CA, USA). The homogenate was centrifuged at 12,000× *g* for 10 min at 4 °C. The extract was neutralized to pH 7.0 using with 7.5 M triethanolamine. Each neutralized solution was divided into 2 assays. One (1 mL) was used to determine the total glutathione content (GSH and GSSG). Another (1 mL) was reacted with 20 μL of 2-vinylpyridine for 60 min at 20 °C to allow the derivatization of GSH and the detection of only GSSG in the subsequent assay. The assay was carried out by adding 50 μL of the sample, 150 μL of 125 mM sodium phosphate buffer (pH 6.5) including 6.3 mM EDTA, 700 μL of 0.3 mM NADPH, 100 μL of 0.6 mM DTNB, and 10 μL of 50 units mL^{-1} GR in a cuvette with a 1 cm light path. The absorbance was monitored spectrophotometrically at 412 nm (Shimadzu UV-1800, Shimadzu, Japan) for 120 s at a temperature of 30 °C. Total glutathione and GSSG contents were estimated from the standard curve of GSH using 25–100 μL.

2.8. Antioxidant Enzyme Extraction and Assay

Enzyme extraction was performed following the method of Ishikawa et al. [30], with certain modifications. Frozen fruit tissue (5 g) was homogenized in a cooled mortar and pestle with 20 mL of cold 50 mM potassium phosphate buffer (pH 7.0) containing 20% sorbitol, 1 mM ascorbate, 1 mM EDTA, and 4% polyvinylpyrrolidone. The homogenate was passed through two layers of Miracloth (Calbiochem, San Diego, CA, USA) and centrifuged at 15,000× *g* for 15 min at 4 °C.

The activity of CAT was determined according to the procedure established by Aebi [31]. H_2O_2 consumption was recorded as the decrease in absorbance at 240 nm (Shimadzu UV-1800, Shimadzu, Japan) at 25 °C for 120 s. The calculation of CAT activity used an extinction coefficient of 39.4 mM^{-1} cm^{-1}.

The activity of APX was estimated according to the procedure of Nakano and Asada [32]. The ascorbate oxidation was recorded as the absorbance decrease at 290 nm (Shimadzu UV-1800, Shimadzu, Japan) at 25 °C for 120 s. The calculation of APX activity used an extinction coefficient of 2.8 mM^{-1} cm^{-1}.

The activity of MDHAR was estimated according to the procedure of Hossain and Asada [33]. The NADH oxidation was recorded as the absorbance decrease at 340 nm (Shimadzu UV-1800, Shimadzu, Japan) at 25 °C for 120 s. The calculation of MDHAR activity used an extinction coefficient of 6.2 mM^{-1} cm^{-1}.

The activity of DHAR was estimated according to the procedure of Hossain and Asada [33]. The generation of ascorbate was recorded as the increase in absorbance at 265 nm (Shimadzu UV-1800, Shimadzu, Japan) at 25 °C for 120 s. The calculation of DHAR activity used an extinction coefficient of 14 mM^{-1} cm^{-1}.

The activity of GR was estimated according to the procedure of Klapheck et al. [34]. The NADPH oxidation was recorded as the absorbance decrease at 340 nm (Shimadzu UV-1800, Shimadzu, Japan) at 25 °C for 120 s. The calculation of GR activity used an extinction coefficient of 6.2 mM^{-1} cm^{-1}.

2.9. Determination of Protein

The protein content in extracts was measured by the method of Bradford [35], using bovine serum albumin as the standard.

2.10. Data Analysis

The data were subjected to two-way analysis of variance (treatments and storage weeks) for a completely randomized design using SPSS (SPSS 16.0 for Windows) statistical software. Mean comparisons were determined by Duncan's multiple range test (DMRT) at a significance level of 0.05 ($p < 0.05$). The data were reported as the means ± S.E. (standard error).

3. Results

3.1. Changes in the Chilling Injury Index and Chilling Injury Incidence

Hot water-treated fruit and untreated fruit (control) presented CI symptoms and incidence by week 2 and developed continuously with storage time (Figures 1 and 2). Fruit heated at 55 °C for 1 min (HWT-1 min) exhibited a delay in the development of CI symptoms and a lower percentage of incidence than HWT-3 min, HWT-5 min, and the control fruit. By 2 weeks of storage, the CI incidence of 60% was shown in HWT-3 min, HWT-5 min, and the control fruit, and it reached 100% in week 3. In week 4, the severity of CI in HWT-1 min fruit was slight (CI index = 2), but damage was moderate to severe (CI index = 4.5–4.8) in control, HWT-3 min, and HWT-5 min fruit.

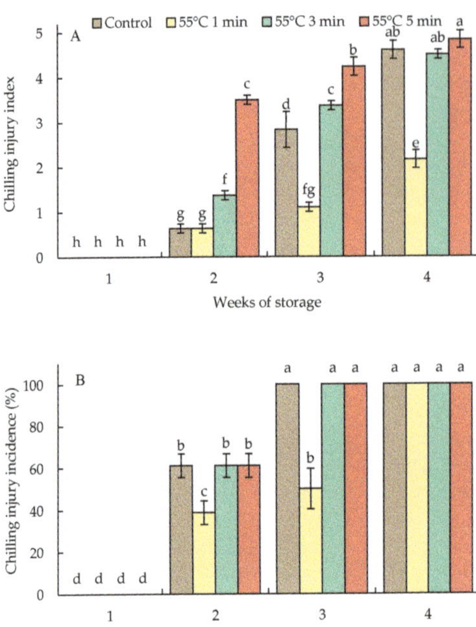

Figure 1. Effect of hot water treatment (55 °C for 1, 3, 5 min) on (**A**) chilling injury index and (**B**) chilling injury incidence in red sweet pepper fruit during storage at 10 °C. Data represent means ± S.E. (n = 3). Different letters on bars are significantly different ($p < 0.05$).

Figure 2. Effect of hot water treatment at 55 °C for (**A**) 0 min (control), (**B**) 1 min, (**C**) 3 min, and (**D**) 5 min on virtual symptoms of CI in red sweet pepper fruit during storage at 10 °C for 4 weeks (1, 2, 3, 4).

3.2. Changes in Quality Attributes

Weight loss continued to increase over the storage period (Figure 3). The accumulated losses did not differ statistically after storage for 1 and 2 weeks. The HWT-3 min and HWT-5 min fruits exhibited significant loss of water by week 4 and had higher losses than HWT-1 min and control fruits.

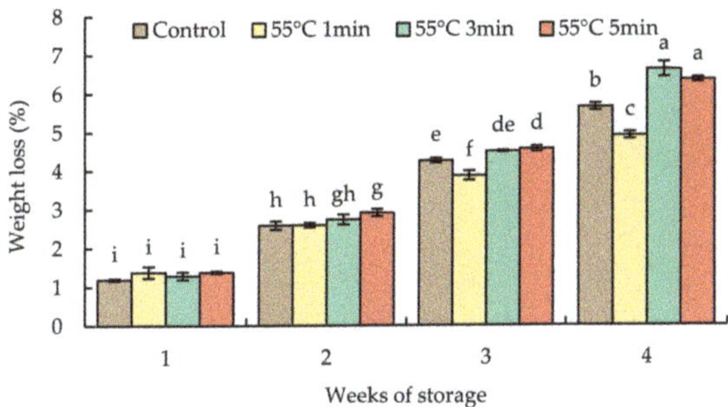

Figure 3. Effect of hot water treatment (55 °C for 1, 3, 5 min) on weight loss in red sweet pepper fruit during storage at 10 °C. Data represent means ± S.E. (n = 3). Different letters on bars are significantly different ($p < 0.05$).

It was found that there was no significant difference of L^*, a^*, or b^* between hot water-treated fruits and the control fruit during storage at 10 °C for 4 weeks ($p > 0.05$) (Figure S1).

The overall quality of sweet pepper fruits declined during prolonged storage in all treatments (Figure 4). HWT-1 min fruits had a higher overall quality score until the end of storage, while HWT- 5 min fruits lost the market potential after storage for 3 weeks; control and HWT-3 min fruits were below the marketable level at week 4 (as the score was less than 3).

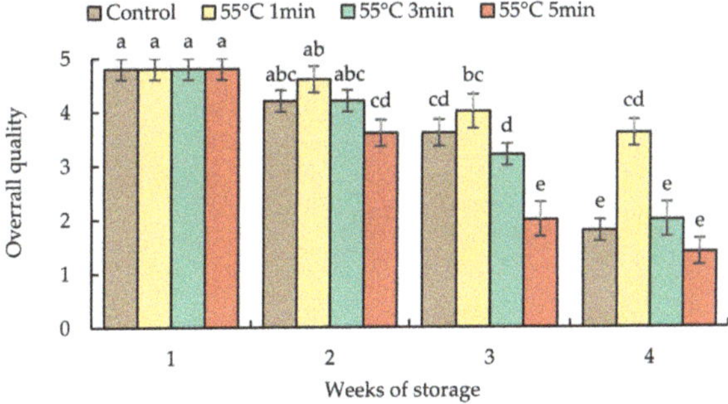

Figure 4. Effect of hot water treatment (55 °C for 1, 3, 5 min) on overall quality in red sweet pepper fruit during storage at 10 °C. Data represent means ± S.E. (n = 3). Different letters on bars are significantly different ($p < 0.05$).

3.3. Changes in Electrolyte Leakage

The value of electrolyte leakage in the initial time was about 30% (Figure 5). After 4 weeks of storage, the electrolyte leakage sharply increased to 67% in HWT-5 min fruit and gradually increased to 50% in HWT-3 min and the control fruit, whereas fruits treated for 1 min had a slight increase (40%). During low-temperature storage, HWT-1 min was given a moderate heat treatment to reduce cold stress, HWT-3 min had no positive effect, and an excessive exposure time—HWT-5 min—triggered cell damage that caused electrolyte leakage that was higher than those in HWT-1 min, HWT-3 min, and the control.

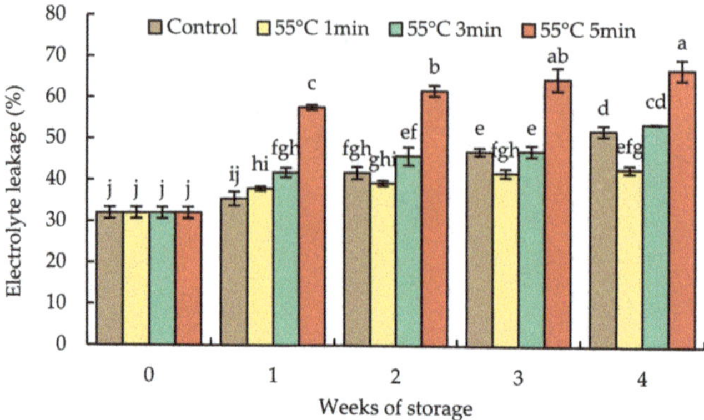

Figure 5. Effect of hot water treatment (55 °C for 1, 3, 5 min) on electrolyte leakage in red sweet pepper fruit during storage at 10 °C. Data represent means ± S.E. (n = 3). Different letters on bars are significantly different ($p < 0.05$).

3.4. Changes in Lipid Peroxidation

The content of MDA in HWT-1 min fruit markedly increased by week 1, declined thereafter, and remained at a lower level than those of HWT-3 min, HWT-5 min, and the control until the end of storage. There was a slight increase and constant level of MDA in HWT-3 min, while the MDA level in HWT-5 min increased higher than those of shorter treatments, and the content of MDA was the same as that of the control during the storage at 10 °C for 4 weeks (Figure 6).

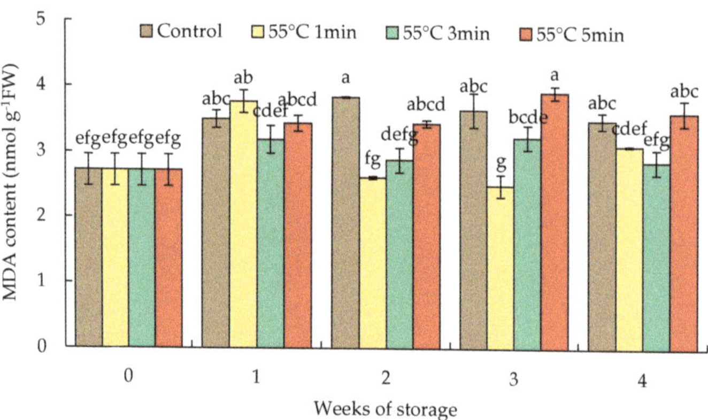

Figure 6. Effect of hot water treatment (55 °C for 1, 3, 5 min) on malondialdehyde (MDA) content in red sweet pepper fruit during storage at 10 °C. Data represent means ± S.E. (n = 3). Different letters on bars are significantly different ($p < 0.05$).

3.5. Changes in Hydrogen Peroxide Content

The hydrogen peroxide level increased slightly until week 1 in all fruits. In week 2, the increase was almost 3.6-fold in HWT-3 min and HWT-5 min fruits and continuously increased until the finale of storage; whereas in HWT-1 min fruit, the content of H_2O_2 was relatively low and stable from week 1 to week 3 but significantly increased by week 4. HWT-1 min showed a lower H_2O_2 content than HWT-3 min, HWT-5 min, and the control (Figure 7).

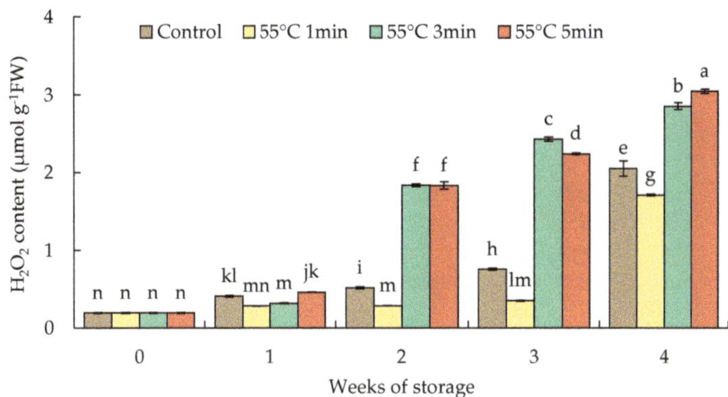

Figure 7. Effect of hot water treatment (55 °C for 1, 3, 5 min) on hydrogen peroxide content in red sweet pepper fruit during storage at 10 °C. Data represent means ± S.E. (n = 3). Different letters on bars are significantly different ($p < 0.05$).

3.6. Changes in Total Ascorbate and Dehydroascorbate Contents

The level of total AsA in HWT-1 min fruit slightly increased by week 1 but gradually decreased thereafter until the end of storage and was at a higher level than those in HWT-3 min, HWT-5 min, and the control. Prolonged heat treatments (3 and 5 min) showed a lower total AsA level and had a trend similar to that of HWT-1 min (Figure 8A). Heat treatment increased the total AsA and DHA in comparison with the control except at longer exposure times (3 and 5 min), in which excessive treatments induced heat damage in red sweet pepper fruits.

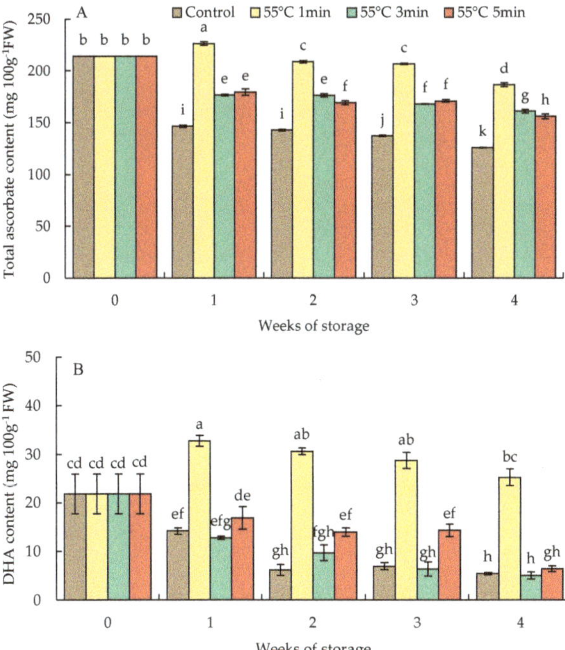

Figure 8. Effect of hot water treatment (55 °C for 1, 3, 5 min) on (**A**) total ascorbate and (**B**) dehydroascorbate (DHA) contents in red sweet pepper fruit during storage at 10 °C. Data represent means ± S.E. (n = 3). Different letters on bars are significantly different ($p < 0.05$).

The content of DHA in HWT-1 min fruit increased sharply, by approximately 1.5 times, at 1 week and then gradually declined; the level was higher than that in HWT-3 min, HWT-5 min, and the control regardless of the storage time (Figure 8B).

3.7. Changes in Reduced Glutathione and Oxidized Glutathione Contents

The content of GSH in HWT-1 min fruit increased after storage for 1 week and then significantly decreased at week 4; the level was higher than those in HWT-3 min, HWT-5 min, and the control fruit. By contrast, HWT-3 min, HWT-5 min, and control fruits indicated a rapid decline of the GSH content after storage for 1 week and were maintained at a lower level during the storage at 10 °C for 4 weeks (Figure 9A).

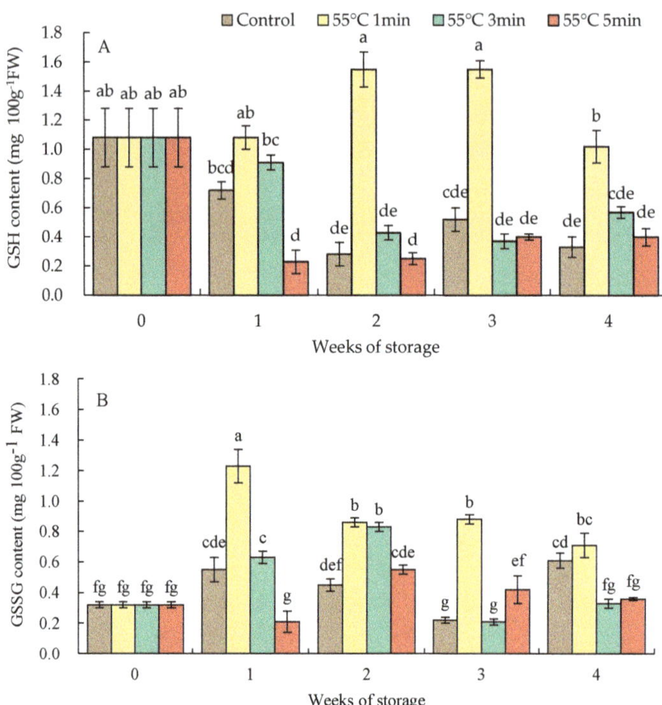

Figure 9. Effect of hot water treatment (55 °C for 1, 3, 5 min) on (**A**) reduced glutathione (GSH) and (**B**) oxidized glutathione (GSSG) contents in red sweet pepper fruit during storage at 10 °C. Data represent means ± S.E. (n = 3). Different letters on bars are significantly different ($p < 0.05$).

The GSSG level had a tendency similar to that of the level of GSH in HWT-1 min fruit, which sharply increased at week 1 and gradually decreased thereafter, and was higher than those in HWT-3 min, HWT-5 min, and control fruits until the final storage period (Figure 9B).

3.8. Changes in Catalase, Ascorbate Peroxidase, Monodehydroascorbate Reductase, Dehydroascorbate Reductase, and Glutathione Reductase Activities

Antioxidant enzymes, such as CAT and AsA-GSH cycle-related enzymes (APX, DHAR, MDHAR, and GR), play important roles to maintain cellular homeostasis by scavenging H_2O_2 and preventing the accumulation of H_2O_2 to toxic levels [14].

The activities of CAT indicated a quick decrease in week 1 and steadily declined until the end of the storage duration in all samples. By week 4, the activities of CAT in HWT-1 min fruit remained at a level higher than those of HWT-3 min, HWT-5 min, and control fruits (Figure 10).

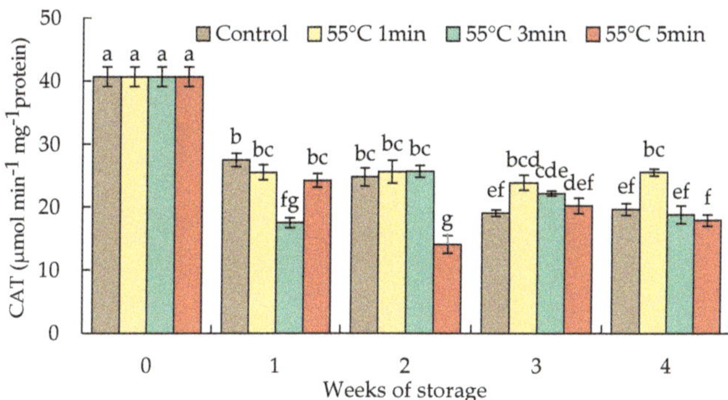

Figure 10. Effect of hot water treatment (55 °C for 1, 3, 5 min) on the activity of catalase (CAT) in red sweet pepper fruit during storage at 10 °C. Data represent means ± S.E. (n = 3). Different letters on bars are significantly different ($p < 0.05$).

Ascorbate peroxidase activities in HWT-1 min fruit slightly decreased in the first week of storage but increased in weeks 2 and 3, then declined thereafter, and remained higher than those of HWT-3 min, HWT-5 min, and control fruits throughout the storage period. HWT-3 min and HWT-5 min fruits presented a significant decrease in APX activities until the end of storage time, showing a level lower than those of HWT-1 min and control fruits (Figure 11A).

Monodehydroascorbate reductase activities in HWT-1 min fruit sharply increased and approximately doubled after 2 weeks but decreased afterward. The MDHAR activities in HWT-1 min fruit were higher than those in HWT-3 min, HWT-5 min, and control fruits, except in week 4, when all fruits showed similar levels (Figure 11B).

Dehydroascorbate reductase activities in HWT-1 min fruit tended to increase during cold storage for 3 weeks and decreased thereafter. In HWT-1 min fruits, DHAR activities were higher than those of HWT-3 min, HWT-5 min, and control fruits, except in week 4, when all fruits showed similar levels (Figure 11C).

Glutathione reductase activities in HWT-1 min fruit sharply increased during the storage at 10 °C for 4 weeks. The activities of GR in HWT-1 min fruit were higher than those of HWT-3 min, HWT-5 min, and control fruits. There was a gradual increase in GR activities in HWT-3 min and HWT-5 min fruits until the end of storage (Figure 11D).

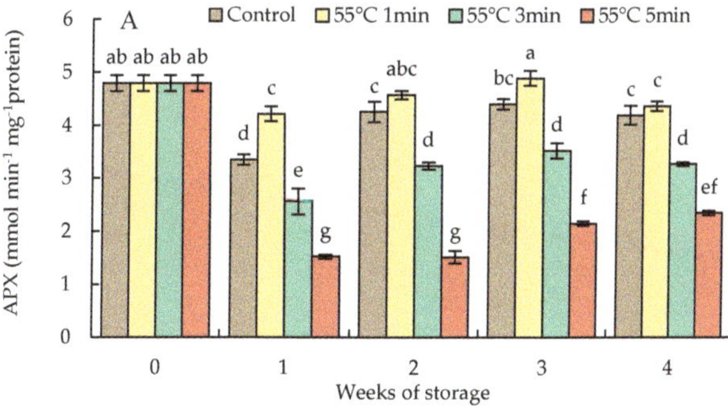

Figure 11. *Cont.*

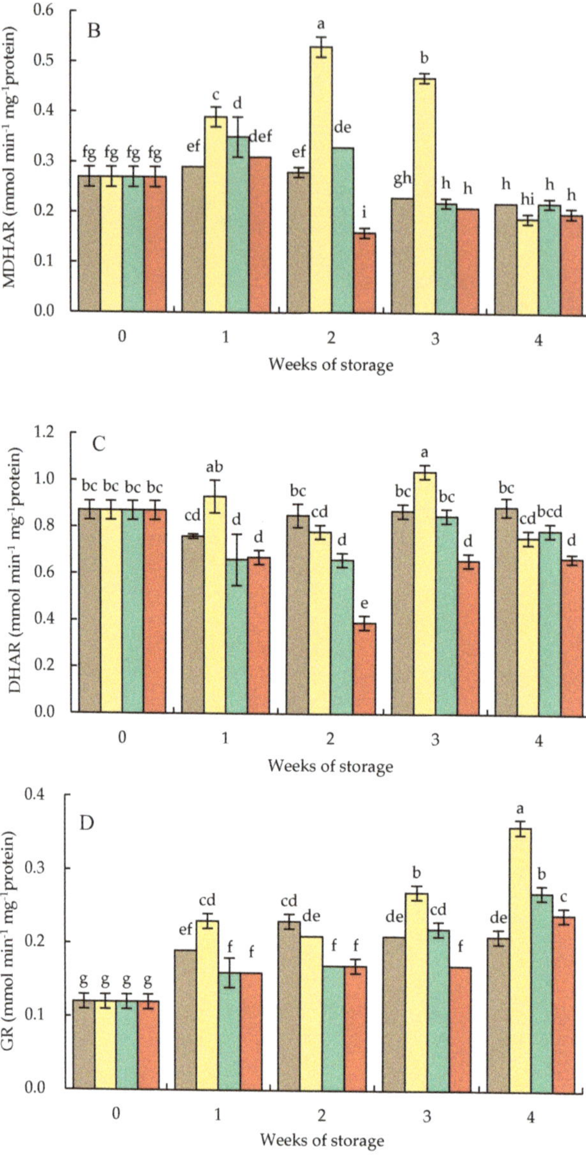

Figure 11. Effect of hot water treatment (55 °C for 1, 3, 5 min) on the activity of (**A**) ascorbate peroxidase (APX), (**B**) monodehydroascobate reductase (MDHAR), (**C**) dehydroascorbate reductase (DHAR), and (**D**) glutathione reductase (GR) in red sweet pepper during storage at 10 °C. Data represent means ± S.E. (n = 3). Different letters on bars are significantly different ($p < 0.05$).

4. Discussion

Low-temperature storage is used extensively to keep the quality and prolong the shelf life of horticultural products after harvest. Nevertheless, because of their sensitivity to cold stress, the utilization of low-temperature storage has some limitations for chilling sensitive horticultural products, such as subtropical and tropical crops [21]. Chilling-induced stress changes the balance between the ROS-forming system and defensive mechanisms, resulting in oxidation-induced CI [10].

In the present study, HWT-1 min was effective in alleviating CI and retarded the development of CI symptoms in red sweet pepper fruit as compared to control and the longer treatment times, HWT-3 min and HWT-5 min. CI symptoms presented in the same period of storage (week 2); however, the level of damage differed. Finally, the CI severity in HWT-1 min fruit was slight; however, in the control, HWT-3 min, and HWT-5 min fruit, damage was moderate to severe (Figures 1 and 2). Previous studies also found that hot water immersion modified responses to cold stress, delaying the onset of CI, and thus has been proven to alleviate CI in fresh produce, such as sweet pepper [13,36], mumes [18], plums [37], and cucumbers [16].

Weight loss in untreated red sweet peppers was less than in those treated with hot water, except for HWT-1 min fruit. Hot water immersion increased fruit weight loss in some fresh produce, such as mandarins [38], which lost more weight than the control during cold storage. From this study, short exposure to hot water (HWT-1 min) reduced weight loss (Figure 3), probably as a result of the stimulated recrystallization or molecular orientation of the waxes on the cuticle layer, which plays a significant role in regulating the water loss and preserving fruit firmness [16,39]. However, red sweet pepper fruits exposed to hot water for a longer time (HWT-3 min and HWT-5 min) lost more fresh weight, which might be due to tissue damage (Figure 2). Although hot treatment may be of benefit to hot water-treated horticultural products, excessive heat, such as revelation of products to lethal temperatures or an extended treatment duration, might induce cell damage [40]. Excess heat stress may cause the destruction of cells, the loss of membrane integrity, and the elimination of the wax coating the outer surface of the cuticle, resulting in higher loss of water.

In previous studies, Fallik et al. [41] found serious heat injury in sweet pepper when hot water treatment at 55 °C was applied, and even observed severe fruit damage when treated at 52 °C for 2 min [4,41]. Thus, they [24] suggested 55 ± 1 °C for only 12 s to maintain fruit quality during prolonged storage. In contrast similar with our results (Figure 4), Shehata et al. [42] found that the high-temperature treatment (55 °C for 1 min) provided a good appearance without visible injury and decay in pepper fruits during cold storage. The difference of heat tolerance in fruits depends on species, genotype, stage of fruit maturity, type and severity of heat treatment applied, and the preconditioning treatments before heat treatment [43].

Cell membrane integrity is most affected by CI. In the cell membrane, the transition from the flexible liquid crystalline phase to a solid gel structure phase is caused at cold storage temperatures, which increases the malfunction of cell membranes [44]. When horticultural product is exposed to damaging temperatures below a certain threshold temperature, cell membranes break, resulting in the leakage of ion, metabolites, and intracellular water, which can be traced as electrolyte leakage [21,45]. Electrolyte leakage in fruits treated with hot water immersion differed depending on the exposure time. During storage at 10 °C, brief hot water immersion (HWT-1 min) showed lower electrolyte leakage because mild heat treatment may reduce chilling stress; however, 5 min (HWT-5 min) was an excessive heating time that triggered cell damage and presented the highest leakage (Figure 5). Similarly, treated cucumbers (55 °C, 5 min) had lower electrolyte leakage than untreated fruit, thus hot water mitigated CI by maintaining membrane integrity [16].

Lipid peroxidation is the initial phenomenon caused by CI. Cold stress alters the structure of the plant cell membrane, causing membrane integrity to deteriorate due to lipid peroxidation, which is measured as the MDA level. MDA is the oxidative secondary metabolite of polyunsaturated fatty acid in cell membranes; it is a suitable indicator of oxidative destruction to cell membrane integrity under temperature stress, and its level is a useful indicator of the oxidative stress level [9,11,21]. Our result found that the MDA level in red sweet pepper fruit immersed with HWT-1 min was lower than that with prolonged heat exposure (3 and 5 min) and in the control (Figure 6). The low level of MDA and electrolyte leakage in HWT-1 min fruit reflected the lower degree of CI and the incidence of CI in red sweet peppers during cold storage for 4 weeks.

Hydrogen peroxide is a strong oxidant, a relatively long-lived molecule, and it is moderately reactive. It is produced by the superoxide dismutase that catalyzes a superoxide radical to H_2O_2. Environmental stresses can cause the production of H_2O_2; however, excessive H_2O_2 can cause oxidative damage in plant cells by disrupting metabolic processes and affecting cell membrane integrity [13,46]. In this study (Figure 7), when comparing different times of heat exposure (55 °C for 1, 3, and 5 min), HWT-1 min fruit had lower H_2O_2 contents; this remained constant until week 3 and then rose significantly by week 4. Prolonged heating times (HWT-3 min and HWT-5 min) tended to overheat red sweet peppers, causing higher oxidative stress, which was demonstrated by a higher accumulation of H_2O_2 from week 2 until the end of storage (Figure 6). H_2O_2 responds to oxidative stress as a signal in the cell compartments that it originates from, leading to an applicable response in the cellular protection system [12]. In low levels, H_2O_2 reacts as a signal molecule involved in acclimatory signaling, triggering tolerance to several stresses; in contrast, in high levels, it acts as a providing factor to stress damages and causes cellular destruction [13,47,48]. HWT-1 min could stimulate H_2O_2 production to a level that triggered the defense mechanism responses, involving enzymatic and non-enzymatic antioxidative scavenging systems, promoting tolerance of subsequent cold storage. Previous study indicated that temperature stresses reacted to antioxidative scavenging systems in plants. Moderate heat treatment of horticultural produce induces mild oxidative stress, which affects the antioxidant condition and induces tolerance to subsequent stress [18]. The oxidative function of H_2O_2 is intimately linked to the redox state in plant tissues. The redox balance is associated with the expression of genes that contribute to stress resistance, stress acclimation, and defense systems [12].

Heat treatment of horticultural crops before storage not only increases their resistance to heat stress but also improves their tolerance to other various stresses [15]. Temperature stress altered homeostasis in plant cell and major processes in its physiological functions, resulting in the accumulation of ROS levels. ROS may potentially play a function in signal transduction processes that activate stress-response pathways and trigger defensive systems [49]. Mild heat treatment causes slight stress in fresh products, stimulating antioxidant responses in both enzymatic and non-enzymatic systems [11]. Non-enzymatic systems include antioxidant substances, such AsA and GSH, whereas antioxidant enzymes involve superoxide dismutase (SOD), CAT, and AsA-GSH cycle-related enzymes (APX, DHAR, MDHAR, and GR) [9,11]. The AsA-GSH cycle consists of an antioxidant and detoxifying system against ROS that has a significant impact on the resistance to chill damage during postharvest storage [50]. Both AsA and GSH are the significant antioxidants in plants that play important roles in stress resistance. High AsA and GSH contents may be responsible for improving performance during postharvest cold storage [51]. AsA is known to be the most powerful antioxidant substance and plays a critical role in maintaining APX activities to detoxify H_2O_2 in the AsA-GSH cycle [48]. In the current study (Figure 8A), the content of AsA in HWT-1 min fruit was gradually decreased but maintained at a higher level than those in HWT-3 min, HWT-5 min, and control fruits during the storage at 10 °C. The decreasing trend of DHA contents in HWT-1 min fruit was similar to that of AsA contents (Figure 9B). Prolonged exposure to hot water (3 min and 5 min) decreased the AsA and DHA content. Hot-water dipping has also been observed to increase the AsA level in other fresh products, including tomatoes [9], zucchini [17], and mumes [18]. The elevated redox potential in AsA may correspond with the chill acclimation of hot water-treated fruit. The increased activity of the AsA metabolism system may indicate a key role in lowering H_2O_2 levels throughout low-temperature storage [18]. In HWT-1 min fruit, the high level of AsA was reflected in the lower content of H_2O_2 during the initial 3 weeks of storage (Figure 7).

Glutathione has a crucial function in the antioxidative defense mechanism, regenerating AsA from its oxidized form, DHA, throughout the AsA-GSH cycle [48]. In this study (Figure 9), the GSH level in HWT-1 min fruit increased after 1 week of storage and significantly decreased at week 4; throughout the storage period, the content remained higher than in fruits treated differently. Hot water treatment for 1 min increased chilling tolerance by the increase of the GSH production and the GR activities.

The accumulation of ROS induced by various stresses is mitigated by enzymatic protection systems, such as SOD, APX, GPX, and CAT, and non-enzymatic protection systems of low molecular substances, such as AsA, GSH, α-tocopherol, carotenoids, and flavonoids. ROS disturbs many functions of cellular metabolism by injuring nucleic acids, oxidizing proteins, and producing lipid peroxidation [48]. Many processes in plant metabolism may indicate different optimal temperatures, and their related enzymes may be less heat unstable, so they respond differently to the stresses of temperature [18]. CAT is one of the main enzymes that scavenges ROS by decomposing H_2O_2 into water and oxygen. The increase in CAT activity is considered to be the adapted characteristic that assists in the disposal of H_2O_2. Abiotic stresses either increase or decrease CAT activity, depending on the strength, period, and kind of stress [12]. In this result (Figure 10), hot water treatments were less effective at stimulating CAT activities. CAT activity in red sweet peppers treated with immersion of hot water (55 °C for 1, 3, and 5 min) and untreated fruit tended to decline during cold storage. Similar trends were indicated in mature green mumes [18], which were treated with immersion of hot water (45 °C for 5 min) during low-temperature storage for 4 weeks. Cold stress might cause the downregulation of abundant CAT enzyme protein and the decrease in CAT activity in red sweet peppers during cold storage.

Ascorbate peroxidase has an important role in the decomposition of H_2O_2 and the control of H_2O_2 levels in cellular compartments. In the present study (Figure 11A), although HWT-1 min did not significantly increase APX activities, they were higher than those in fruits treated for prolonged times (3 and 5 min) and the control. HWT-1 min might be a modest heat exposure that could induce APX activity to maintain low levels of H_2O_2. APX is considered to play a role in the defensive mechanism against CI development during cold storage, and to enhance the tolerance to cold storage, as well as having a more important action in the ROS control or being responsible for regulating ROS signaling [13].

Ascorbate is a powerful antioxidant that can directly detoxify free radicals. It comes in two oxidized forms: MDHA and DHA. In living organisms, the AsA recycling metabolic system is significant for preserving AsA homeostasis against exogenous stimuli. DHAR and MDHAR are two major AsA recycling enzymes and are members of the AsA-GSH cycle [52]. DHAR catalyzes the recycle action of DHA to AsA, utilizing GSH as a hydrogen donor prior to the spontaneous hydrolysis of DHA to irreversibly form 2,3-diketogulonic acid. On the other hand, MDHAR uses reduced ferredoxin or NAD(P)H as an electron donor to recycle MDHA prior to the spontaneous oxidation of MDHA to form DHA [44]. The activity of MDHAR in HWT-1 min fruit highly enhanced and about doubled after 2 weeks of storage but reduced subsequently (Figure 11B). DHAR activity in HWT-1 min fruit slightly increased throughout the first 3 weeks of storage and reduced thereafter (Figure 11C). The levels of both DHAR and MDHAR activities in HWT-1 min fruit were higher than those in HWT-3 min, HWT-5 min, and control fruits. The increase of DHAR and MDHAR activities indicates accumulative responses to cold stress. In a previous study, the higher H_2O_2 level produced by heat treatment promoted DHAR and MDHAR activities [13]. Temporally, an increase in gene expression of DHAR and MDHAR was induced in H_2O_2 accumulation caused by heat treatment [53]. H_2O_2 responds as a molecular-signaling secondary messenger of metabolic control, and regulator of gene expression, increasing responses and activation of protection pathways to various stresses. Thus, the higher activities of DHAR and MDHAR enzymes play a significant role in alleviating CI.

Glutathione reductase is a key enzyme in the AsA-GSH cycle and catalyzes the conversion of GSSH to GSH by utilization of NADPH as an electron donor. It also plays a crucial role in the defense against ROS by maintaining the reduction state of GSH and the AsA pools [13]. GR activity in HWT-1 min fruit sharply increased and was higher than those in HWT-3 min, HWT-5 min, and control fruits throughout the storage period (Figure 11D). High GR activity induced a larger GSH pool and maintained the AsA-GSH cycle, and consequently enhanced the AsA pool [54]. Therefore, higher levels of GSH content and GR activity imply an influence on chilling tolerance and cold acclimation. The GR has a significant action in resistance systems for chilling stress.

5. Conclusions

The findings of our study revealed that pre-treatment with immersion of hot water at 55 °C for 1 min alleviated CI in red sweet pepper fruit during the chilling storage duration, which may be due to the increase of the AsA-GSH cycle by enhanced antioxidant contents of AsA and GSH and the activity of antioxidant-related enzymes. During cold storage, chilling resistance was improved, and the onset of CI was delayed and mitigated in HWT-1 min fruit, as illustrated by low H_2O_2 levels, whereas prolonged exposure to hot water (3 and 5 min) caused cellular damage, as showed by increases in weight loss, the CI index, the level of electrolyte leakage, and the MDA level. The non-damaging heat condition of hot water at 55 °C for 1 min can allow cold storage at storage temperatures lower than the optimal temperature. This hot water treatment in the present study is a safe treatment, technologically easier, cheaper, and more feasible for suitable postharvest treatment.

Supplementary Materials: The following are available online at https://www.mdpi.com/article/10.3390/foods10123031/s1, Figure S1: Effect of hot water treatment (55 °C for 1, 3, 5 min) on L^*, a^*, b^* color in red sweet pepper fruit during storage at 10 °C.

Author Contributions: Conceptualization, Y.I. and J.K.; methodology, Y.I.; formal analysis, J.K.; investigation, Y.I.; data curation: Y.I. and J.K.; writing—original draft preparation, J.K.; writing—review and editing, Y.I. and J.K.; supervision, Y.I. All authors have read and agreed to the published version of the manuscript.

Funding: This research received no external funding.

Conflicts of Interest: The authors declare that they have no conflicts of interest for this article.

References

1. FAO: Chillies and Peppers. Available online: http://www.fao.org/faostat/en/#data/QC (accessed on 18 April 2021).
2. Howard, L.R.; Talcott, S.T.; Brenes, C.H.; Villalon, B. Changes in phytochemical and antioxidant activity of selected pepper cultivars (*Capsicum* Species) as influenced by maturity. *J. Agric. Food Chem.* **2000**, *48*, 1713–1720. [CrossRef]
3. Giuffrida, D.; Dugo, P.; Torre, G.; Bignardi, C.; Cavazza, A.; Corradini, C.; Dugo, G. Characterization of 12 *Capsicum* varieties by evaluation of their carotenoid profile and pungency determination. *Food Chem.* **2013**, *140*, 794–802. [CrossRef]
4. Fallik, E.; Bar-Yosef, A.; Alkalai-Tuvia, S.; Aharon, Z.; Perzelan, Y.; Ilić, Z.; Lurie, S. Prevention of chilling injury in sweet bell pepper stored at 1.5 °C by heat treatments and individual shrink packaging. *Folia Hortic.* **2009**, *21*, 87–97. [CrossRef]
5. Fallik, E.; Perzelan, Y.; Alkalai-Tuvia, S.; Nemny-Lavy, E.; Nestel, D. Development of cold quarantine protocols to arrest the development of the Mediterranean fruit fly (*Ceratitis capitata*) in pepper (*Capsicum annuum* L.) fruit after harvest. *Postharvest Biol. Technol.* **2012**, *70*, 7–12. [CrossRef]
6. Özden, Ç.; Bayindirli, L. Effects of combinational use of controlled atmosphere, cold storage and edible coating applications on shelf life and quality attributes of green peppers. *Eur. Food Res. Technol.* **2002**, *214*, 320–326. [CrossRef]
7. González-Aguilar, G.A. Pepper. In *The Commercial Storage of Fruits, Vegetables, and Florist and Nursery Stocks (Agriculture Handbook 66)*; Gross, K.C., Wang, C.Y., Saltveit, M., Eds.; U.S. Department of Agriculture, Agricultural Research Service: Washington, DC, USA, 2016; pp. 481–484.
8. Saltveit, M.E., Jr.; Morris, L.L. Overview on chilling injury of horticultural crops. In *Chilling Injury of Horticultural Crops*; Wang, C.Y., Ed.; CRC Press: Boca Raton, FL, USA, 1990; pp. 3–15.
9. Imahori, Y.; Bai, J.; Baldwin, E. Antioxidative responses of ripe tomato fruit to postharvest chilling and heating treatments. *Sci. Hort.* **2016**, *198*, 398–406. [CrossRef]
10. Hodges, D.M.; DeLong, J.M.; Forney, C.F.; Prange, R.K. Improving the thiobarbituric acid-reactive-substances assay for estimating lipid peroxidation in plant tissues containing anthocyanin and other interfering compounds. *Planta* **1999**, *207*, 604–611. [CrossRef]
11. Imahori, Y. Role of ascorbate peroxidase in postharvest treatments of horticultural crops. In *Oxidative Damage to Plants*; Ahmad, P., Prasad, M.N.V., Eds.; Academic Press: San Diego, CA, USA, 2014; pp. 425–451.
12. Imahori, Y.; Takemura, M.; Bai, J. Chilling-induced oxidative stress and antioxidant responses in mume (*Prunus mume*) fruit during low temperature storage. *Postharvest Biol. Technol.* **2008**, *49*, 54–60. [CrossRef]
13. Endo, H.; Miyazaki, K.; Ose, K.; Imahori, Y. Hot water treatment to alleviate chilling injury and enhance ascorbate-glutathione cycle in sweet pepper fruit during postharvest cold storage. *Sci. Hort.* **2019**, *257*, 108715. [CrossRef]

14. Yao, M.; Ge, W.; Zhou, Q.; Zhou, X.; Luo, M.; Zhao, Y.; Wei, B.; Ji, S. Exogenous glutathione alleviates chilling injury in postharvest bell pepper by modulating the ascorbate-glutathione (AsA-GSH) cycle. *Food Chem.* **2021**, *352*, 129458. [CrossRef]
15. Imahori, Y. Postharvest stress treatments in fruits and vegetables. In *Abiotic Stress Responses in Plants: Metabolism, Productivity and Sustainability*; Parvaiz, A., Prasad, M.N.V., Eds.; Springer: Berlin/Heidelberg, Germany, 2012; pp. 347–358.
16. Nasef, I.N. Short hot water as safe short hot water as safe treatment induces chilling tolerance and antioxidant enzymes, prevents decay and maintains quality of cold-stored cucumbers. *Postharvest Biol. Technol.* **2018**, *138*, 1–10. [CrossRef]
17. Zhang, M.; Liu, W.; Li, C.; Shao, T.; Jiang, X.; Zhao, H.; Ai, W. Postharvest hot water dipping and hot water forced convection treatments alleviate chilling injury for zucchini fruit during cold storage. *Sci. Hort.* **2019**, *249*, 219–227. [CrossRef]
18. Endo, H.; Ose, K.; Bai, J.; Imahori, Y. Effect of hot water treatment on chilling injury incidence and antioxidative responses of mature green mume (*Prunus mume*) fruit during low temperature storage. *Sci. Hort.* **2019**, *246*, 550–556. [CrossRef]
19. Ummarat, N.; Matsumoto, T.K.; Wall, M.M.; Seraypheap, K. Changes in antioxidants and fruit quality in hot water-treated 'Hom Thong' banana fruit during storage. *Sci. Hort.* **2011**, *130*, 801–807. [CrossRef]
20. Bassal, M.; El-Hamahmy, M. Hot water dip and preconditioning treatments to reduce chilling Injury and maintain postharvest quality of Navel and Valencia oranges during cold quarantine. *Postharvest Biol. Technol.* **2011**, *60*, 186–191. [CrossRef]
21. Aghdam, M.S.; Sevillano, L.; Flores, F.B.; Bodbodak, S. Heat shock proteins as biochemical markers for postharvest chilling stress in fruits and vegetables. *Sci. Hort.* **2013**, *160*, 54–64. [CrossRef]
22. Lurie, S. Postharvest heat treatments. *Postharvest Biol. Technol.* **1998**, *14*, 257–269. [CrossRef]
23. Fallik, E. Prestorage hot water treatments (immersion, rinsing and brushing). *Postharvest Biol. Technol.* **2004**, *32*, 125–134. [CrossRef]
24. Fallik, E.; Grinberg, S.; Alkalai, S.; Yekutieli, O.; Wiseblum, A.; Regev, R.; Beres, H.; Bar-Lev, E. A Unique rapid hot water treatment to improve storage quality of sweet pepper. *Postharvest Biol. Technol.* **1999**, *15*, 25–32. [CrossRef]
25. Wang, Y.; Gao, L.; Wang, Q.; Zuo, J. Low temperature conditioning combined with methyl jasmonate can reduce chilling injury in bell pepper. *Sci. Hort.* **2019**, *243*, 434–439. [CrossRef]
26. Dipierro, S.; De Leonardis, S. The ascorbate system and lipid peroxidation in stored potato (*Solanum tuberosum* L.) tubers. *J. Exp. Bot.* **1997**, *48*, 779–783. [CrossRef]
27. Veljovic-Jovanovic, S.; Noctor, G.; Foyer, C.H. Are leaf hydrogen peroxide concentrations commonly overestimated? The potential influence of artefactual interference by tissue phenolics and ascorbate. *Plant Physiol. Biochem.* **2002**, *40*, 501–507. [CrossRef]
28. Stevens, R.; Buret, M.; Garchery, C.; Carretero, Y.; Causse, M. Technique for rapid, small-scale analysis of vitamin C levels in fruit and application to a tomato mutant collection. *J. Agric. Food Chem.* **2006**, *54*, 6159–6165. [CrossRef]
29. Griffith, O.W. Determination of glutathione and glutathione disulfide using glutathione reductase and 2-vinylpyridine. *Anal. Biochem.* **1980**, *106*, 207–212. [CrossRef]
30. Ishikawa, T.; Takeda, T.; Shigeoka, S. Purification and characterization of cytosolic ascorbate peroxidase from komatsuna (*Brassica rapa*). *Plant Sci.* **1996**, *120*, 11–18. [CrossRef]
31. Aebi, H. Catalase in vitro. *Meth. Enzymol.* **1984**, *105*, 121–126. [CrossRef]
32. Nakano, Y.; Asada, K. Purification of ascorbate peroxidase in spinach chloroplasts; its inactivation in ascorbate-depleted medium and reactivation by monodehydroascorbate radical. *Plant Cell Physiol.* **1987**, *28*, 131–140. [CrossRef]
33. Hossain, M.A.; Asada, K. Purification of dehydroascorbate reductase from spinach and its characterization as a thiol enzyme. *Plant Cell Physiol.* **1984**, *25*, 85–92. [CrossRef]
34. Klapheck, S.; Zimmer, I.; Cosse, H. Scavenging of hydrogen peroxide in the endosperm of *Ricinus communis* by ascorbate peroxidase. *Plant Cell Physiol.* **1990**, *31*, 1005–1013. [CrossRef]
35. Bradford, M.M. A rapid and sensitive method for the quantitation of microgram quantities of protein utilizing the principle of protein-dye binding. *Anal. Biochem.* **1976**, *72*, 248–254. [CrossRef]
36. Rehman, R.N.U.; Malik, A.U.; Khan, A.S.; Hasan, M.U.; Anwar, R.; Ali, S.; Haider, M.W. Combined application of hot water treatment and methyl salicylate mitigates chilling injury in sweet pepper (*Capsicum annuum* L.) fruits. *Sci. Hort.* **2021**, *283*, 110113. [CrossRef]
37. Sun, J.; Chen, J.; Kuang, J.; Chen, W.; Lu, W. Expression of sHSP genes as affected by heat shock and cold acclimation in relation to chilling tolerance in plum fruit. *Postharvest Biol. Technol.* **2010**, *55*, 91–96. [CrossRef]
38. Schirra, M.; D'hallewin, G. Storage performance of Fortune mandarins following hot water dips. *Postharvest Biol. Technol.* **1997**, *10*, 229–238. [CrossRef]
39. Kissinger, M.; Tuvia-Alkalai, S.; Shalom, Y.; Fallik, E.; Elkind, Y.; Jenks, M.A.; Goodwin, M.S. Characterization of physiological and biochemical factors associated with postharvest water loss in ripe pepper fruit during storage. *J. Am. Soc. Hortic. Sci.* **2005**, *130*, 735–741. [CrossRef]
40. Lu, J.; Vigneault, C.; Charles, M.; Raghavan, V. Heat treatment application to increase fruit and vegetable quality. *Stewart Postharvest Rev.* **2007**, *3*, 1–7. [CrossRef]
41. Fallik, E.; Grinberg, S.; Alkalai, S.; Lurie, S. The effectiveness of postharvest hot water dipping on the control of grey and black moulds in sweet red pepper (*Capsicum annuum*). *Plant Pathol.* **1996**, *45*, 644–649. [CrossRef]
42. Shehata, S.; Ibrahim, M.I.A.; El-Mogy, M.; Gawad, K. Effect of hot water dips and modified atmosphere packaging on extend the shelf life of bell pepper fruits. *Wulfenia* **2013**, *20*, 315–328.

43. Gonzalez-Aguilar, G.A.; Villa-Rodriguez, J.A.; Ayala-Zavala, J.F.; Yahia, E.M. Improvement of the antioxidant status of tropical fruits as a secondary response to some postharvest treatments. *Trends Food Sci. Technol.* **2010**, *21*, 475–482. [CrossRef]
44. Madani, B.; Mirshekari, A.; Imahori, Y. Physiological responses to stress. In *Postharvest Physiology and Biochemistry of Fruits and Vegetables*; Yahia, E.M., Ed.; Woodhead Publishing: Duxford, UK, 2019; pp. 405–423.
45. Sharom, M.; Willemot, C.; Thompson, J.E. Chilling injury induces lipid phase changes in membranes of tomato fruit. *Plant Physiol.* **1994**, *105*, 305–308. [CrossRef]
46. Van Breusegem, F.; Vranová, E.; Dat, J.F.; Inzé, D. The role of active oxygen species in plant signal transduction. *Plant Sci.* **2001**, *161*, 405–414. [CrossRef]
47. Quan, L.-J.; Zhang, B.; Shi, W.-W.; Li, H.-Y. Hydrogen peroxide in plants: A versatile molecule of the reactive oxygen species network. *J. Integr. Plant Biol.* **2008**, *50*, 2–18. [CrossRef]
48. Gill, S.S.; Tuteja, N. Reactive oxygen species and antioxidant machinery in abiotic stress tolerance in crop plants. *Plant Physiol. Biochem.* **2010**, *48*, 909–930. [CrossRef] [PubMed]
49. Suzuki, N.; Mittler, R. Reactive oxygen species and temperature stresses: A delicate balance between signaling and destruction. *Physiol. Plant* **2006**, *126*, 45–51. [CrossRef]
50. Constán-Aguilar, C.; Leyva, R.; Blasco, B.; Sánchez-Rodríguez, E.; Soriano, T.; Ruiz, J.M. Biofortification with potassium: Antioxidant responses during postharvest of cherry tomato fruits in cold storage. *Acta Physiol. Plant* **2014**, *36*, 283–293. [CrossRef]
51. Davey, M.W.; Keulemans, J. Determining the potential to breed for enhanced antioxidant status in malus: mean inter- and intravarietal fruit vitamin C and glutathione contents at harvest and their evolution during storage. *J. Agric. Food Chem.* **2004**, *52*, 8031–8038. [CrossRef] [PubMed]
52. Noctor, G.; Foyer, C.H. Ascorbate and glutathione: Keeping active oxygen under control. *Annu. Rev. Plant. Physiol. Plant. Mol. Biol.* **1998**, *49*, 249–279. [CrossRef] [PubMed]
53. Wang, K.; Shao, X.; Gong, Y.; Xu, F.; Wang, H. Effects of postharvest hot air treatment on gene expression associated with ascorbic acid metabolism in peach fruit. *Plant Mol. Biol. Rep.* **2014**, *32*, 881–887. [CrossRef]
54. Imahori, Y.; Bai, J.; Ford, B.L.; Baldwin, E.A. Effect of storage temperature on chilling injury and activity of antioxidant enzymes in carambola "Arkin" fruit. *J. Food Process. Preserv.* **2021**, *45*. [CrossRef]

 foods

MDPI

Article

Improvement of the Performance of Chitosan—*Aloe vera* Coatings by Adding Beeswax on Postharvest Quality of Mango Fruit

Usman Amin [1,2], Muhammad Kashif Iqbal Khan [1,2], Muhammad Usman Khan [3], Muhammad Ehtasham Akram [2], Mirian Pateiro [4,*], José M. Lorenzo [4,5] and Abid Aslam Maan [1,*]

1 National Institute of Food Science and Technology, University of Agriculture, Faisalabad 38000, Punjab, Pakistan; uamin2@ncsu.edu (U.A.); Kashif.khan@uaf.edu.pk (M.K.I.K.)
2 Department of Food Engineering, University of Agriculture, Faisalabad 38000, Punjab, Pakistan; ehtasham.akram@uaf.edu.pk
3 Department of Energy Systems Engineering, University of Agriculture, Faisalabad 38000, Punjab, Pakistan; usman.khan@uaf.edu.pk
4 Centro Tecnológico de la Carne de Galicia, Avd. Galicia No. 4, Parque Tecnológico de Galicia, San Cibrao das Viñas, 32900 Ourense, Spain; jmlorenzo@ceteca.net
5 Área de Tecnología de los Alimentos, Facultad de Ciencias de Ourense, Universidad de Vigo, 32004 Ourense, Spain
* Correspondence: mirianpateiro@ceteca.net (M.P.); abid.maan@uaf.edu.pk (A.A.M.)

Abstract: The effect of the application of chitosan–*Aloe vera* coatings emulsified with beeswax (0, 0.5, 1, 1.5 and 2%) during storage of *Mangifera indica* L. (cv Anwar Ratol) was investigated. Particle size of emulsions was reduced significantly with an increase in beeswax concentration. Water vapor permeability of the coatings was reduced by 43.7% with an increase in concentration of beeswax to 2%. The coated mangoes (at all concentrations of beeswax) exhibited reduced weight loss, delayed firmness loss, minimized pH change, maintained the total soluble solid contents, and retained free radical scavenging activity and total phenolic contents when stored at 18 °C and 75 ± 5% R.H. The best results were produced with a formulation containing 2.0% beeswax. Antimicrobial properties of chitosan and *Aloe vera* coatings were also improved with an increase in beeswax concentration and remarkably reduced the disease incidence in mangoes. In conclusion, beeswax-emulsified chitosan–*Aloe vera* coatings can be effectively used to increase the shelf life and marketable period of mangoes.

Keywords: mango; bioactive; coatings; biodegradable; *Aloe vera*; chitosan

Citation: Amin, U.; Khan, M.K.I.; Khan, M.U.; Ehtasham Akram, M.; Pateiro, M.; Lorenzo, J.M.; Maan, A.A. Improvement of the Performance of Chitosan—*Aloe vera* Coatings by Adding Beeswax on Postharvest Quality of Mango Fruit. *Foods* **2021**, *10*, 2240. https://doi.org/10.3390/foods10102240

Academic Editors: Eleni Tsantili and Jinhe Bai

Received: 20 August 2021
Accepted: 17 September 2021
Published: 22 September 2021

1. Introduction

Mango (*Mangifera indica* L.), commonly known as the "king of fruits", is the most popular fruit in the world due to its attractive color, taste, nutritional value and health benefiting impact [1]. After citrus, mango is the second largest fruit crop in Pakistan, making it the fifth largest producer among 87 countries. This fruit provides a significant socio-economic contribution to the economy [2,3]. Several postharvest changes i.e., continuous respiration, disease incidence, weight loss etc., end in loss of quality of mango fruits [4]. Various strategies have been developed to reduce waste. Among them, physical control (i.e., controlled atmospheric (CA) storage), chemical control (i.e., chemical preservatives) and fungicides have been used as a common practice to extend the shelf life of fruits and vegetables [5]. Changes in relative humidity, gas composition and temperature fluctuations during CA storage result in loss of weight, fruit texture, chlorophyll decay, respiration and microbial infection. Similarly, chemical preservatives pose a threat to the environment and human health. Thus, there is a need to develop human environment friendly alternates that may extend the shelf life of fruits.

In this regards, biodegradable coatings have been considered as an assistive approach to increase the shelf life and acceptability of stored fruits [6]. The term "biodegradable" refers to the condition that these coatings are prepared from natural polymers and can be decomposed by microorganisms after consumption of fruits [7]. Thus, they are safe, sustainable and environment friendly tool to assist the preservation of fruits and vegetables [8]. They have also been used as a carrier to several functional ingredients i.e., antioxidants and antimicrobial agents, to prevent the growth of microorganisms and oxidation reactions in fruits and vegetables [9,10]. These types of coatings are commonly known as active coatings. Performance of biodegradable coatings depends on type of polymer, bioactive agent, and lipid used, as lipids play an essential role to prevent moisture loss and preserve the freshness of the food product. In the past, scientists focused only on the development of these types of coatings. Therefore, the literature reports limited application of these coatings to enhance the shelf life of food products.

Chitosan, an amino-polysaccharide, is a non-toxic, biodegradable, edible and excellent film forming material [11]. Positively charged amino groups of chitosan interact with negatively charged cell walls of microorganisms, exhibiting antimicrobial potentials of chitosan [12]. Incorporation of *Aloe vera* in combination with chitosan has been reportedly used in the literature to improve the functional properties of biodegradable coatings [9]. Combination of chitosan and *Aloe vera* has been used to enhance the shelf life and disease control in cucumber [13], blueberries [14], strawberries [15], papaya [16], fresh-cut kiwifruit slices [17], fresh-cut red bell pepper [18] and several other applications.

To further improve the barrier properties and functionality of biodegradable coatings, beeswax has been reportedly used in coating solutions. Previously, beeswax coatings have been applied on several fruits such as kinnow [19], guava [20], blackberries [21], sapodilla fruits [12] and other similar fruits. These coatings had successfully minimized weight loss and maintained total soluble solids, skin color, and bioactive components i.e., vitamins, antioxidants, flavonoids.

Chitosan–*Aloe vera* coatings and beeswax has been separately evaluated on various fruits, however, no application of Chitosan–*Aloe vera* coatings emulsified with beeswax have been reported on any of the fruits and vegetables, to the best of our knowledge. Moreover, authors previously developed edible films using different concentrations of these materials [22]. Films revealed excellent physicochemical and mechanical properties, which strengthened their potential for food packaging application. Therefore, the purpose of current study was the development and characterization of beeswax-emulsified chitosan–*Aloe vera* coatings. Furthermore, effects of application of these coatings on physicochemical, mechanical and functional properties of mango fruits have been studied.

2. Materials and Methods

2.1. Selection of Healthy Mango Fruits

Fully ripened mango (*Magnifera Indica* L. cv. Anwar Ratol) fruits were collected from the local market of Faisalabad, Pakistan. Fruits with uniform shape, color, size were selected, while diseased or blemished fruits were discarded. Sorted fruits were first washed with distilled water and then dipped in NaOCl (3%) solution for 3 min to sterilize the mango fruits. Afterward, fruits were washed and cleaned with soft tissues to remove small water droplets. Afterward, fruits were randomly distributed into five different groups (each containing 20 fruits) prior to treatments. Each fruit was assigned a specific name ($Control_i$, $S1_i$, $S2_i$, $S3_i$, $S4_i$; where $i = 1, 2, 3, 4, \ldots$ n) and stored at 8 °C prior to coating. Uncoated samples were considered as control samples.

2.2. Procurement of Coating Material

Chitosan (origin: crab shells), beeswax (yellow soft), gallic acid (MW: 170.12), Folin-Ciocalteu (2 N), glycerol (MW: 92.09), tween 20 (conductivity: 50 µS/cm), acetic acid (≥99.5%) and NaOCl (4–4.99% available chlorine) were purchased form Sigma Aldrich, USA. Fresh *Aloe vera* plants were harvested from a local nursery in Faisalabad, Pakistan.

2.3. Preparation of Coating Solutions

Preparation of coatings was performed in three steps. First, chitosan solution was prepared by slowly dissolving 1 g of chitosan in 100 mL of 2% acetic acid solution (*v*/*v*) on a magnetic stirrer (350 rpm) to get 1% chitosan solution (*w*/*v*) [23]. For *Aloe vera* gel preparation, *Aloe vera* leaves were peeled with stainless steel knife and gel was extracted. It was then blended and homogenized in a blender and filtered to remove impurities and debris material. A 20% *Aloe vera*–chitosan solution was prepared by mixing 20 mL of *Aloe vera* with 80 mL of prepared chitosan solution (1%). This combination was selected because films of these materials revealed superior barrier and mechanical strength in our previous study [22]. Afterward, glycerol (0.2 g) and tween 20 (0.2 g) were added in chitosan–*Aloe vera* solution. Finally, beeswax was melted at 70 °C and mixed with the chitosan–*Aloe vera* solution in the proportions described in Table 1. Coating formulations were homogenized at 13,500 rpm for 3–4 min using ultra-turrax homogenizer (IKA-T18-Germany).

Table 1. Proportions of the emulsions used to formulate the different treatments.

Treatment	*Aloe vera*—Chitosan Solution (mL)	Beeswax (Melted) (mL)
S1	99.5	0.5
S2	99.0	1.0
S3	98.5	1.5
S4	98.0	2.0

2.4. Application of Coatings on Mango Fruits

Coatings (slightly yellowish in color) were applied on mango fruits by dipping the mango fruits in the coating solution. Mango fruits in each group were dipped in the coating formulations, separately, for the same time under ambient conditions (20 °C, R.H. 70%). After drying of coatings, all the samples were stored at 18 °C and 75 ± 5% R.H. These conditions were selected to mimic the market environment, as this study attempted to increase the marketable life of mangoes. For subsequent physicochemical analyses, each sample was marked with the group name and fruit number for an effective storage study (e.g., sample S1 as $S1_1$, $S1_2$, $S1_3$, etc.). Then, each sample was evaluated for physicochemical characteristics in 7 day intervals for three weeks.

2.5. Characterization of Coating Emulsions

2.5.1. Particle Size Analysis

Prepared emulsions were subjected to particle size analysis using particle size analyzer (Bettersizer ST, Bettersize Instruments Ltd., Dandong, Liaoning, China).

2.5.2. Emulsion Stability

Stability of the emulsions was determined by following the method reported by Amin et al. [22] with little modifications. Briefly, 6 mL from each coating formulation was taken in a test tube on test tube rack and placed under ambient condition for three days. Stability of the emulsion was measured by the following equation:

$$\text{Emulsion stability} = \frac{h_o - h_t}{h_o} \quad (1)$$

where h_o and h_t is the initial and final height of the emulsion in the test tube after 3 days' time, respectively.

2.5.3. Water Vapor Transmission Rate

Water vapor permeability of the coatings was measured according to ASTM method [24] with few modifications. Coating solutions were casted into films and dried at 23 ± 2 °C on a smooth horizontal surface until the coating solutions shaped into a thin sheet or films. Glass tubes (5 × 10 cm) were filled with desiccant (silica gel) and dried films were fixed over the face of the tube (by using gel). Films were placed in the climate chamber (POL-EKO-APARATURA-KK-350, Poland) and maintained at 38 ± 0.6 °C and 90 ± 2% R.H. Change in weight of tubes was measured periodically for 7 days to obtain moisture transfer rate. Water vapor permeability (WVP) was calculated by the following equation:

$$\text{WVP} = \frac{\Delta m \times T}{A \times t \times \Delta P} \tag{2}$$

Here, Δm is the change in mass (g) before and after the specific time, A is the area (m^2), t is the time (hours), T is the thickness of films and ΔP is the difference in pressure at saturated pressure and pressure under the testing conditions.

2.5.4. Diffusion Coefficient

The moisture loss in term of diffusion through the coatings was determined by using Fick's law of diffusion, written as:

$$MR = \frac{M_t - M_e}{M_o - M_e} = \frac{8}{\pi^2} exp\left(\frac{-\pi^2 Dt}{4L^2}\right) \tag{3}$$

Here, MR, D, t and L are the moisture ratio, effective diffusion coefficient (m^2/s), time (s) and thickness (m) of the sample layer distributed in the drying chamber, respectively. Moreover, M_t, M_e, M_o represents moisture levels at time interval of t, initial and equilibrium, respectively. The diffusion coefficient can be calculated from the slope of line (α) drawn between ln(MR) and time as:

$$\alpha = \left(\frac{\pi^2 Dt}{4L^2}\right) \tag{4}$$

2.6. *Physicochemical Analysis of Coated and Uncoated Mango Fruits*

2.6.1. Percentage Weight Loss

Percentage weight loss of fruits was measured by using digital weight balance. The weight of each marked sample was measured before the start of experiment. Change in weight of stored samples was observed and percentage weight loss was calculated by the following equation [19]:

$$\text{Weight loss } (\%) = \frac{W_i - W_f}{W_\text{i}} \tag{5}$$

where W_i is the initial weight of the coated fruit and W_f is the weight after specific time interval during storage.

2.6.2. Firmness

Firmness of mangoes was measured by using a Texture Analyzer (TA XT Plus, Stable Micro Systems, Godalming, Surrey, UK) as performed by Meindrawan et al. [25] with slight modification in speed and replications. Three samples from each group of S1, S2, S3, and S4 were selected randomly after 0, 7, 14, 21 days and repeatedly placed on the stage of Texture analyzer. Sampling probe was inserted inside the fruit sample at a speed of 3 mm/s and repeated three times on different areas of a sample. The penetration force was expressed in grams the fruits can bear.

2.6.3. Total Soluble Solids

The total soluble solid contents of the fruits were determined using digital refractometer (DR201-95, KRÜSS Optronic GmbH, Hamburg, Germany). The pulp of the mangoes was extracted and homogenized for uniform distribution of solid particles in the solution. A drop from the pulp was placed on the sample stage of the refractometer, and the results were expressed as °Brix.

2.6.4. pH

Homogenized fruit pulp was subjected to pH measurement using pH meter (ST5000, OHAUS, Parsippany, NJ, USA) through the method described by Eshetu et al. [26]. pH meter was first calibrated, and the probe was inserted into the homogenized sample. The value of pH was noted as the concentration of hydronium ion on pH scale 0–14.

2.6.5. Total Phenolic Contents (TPC)

Total phenolic contents were measured by the Folin-Cicalteau method [8]. Briefly, 0.3 mL of the extract prepared for antioxidant activity was mixed with 1.2 mL of 7% sodium carbonate solution. The mixture was then mixed with 1.5 mL of Folin reagent and placed on shaker for 1.5 to 2 h. The absorbance was measured at 765 nm by UV/VIS spectrophotometer (T80). Gallic acid calibration curve (10 ppm–1000 ppm) was used to measure TPC, and results were expressed as mg gallic acid (GAE)/100g MP.

2.6.6. Total Antioxidant Capacity

The antioxidant activity was assayed by the method adopted by Ebrahimi et al. [8]. To prepare the extract, homogenized fruit pulp was added in 80% methanol and placed on the shaker for 3–4 h. After this, DPPH solution was prepared by adding 0.025 g of DPPH in 10 mL of 85% ethanol. A volume of 50 µL of the prepared extract was mixed well with 950 µL of DPPH solution. The samples were then placed in the dark room for 30 min. Absorbance by the samples was measured at 517 nm using UV/VIS spectrophotometer (T80). Free radical scavenging activity (%) was measured by using the following equation:

$$\text{Free redical Scavenging activity } (\%) = \frac{(\text{Absorbance})_{\text{DPPH}} - (\text{absorbance})_{\text{Sample}}}{(\text{Absorbance})_{\text{DPPH}}} \quad (6)$$

2.6.7. Decay Incidence

Decay incidence on mango fruits was assessed through the method reported by Khaliq et al. [27] and Eshetu et al. [26]. Fruits were carefully observed for fungal and bacterial infection after 7, 14 and 21 days. Scale was observed as 1 = no decay, 2 = 0–5% decay, 3 = 5–25%, 4 = 25–50%, 5 = 50–75%, 6 = >75%. Percentage of fungal and bacterial decay was calculated by the following equation:

$$\text{Decay incidence} = \frac{\sum(\text{decay level } \times \text{ Number of fruit at that level})}{\text{Total number of fruits } \times \text{ Maximum decay level}} \times 100 \quad (7)$$

2.7. Statistical Analysis

Data was obtained in triplicate and subjected to statistical analysis at 5% confidence interval. Characteristics of emulsion were analyzed through ANOVA and differences between the means was analyzed through pairwise comparison using SAS software package (SAS institute, Cary, NC, USA). Shelf life study parameters were analyzed using MATLAB curve fitting tool (MATLAB 2016a, Natick, MA, USA).

3. Results and Discussions

3.1. Characterization of Emulsion Coatings

3.1.1. Particle Size and Emulsion Stability

Particle size and size distribution are the most important parameters of emulsions, determining their characteristics such as rheology, appearance, stability, etc. [28]. The effect of beeswax concentration on logarithm distribution of particles in emulsions is shown in Figure 1. A non-significant difference was observed in particle size distribution with increase in concentration of beeswax in chitosan—*Aloe vera* emulsions. The average particle diameter for all the emulsions was observed to be around 4 µm. Xie et al. [29] observed similar behavior for emulsions of beeswax in carboxymethyl chitosan and cellulose nanofibrils. Contraction in particle size range with increased beeswax concentration might have produced more shear during homogenization and produced more rod-like structures during hardening of beeswax upon cooling [30]. Moreover, emulsions remained stable for all concentrations of beeswax after 72 h, without any separation of dispersed phase. This reveals better dispersibility of beeswax in chitosan—*Aloe vera* emulsions.

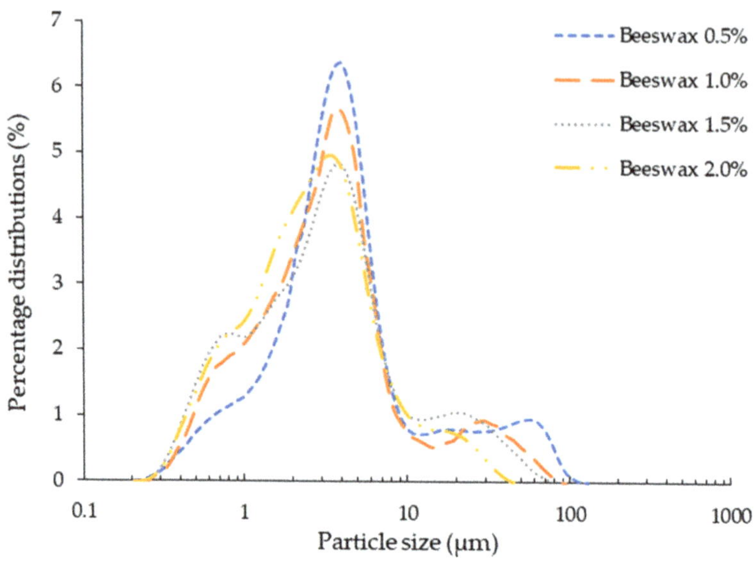

Figure 1. Particle size distribution of emulsion.

3.1.2. Water Vapor Permeability and Diffusivity

Results for water vapor permeability of films is shown in Figure 2. The highest value of WVP was observed for 0.5% beeswax and the lowest for 2% concentration of beeswax. Thus, increasing the concentration of beeswax from 0.5 to 2% resulted in 43.78% reduction in water vapor permeability of films. This is obvious due to increase concentration of wax in the coatings. Pérez-Vergara et al. [31] attributed the reduction in moisture transfer to a large number of long-chain fatty alcohols and alkanes present in the orthorhombic structure of beeswax. Zhang et al. [32] have reported that synergistic interaction between chitosan and beeswax films exhibits better water barrier properties compared to pure beeswax films.

Similarly, diffusion of moisture through the films was also calculated and drawn in Figure 2. Results of diffusion coefficient revealed that an increased concentration of beeswax in *Aloe vera* and chitosan blend had significantly reduced moisture diffusion through the coatings. Initially, moisture diffusion for uncoated mango fruits was observed to be 1.5×10^{-10} m^2/s, which was reduced to 7.94×10^{-11} m^2/s as the concentration of beeswax increased to 2% in the coating formulations.

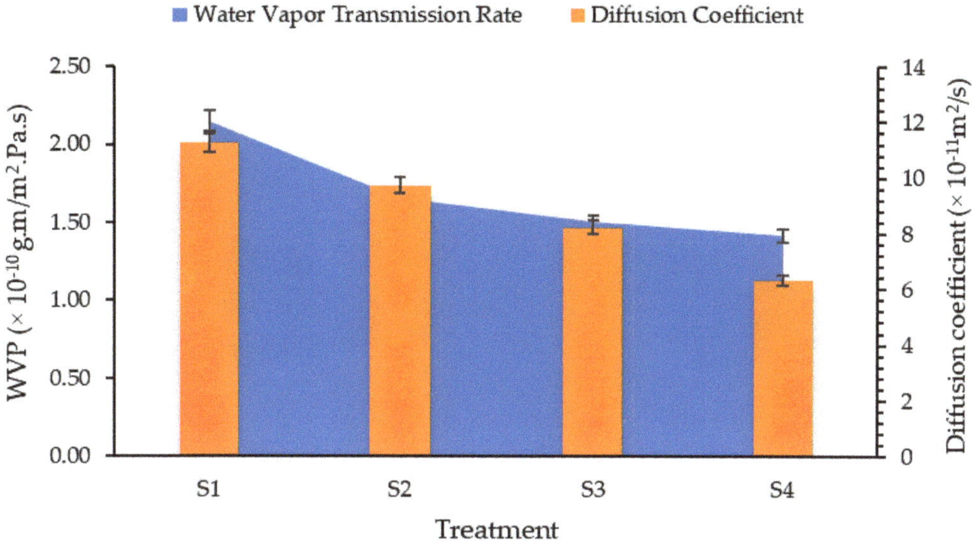

Figure 2. Water vapor permeability and diffusion coefficient of chitosan–*Aloe vera* gel coatings with different concentrations of beeswax.

3.2. Physicochemical Characteristics of Coated and Uncoated Mangoes

3.2.1. Weight Loss

Fruits and vegetables are mainly composed of water, which is known as the indicator of freshness of fruits and vegetables. Fluctuation in storage temperature and relative humidity of the storage environment may trigger respiration and transpiration rate, which may cause weight loss in fruits and vegetables [26]. In the current study, weight loss of mangoes was significantly reduced by all coating solutions as shown in Figure 3a. However, weight loss was observed to increase throughout storage. Mangoes coated with treatment S4 (containing 2% beeswax) exhibited minimum weight loss during storage periods. Increased concentration of beeswax increased the hydrophobicity of coatings, which conversely reduced the moisture removal, and consequently, the weight loss from the fruits. This reduction of weight loss in polysaccharide coatings may be associated to better crosslinking between hydroxyl group of polysaccharide and hydrophilic substance (such as phenols) in coatings through hydrogen bonding [33]. The results are supported by the findings of Nasrin et al. [34], who used beeswax coatings in combination with coconut oil and observed the similar behavior of weight loss in lemons. Eshetu et al. [26] also reported that 2% beeswax in chitosan coatings preserved moisture inside the mangoes.

3.2.2. Total Soluble Solids

Figure 3b exhibits the effect of beeswax concentration and storage time on total soluble contents of mango fruits. Concentration of beeswax in coating solutions significantly affected total soluble contents in mango fruits. A rapid change in degree brix was observed in the control sample, which increased from 60.08 to 65.13 °Brix during storage time. However, a non-significant change in brix was observed for coated fruits. This may be due to the reduced conversion of acid into sugar during 21 days of storage. Chitosan–*Aloe vera* coatings with 2% beeswax revealed minimum change in TSS (from 60.08 to 61.74 °Brix). Moalemiyan et al. [35] reported that an increase in TSS may be associated with weight loss causing an increase in sugar concentration. Moreover, conversion of complex carbohydrates present in the mangoes into soluble sugars also increased the TSS contents. Coatings provide an efficient cover from the environmental oxygen and hinder the metabolic activities responsible for rapid conversions of acids into sugar [19]. As explained earlier, a slight change in weight loss was observed for coatings containing 2% beeswax, which may be responsible for a slight increase in TSS contents in mangoes treated with S4.

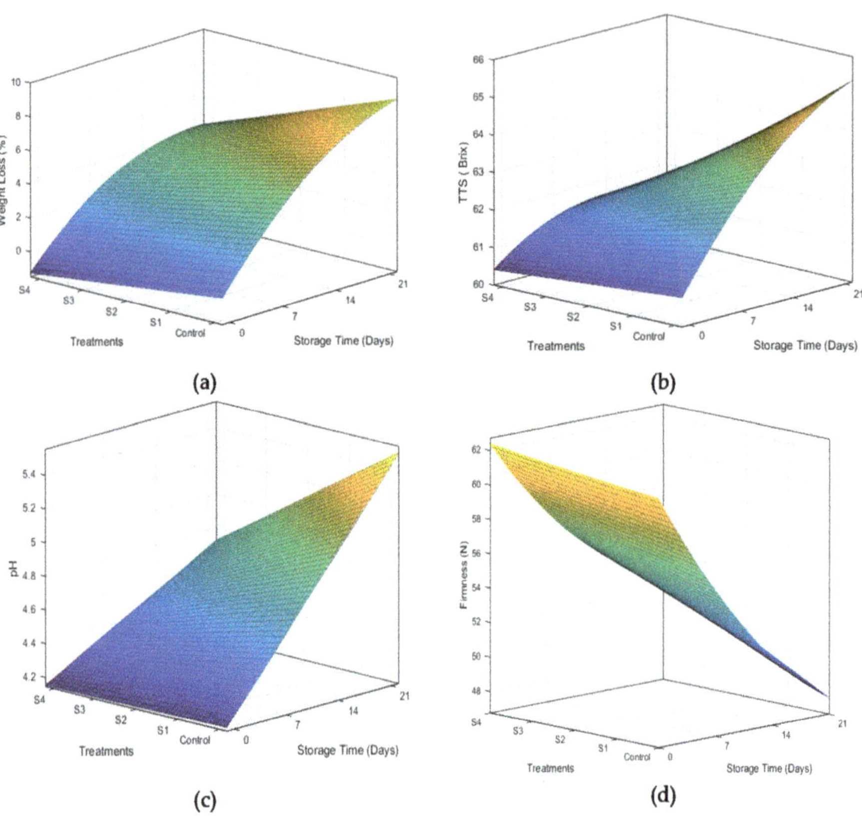

Figure 3. Effect of storage time and beeswax concentration on the various properties of coated mangoes, i.e., weight loss (**a**), total soluble solids (**b**), pH (**c**) and firmness (**d**) during storage at 18 °C and 75 ± 5% R.H.

3.2.3. pH

Concentration of beeswax in the coating solution significantly affected the pH of mangoes during storage, as shown in Figure 3c. A linear and rapid increase in concentration of hydronium ion was observed for uncoated fruits. As the storage time was increased, their change in pH became more pronounced. All the coating formulations significantly hindered the change in pH as compared to the uncoated treatments. However, treatment S4 was the most efficient in reducing the rate of change in pH. Increase in pH values is attributed to an increase in conversion of organic acids into sugars [35]. As explained in the section above, TSS values changed slightly for mango coated with 2% beeswax (S4). Therefore, minimum variations in pH were observed for samples treated with S4 as compared to control samples. Findings of Velickova et al. [23] support the results obtained in the current study, who applied chitosan–beeswax coatings on strawberries.

3.2.4. Firmness

Firmness of fruit is an important indicator of freshness and quality. Firmness of coated mango fruits was tested and compared with uncoated mangoes. Figure 3d reveals the effect of beeswax concentration and storage time on firmness of mango fruits. Firmness of mango fruits was observed to decrease linearly with an increase in storage time for all coating treatments. However, the rate of firmness loss was consistently reduced with addition of beeswax up to 2%. Application of hydroxypropyl and carnauba wax coatings [36] also revealed similar results. Klangmuang and Sothornvit [37] reported that loss in firmness of mango fruits is attributed to cell wall digestion by different enzymes during ripening. Enzymes such as pectin esterase and polyglacturonase trigger the hydrolysis of starch and pectin present in the cell wall. This causes the loss in cellular turgor and subsequent loss of firmness of mango fruits. Emulsified coatings slowed down the metabolic activity in the fruits by controlling the environment of coated fruits, which maintained the firmness in coated samples [38].

3.2.5. Free Radical Scavenging Activity

Antioxidant activity of mangoes has been attributed to a number of compounds, i.e., organic acids, chlorophyl, phenolic compounds etc. The role of coatings in maintaining antioxidant activity during storage is shown in Figure 4a. Uncoated samples exhibited maximum reduction in antioxidant activity (84.89% to 80.66%) during the storage of 21 days. The loss of antioxidant activity was reduced with an increase in beeswax concentration, with minimum reduction observed in S4. Reduction in free radical scavenging activity was more pronounced at the end of the first week for all samples. These results are in correspondence with Ebrahimi et al. [8], who applied guar gum–*Aloe vera* coatings containing *Spirulina platensis* on mango fruits.

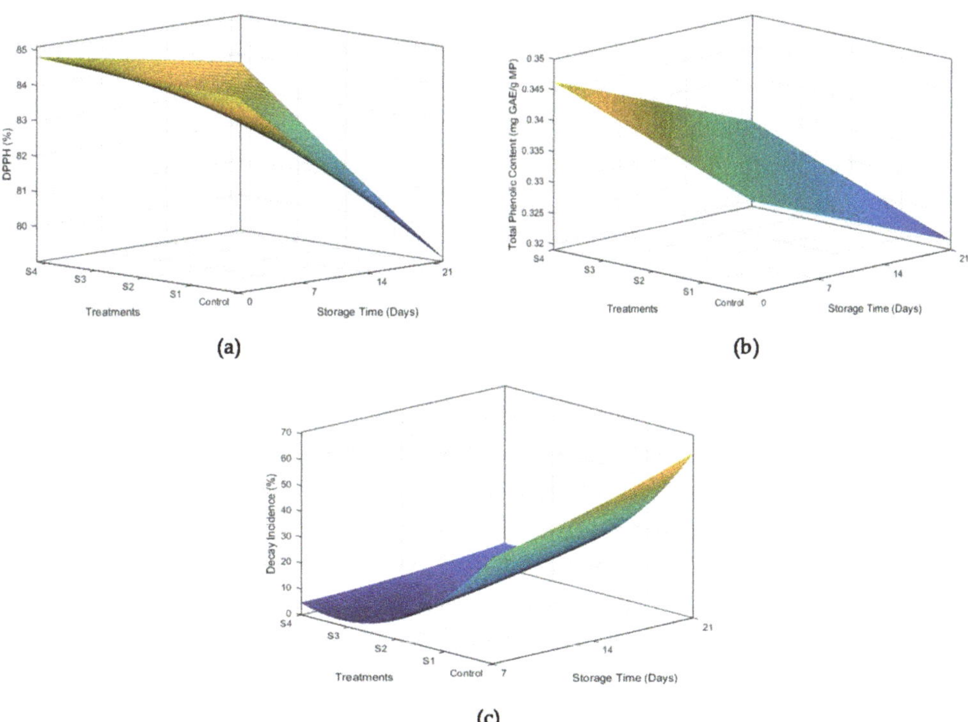

Figure 4. Effect of time and beeswax concentration on DPPH activity (**a**), total phenolic contents (**b**), and decay incidence (**c**) of mango fruits stored for 21 days at 18 °C and 75 ± 5% R.H.

3.2.6. Total Phenolic Contents (TPC)

Phenolic compounds are present in almost all plants as essential secondary metabolites. They also provide protective mechanisms against different pathogens. However, phenolic compounds start to decrease with an increase in storage time in fruits and vegetables [27]. The effect of coatings on phenolic contents of coated and uncoated mangoes during storage is presented in Figure 4b. All the coating formulations were observed to efficiently retain the TPC throughout the storage period, with S4 being the most efficient. TPC in the control sample reduced from 0.342 to 0.3201 mg GAE/g of mango pulp. Maintenance of phenolic compounds with increase in beeswax concentration can be attributed to the low oxygen permeability. Hinderance in oxygen availability to mango fruits through coatings slowed down the enzymatic degradation and oxidation in coated samples [33].

3.2.7. Decay Incidence

Percentage decay incidence was observed throughout the storage of 21 days at 8 °C. An increase in decay of the samples became more pronounced between 7–14 and 14–21 days interval. All the fruits in the control treatment showed microbial infection with different levels of severity on the disease incidence scale (Figure 4c). Maximum disease incidence of 67.33% was observed for uncoated samples during the first week of storage and corresponded with the results of Klangmuang and Sothornvit [37]. The decay incidence decreased with an increase in beeswax concentration, with S4 being the most efficient in reducing fruit decay. This may be attributed to the antimicrobial properties and inhibitory action of chitosan and *Aloe vera*, which helped to maintain the cellular integrity. Amino group of chitosan attaches to the cell membrane of bacteria through electrostatic forces, which permanently destroys the bacterial cell [10]. Additionally, *Aloe vera* extract used in coatings might have released phenolic compounds to thwart microbial growth [27]. Hinderance in the transport of oxygen from the environment to the surface of fruit might also have contributed to reduced microbial growth [9].

4. Conclusions

In this study, the potential of beeswax-emulsified chitosan–*Aloe vera* coatings was explored for the preservation of mangoes. Increased concentration of beeswax (0.5 to 2%) in coating formulations produced significantly stable emulsions with decreased polydispersity and average particle size of 4 µm. Hydrophobicity of beeswax and better cross-linkages between polymer and bioactive compounds in *Aloe vera* extract reduced the water transmission and diffusion coefficient to 33.83% and 43.78%, respectively. Furthermore, coatings having variable beeswax concentration significantly conserved the physicochemical characteristics of fruits, i.e., weight loss, firmness, pH, and total soluble solid contents. Coatings also reduced the oxygen permeability, which minimized the oxidation of phenolic compounds in the coated fruits. A significant reduction in decay incidence (4.2% on coated fruits with S4 formulation as compared to 62.1% in uncoated fruits) reveals their potential for practical application to decrease senescence and ripening of fruits during their exposure to market conditions.

Author Contributions: Conceptualization, M.U.K., M.E.A. and A.A.M.; writing—original draft preparation, U.A. and M.K.I.K.; writing—review and editing, M.U.K., M.E.A., M.P., J.M.L. and A.A.M. All authors have read and agreed to the published version of the manuscript.

Funding: This research received no external funding.

Institutional Review Board Statement: Not applicable.

Informed Consent Statement: Not applicable.

Data Availability Statement: The data presented in this study are available on request from the corresponding authors.

Acknowledgments: Thanks to GAIN (Axencia Galega de Innovación) for supporting this research (grant number IN607A2019/01).

Conflicts of Interest: The authors declare no conflict of interest.

References

1. Sudheeran, P.K.; Love, C.; Feygenberg, O.; Maurer, D.; Ovadia, R.; Oren-Shamir, M.; Alkan, N. Induction of red skin and improvement of fruit quality in 'Kent', "Shelly" and 'Maya' mangoes by preharvest spraying of prohydrojasmon at the orchard. *Postharvest Biol. Technol.* **2019**, *149*, 18–26. [CrossRef]
2. Badar, H.; Ariyawardana, A.; Collins, R. Dynamics of mango value chains in Pakistan. *Pak. J. Agric. Sci.* **2019**, *56*, 523–530.
3. Musharraf, S.G.; Uddin, J.; Siddiqui, A.J.; Akram, M.I. Quantification of aroma constituents of mango sap from different Pakistan mango cultivars using gas chromatography triple quadrupole mass spectrometry. *Food Chem.* **2016**, *196*, 1355–1360. [CrossRef] [PubMed]
4. Anusuya, P.; Nagaraj, R.; Janavi, G.J.; Subramanian, K.S.; Paliyath, G.; Subramanian, J. Pre-harvest sprays of hexanal formulation for extending retention and shelf-life of mango (*Mangifera indica* L.) fruits. *Sci. Hortic.* **2016**, *211*, 231–240. [CrossRef]
5. Rosalie, R.; Léchaudel, M.; Dhuique-Mayer, C.; Dufossé, L.; Joas, J. Antioxidant and enzymatic responses to oxidative stress induced by cold temperature storage and ripening in mango (*Mangifera indica* L. cv. 'Cogshall') in relation to carotenoid content. *J. Plant Physiol.* **2018**, *224–225*, 75–85. [CrossRef] [PubMed]
6. Cosme Silva, G.M.; Silva, W.B.; Medeiros, D.B.; Salvador, A.R.; Cordeiro, M.H.M.; da Silva, N.M.; Santana, D.B.; Mizobutsi, G.P. The chitosan affects severely the carbon metabolism in mango (*Mangifera indica* L. cv. Palmer) fruit during storage. *Food Chem.* **2017**, *237*, 372–378. [CrossRef]
7. Amin, M.U.; Khurram, M.; Khattak, B.; Khan, J. Antibiotic additive and synergistic action of rutin, morin and quercetin against methicillin resistant *Staphylococcus aureus*. *BMC Complement. Altern. Med.* **2015**, *15*, 1–12. [CrossRef]
8. Ebrahimi, F.; Rastegar, S. Preservation of mango fruit with guar-based edible coatings enriched with Spirulina platensis and Aloe vera extract during storage at ambient temperature. *Sci. Hortic.* **2020**, *265*, 109258. [CrossRef]
9. Shah, S.; Hashmi, M.S. Chitosan–aloe vera gel coating delays postharvest decay of mango fruit. *Hortic. Environ. Biotechnol.* **2020**, *61*, 279–289. [CrossRef]
10. Xing, Y.; Yang, H.; Guo, X.; Bi, X.; Liu, X.; Xu, Q.; Wang, Q.; Li, W.; Li, X.; Shui, Y.; et al. Effect of chitosan/Nano-TiO_2 composite coatings on the postharvest quality and physicochemical characteristics of mango fruits. *Sci. Hortic.* **2020**, *263*, 109135. [CrossRef]
11. Khatri, D.; Panigrahi, J.; Prajapati, A.; Bariya, H. Attributes of Aloe vera gel and chitosan treatments on the quality and biochemical traits of post-harvest tomatoes. *Sci. Hortic.* **2020**, *259*, 108837. [CrossRef]
12. Foo, S.Y.; Nur Hanani, Z.A.; Rozzamri, A.; Ibadullah, W.Z.W.; Ismail-Fitry, M.R. Effect of Chitosan–Beeswax Edible Coatings on the Shelf-life of Sapodilla (*Achras zapota*) Fruit. *J. Packag. Technol. Res.* **2019**, *3*, 27–34. [CrossRef]
13. Adetunji, C.; Fadiji, A.; Aboyeji, O. Effect of chitosan coating combined Aloe vera gel on cucumber (*Cucumis Sativa* L.) post-harvest quality during ambient storage. *J. Emerg. Trends Eng. Appl. Sci.* **2014**, *5*, 391–397.
14. Vieira, J.M.; Flores-López, M.L.; de Rodríguez, D.J.; Sousa, M.C.; Vicente, A.A.; Martins, J.T. Effect of chitosan–Aloe vera coating on postharvest quality of blueberry (*Vaccinium corymbosum*) fruit. *Postharvest Biol. Technol.* **2016**, *116*, 88–97. [CrossRef]
15. Pinzon, M.I.; Sanchez, L.T.; Garcia, O.R.; Gutierrez, R.; Luna, J.C.; Villa, C.C. Increasing shelf life of strawberries (*Fragaria* ssp) by using a banana starch-chitosan-Aloe vera gel composite edible coating. *Int. J. Food Sci. Technol.* **2020**, *55*, 92–98. [CrossRef]
16. Monzón-Ortega, K.; Salvador-Figueroa, M.; Gálvez-López, D.; Rosas-Quijano, R.; Ovando-Medina, I.; Vázquez-Ovando, A. Characterization of Aloe vera-chitosan composite films and their use for reducing the disease caused by fungi in papaya Maradol. *J. Food Sci. Technol.* **2018**, *55*, 4747–4757. [CrossRef]
17. Benítez, S.; Achaerandio, I.; Pujolà, M.; Sepulcre, F. Aloe vera as an alternative to traditional edible coatings used in fresh-cut fruits: A case of study with kiwifruit slices. *LWT—Food Sci. Technol.* **2015**, *61*, 184–193. [CrossRef]
18. Sathiyaseelan, A.; Saravanakumar, K.; Mariadoss, A.V.A.; Ramachandran, C.; Hu, X.; Oh, D.H.; Wang, M.H. Chitosan-tea tree oil nanoemulsion and calcium chloride tailored edible coating increase the shelf life of fresh cut red bell pepper. *Prog. Org. Coat.* **2021**, *151*, 106010. [CrossRef]
19. Baswal, A.K.; Dhaliwal, H.S.; Singh, Z.; Mahajan, B.V.C.; Kalia, A.; Gill, K.S. Influence of carboxy methylcellulose, chitosan and beeswax coatings on cold storage life and quality of Kinnow mandarin fruit. *Sci. Hortic.* **2020**, *260*, 108887. [CrossRef]
20. Oliveira, V.R.L.; Santos, F.K.G.; Leite, R.H.L.; Aroucha, E.M.M.; Silva, K.N.O. Use of biopolymeric coating hydrophobized with beeswax in post-harvest conservation of guavas. *Food Chem.* **2018**, *259*, 55–64. [CrossRef]
21. Pérez-Gallardo, A.; García-Almendárez, B.; Barbosa-Cánovas, G.; Pimentel-González, D.; Reyes-González, L.R.; Regalado, C. Effect of starch-beeswax coatings on quality parameters of blackberries (*Rubus* spp.). *J. Food Sci. Technol.* **2014**, *52*, 5601–5610. [CrossRef] [PubMed]
22. Amin, U.; Khan, M.A.; Akram, M.E.; Al-Tawaha, A.R.M.S.; Laishevtcev, A.; Shariati, M.A. Characterization of compisote edible films from aloe vera gel, beeswax and chitosan. *Potravin. Slovak J. Food Sci.* **2019**, *13*, 854–862. [CrossRef]
23. Velickova, E.; Winkelhausen, E.; Kuzmanova, S.; Alves, V.D.; Moldão-Martins, M. Impact of chitosan-beeswax edible coatings on the quality of fresh strawberries (*Fragaria ananassa* cv Camarosa) under commercial storage conditions. *LWT—Food Sci. Technol.* **2013**, *52*, 80–92. [CrossRef]

24. ASTM. *ASTM D1653-13: Standard Test Methods for Water Vapor Transmission of Organic Coating Films*; ASTM International: West Conshohocken, PA, USA, 2013.
25. Meindrawan, B.; Suyatma, N.E.; Wardana, A.A.; Pamela, V.Y. Nanocomposite coating based on carrageenan and ZnO nanoparticles to maintain the storage quality of mango. *Food Packag. Shelf Life* **2018**, *18*, 140–146. [CrossRef]
26. Eshetu, A.; Ibrahim, A.M.; Forsido, S.F.; Kuyu, C.G. Effect of beeswax and chitosan treatments on quality and shelf life of selected mango (*Mangifera indica* L.) cultivars. *Heliyon* **2019**, *5*, e01116. [CrossRef]
27. Khaliq, G.; Ramzan, M.; Baloch, A.H. Effect of Aloe vera gel coating enriched with Fagonia indica plant extract on physicochemical and antioxidant activity of sapodilla fruit during postharvest storage. *Food Chem.* **2019**, *286*, 346–353. [CrossRef]
28. Maan, A.A.; Schroën, K.; Boom, R. Spontaneous droplet formation techniques for monodisperse emulsions preparation—Perspectives for food applications. *J. Food Eng.* **2011**, *107*, 334–346. [CrossRef]
29. Xie, B.; Zhang, X.; Luo, X.; Wang, Y.; Li, Y.; Li, B.; Liu, S. Edible coating based on beeswax-in-water Pickering emulsion stabilized by cellulose nanofibrils and carboxymethyl chitosan. *Food Chem.* **2020**, *331*, 127108. [CrossRef]
30. Han, J.H.; Seo, G.H.; Park, I.M.; Kim, G.N.; Lee, D.S. Physical and Mechanical Properties of Pea Starch Edible Films Containing Beeswax Emulsions. *J. Food Sci.* **2006**, *71*, E290–E296. [CrossRef]
31. Pérez-Vergara, L.D.; Cifuentes, M.T.; Franco, A.P.; Pérez-Cervera, C.E.; Andrade-Pizarro, R.D. Development and characterization of edible films based on native cassava starch, beeswax, and propolis. *NFS J.* **2020**, *21*, 39–49. [CrossRef]
32. Zhang, W.; Xiao, H.; Qian, L. Enhanced water vapour barrier and grease resistance of paper bilayer-coated with chitosan and beeswax. *Carbohydr. Polym.* **2014**, *101*, 401–406. [CrossRef]
33. Riaz, A.; Aadil, R.M.; Amoussa, A.M.O.; Bashari, M.; Abid, M.; Hashim, M.M. Application of chitosan-based apple peel polyphenols edible coating on the preservation of strawberry (*Fragaria ananassa* cv Hongyan) fruit. *J. Food Process. Preserv.* **2021**, *45*. [CrossRef]
34. Nasrin, T.A.A.; Rahman, M.A.; Arfin, M.S.; Islam, M.N.; Ullah, M.A. Effect of novel coconut oil and beeswax edible coating on postharvest quality of lemon at ambient storage. *J. Agric. Food Res.* **2020**, *2*, 100019. [CrossRef]
35. Moalemiyan, M.; Ramaswamy, H.S.; Maftoonazad, N. Pectin-based edible coating for shelf-life extension of Ataulfo mango. *J. Food Process Eng.* **2012**, *35*, 572–600. [CrossRef]
36. Baldwin, E.A.; Burns, J.K.; Kazokas, W.; Brecht, J.K.; Hagenmaier, R.D.; Bender, R.J.; Pesis, E. Effect of two edible coatings with different permeability characteristics on mango (*Mangifera indica* L.) ripening during storage. *Postharvest Biol. Technol.* **1999**, *17*, 215–226. [CrossRef]
37. Klangmuang, P.; Sothornvit, R. Active hydroxypropyl methylcellulose-based composite coating powder to maintain the quality of fresh mango. *LWT* **2018**, *91*, 541–548. [CrossRef]
38. Sousa, F.F.; Pinsetta Junior, J.S.; Oliveira, K.T.E.F.; Rodrigues, E.C.N.; Andrade, J.P.; Mattiuz, B.H. Conservation of 'Palmer' mango with an edible coating of hydroxypropyl methylcellulose and beeswax. *Food Chem.* **2021**, *346*, 128925. [CrossRef]

Article

Molecular Mechanisms through Which Short-Term Cold Storage Improves the Nutritional Quality and Sensory Characteristics of Postharvest Sweet Potato Tuberous Roots: A Transcriptomic Study

Shuqian Zhou, Lu Chen, Gang Chen, Yongxin Li and Huqing Yang *

School of Food and Health, Zhejiang Agricultural & Forestry University, Wusu Street #666, Lin'an District, Hangzhou 311300, China; shuqianzhou@126.com (S.Z.); lu_chen0829@126.com (L.C.); chenggangdeng@126.com (G.C.); lei_ru@126.com (Y.L.)
* Correspondence: yanghq@zafu.edu.cn; Tel.: +86-571-6374-1276

Abstract: Sweet potato (*Ipomoea batatas* (L.) Lam.) is a commercially relevant food crop with high demand worldwide. This species belongs to the Convolvulaceae family and is native to tropical and subtropical regions. Storage temperature and time can adversely affect tuberous roots' quality and nutritional profile. Therefore, this study evaluates the effect of storage parameters using physicochemical and transcriptome analyses. Freshly harvested tuberous roots (Xingxiang) were stored at 13 °C (control) or 5 °C (cold storage, CS) for 21 d. The results from chilling injury (CI) evaluation demonstrated that there was no significant difference in appearance, internal color, weight, and relative conductivity between tuberous roots stored at 13 and 5 °C for 14 d and indicated that short-term CS for 14 d promoted the accumulation of sucrose, chlorogenic acid, and amino acids with no CI symptoms development. This, in turn, improved sweetness, antioxidant capacity, and nutritional value of the tuberous roots. Transcriptome analyses revealed that several key genes associated with sucrose, chlorogenic acid, and amino acid biosynthesis were upregulated during short-term CS, including sucrose synthase, sucrose phosphate synthase, phenylalanine ammonia-lyase, 4-coumarate-CoA ligase, hydroxycinnamoyl-CoA quinate hydroxycinnamoyltransferase, serine hydroxymethyltransferase, alanine aminotransferase, arogenate dehydrogenase, and prephenate dehydratase. These results indicated that storage at 5 °C for 14 d could improve the nutritional quality and palatability of sweet potato tuberous roots without compromising their freshness.

Keywords: sweet potato; postharvest treatment; edible quality; chilling injury; transcriptome

Citation: Zhou, S.; Chen, L.; Chen, G.; Li, Y.; Yang, H. Molecular Mechanisms through Which Short-Term Cold Storage Improves the Nutritional Quality and Sensory Characteristics of Postharvest Sweet Potato Tuberous Roots: A Transcriptomic Study. *Foods* **2021**, *10*, 2079. https://doi.org/10.3390/foods10092079

Academic Editors: Eleni Tsantili and Jinhe Bai

Received: 30 July 2021
Accepted: 31 August 2021
Published: 2 September 2021

1. Introduction

Sweet potato (*Ipomoea batatas*) is an important crop that has recently been recognized as a functional food due to its health-promoting properties and nutraceutical components [1]. Both the leaves and tuberous roots of sweet potato are consumed, providing carbohydrates, fibers, carotenes, thiamine, riboflavin, niacin, minerals, vitamins A and C, and protein [2,3]. In 2019, the global production of sweet potato reached 91,820,929 t, with a plantation area of 7,768,870 ha (Food and Agriculture Organization of the United Nations; http://www.fao.org (accessed on 16 August 2020) [4]. China is the largest producer of sweet potatoes in the world, with an annual production of 51,992,156 t (56.6% of the world's production) in 2019 [4].

In China, a large portion of harvested sweet potato tuberous roots is shipped for fresh consumption. Sweetness and nutritional value are fundamental quality factors for sweet potato tuberous roots. Particularly, low sweetness is an undesirable characteristic of freshly harvested tuberous roots, which negatively impacts marketability. Therefore, some sweet potato tuberous roots are routinely submitted to a postharvest sweetening process prior to their marketing. For instance, Masuda et al. [5] reported that 3–4 months of storage at 13 °C is required to sweeten "Kokei 14" sweet potato tuberous roots before meeting the sweetness requirements of the fresh market. Previous studies have demonstrated that the sugar content and sweetness of postharvest sweet potato tuberous roots could be enhanced via cold storage at 4–5 °C for more than 20 d; however, the root suffered serious CI, thus affecting its overall quality and producing off-flavors [6]. CI of chilling-sensitive produce is generally thought to be the consequence of oxidative stress caused by excessive accumulation of reactive oxygen species (ROS) under low temperature [7]. Cold-induced ROS production leads to increased electrolyte leakage and compromises membrane integrity. CI is characterized by surface pitting, dark watery patches, and internal tissue browning [8,9]. Therefore, cold storage conditions must be optimized to shorten the sweetening period while also avoiding CI.

Besides sugars, chlorogenic acid is one of the key components in sweet potato tuberous roots and an important factor leading to the quality of sweet potato tuberous roots. This functional compound possesses a wide variety of health-promoting properties, including antioxidant, antimicrobial, anti-inflammatory, and antitumor activities [10–12]. Therefore, monitoring its changes during the postharvest process would provide a useful indicator of the nutrient composition of the sweet potato tuberous roots.

Therefore, this study sought to develop a quick and safe method to improve the quality of freshly harvested "Xinxiang" sweet potato tuberous roots. Tuberous roots were stored at 13 °C (control) and 5 °C (cold storage, CS) for 3 weeks. Afterward, the effects of different storage temperature and time combinations on quality parameters such as mass loss, flesh color, relative conductivity, and sugar, chlorogenic acid, and amino acid contents during storage were investigated. Furthermore, transcriptomic analyses were also performed to understand the molecular mechanisms that drive the aforementioned processes, and key genes associated with quality changes were identified.

2. Materials and Methods

2.1. Plant Materials

"Xinxiang" sweet potato tuberous roots were harvested at optimum maturity (135 d) from the Lingxi sweet potato professional cooperative (Hangzhou City, China). Similarly sized tuberous roots without mechanical or biological damage were selected. The selected tuberous roots were randomly divided into two groups, each with 240 tuberous roots (3 replicates; $n = 80$). Two groups of tuberous roots were packed in non-woven bags and stored at 13 °C (control) or 5 °C (cold storage, CS), respectively, for 21 d under 80%–85% relative humidity. After 14 and 21 d, some samples were transferred and stored at 20 °C for another 3 d to simulate shelf display. Shelf display is reflected in whether there will be different changes between the exterior and interior when it is transferred to normal temperature after cold storage. Three tuberous roots were randomly selected at 0, 14 + 3, and 21 + 3 d for visual assessment of CI and flesh color. Tuberous root tissue samples from under the skin (0.5–1 cm) were collected on 0, 7, 14, and 21 d of storage frozen in liquid nitrogen and stored at −80 °C for further phytochemical and molecular analyses.

2.2. Relative Conductivity and Mass Loss

Relative conductivity was measured as described by Li et al. [9] with some modifications. Thirty tuberous root disks (10 mm diameter and 2 mm thickness) were put in 100 mL of deionized water and shake at 150 cycles min^{-1} for 1 h on a lab plate shaker. The solution conductivity was measured using a Model EC 215 conductivity meter (HANNA Instruments, Beijing, China). The total conductivity was obtained after each sample was boiled for 15 min and the relative electrolyte leakage was expressed as a percentage of the total conductivity. The results were calculated using the following Equation (1):

$$\text{Relative conductivity} = (P_1/P_0) \times 100\% \quad (1)$$

where P_1 is the conductivity after vibration; P_0 is the conductivity after boiling.

Ten marked tuberous roots were weighted, and the mass loss was calculated as follows Equation (2):

$$\text{Mass loss} = [(W_1 - W_2)/W_1] \times 100\% \quad (2)$$

where W_1 is the fresh mass; W_2 is the mass after storage.

2.3. Sugar Content, Sweetness Index, and Taste Evaluation

The sugar content was evaluated as described by Li et al. [9] with some modifications. Briefly, 2.0 g of sample was ground in 15 mL of 40% acetonitrile in a mortar on ice. The homogenate was then transferred to a centrifuge tube and sonicated for 20 min before centrifuging at 12,000× *g* for 15 min. The precipitate was reextracted twice with 2 × 15 mL extracting solvent. All supernatants were then collected to a final 50 mL volume, after which the sample was filtered through a 0.45 µm membrane filter prior to assessment.

Next, 10 µL of the filtrate was injected into a high-performance liquid chromatography (HPLC) system (Agilent 1200, Agilent Technologies, Palo Alto, CA, USA) fitted with a refractive index detector (RID-1260, Agilent) and an Agilent ZORBAX Carbohydrate Analysis Column (4.6 mm × 250 mm, Agilent). Soluble sugars were separated in 75% (*v*/*v*) acetonitrile at a 1 mL min^{-1} flow rate at room temperature. Target peaks were identified by comparing the retention times of the compounds in the sample solutions to those of a standard mixture, and compound concentrations were determined via the area percentage method. All results were expressed as mg kg^{-1} on a dry mass basis.

The sweetness index, an estimate of total sweetness perception, is calculated based on the amount and sweetness properties of individual carbohydrates in fruits and vegetables [13]. The sweetness contribution of each carbohydrate was calculated based on the fact that fructose and sucrose are 2.30 and 1.35 times sweeter than glucose, respectively [14]. The sweetness index was calculated using the following Equation (3):

$$\text{Sweetness index} = 1.00 \times \text{glucose} + 2.30 \times \text{fructose} + 1.35 \times \text{sucrose} \quad (3)$$

The sensory quality of sweet potato tuberous roots was evaluated by measuring the sensory quality index. Evenly sized tuberous roots were selected, cleaned, weighed, and boiled for 40 min under atmospheric pressure. After cooking, the tuberous roots were allowed to cool for 30 min prior to sensory evaluation. A group of 10 people of different ages, sex, and education levels participated in the assessment. A five-point hedonic scale: 2 = "extremely dislike"; 4 = "dislike"; 6 = "neither like nor dislike"; 8 = "like"; and 10 = "extremely like" was used to assess the sensory qualities of the boiled tuberous roots. The sensory attributes evaluated were general appearance, smell, sweetness, and overall acceptability.

2.4. Total Phenolics, Individual Phenolic Acid Content, and Antioxidant Capacity

Sample extracts were prepared as described by Wang et al. [15] with some modifications. Briefly, 3.0 g of sample was ground in 15 mL of 80% methanol in a mortar on ice. The homogenate was then sonicated for 20 min and centrifuged at 12,000× *g* for 15 min at 4 °C. The residue was reextracted twice with 2 × 15 mL of extracting solvent. The supernatants were then pooled, and vacuum evaporated at 40 °C to remove all methanol residues. The remaining extract was adjusted to pH 1.5 with 6 mol L^{-1} HCl and centrifuged at 12,000× *g* for 15 min at 4 °C. Afterward, the sample was extracted five additional times with an equal volume of ethyl acetate-ether (*v*/*v*; 1:1), then pooled. The supernatant was dried with anhydrous sodium sulfate, filtered with filter paper, and vacuum evaporated at 35 °C until dry. The residue was dissolved in 5 mL methanol, and the sample was filtered through a 0.45 µm membrane prior to analysis.

A total phenolic standard curve was constructed as described by Kalt et al. [16]. The total phenolic content (TPC) of the sweet potato tuberous roots extracts were determined spectrophotometrically using Folin–Ciocalteu's reagent, and the results were expressed as gallic acid equivalents per gram of dry mass (mg GAE kg^{-1} DW).

The levels of individual phenolic acids in tuberous roots were determined via HPLC analyses. Phenolic acid standards were prepared in methanol (m/v; 0.01:10) and stored in brown reagent bottles before use. Afterward, 10 μL of the filtrate was injected into an HPLC system fitted with a refractive index detector (RID-1260, Agilent) and an Agilent ZORBAX Eclipse Plus C18 column (4.6 mm × 250 mm, Agilent). The mobile phase was composed of (A) 2% acetic acid (aqueous) and (B) acetonitrile, and gradient elution was performed as follows: 0 min, 97:3; 20 min, 95:5; 35 min, 85:15; 65 min, 70:30; and 65 min, 0:100. The mobile phase was vacuum filtered through a 0.45 μm membrane filter before use. All HPLC analyses were conducted at a 1 mL min^{-1} flow rate at room temperature, and the results were expressed as mg kg-1 DW.

DPPH radical-scavenging activity was determined as described by Oliveira et al. [17] with some modifications. Scavenging activity was calculated using the following Equation (4):

$$\text{Scavenging activity} = [(A_{517} \text{ of control} - A_{517} \text{ of sample})/A_{517} \text{ of control}] \times 100 \quad (4)$$

DPPH scavenging activity values were expressed as mg Trolox equivalent antioxidant capacity·kg^{-1} (mg TEAC kg^{-1} DW).

Ferric reducing antioxidant power (FRAP) assays were conducted as described by Apati et al. [18] with some modifications. Briefly, 1 mL sample solution was mixed with 2.5 mL of phosphate buffer (0.2 mol L^{-1}, pH 6.6) and 2.5 mL of $K_3Fe(CN)_6$ (v/v; 1:100).The mixture was incubated at 50 °C for 20 min and added 2.5 mL of trichloroacetic acid (v/v; 1:10), which was then centrifugated at 3000 rpm for 10 min. The supernatants were sucked 2.5 mL and mixed with 2.5 mL of bidistilled water and 0.5 mL of $FeCl_3$ (v/v; 0.1:100). The absorbance was measured at 700 nm. The results were expressed as mg TEAC kg^{-1} DW.

2.5. *Free Amino Acid Content Determination*

A 0.5 g sample was ground in a mortar on ice with 5 mL of extracting solution containing 50% alcohol and 0.01 mol L^{-1} of HCl. The homogenate was transferred to a centrifuge tube and sonicated for 30 min at low temperature, then centrifuged at 12,000× g for 5 min. Next, 1 mL of supernatant was recovered, freeze-dried, and dissolved in 1 mL of amino acid diluent. The sample was filtered through a 0.22 μm membrane filter before use. Finally, 50 μL of the filtrate was assessed in an automatic amino acid analyzer (SykamS433D, SYKAM Vertriebs GmbH, Fürstenfeldbruck, Germany) with an analysis column (LCA K07/Li, Sykam). The results were expressed as mg kg^{-1} DW.

2.6. *RNA Isolation and Sequencing*

Total RNA was extracted from control samples on days 0 and 14 and from CS samples on day 14 using the RNAprep Pure Plant Kit (Tiangen Biotech Co., Ltd., Beijing, China). Agarose gel electrophoresis was used to evaluate the RNA integrity. RNA sequencing (RNA-Seq) was then conducted by BmK Biotechnology Co., Ltd. (Beijing, China). Adapter sequences and low-quality sequences were then filtered out to ensure that all downstream analyses were conducted using clean and high-quality data. The hisat2 software was then used to map the data to the wild sweet potato reference genome (http://sweetpotato.plantbiology.msu.edu/ (accessed on 15 October 2020)). The hisat2 software uses an indexing scheme based on the Burrows-Wheeler transform and the Ferragina-Manzini (FM) index, employing two types of indexes for alignment: a whole-genome FM index to anchor each alignment and numerous local FM indexes for very rapid extensions of these alignments [19]. Functional annotation, cluster analysis, protein-protein interaction network analysis, and more in-depth mining analysis were conducted thereafter.

2.7. *Quantitative Real-Time PCR Validation*

Representative differential expressed genes (DEGs) identified by RNA-Seq were selected for experimental quantitative real-time PCR (qRT-PCR) validation. Gene-specific primers were synthesized by Beijing Qingke Biotechnology Co., Ltd. (Beijing, China) (Table S1). The RNA was reverse transcribed into complementary DNA (cDNA) using a PrimeScript™ RT reagent Kit with gDNA Eraser (Takara Bio Inc., Kusatsu (Shiga), Japan). qRT-PCR was performed using a TB Green® Premix Ex Taq™ II (Tli RNaseH Plus) (Takara Bio Inc., Kusatsu (Shiga), Japan) under the following conditions: 95 °C for 30 s, followed by 40 cycles of 95 °C for 5 s and 60 °C for 30 s. Relative gene expression was calculated using the $_{\Delta\Delta}$Ct method and normalized to the expression levels of *α-tubulin*. The qRT-PCR experiments were performed using at least three biological replicates, a negative control, and two technical replicates.

2.8. Statistical Analyses

The data were subjected to ANOVA analysis, and significant differences between means were determined using Duncan's multiple range test at a $p < 0.05$ probability (level). All results were reported as the mean of three replicates.

3. Results

3.1. Chilling Injury, Relative Conductivity, and Mass Loss in Sweet Potato Tuberous Roots

The typical symptoms of CI in sweet potato tuberous roots are pitting, water-soaked areas, and skin depression. Moreover, fungal infestation may be observed in damaged tissues in severe cases. The tuberous root in the CS group had no CI symptoms after 14 d at 5 °C plus 3 d of shelf display, whereas slight surface pitting and internal browning developed after 21 d at 5 °C and 3 d at 20 °C. Tuberous roots stored at 13 °C did not develop CI even after 21 d of storage (Figure 1A,B).

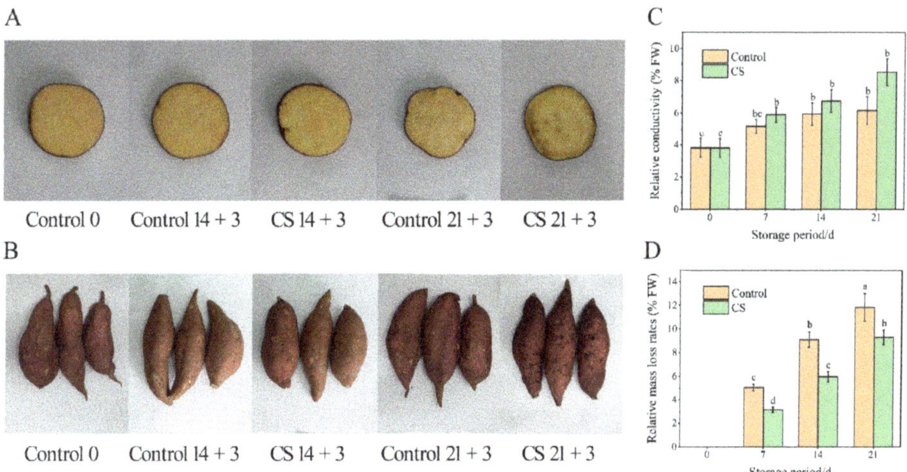

Figure 1. The color changes, relative conductivity, and mass loss of sweet potato tuberous roots during 21 d storage at the control and CS group. (**A**) color changes on the cutting surface, (**B**) color changes on the skin, (**C**) relative conductivity, (**D**) relative mass-loss rates. Results are expressed as the mean ± standard error of three replicates. Different characters indicate significant differences between treatment means at the $p < 0.05$ level.

As shown in Figure 1C, the relative conductivity of either control or CS tuberous roots increased throughout storage. No significant difference was detected between the control or CS group until day 21, where the relative conductivity was 8.5% higher than that of the control group.

Excessive mass loss can compromise the quality of fresh produce. As shown in Figure 1D, the two groups underwent substantial loss after 21 d of storage, reaching up to

11.8% and 9.3% at 13 and 5 °C, respectively. The mass loss at 13 °C was significantly higher ($p < 0.05$) than that at 5 °C.

3.2. Changes in Soluble Sugars, Sweetness Index, and Sensory Qualities

Soluble sugars are the main nutritional components of sweet potato tuberous roots, among which fructose, glucose, and sucrose are the main free sugars that determine their sweetness. Particularly, sucrose content was significantly higher in tuberous roots of the CS group compared with the control ($p < 0.05$) after 14 d of storage. In contrast, glucose and fructose contents did not change significantly in tuberous roots of either CS or control groups during storage (Figure 2A–C).

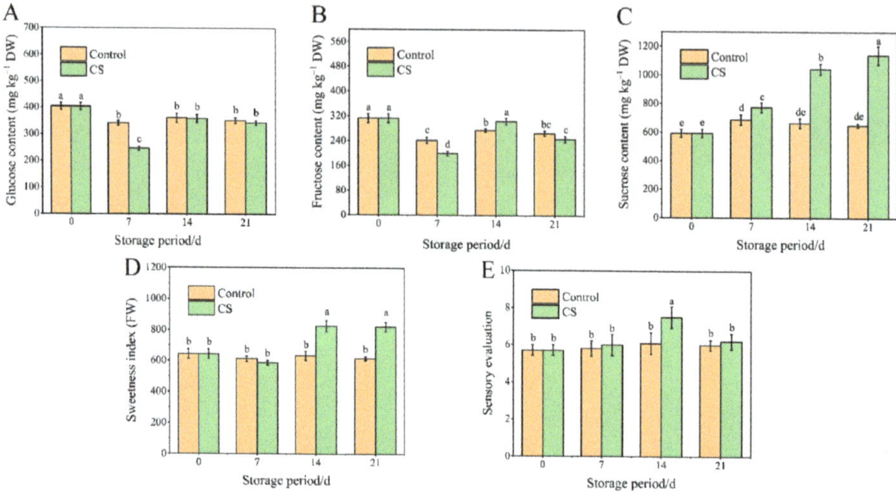

Figure 2. Contents of (**A**) glucose, (**B**) fructose, (**C**) sucrose, and (**D**) sweetness index and sensory score (**E**) of sweet potato tuberous roots of control and CS group during 21 d storage. Results are expressed as the mean ± standard error of three replicates. Data with different characters are significantly different from each other at the $p < 0.05$ level.

As shown in Figure 2D, the sweetness index in CS group tuberous roots peaked at day 14, reaching a level that was significantly higher than that of the control group. Therefore, sweet potato tuberous roots stored at 5 °C successfully achieved low-temperature sweetening. However, the sweetness index of sweet potato tuberous roots not further improved after 14 d of cold storage. Results from the sensory evaluation indicated that rates of consumer acceptance of tuberous roots did not change after 21 d of storage at 13 °C. In contrast, the sensory score of tuberous roots stored at 5 °C for 14 d was significantly higher than that of the control tuberous roots; however, these scores decreased in tuberous roots stored for 21 d at 5 °C due to flesh browning and CI-induced off-flavors after cooking (Figure 2E).

3.3. Changes in Total Phenolics, Free Phenolic Acid Content, and Antioxidant Capacity

The total phenol contents in tuberous roots of both groups increased after 21 d of storage (Figure 3A). Further, the total phenol content in tuberous roots of the CS group was significantly higher than that in the control group ($p < 0.05$) after 14 d of storage. Chlorogenic acid is the main phenolic compound in sweet potatoes and thus might have contributed to the increase in total phenolic. The chlorogenic acid content in CS tuberous roots was significantly higher ($p < 0.05$) than that in the control group after 14 d (Figure 3B).

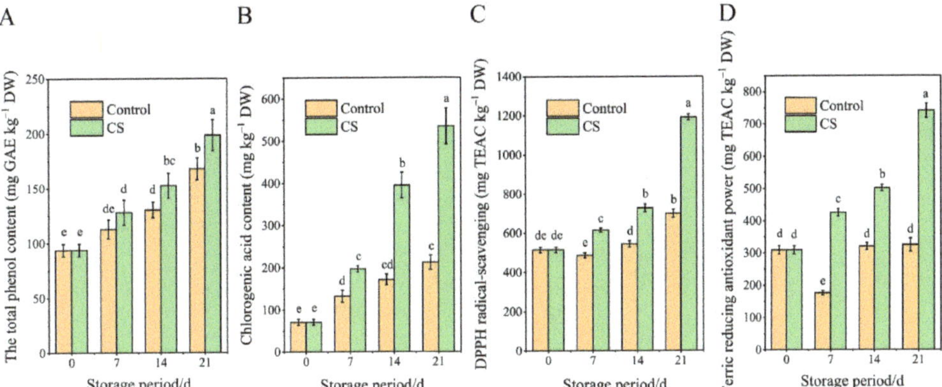

Figure 3. (**A**) Total phenol and (**B**) chlorogenic acid contents, (**C**) 2,2-Diphenyl-1-Picrylhydrazyl (DPPH) radical-scavenging activity, and (**D**) ferric reducing antioxidant power (FRAP) of sweet potato tuberous roots during 21 d storage control and CS groups. Data are the mean ± standard error of three replicates. Data with different characters indicate significant differences between treatments at the $p < 0.05$ level.

Moreover, the DPPH scavenging activity in tuberous roots of the CS group was significantly ($p < 0.05$) higher than that in the control group after 7 d of storage (Figure 3C). As shown in Figure 3D, the FRAP in the CS group also showed a significant upward trend during storage, with CS tuberous roots showing significantly higher FRAP levels than the control group ($p < 0.05$).

3.4. Changes in Free Amino Acid Content in Sweet Potato Tuberous Roots during Storage

The free amino acids analyzed herein were glycine, phenylalanine, tyrosine, and propanine. The initial phenylalanine content was approximately 90 mg kg^{-1} FW, followed by alanine and tyrosine. The glycine content was the lowest among the four amino acids tested. Almost all of these free amino acids tended to increase continuously between two and six times during storage at temperatures (Figure 4A–D).

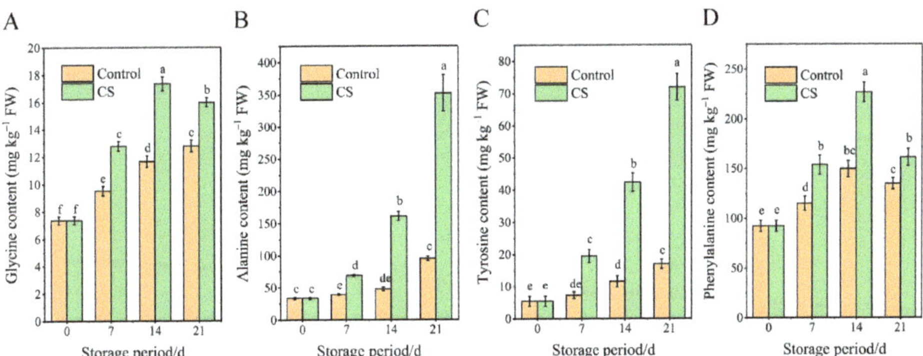

Figure 4. Contents of (**A**) glycine, (**B**) alanine, (**C**) tyrosine, and (**D**) phenylalanine of sweet potato tuberous roots of control and CS group during 21 d storage. Results are expressed as the mean ± standard error of three replicates. Different characters indicate significant differences between means of treatments at the $p < 0.05$ level.

3.5. *Sweet Potato Tuberous Roots Transcriptome Analysis at Different Storage Temperatures*

3.5.1. Data Quality Evaluation and Analysis

To gain a better understanding of the physiological changes of sweet potato tuberous roots under different storage temperatures, RNA samples were collected on days 0 (control, C0) and 14 (control and CS group, C14, and CS14). A total of 6.44 Gb to 9.51 Gb clean reads were obtained, and more than 94.0% of the reads had a Q30 quality score (Table S2). The sequencing data were therefore deemed adequate for correlation analysis. Hisat2 was used to align the clean reads with the sweet potato reference genome (https://ipomoea-genome.org/ (accessed on 15 October 2020)), achieving a 77.64–82.57% alignment efficiency. The alignment results are shown in Table S3.

Pearson correlation analysis was performed to validate the gene expression profiles based on the transcripts from nine different samples (Figure 5A). As expected, the results indicated that biological replicates from the same treatment group were highly correlated. Particularly, C0 and C14 were highly correlated, whereas CS14 and C0 were less correlated. The principal component analysis (PCA) (Figure 5B) results coincided with the correlation analysis findings. These findings indicated that the gene expression profile of the CS14 and C0 groups exhibited the most significant differences.

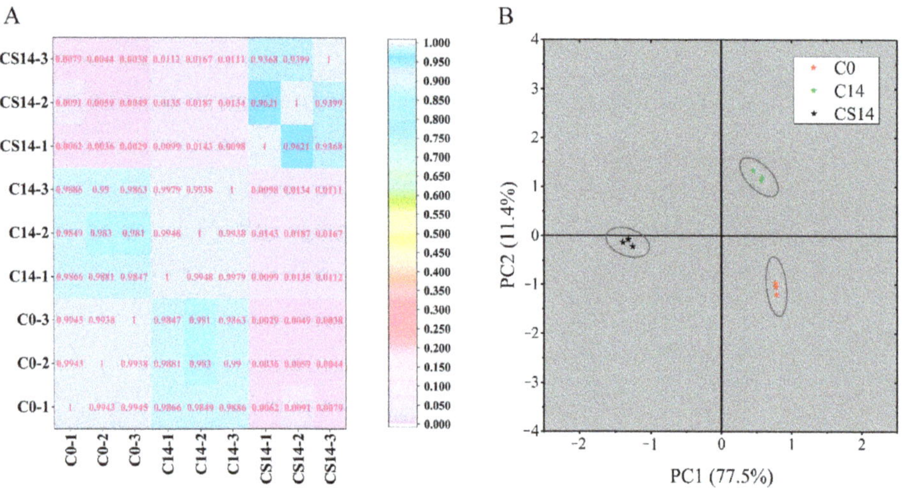

Figure 5. Global transcriptomic changes at the control and CS groups after 14 d of storage (C14 and CS14). (**A**) 2D hierarchical clustering, blue indicates a high correlation, purple represents low correlation. (**B**) Principal component analysis (PCA) of the RNA sequencing data of sweet potato tuberous roots samples.

3.5.2. Unigene Annotation Statistics

Unigene function annotation was conducted using the COG, GO, KEGG, KOG, Pfam, Swissprot, eggnog, and NR databases (Table S4). A total of 24,045 single genes were annotated in eight databases, of which the NR database had the highest annotation rate, with 31,000 annotations, accounting for 99.52% of the total annotated genes.

3.5.3. DEG Screening and Annotation Analysis

Comparison of Two DEG Groups

Degseq was used to identify differentially expressed genes (DEGs) in C0 vs. C14 and C0 vs. CS14. The screening criteria were fold change ≥ 2 and FDR < 0.01. Compared with the C0, 602 unigenes were upregulated, and 466 were downregulated in C14, whereas more genes were differentially expressed in the CS14 group (4850 upregulated and 5086 downregulated) (Table 1). The two groups of DEGs were annotated using the COG, eggnog, NR, Pfam, Swissprot, GO, and KEGG databases. The number of annotated genes is shown in Table S5, and the annotation ratio of the NR database was the highest (>95%).

Table 1. Statistics of the number of deferentially expressed genes.

DEG Set	DEG NUMBER	Upregulated	Downregulated
C0 vs. C14	1068	602	466
C0 vs. CS14	9936	4850	5086

GO Database Annotation Analysis

The GO annotation system includes three main branches: cellular component, molecular function, and biological process. The two groups of DEGs were classified into 53 functional subgroups (Figure S1). As indicated in Table S6, in terms of cellular components, the DEGs were mainly associated with seven functional sub-categories: "cell", "membrane", "macroscopic complex", "organelle", "organelle part", "membrane part", and "cell part". Regarding molecular function, DEGs were mainly associated with the two functional subclasses of "catalytic activity" and "binding". For the biological process classification, the DEGs were mainly enriched in pathways associated with "biological process", "cellular process", "single organization process", "response to stimulus", "localization", "biological regulation", and "cellular component organization or biogenesis". Regarding the differences in DEG ratios between the two groups, the "cell", "membrane-enclosed lumen", "macroscopic complex", "organelle", "organelle part", "cell part", "structural molecular activity", "growth and cellular component organization or biogenesis" process ratios exhibited marked differences (e.g., up to 40-fold differences). Moderately low temperatures had an important effect on the aforementioned processes at the transcriptome level in sweet potato tuberous roots stored for 14 d.

KEGG Functional Annotation

The enriched KEGG pathways could be divided into five categories, including "cellular processes", "environmental information processing", "genetic information processing", "metabolism", and "organic systems". According to the KEGG pathway classification (Figure S2A,C) and enrichment analysis (Figure S2B,D), the DEGs were mainly associated with metabolism pathways. As shown in Table S7, the C0 vs. C14 and C0 vs. CS14 comparisons indicated that "plant hormone signal transduction", "protein processing in endoplasmic reticulum", "ribosome", "phenylpropanoid biosynthesis", "starch and sucrose metabolism", "biosynthesis of amino acids", and "carbon metabolism" were the main metabolic pathways affected by different low-temperature storage conditions. Among them, the "phenylpropanoid biosynthesis", "starch and sucrose metabolism", and "biosynthesis of amino acids" pathways are associated with the synthesis of partial free phenolic acids, sugars, and free amino acids, respectively. The present study mainly focused on these three metabolic pathways by screening key genes linked to metabolic changes.

3.5.4. Key Genes Involved in Sweet Potato Tuberous Roots Quality

As shown in Table 2, the key genes related to sucrose synthesis were sucrose synthase and sucrose phosphate synthase, among which sucrose synthase exhibited differential expression and sucrose phosphate synthase was downregulated at low temperatures. Further, upregulated genes were more abundant than downregulated genes in sucrose synthase.

Table 2. Selected genes associated with the accumulation of sucrose, amino acid, and chlorogenic acid in the biosynthesis pathways.

Name	Gene ID	FPKM			Enzyme
		C0	C14	CS14	
sucrose synthase	itf11g07860	88.50	67.61	912.29	EC:2.4.1.13
	itf06g18950	4.20	7.05	113.25	EC:2.4.1.13
	itf02g07130	555.91	451.65	72.19	EC:2.4.1.13
sucrose phosphate synthase	itf03g21140	84.66	96.66	19.06	EC:2.4.1.14
phenylalanine ammonia-lyase	itf09g14800	10.94	51.55	231.74	EC:4.3.1.24
	itf09g14820	6.15	30.92	138.57	EC:4.3.1.24
	itf15g00190	1.69	24.46	56.15	EC:4.3.1.24
4-coumarate-CoA ligase	itf11g10280	61.32	70.53	166.77	EC:6.2.1.12
hydroxycinnamoyl-CoA quinate hydroxycinnamoy- ltransferase	itf07g23450	100.19	220.02	448.15	EC:2.3.1.133
serine hydroxymethyltransferase	itf09g04020	26.86	32.19	60.77	EC:2.1.2.1
alanine aminotransferase	itf11g08270	44.75	47.50	283.38	EC:2.6.1.2
arogenate dehydrogenase	itf10g08090	3.86	5.57	33.58	EC:1.3.1.78
	itf12g23730	1.77	2.48	19.53	EC:1.3.1.78
arogenate dehydratase/ prephenate dehydratase	itf07g12100	0.16	3.86	12.61	EC:4.2.1.91 4.2.1.51
	itf14g14710	11.57	14.87	27.00	EC:4.2.1.91 4.2.1.51
tyrosine aminotransferase	itf02g15070	38.52	20.45	7.71	EC:2.6.1.5

The phenylalanine ammonia-lyase, 4-coumarate-coa ligase, and hydroxycinnamoyl-CoA quinate hydroxycinnamoyltransferase genes were upregulated in the phenylalanine metabolism pathway, which has been associated with chlorogenic acid accumulation in sweet potato tuberous roots at low temperatures. In the amino acid biosynthesis pathway, low-temperature storage upregulated the alanine aminotransferase, serine hydroxymethyltransferase, aromatate dehydrogenase, and aromatate dehydratase/precursor dehydratase genes. Therefore, the lower storage temperature translated to higher expression levels, which are highly associated with the accumulation of glycine, alanine, tyrosine, and phenylalanine in sweet potato tuberous roots.

3.6. Validation of RNA-Seq Results via qRT-PCR

The results from the transcriptome analysis were validated in a biologically independent experiment using qRT-PCR. A total of 7 DEGs linked to sweet potato tuberous roots quality were selected and analyzed, and *α-tubulin* was used as a reference gene. As illustrated in Figure 6, the relative expression levels of sucrose synthase (itf11g07860), phenylalanine ammonia-lyase (itf09g14800), 4-coumarate-CoA ligase (itf11g10280), serine hydroxymethyltransferase (itf09g04020), alanine aminotransferase (itf11g08270), arogenate dehydrogenase (itf10g08090), and arogenate dehydratase/prephenate dehydratase (itf07g12100) were all upregulated by low-temperature storage, which was consistent with our RNA-Seq results.

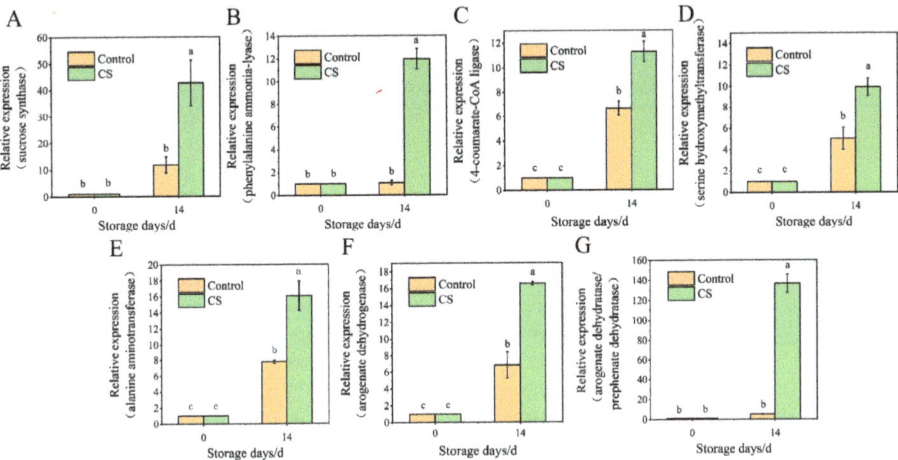

Figure 6. Validation of differentially expressed genes (DEGs) by quantitative real-time PCR. All data were assessed by ANOVA, and the results are expressed as the mean ± standard error of three replicates. Data with different characters indicate significant differences between treatments ($p < 0.05$).

4. Discussion

Sweet potato is now globally recognized as a functional food due to its preventive and therapeutic effects against chronic diseases [1]. However, the sweetness of freshly harvested sweet potato tuberous roots is not enough, which limits the acceptance of consumers. Cold storage can improve the sweetness of sweet potato tuberous roots, but long-term storage will cause CI. The development and likelihood of CI can also depend on the sweet potato origin, variety, cultivar, harvest season, temperature, and chilling stress duration [20]. Therefore, precise control of storage temperature and time may improve tuberous root sweetness without resulting in CI. Our study demonstrated that the sweetness of mature "Xinxiang" sweet potato tuberous roots increased 2-fold after 14 d of storage at 5 °C, which significantly improved their taste according to our sensory tests. Moreover, no CI was detected at this time point. During storage at 13 °C, the soluble sugar and sweetness index of the sweet potato tuberous roots did not change significantly, which contrasted with the findings of Ru et al. [21], who demonstrated that the levels of sugars in tuberous roots stored at both 13 and 4 °C increased after 15 d. We speculate that these differences were related to the maturity of the sweet potatoes. In this study, sweet potatoes were harvested 135 d after planting, which was different from our previous study (Ru et al.) [21]. More mature tuberous roots appeared to be less sensitive to low-temperature stress [21]. Therefore, there were no observable differences in the soluble sugar contents of tuberous roots stored at 13 °C even after 21 d.

Phenolic compounds are naturally occurring chemicals that defend sweet potatoes and other plants against biotic and abiotic stressors [22–24]. Chlorogenic acid (5-O-caffeoyliquinic acid, CGA) is the predominant component of phenolic acids in sweet potato tuberous roots and is the primary contributor to its antioxidative, antimutagenic, and radical-scavenging properties [25–27]. The increase in phenolic compound contents due to low-temperature exposure may enhance the nutraceutical value of sweet potato tuberous roots. The chlorogenic acid content in "Xinxiang" sweet potato tuberous roots increased markedly after 14 d of storage at 5 °C. This is consistent with the findings of Ishiguro et al. [28], who detected a significant increase in total phenolic content in sweet potato tuberous roots after two weeks of low-temperature exposure.

Low temperature also increases the concentration of some free amino acids in sweet potato tuberous roots, thus enhancing the nutritional value of cold-stored tuberous roots.

However, tyrosine is one of the main substrates of enzymatic browning. Specifically, the enzyme tyrosinase oxidizes tyrosine to produce quinones. This, in turn, leads to melanin production, resulting in browning [29]. Therefore, a large amount of tyrosine accumulated during long-term low-temperature storage can lead to browning/blackening in fruits and vegetables [30]. This is consistent with the internal browning of sweet potato tuberous roots with surface cuts after 21 d of storage at 5 °C and 3 d at room temperature.

Several studies have assessed the effects of low temperature on transcript profiles in stored sweet potato tuberous roots. For instance, Ji et al. [8] analyzed the transcriptome changes of sweet potato tuberous roots stored at an optimal (13 °C) and low temperature (4 °C) for 6 weeks and found that the tuberous roots could resist CI by inducing genes related to the biosynthesis of unsaturated fatty acids, pathogen defense, and phenylalanine metabolism. These results indicate that phenylalanine metabolism can affect the production of phenolic compounds at low temperatures, which is consistent with the results of the present study. Additionally, their study demonstrated that membrane damage may be the main cause of chilling injury, thus highlighting the critical importance of lipid metabolism to improve the stress resistance of tuberous roots under low-temperature storage conditions. Xie et al. [31] also conducted transcriptomic analyses on "xushu18" sweet potato tuberous roots and found that the expression of genes associated with carbohydrate metabolism was regulated during cold storage, leading to the accumulation of sucrose in sweet potato tuberous roots, which was consistent with the results of this study. There have been reports on long-term cold storage of sweet potato tuberous roots to study the molecular mechanism of CI in tuberous roots, but there are no transcriptome reports on short-term cold storage of sweet potato tuberous roots. In this study, RNA-Seq was used to explore the molecular mechanisms through which short-term low-temperature storage enhances sweet potato tuberous roots' quality. The concentrations of soluble sugars, phenolic compounds, and amino acids increased significantly after short-term low-temperature storage.

Sucrose synthase (SUS; EC 2.4.1.13), sucrose phosphate synthase (SPS; EC 2.4.1.14), and acid convertase (AI; EC 3.2.1.26) are key enzymes in sucrose metabolism (Figure 7). Among them, sucrose synthase can catalyze the reversible reaction of sucrose synthesis and decomposition [32]. Previous studies have reported that low temperatures increase the activity of sucrose synthase in wheat [33]. In this study, low-temperature storage upregulated the expression of two sucrose synthase genes. SPS is a rate-limiting enzyme in the synthesis of sucrose [34]. In this study, SPS expression was downregulated in the CS treatment, suggesting that the regulation of SPS activity is highly complex and may not be affected at the transcript level [21].

In plants, CGA biosynthesis occurs downstream of the phenylpropanoid pathway (red and blue color represents upregulation and downregulation, respectively) (Figure 7). Phenylalanine generates P-coumaroyl-CoA, which in turn is catalyzed by key enzymes in the biosynthesis of chlorogenic acid, including phenylalanine ammonia-lyase (PAL; EC 4.3.1.24), cinnamic acid 4-hydroxylase (C4H; EC 1.14.14.91), 4-coumarate-CoA ligase (4CL; EC 6.2.1.12), hydroxycinnamoyl-CoA shikimate/quinate hydroxycinnamoyltransferase (HCT; EC 2.3.1.133), β-coumaroylester 3-hydroxylases (C3H; EC 1.14.14.96), and hydroxycinnamoyl-CoA quinate hydroxycinnamoyltransferase (HQT; EC 2.3.1.99) [35,36]. In this study, several key genes belonging to the chlorogenic acid synthesis pathway, such as 3 PAL, 1 4CL, and 1 HQT, were regulated by low temperature. The expression levels of PAL in the control and CS groups increased by 5.69- and 22.71-fold, 4CL increased by 1.17- and 2.72-fold, respectively, and HQT increased by 2.20- and 4.47-fold, respectively. Further, some key genes in the shikimic acid pathway were also upregulated by low temperature, which might have contributed to the accumulation of tyrosine in CS tuberous roots.

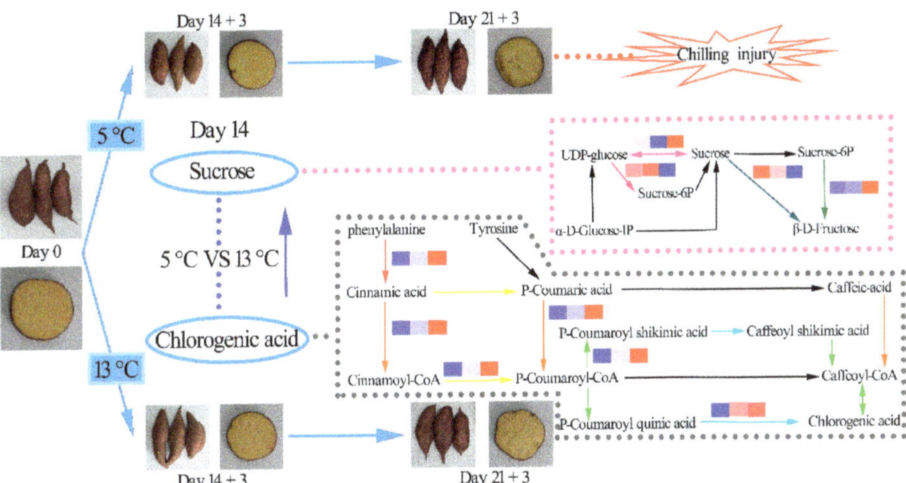

Figure 7. Color, sucrose content, and chlorogenic acid content changes in sweet potato tuberous roots during 14 d of storage. The figure contains the metabolite map of the sucrose and chlorogenic acid synthesis pathway. The relative expression levels of C0, C14, and CS14 are shown as heat maps.

5. Conclusions

Our study demonstrated that the sweetness index, antioxidant activity, amino acid content, and sensory characteristics of postharvest mature sweet potato tuberous roots were significantly improved after 14 d of storage at 5 °C without any indications of CI, but a continued extension of CS time would lead to cold injury, resulting in the decline of storage tolerance. The key genes associated with sweet potato quality parameters were also identified through RNA-Seq and verified using qPCR, and exhibited the involvement of sucrose synthase and sucrose phosphate synthase in starch and sucrose metabolism pathway, phenylalanine ammonia-lyase, 4-coumarate-CoA ligase, and hydroxycinnamoyl-CoA quinate hydroxycinnamoyltransferase in phenylpropane biosynthesis pathway and serine hydroxymethyltransferase, alanine aminotransferase, arogenate dehydrogenase and prephenate dehydratase in the biosynthesis of amino acids pathway, thus providing insights into the molecular mechanisms by which short-term cold storage enhances sweet potato quality and nutritional profile. Therefore, short-term CS treatment is an effective method to improve the nutritional and sensory quality of sweet potato tuberous roots.

Supplementary Materials: The following are available online at https://www.mdpi.com/article/10.3390/foods10092079/s1, Table S1: Primers for Quantitative Real-time PCR, Table S2: Sample sequencing data evaluation statistics table, Table S3: Statistical table of comparison results between the second-generation sequencing data and the reference genome, Table S4: Annotation statistics table of single gene sequence, Table S5: Statistics of the number of annotations for differentially expressed genes, Table S6: Comparison table of GO function annotations of differentially expressed genes, Table S7: Function annotation comparison table of differentially expressed gene KEGG, Figure S1: Statistical chart of GO annotation classification of DEGs, Figure S2: KEGG classification map and KEGG enrichment map of DEGs.

Author Contributions: Conceptualization, Y.L. and H.Y.; Data curation, G.C.; Investigation, S.Z. and L.C.; Writing—original draft, S.Z. and L.C. S.Z. and L.C. contributed equally to this work. All authors have read and agreed to the published version of the manuscript.

Funding: This research was funded by the National Natural Science Foundation of China (No. 31871857) and the Zhejiang Provincial Natural Science Foundation of China (No. LQ20C200002).

Data Availability Statement: The data presented in this study are available on request from the corresponding author.

Acknowledgments: All authors thank Yanrong Huo and Peng Cui for helping to illustrate the graphical abstract.

Conflicts of Interest: The authors declare no conflict of interest.

References

1. Ayeleso, T.B.; Ramachela, K.; Mukwevho, E. A review of therapeutic potentials of sweet potato: Pharmacological activities and influence of the cultivar. *Trop. J. Pharm. Res.* **2016**, *15*, 2751–2761. [CrossRef]
2. Bovell-Benjamin, A.C. Sweet potato: A review of its past, present, and future role in human nutrition. *Adv. Food Nutr. Res.* **2007**, *52*, 1–59. [CrossRef] [PubMed]
3. Sheikha, A.; Ray, R.C. Potential impacts of bio-processing of sweet potato: Review. *Crit. Rev. Food Sci. Nutr.* **2015**, *57*, 455–471. [CrossRef]
4. Food and Agriculture Organization of the United Nations (FAO). FAOSTAT, Crops and Livestock Products. Available online: http://www.fao.org/faostat/en/#data/QCL (accessed on 1 July 2021).
5. Masuda, D.; Fukuoka, N.; Goto, H.; Kano, Y. Effect of cold treatment after harvest on sugar contents and storability in sweet potato (*Ipomoea batatas* L.). *Hortic. Res.* **2007**, *6*, 597–601. [CrossRef]
6. Sakamoto, T.; Masuda, D.; Nishimura, K.; Ikeshita, Y. Relationship between invertase gene expression and sucrose concentration in the tuberous roots of sweet potato (*Ipomoea batatas* L. Lam.) during cold storage. *J. Pomol. Hortic. Sci.* **2014**, *89*, 229–235. [CrossRef]
7. Lyons, J.M. Chilling injury in plants. *Annu. Rev. Plant Physiol.* **1973**, *24*, 445–466. [CrossRef]
8. Ji, C.Y.; Chung, W.H.; Kim, H.S.; Jung, W.Y.; Kang, L.; Jeong, J.C.; Kwak, S.S. Transcriptome profiling of sweetpotato tuberous roots during low temperature storage. *Plant Physiol. Biochem.* **2017**, *112*, 97–108. [CrossRef]
9. Li, X.; Yang, H.Q.; Lu, G.Q. Low-temperature conditioning combined with cold storage inducing rapid sweetening of sweetpotato tuberous roots (*Ipomoea batatas* (L.) Lam) while inhibiting chilling injury. *Postharvest Biol. Technol.* **2018**, *142*, 1–9. [CrossRef]
10. Liang, N.J.; Kitts, D.D. Role of chlorogenic acids in controlling oxidative and inflammatory stress conditions. *Nutrients* **2016**, *8*, 16. [CrossRef]
11. Liu, Y.J.; Zhou, C.Y.; Qiu, C.H.; Lu, X.M.; Wang, Y.T. Chlorogenic acid induced apoptosis and inhibition of proliferation in human acute promyelocytic leukemia HL-60 cells. *Mol. Med. Rep.* **2013**, *8*, 1106–1110. [CrossRef] [PubMed]
12. Xu, J.G.; Hu, Q.P.; Liu, Y. Antioxidant and DNA-protective activities of chlorogenic acid isomers. *J. Agric. Food Chem.* **2012**, *60*, 11625–11630. [CrossRef]
13. Magwaza, L.; Opara, U.L. Analytical methods for determination of sugars and sweetness of horticultural products—A review. *Sci. Hortic.* **2015**, *184*, 179–192. [CrossRef]
14. Keutgen, A.; Pawelzik, E. Modifications of taste-relevant compounds in strawberry fruit under NaCl salinity. *Food Chem.* **2007**, *105*, 1487–1494. [CrossRef]
15. Wang, Y.C.; Chuang, Y.C.; Ku, Y.H. Quantitation of bioactive compounds in citrus fruits cultivated in Taiwan. *Food Chem.* **2007**, *102*, 1163–1171. [CrossRef]
16. Kalt, W.; Forney, C.F.; Martin, A.; Prior, R.L. Antioxidant capacity, vitamin C, phenolics and anthocyanins after fresh storage of small fruits. *J. Agric. Food Chem.* **1999**, *47*, 4638–4644. [CrossRef]
17. Oliveira, I.; Sousa, A.; Ferreira, I.C.; Bento, A.; Estevinho, L.; Pereira, J.A. Total phenols, antioxidant potential and antimicrobial activity of walnut (Juglans regia L.) green husks. *Food Chem. Toxicol.* **2008**, *46*, 2326–2331. [CrossRef] [PubMed]
18. Apáti, P.; Szentmihályi, K.; Kristó, S.T.; Papp, I.; Vinkler, P.; Szoke, E.; Kéry, A. Herbal remedies of solidago—correlation of phytochemical characteristics and antioxidative properties. *J. Pharm. Biomed. Anal.* **2003**, *32*, 1045–1053. [CrossRef] [PubMed]
19. Kim, D.; Langmead, B.; Salzberg, S.l. HISAT: A fast spliced aligner with low memory requirements. *Nat. Methods* **2015**, *12*, 357–360. [CrossRef]
20. Sevillano, L.; Sanchez-Ballesta, M.T.; Romojaro, F.; Flores, F.B. Physiological, hormonal and molecular mechanisms regulating chilling injury in horticultural species. Postharvest technologies applied to reduce its impact. *J. Sci. Food Agric.* **2010**, *89*, 555–573. [CrossRef]
21. Ru, L.; Chen, B.W.; Li, Y.X.; Wills, R.B.M.; Lv, Z.F.; Lu, G.Q.; Yang, H.Q. Role of sucrose phosphate synthase and vacuolar invertase in postharvest sweetening of immature sweetpotato tuberous roots (*Ipomoea batatas* (L.) Lam cv 'Xinxiang'). *Sci. Hortic.* **2021**, *282*, 110007. [CrossRef]
22. Walter, W.M.; Schadel, W.E. Distribution of phenols in "Jewel" sweet potato [*Ipomoea batatas* (L.) Lam.] roots. *J. Agric. Food Chem.* **1981**, *29*, 904–906. [CrossRef]
23. Harrison, H.F.; Peterson, J.K.; Snook, M.E.; Bohac, J.R.; Jackson, D.M. Quantity and potential biological activity of caffeic acid in sweet potato [*Ipomoea batatas* (L.) Lam.] storage root periderm. *J. Agric. Food Chem.* **2003**, *51*, 2943–2948. [CrossRef]
24. Padda, M.S.; Picha, D.H. Effect of low temperature storage on phenolic composition and antioxidant activity of sweetpotatoes. *Postharvest Biol. Technol.* **2008**, *47*, 176–180. [CrossRef]

25. Islam, S. Sweetpotato (*Ipomoea batatas* L.) Leaf: Its potential effect on human health and nutrition. *J. Food Sci.* **2006**, *71*, 13–21. [CrossRef]
26. Padda, M.S.; Picha, D.H. Methodology optimization for quantification of total phenolics and individual phenolic acids in sweetpotato (*Ipomoea batatas* L.) roots. *J. Food Sci.* **2010**, *72*, 412–416. [CrossRef]
27. Mohanraj, R.; Sivasankar, S. Sweet potato (*Ipomoea batatas* [L.] Lam)-A valuable medicinal food: A review. *J. Med. Food* **2014**, *17*, 733–741. [CrossRef]
28. Ishiguro, K.; Yahara, S.; Yoshimoto, M. Changes in polyphenolic content and radical-scavenging activity of sweet potato (*Ipomoea batatas* L.) during storage at optimal and low temperatures. *J. Agric. Food Chem.* **2007**, *55*, 10773–10778. [CrossRef]
29. Moon, K.M.; Kwon, E.B.; Lee, B.; Kim, C.Y. Recent trends in controlling the enzymatic browning of fruit and vegetable products. *Molecules* **2020**, *25*, 2754. [CrossRef]
30. Moon, K.M.; Lee, B.; Cho, W.K.; Lee, B.S.; Kim, C.Y.; Ma, J.Y. Swertiajaponin as an anti-browning and antioxidant flavonoid. *Food Chem.* **2018**, *252*, 207. [CrossRef]
31. Xie, Z.Y.; Zhou, Z.L.; Li, H.M.; Yu, J.J.; Jiang, J.J.; Tang, Z.H.; Ma, D.F.; Zhang, B.H.; Han, Y.H.; Li, Z.Y. High throughput sequencing identifies chilling responsive genes in sweetpotato (*Ipomoea batatas* Lam.) during storage. *Genomics* **2019**, *111*, 1006–1017. [CrossRef]
32. Lothar, E. Effect of metal ions on sucrose synthase from rice grains—A study on enzyme inhibition and enzyme topography. *Glycobiology* **1995**, *5*, 201–206. [CrossRef]
33. Santoiani, C.S.; Tognetti, J.A.; Pontis, H.G.; Salerno, G.L. Sucrose and fructan metabolism in wheat roots at chilling temperatures. *Physiol. Plant.* **2010**, *87*, 84–88. [CrossRef]
34. Huber, S.C.; Huber, J.L. Role and regulation of sucrose-phosphate synthase in higher plants. *Annu. Rev. Plant Physiol. Plant Mol. Biol.* **1996**, *47*, 431–444. [CrossRef] [PubMed]
35. Kojima, M.; Uritani, I. Studies on chlorogenic acid biosynthesis in sweet potato root tissue using trans-cinnamic acid-2-14C and quinic acid-G-3H. *Plant Cell Physiol.* **1972**, *13*, 311–319. [CrossRef]
36. Yu, Y.; Zhang, Q.; Liu, S.; Ma, P.Y.; Jia, Z.D.; Xie, Y.Z.; Bian, X.F. Effects of exogenous phytohormones on chlorogenic acid accumulation and pathway-associated gene expressions in sweetpotato stem tips. *Plant Physiol. Biochem.* **2021**, *164*, 21–26. [CrossRef] [PubMed]

MDPI

Article

Susceptibility of Impact Damage to Whole Apples Packaged Inside Molded Fiber and Expanded Polystyrene Trays

Kyle Dunno *, **Isabel Stoeckley** and **Matthew Hofmeister**

Department of Packaging Science, Rochester Institute of Technology, 78 Lomb Memorial Drive, Rochester, NY 14623, USA; irs2530@g.rit.edu (I.S.); mxh6953@g.rit.edu (M.H.)
* Correspondence: kddipk@rit.edu

Abstract: Postharvest damage, leading to loss and waste, continues to be a significant problem in the fresh produce industry. Trays, designed to reduce fruit-to-fruit contact, are utilized by the apple industry to minimize bruising of whole apples. During distribution, packaged apples are subjected to various supply chain hazards, which may lead to bruising damage. Currently, molded fiber (MF) and expanded polystyrene (EPS) trays transport whole apples from the packhouse to the retail outlet. Mechanical shock, by free-fall drop method, was used to evaluate the performance differences between the two trays and quantify the bruising characteristics of the apples. Results showed that the EPS trays provided better shock protection to the apple as compared to the MF tray, reducing the impact acceleration by more than 70%. Additionally, the bruise susceptibility was 40% less for the apples packaged inside the EPS trays, regardless of drop height. However, apples packaged in the middle layer trays were most susceptible to bruising damage, regardless of tray type.

Keywords: bruise susceptibility; apples; mechanical shock; transportation; molded fiber; expanded polystyrene

Citation: Dunno, K.; Stoeckley, I.; Hofmeister, M. Susceptibility of Impact Damage to Whole Apples Packaged Inside Molded Fiber and Expanded Polystyrene Trays. *Foods* **2021**, *10*, 1980. https://doi.org/10.3390/foods10091980

Academic Editors: Eleni Tsantili and Jinhe Bai

Received: 27 July 2021
Accepted: 23 August 2021
Published: 25 August 2021

1. Introduction

Fresh produce travels through a demanding food supply chain to reach the consumer. During this journey, products are exposed to a variety of supply chain hazards, such as mechanical shock, vibration, and compression. In apples, these hazards can result in bruising to the fruit, altering their quality and perceived deterioration by consumers, especially during bulk display at retail [1–6]. In today's market, consumers demand fresh produce to be free from visible defects such as bruising. Excessive defects in the apple will deter consumers, resulting in them selecting other items causing product shrinkage [4]. Although the bruise size is a function of the mechanical properties of the apple flesh, a visible bruise greater than 100 mm^2 will typically result in the apple being discarded as waste [7,8]. Peggie [9] reported that approximately 8–10% of apples harvested were discarded mainly due to bruise damage. However, Lewis et al. [4] reported data from apple distributors that indicated apple waste could be 50% or higher due to bruising. As a result, the product's visual appearance is critical to the purchasing instincts of the consumer.

Packaging performs a variety of functions, one of which is to protect the product during transport. For apples, the type of packaging system employed is dependent on its position in the postharvest supply chain. For example, bulk bins are used during post-harvesting to move the apples from the grower to the packhouse, while apples traveling to the retail outlet are packaged inside corrugated containers containing bags, pouches, or trays of apples. Unfortunately, although a wide range of packaging formats are available, most of them are not designed to adequately protect the apples during transit, resulting in bruise damage still being a frequent quality problem for growlers and retailers [1].

Limiting apple-to-apple contact during transport is desirable as this minimizes the opportunity of the fruit to bruise. To accomplish this, apples traveling to retail are often packaged into trays with individual cells. The two predominant tray materials utilized by the apple industry are molded fiber (MF) and expanded polystyrene (EPS). MF, produced from pulp slurry, can be appealing due to its end-of-life impact but is susceptible to moisture gain and swelling, which could be a deterrent for some applications. EPS is a closed-cell foam consisting of 98% air, making it incredibly lightweight while also providing excellent energy absorption and dissipation during impacts. However, in recent years there have been restrictions put in place regarding the use of EPS due to its environmental impact including legislation passed by local and state governments [10]. Food losses and waste (FLW) are globally becoming a top priority in food management in order to increase food security, while also striving to reduce environmental impacts [11]. Therefore, it is imperative to understand not only the environmental impact of the package material, but also understand its influence on food loss. Additional packaging solutions exist for the transport of fresh fruit, such as corrugated sleeves, mesh netting, and stand-up pouches and bags, but these solutions are not commercially used in the transport of apples. Although both MF and EPS have been used extensively by the apple industry to move whole apples to retail, limited data exists comparing these two materials to understand which tray provides better protection against bruising.

Previous research has evaluated the influence of packaging materials on the bruising characteristics of whole apples undergoing transportation simulations, but the vast majority of research available examines impacting the fruit using a pendulum method [5,6,12–15]. Singh et al. [16] and Singh and Xu [17] examined different packaging materials, including MF and EPS, focused on the damage resulting from laboratory vibration simulations. The results, however, indicated the fruit packaged inside the EPS trays had less damage than those inside the MF trays. Fadiji et al. [1] reported that apples were more susceptible to bruising when packaged inside plastic bags as compared to trays during multiple impacts. By placing the apples inside the individual cells, the fruit-to-fruit impact was reduced [18]. Batt et al. [19] investigated the performance of MF and EPS tray types during simulated transport conditions and noted there were no significant differences in apple bruise frequency or size. However, this project examined only one impact from 61 cm, unlike Fadiji et al. [1], which examined apple bruising using multiple drops from shorter drop heights which are more common for fruit packaged at this stage in the supply chain.

Numerous studies focusing on the design of the ventilated corrugated container for apples and other fruits have been published [20–23]; however, little is known about the performance of the fruit trays designed to prevent fruit damage. The objective of this research was to investigate the bruise susceptibility of whole apples during mechanical shock inside ventilated corrugated containers when packaged using either MF or EPS trays, including the bruise frequency and impact acceleration experienced by the fruit.

2. Materials and Methods

2.1. Apple Variety

'Minneiska' (SweeTango®) apples packaged using two different tray designs (MF and EPS) were acquired during commercial harvest, in September, from a packhouse in Wolcott, NY, USA. Fruit of uniform size and maturity based on color, firmness, and free from physical defects were used for the experiments. The mean diameter of the apple was 74.4 ± 8.8 mm. The average mass of the apple was 197 ± 25 g. The packaged apples were stored at refrigerated conditions (4 °C) for at least 48 h before the experiment.

2.2. Packaging Materials

Selected for this study were two interior trays types used for whole apple transport, molded fiber (MF) and expanded polystyrene (EPS). The trays used for this study were standard apple trays designed to hold 88 apples per ventilated corrugated container. Each tray held 22 apples, and there are four trays of apples in each container. The trays were numbered sequentially as Tray 1–4 from bottom to top, starting with Tray 1 located at the bottom of the container. The outer dimensions of the ventilated corrugated container were 514 mm × 327 mm × 635 mm. The final mass of the filled corrugated containers with whole apples was 18.81 ± 0.05 kg and 18.31 ± 0.17 kg for the containers with MF and EPS trays, respectively. The apples were carefully inspected and arranged in the trays with the flower stalk axis horizontal and running parallel with the molded pockets of the trays (Figure 1).

Figure 1. Arrangement of apples in molded pockets prior to testing.

2.3. Free-Fall Drop Test

A Lansmont Model PDT 227 Precision Drop Tester (Lansmont Corporation, Monterey, CA, USA) was used to perform the free-fall drop events (Figure 2). Impact bruises were produced by dropping each corrugated container five times from a predetermined drop height onto the concrete floor surface of the drop test equipment. In this study, the packages were dropped onto the bottom panel of the package from two drop heights, 30 cm and 50 cm. The testing was performed in duplicate for each package configuration at the two different drop heights.

Figure 2. Packaged product on the Lansmont PDT 227.

To record the impact acceleration of the apple from the free-fall drop event, a model HT356B21 triaxial accelerometer (PCB Piezotronics, Inc., Depew, NY, USA) was attached to the apple located in the corner cell of the top tray. The sensor was placed in this position to ensure it would not shear off during the impacts with surrounding apples. The Lansmont Test Partner 3 (Lansmont Corporation, Monterey, CA, USA) processed the signal events and the resultant acceleration from each impact recorded.

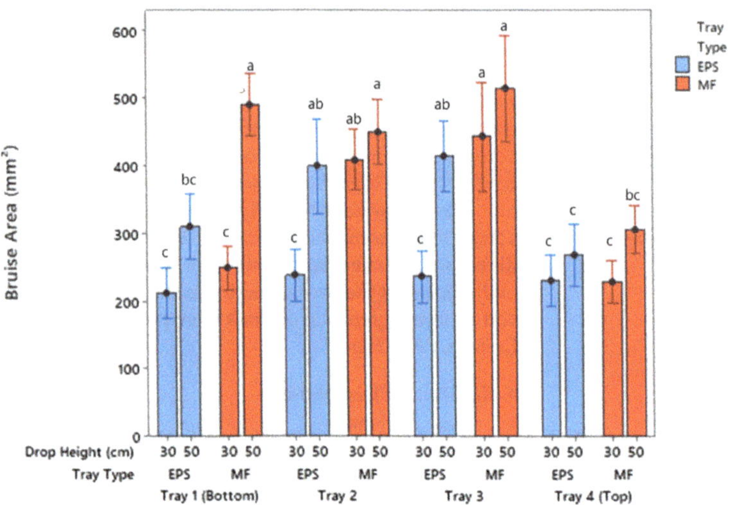

Figure 4. Bruise area of apples by tray location packaged using MF and EPS trays from 30 and 50 cm drop height (mean ± SD, $n = 2$). Different letters indicate statistically significant differences at $p < 0.05$. Bars with no common letters are significantly different ($p < 0.05$).

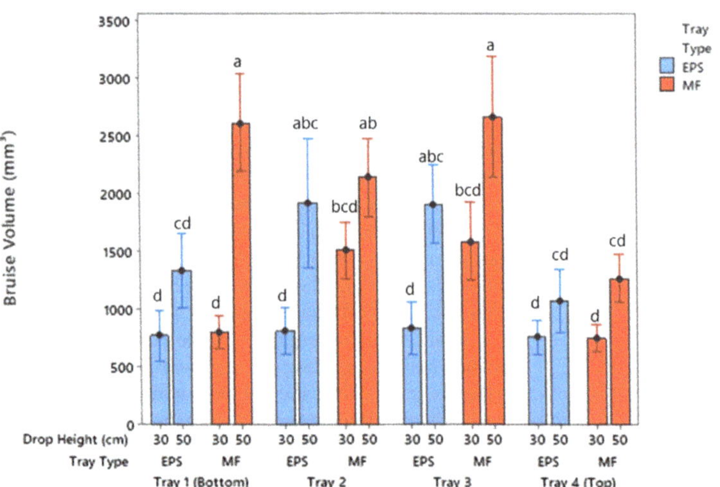

Figure 5. Bruise volume of apples by tray location packaged using MF and EPS trays from 30 and 50 cm drop height (mean ± SD, $n = 2$). Different letters indicate statistically significant differences at $p < 0.05$. Bars with no common letters are significantly different ($p < 0.05$).

Additionally, the frequency, or occurrence, of bruising was observed during the analysis. Figures 6 and 7 display bruise frequency based on the drop height and tray material. Results from the 30 cm and 50 cm drops show EPS trays reported a lower total count of bruises than the MF trays. The MF trays from 30 cm show an increasing trend in the total number of bruising from the bottom to top trays, whereas the remaining treatments have the greatest number of bruises occurring to the middle trays.

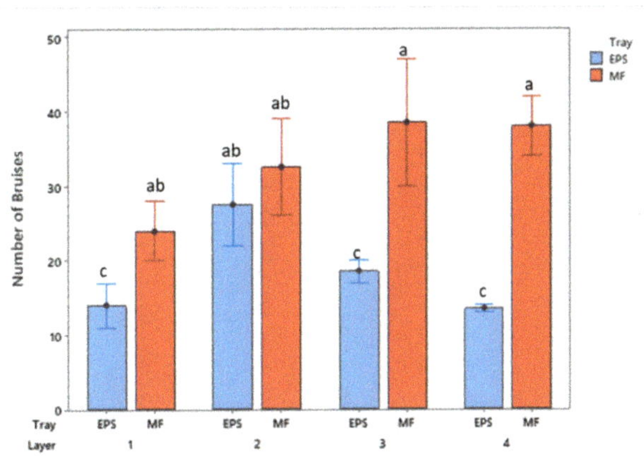

Figure 6. Bruise frequency of apples by tray location packaged using MF and EPS trays from 30 cm drop height (mean ± SD, $n = 2$). Different letters indicate statistically significant differences at $p < 0.05$. Bars with no common letters are significantly different ($p < 0.05$).

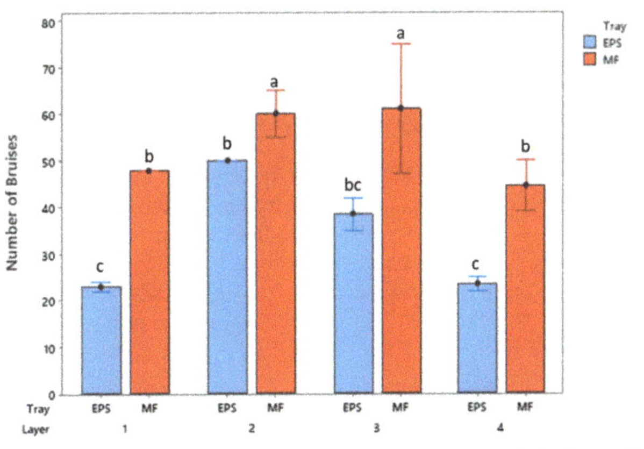

Figure 7. Bruise frequency of apples by tray location packaged using MF and EPS trays from 50 cm drop height (mean ± SD, $n = 2$). Different letters indicate statistically significant differences at $p < 0.05$. Bars with no common letters are significantly different ($p < 0.05$).

3.2. Bruise Susceptibility

Table 1 displays the impact energy associated with each package configuration based on the mass of the package system and the drop height. The impact energies at each of the drop heights were effectively the same, noting the difference being the lighter mass of the EPS tray compared to the MF tray. Although the impact energies of the two package systems were similar, the bruise susceptibility, in terms of the ratio of the bruise volume to the impact energy [7], was greater for apples packaged using MF trays (Figure 8). These results indicate the EPS trays were able to absorb more of the impact energy, reducing the amount of energy transferred to the apple during the impact event. Although the MF tray reduces the lateral movement of the apples, it provides minimal energy absorption during vertical impacts, resulting in the apples being exposed to greater impact energy and subsequent bruising. The apple package system absorbs the energy through stretching of the trays, container sidewall buckling, and compression between the apple contact surfaces [28,32]. The most significant difference in the two package systems evaluated during this study was the compression of the apple contact surfaces. The EPS trays were able to reduce the compression between apples better than the MF trays.

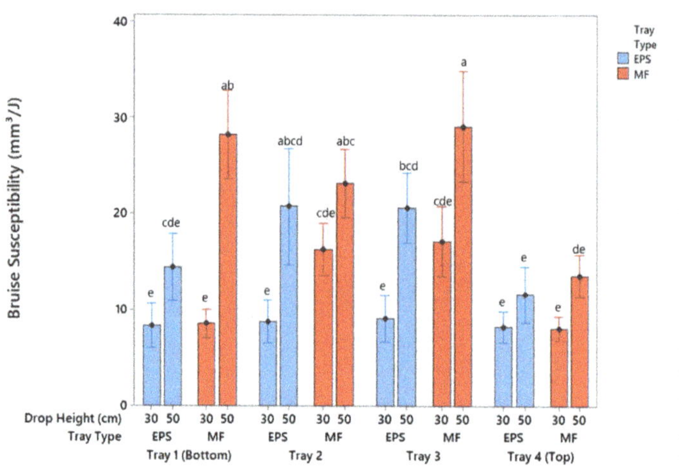

Figure 8. Bruise susceptibility of apples by tray location packaged using MF and EPS trays from 30 and 50 cm drop height (mean ± SD, $n = 2$). Different letters indicate statistically significant differences at $p < 0.05$. Bars with no common letters are significantly different ($p < 0.05$).

To compare the protective cushion properties of the two trays, the resultant acceleration was recorded for each free-fall event. Figure 9 displays an example acceleration versus time curve from the individual impacts showing the apples' response to the free-fall impact. The impact accelerations were averaged for comparison for each treatment type (drop height and material type). As shown in Figure 10, the average acceleration of the apples packaged with the EPS tray was significantly less than that of the MF tray for both the 30 cm and 50 cm drops. The impact duration of the apples in the EPS were twice as long on average compared to MF trays. Comparing the trays from the 30 cm drop tests, the acceleration levels experienced by the apples increased by 30% for those packaged with MF trays. Additionally, examining the results from the 50 cm drop tests, the acceleration levels increased by 71% for apples packaged inside the MF as compared to those of the apples packaged inside the EPS trays. These results indicate that the EPS trays could absorb the impact energy from the event, resulting in less bruising as noted in previous sections.

3.3. Tray Damage

For packaging to protect the product, the material must absorb the mechanical shock event to mitigate the event from transferring to the product. After drop testing, the trays were graded based on the level of damage observed using the scale outlined in Section 2.5. The MF trays from the 30 cm and 50 cm drops were graded as good condition. Regardless of drop height, the MF trays showed occasional small tears (<25 cm) and minor creasing. The EPS trays from both the 30 cm and 50 cm drops experience heavy damage, and were graded as poor condition. Figure 11 illustrates the type of damage experienced by the EPS trays as result of the drop testing. This absorption of shock will often damage the packaging materials, especially at refrigerated conditions where the materials, specifically plastics, are more brittle than fiber-based trays [33]. However, although the EPS trays were damaged, the apple bruising was not as prevalent as the MF tray material. This indicates that the EPS tray material, although more susceptible to fracturing due to the storage conditions, absorbed more impact energy during the drop testing as compared to MF. For all of the drop tests, both with the MF and EPS trays, no damage was observed to any of the corrugated containers used.

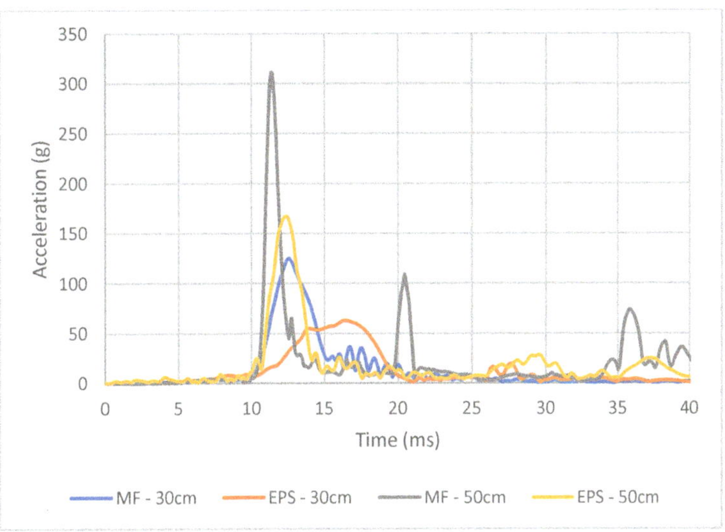

Figure 9. Example of acceleration vs. time curve during impact response from each of the experimental treatments.

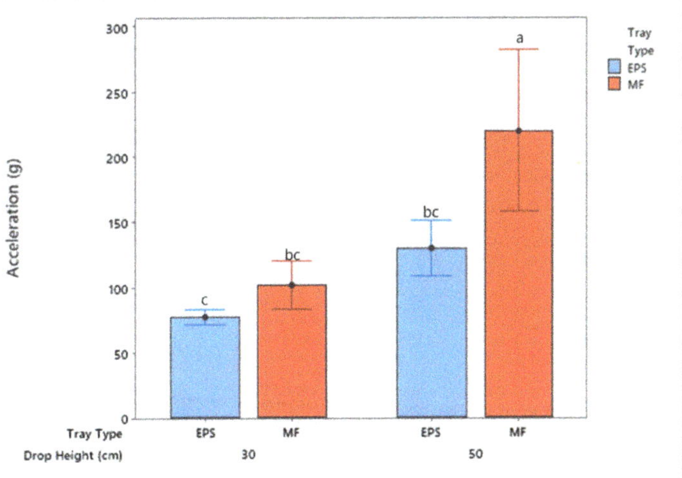

Figure 10. Impact acceleration from the free-fall drop events of apples by tray location packaged using MF and EPS trays from 30 and 50 cm drop height (mean ± SD, $n = 2$). Different letters indicate statistically significant differences at $p < 0.05$. Bars with no common letters are significantly different ($p < 0.05$).

(a)

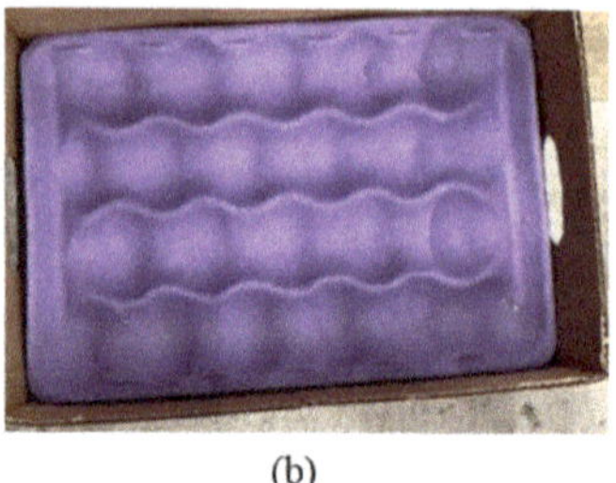
(b)

Figure 11. Example of tray damage, (**a**) EPS and (**b**) MF.

4. Conclusions

Examined in this research project were two packaging tray materials, molded fiber and expanded polystyrene, commonly used to transport whole apples. To evaluate the protective nature of these trays, packaged product systems were subjected to mechanical drops by free-fall, promoting bruise damage and reducing fruit quality. This study indicates the EPS trays decreased the bruise susceptibility of whole apples compared to the MF trays, regardless of drop height. This was confirmed in both bruise analysis performed on the individual apples as well as comparing the impact accelerations of the apples packaged in these two configurations. The apples packaged using EPS trays experienced significantly less impact acceleration as those packaged inside MF trays, indicating the EPS trays mitigated the shock, thus reducing the severity and level of bruising to the apples. Fruit damage was more prevalent to the middle layers of the packages, noting additional cushioning material may be desirable to reduce bruising in those areas. Based on the data from this study, although both materials prevent direct fruit-to-fruit contact during handling, they do not provide the same level of performance in reducing bruise damage. Therefore, it is imperative to select a packaging material, which can decrease the likelihood of bruising to the apple as a result of mechanical forces experienced during transport and handling. Results from this study would be of great benefit to apple growers and packaging engineers who are seeking to reduce or minimize the effects of bruising to apples during transport.

Author Contributions: Conceptualization, K.D.; methodology, K.D., I.S. and M.H.; validation, K.D., I.S. and M.H.; formal analysis, I.S., M.H.; investigation, K.D., I.S. and M.H.; writing—original draft preparation, K.D.; writing—review and editing, K.D., I.S. and M.H.; visualization, I.S., M.H.; supervision, K.D.; project administration, K.D.; funding acquisition, K.D. All authors have read and agreed to the published version of the manuscript.

Funding: This project was supported financially by the Sponsored Research Services of Rochester Institute of Technology (Rochester, NY, USA). The authors declare that they have no known competing financial interests or personal relationships that could have appeared to influence the work reported in this paper.

Institutional Review Board Statement: Not applicable.

Informed Consent Statement: Not applicable.

Data Availability Statement: The data presented in this study are available on request from the corresponding author.

Conflicts of Interest: The authors declare no conflict of interest.

References

1. Fadiji, T.; Coetzee, C.; Pathare, P.; Opara, U.L. Susceptibility to impact damage of apples inside ventilated corrugated paperboard packages: Effects of package design. *Postharvest Biol. Technol.* **2016**, *111*, 286–296. [CrossRef]
2. Opara, L.U. Bruise susceptibilities of "Gala" apples as affected by orchard management practices and harvest date. *Postharvest Biol. Technol.* **2007**, *43*, 47–54. [CrossRef]
3. Mukama, M.; Ambaw, A.; Opara, U.L. Advances in design and performance evaluation of fresh fruit ventilated distribution packaging: A review. *Food Packag. Shelf Life* **2020**, *24*, 100472. [CrossRef]
4. Lewis, R.; Yoxall, A.; Marshall, M.B.; Canty, L.A. Characterising pressure and bruising in apple fruit. *Wear* **2008**, *264*, 37–46. [CrossRef]
5. Jarimopas, B.; Singh, S.P.; Sayasoonthorn, S.; Singh, J. Comparison of package cushioning materials to protect post-harvest impact damage to apples. *Packag. Technol. Sci.* **2007**, *20*, 315–324. [CrossRef]
6. Stopa, R.; Szyjewicz, D.; Komarnicki, P.; Kuta, Ł. Determining the resistance to mechanical damage of apples under impact loads. *Postharvest Biol. Technol.* **2018**, *146*, 79–89. [CrossRef]
7. Pang, D.W.; Studman, C.J.; Banks, N.H. Apple bruising thresholds for an instrumented sphere. *Trans. ASAE USA* **1994**, *37*, 893–897. [CrossRef]
8. Bollen, A.F.; Cox, N.R.; Dela Rue, B.T.; Painter, D.J. PH—Postharvest Technology: A descriptor for damage susceptibility of a population of produce. *J. Agric. Eng. Res.* **2001**, *78*, 391–395. [CrossRef]
9. Peggie: Technical Problems in the Retail Marketing. Available online: https://scholar.google.com/scholar_lookup?title=Technical%20problems%20in%20the%20retail%20marketing%20of%20fruit%20and%20vegetables&publication_year=1968&author=I.%20Peggie (accessed on 22 June 2020).
10. Wagner, T.P. Policy instruments to reduce consumption of expanded polystyrene food service ware in the USA. *Detritus* **2020**, *9*, 11–26. [CrossRef]
11. Winans, K.; Marvinney, E.; Gillman, A.; Spang, E. An evaluation of on-farm food loss accounting in Life-Cycle Assessment (LCA) of Four California Specialty Crops. *Front. Sustain. Food Syst.* **2020**, *4*, 10. [CrossRef]
12. Fu, H.; He, L.; Ma, S.; Karkee, M.; Chen, D.; Zhang, Q.; Wang, S. Bruise responses of apple-to-apple impact. *IFAC-Pap.* **2016**, *49*, 347–352. [CrossRef]
13. Van Zeebroeck, M.; Van linden, V.; Darius, P.; De Ketelaere, B.; Ramon, H.; Tijskens, E. The effect of fruit factors on the bruise susceptibility of apples. *Postharvest Biol. Technol.* **2007**, *46*, 10–19. [CrossRef]
14. Zhu, Q.; Guan, J.; Huang, M.; Lu, R.; Mendoza, F. Predicting bruise susceptibility of "golden delicious" apples using hyperspectral scattering technique. *Postharvest Biol. Technol.* **2016**, *114*, 86–94. [CrossRef]
15. Stropek, Z.; Gołacki, K. The effect of drop height on bruising of selected Apple Varieties. *Postharvest Biol. Technol.* **2013**, *85*, 167–172. [CrossRef]
16. Singh, S.P.; Burgess, G.; Xu, M. Bruising of apples in four different packages using simulated truck vibration. *Packag. Technol. Sci.* **1992**, *5*, 145–150. [CrossRef]
17. Singh, S.P.; Xu, M. Bruising in apples as a function of truck vibration and packaging. *Appl. Eng. Agric.* **1993**, *9*, 455–460. [CrossRef]
18. Peleg, K. Biomechanics of fruits and vegetables. *J. Biomech.* **1985**, *18*, 843–862. [CrossRef]
19. Batt, G.S.; Lussier, M.; Cooksey, K.; Northcutt, J. Transportation, handling, and microbial growth performance of molded fiber and expanded polystyrene apple trays. *Packag. Technol. Sci.* **2019**, *32*, 49–56. [CrossRef]
20. Ambaw, A.; Fadiji, T.; Opara, U.L. Thermo-mechanical analysis in the fresh fruit cold chain: A review on recent advances. *Foods* **2021**, *10*, 1357. [CrossRef]
21. Fadiji, T.; Coetzee, C.J.; Berry, T.M.; Opara, U.L. Investigating the role of geometrical configurations of ventilated fresh produce packaging to improve the mechanical strength—Experimental and numerical approaches. *Food Packag. Shelf Life* **2019**, *20*, 100312. [CrossRef]
22. Fadiji, T.; Coetzee, C.J.; Opara, U.L. Analysis of the creep behaviour of ventilated corrugated paperboard packaging for handling fresh produce—An experimental study. *Food Bioprod. Process.* **2019**, *117*, 126–137. [CrossRef]
23. Fadiji, T.; Coetzee, C.J.; Opara, U.L. Evaluating the displacement field of paperboard packages subjected to compression loading using Digital Image Correlation (DIC). *Food Bioprod. Process.* **2020**, *123*, 60–71. [CrossRef]
24. Bollen, A.F.; Nguyen, H.X.; Dela Rue, B.T. Comparison of methods for estimating the bruise volume of apples. *J. Agric. Eng. Res.* **1999**, *74*, 325–330. [CrossRef]
25. Lu, F.; Ishikawa, Y.; Kitazawa, H.; Satake, T. Measurement of impact pressure and bruising of apple fruit using pressure-sensitive film technique. *J. Food Eng.* **2010**, *96*, 614–620. [CrossRef]
26. Opara, U.L.; Pathare, P.B. Bruise damage measurement and analysis of fresh horticultural produce—A review. *Postharvest Biol. Technol.* **2014**, *91*, 9–24. [CrossRef]
27. Schoorl, D.; Holt, J.E. Impact bruising in 3 apple pack arrangements. *J. Agric. Eng. Res.* **1982**, *27*, 507–512. [CrossRef]
28. Holt, J.E.; Schoorl, D. Package protection and energy dissipation in apple packs. *Sci. Hortic.* **1984**, *24*, 165–176. [CrossRef]
29. Zarifneshat, S.; Ghassemzadeh, H.R.; Sadeghi, M.; Abbaspour-Fard, M.H.; Ahmadi, E.; Javadi, A.; Shervani-Tabar, M.T. Effect of impact level and fruit properties on golden delicious apple bruising. *Am. J. Agric. Biol. Sci.* **2010**, *5*, 114–121. [CrossRef]
30. Ahmadi, E. Bruise susceptibilities of kiwifruit as affected by impact and fruit properties. *Res. Agric. Eng.* **2012**, *58*, 107–113. [CrossRef]

31. Ahmadi, E.; Ghassemzadeh, H.R.; Sadeghi, M.; Moghaddam, M.; Neshat, S.Z. The effect of impact and fruit properties on the bruising of peach. *J. Food Eng.* **2010**, *97*, 110–117. [CrossRef]
32. Van Zeebroeck, M.; Van linden, V.; Ramon, H.; De Baerdemaeker, J.; Nicolaï, B.M.; Tijskens, E. Impact damage of apples during transport and handling. *Postharvest Biol. Technol.* **2007**, *45*, 157–167. [CrossRef]
33. Marsh, K.; Bugusu, B. Food packaging—Roles, materials, and environmental issues. *J. Food Sci.* **2007**, *72*, R39–R55. [CrossRef] [PubMed]

MDPI

Article

Effects of Harvest Maturity, Refrigeration and Blanching Treatments on the Volatile Profiles of Ripe "Tasti-Lee" Tomatoes

Yu Xi [1], Qing Li [1], Jiaqi Yan [2], Elizabeth Baldwin [3], Anne Plotto [3], Erin Rosskopf [3], Jason C. Hong [3], Jinhua Zuo [1,4], Jinhe Bai [3,*] and Jian Li [1,*]

1 Beijing Advanced Innovation Center for Food Nutrition and Human Health, Beijing Technology and Business University, NO. 33 Fucheng Road, Beijing 100048, China; xiyu@btbu.edu.cn (Y.X.); liqing1232021@126.com (Q.L.); zuojinhua@126.com (J.Z.)

2 College of Horticulture, China Agricultural University, NO. 2 Yuanmingyuan West Road, Beijing 100193, China; yanjiaqi@cau.edu.cn

3 U.S. Horticultural Research Laboratory, 2001 South Rock Road, Ft. Pierce, FL 34945, USA; baldwinliz052@gmail.com (E.B.); anne.plotto@usda.gov (A.P.); erin.rosskopf@usda.gov (E.R.); Jason.hong@usda.gov (J.C.H.)

4 Beijing Key Laboratory of Fruits and Vegetable Storage and Processing, National Engineering Research Center for Vegetables, Beijing Academy of Agriculture and Forestry Sciences, Beijing 100097, China

* Correspondence: jinhe.bai@usda.gov (J.B.); lijian@btbu.edu.cn (J.L.)

Abstract: The interactive effects of six maturity stages and refrigerated storage (chilling)/blanching (heating) treatments on the volatile profiles of ripe tomatoes were studied. A total of 42 volatiles were identified, of which 19 compounds had odor activity values equal to or greater than 1. Of those, "green" and "leafy" aroma volatiles were most abundant. Chilling and heating treatments both suppressed overall volatile production, with chilling having the greater impact, regardless of harvest maturity. However, fruit harvested at the turning stage had the least volatile suppression by chilling and heating treatments in comparison with fruit harvested earlier or later, mostly in the fatty acid- and phenylalanine-derived volatiles. Volatiles derived from amino acids were promoted by heat treatment for fruit harvested at all maturities, and those derived from carotenoid and phenylalanine pathways and harvested at advanced harvest maturities were stimulated by chilling treatment. Volatile production is generally believed to be improved by delayed harvest, with vine-ripe being optimum. However, opposite results were observed possibly because the later-harvested fruit had longer exposure to open-field weather stress. The best harvest maturity recommendation is the turning stage where fruit developed abundant volatiles and were least impacted by chilling and heating treatments.

Keywords: *Solanum lycopersicum*; aroma; blanching; chilling; synthetic pathway; volatile; maturity; tomato; flavor; postharvest

Citation: Xi, Y.; Li, Q.; Yan, J.; Baldwin, E.; Plotto, A.; Rosskopf, E.; Hong, J.C.; Zuo, J.; Bai, J.; Li, J. Effects of Harvest Maturity, Refrigeration and Blanching Treatments on the Volatile Profiles of Ripe "Tasti-Lee" Tomatoes. *Foods* **2021**, *10*, 1727. https://doi.org/10.3390/foods10081727

Academic Editor: Onofrio Corona

Received: 24 June 2021
Accepted: 19 July 2021
Published: 26 July 2021

1. Introduction

Tomato is widely consumed worldwide due to its nutrition, flavor, and processing properties [1]. It has been determined that tomatoes contain large amounts of vitamins, carotene, lycopene, and other antioxidants which are beneficial to human health [2]. Moreover, about 400 volatile compounds have been detected in the ripe tomato that contribute to fruit palatability [3–5]. These volatile substances are derived from different pathways such as the amino acid pathway, phenylalanine pathway, fatty acid pathway, and carotenoid pathway [6]. However, many consumers complain that the flavor of modern commercial tomatoes is lacking typical tomato flavor [7]. In fact, the formation of tomato flavor is a complicated process, which is related to many factors, such as variety, pre-harvest environment, and cultural practices; harvest maturity; and postharvest conditions, such as storage temperature, and blanching. In this research we focused on the effect of harvest maturity and postharvest treatments on the volatile profiles of "Tasti-Lee" tomatoes.

Under the current tomato production system, tomatoes are often harvested before full ripeness to avoid postharvest losses and extend storage life [8,9]. Results have not been consistent on how harvest maturity affects the flavor quality of fruit after reaching full red in modern commercial cultivars [5,10,11]. However, it is thought that poor flavor quality often results from harvesting tomatoes prior to breaker stage [8,12]. Xu et al. [13] found that a Florida "heirloom" cultivar had substantially lower quality after reaching full ripe if harvested before the breaker stage, however for commercial cultivars, harvest maturity after mature green did not affect fruit quality in terms of soluble solids content (SSC) and titratable acidity (TA), although differences between harvest maturities for aroma volatiles were detected by an electronic tongue device. "Tasti-Lee" is a hybrid with the crimson gene, high lycopene content, and flavor quality [14,15].

Tomatoes are chilling sensitive at temperatures below 10 °C, and longer exposure time and lower temperature aggravate this sensitivity [16,17]. In Florida, tomatoes are harvested green, gassed with ethylene to initiate and synchronize ripening, then stored at 12–13 °C to slow ripening during packing, repacking, shipping, and marketing [8]. However, on the consumer side, ripe fruit are often stored in the refrigerator (about 5 °C) for several days before cooking and consumption, which has been confirmed to suppress tomato volatiles [18]. However, few studies have been conducted on the interactive effects of the chilling treatment and harvest maturity on tomato flavor.

Blanching (heating) treatment is another method widely used in kitchen and foodservice operations, which can reduce foodborne microorganisms and remove the epidermis structure [19–21]. Heating treatment causes volatile flavor loss, which has been reported in many fruits, including in tomatoes [18,22,23], although the volatile production could be partially recovered after a mild high temperature treatment [18,22]. Blanching uses boiling water to treat fruit, and the treated fruit are then immediately cooked or processed. Thus, it is hypothesized there is no time for the fruit to form a visible physiological response after treatment.

The objectives of this research were to analyze the independent and interactive effects of harvest maturity and temperature treatments on volatile profiles of "Tasti-Lee" tomatoes, and to discuss the effects on precursor pathways for aroma volatiles.

2. Materials and Methods

2.1. Plant Materials

"Tasti-Lee" tomato plants were selected from a tomato research block at the USDA Horticultural Research Laboratory Picos Farm in Fort Pierce, Florida, USA. The harvest maturity of the fruit in the field ranged from immature to full ripe (Figure 1). Sixty fruit of similar size (about 210 g) at each of six mature stages (red, light red, pink, turning, breaker, and mature green) were harvested on 27 December 2015 and stored in a 20 °C dark storage room (humidity: 90%) until they reached the full (red) ripe stage. For each harvest maturity group, when 45 fruits reached red stage (color a* value $\approx$ 20), the fruits were randomly divided into three temperature treatment sub-groups: non-treated control, refrigeration (chilling; 5 °C air for 4 days), and blanching (heating; 100 °C water for 1 min, then ice water used to cool the fruits to room temperature within 3 min). The changes in core temperature were monitored after blanching. Immediately after the temperature treatment, fruits were further divided into five replicates with three fruits each. The experimental design is depicted in Figure 1.

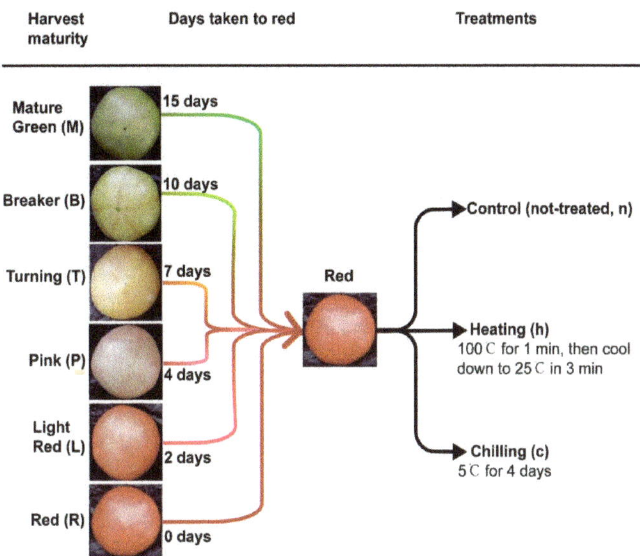

Figure 1. Diagrammatic sketch of overall experimental design. Fruit harvested at six maturities were ripened at 20 °C and then treated by refrigeration (chilling, c), blanching (heating, h), or non-treated (n) as control.

2.2. Sample Preparation

Each replicate sample of three fruits was blended with a homogenizer (Model VM0101, Vita-mix Corp., Cleveland, OH, USA) for one minute. The homogenate, 4.3 g, was pipetted into a 20 mL vial, mixed with 1.7 mL of a saturated $CaCl_2$ solution [24], and crimp-capped with a silicone septum (Gerstel Inc., Linthicum, MD, USA). Samples were vortexed, flash frozen in liquid nitrogen, and stored at −80 °C until analysis.

2.3. Volatile Analysis

Volatiles were analyzed from the vial headspace using solid-phase micro-extraction, and gas chromatography–mass spectrometry (HS-SPME-GC-MS), as described by Wang et al. [25] with modifications. The volatile substances were extracted using a 2 cm SPME fiber (50/30 µm DVB/Carboxen/PDMS; Supelco, Bellefonte, PA, USA). The homogenate mixture was incubated for 30 min at 40 °C and another 30 min was needed for exposure to the fiber.

The volatile compounds were desorbed in a GC-MS (Model 7890 GC coupled with a 5975 MS detector; Agilent, Santa Clara, CA, USA) by using the HP-5 column (50 m × 0.32 mm × 1.05 µm, J&W Scientific, Agilent, Santa Clara, CA, USA). The GC oven temperature was programmed to increase from 40 to 230 °C at a rate of 4 °C min^{-1} and the holding time was 11.70 min. Helium (37 kPa) was used as the carrier gas at a constant flow of 1.5 mL min^{-1}. For the MS system, the temperatures of the inlet, ionizing source, and transfer line were 250, 230, and 280 °C, respectively; electron impact mass units were recorded at 70 eV ionization voltages.

Volatile compounds were tentatively identified by matching their mass spectra to entries in the NIST 14 library (National Institute of Standards and Technology, Gaithersburg, MA, USA) and the retention indexes were compared with the standard volatile compounds. A calibration curve (peak area vs. concentration of reference standards) was prepared from the serial dilutions of the standard and used for sample quantification. Standard aroma compounds (all volatiles identified in this study) were purchased from Sigma-Aldrich (St. Louis, MO, USA) and Fluka Chemical Corporation (Buchs, Switzerland).

2.4. Weather Data Collection

Daily maximum and minimum temperatures and precipitation data for 1–14 days before harvest in 2015 were collected from the U.S. Climate Data (Station of Ft. Pierce Florida, 10 km from the farm) (https://www.usclimatedata.com/climate/fort-pierce/florida/united-states/usfl0156, accessed on 10 May 2021).

2.5. Statistical Analysis

Data were evaluated by analysis of variance (ANOVA) with the statistical analysis system of SPSS Statistics 17.0 (SPSS Inc., Chicago, IL, USA). Two-way ANOVA was carried out to determine the harvest maturity and treatment interaction (maturity × treatment) on each of the assays. Significant differences across maturity × temperature treatment combined were performed by Duncan's new multiple range tests, where differences at $p < 0.05$ were considered significant. Principal component analysis (PCA) was performed by Origin 2019b (Microcal Software Inc., Northampton, MA, USA) on the covariance for analyzing the significant differences and relationship of the volatile organic compounds among different treatments.

3. Results and Discussion

3.1. Characterization of the Volatile Compounds and Effect of Harvest Maturity on Volatile Profiles

A total of 42 volatile compounds were screened by GC-MS in the tomato samples, consisting of 18 aldehydes (methacrolein, butanal, 3-methylbutanal, 2-methylbutanal, 2-methyl-2-butenal, tiglic aldehyde, trans-2-penten-1-al, cis-3-hexenal, hexanal, trans-2-hexenal, heptanal, trans, trans-2, 4-hexadienal, benzaldehyde, octanal, benzeneacetaldehyde, 2-octenal, nonanal, and neral), 4 hydrocarbons (α-pinene, cymene, limonene, and terpinolene), 7 alcohols (2-methylpropanol, 1-penten-3-ol, 3-methylbutanol, 2-methylbutanol, 1-pentanol, 4-methylpentanol, and 3-methylpentanol), 5 ketones (acetone, 2-butanone, 1-penten-3-one, 6-methyl-5-hepten-2-one, and geranyl acetone), 3 oxygen-containing heterocyclic compounds (2-methylfuran, 2-ethyl furan, and 2-pentyl furan), 3 esters (butyl acetate, 2-methylbutyl acetate, and methyl salicylate), 1 sulfur compound (dimethyl disulfide), and 1 sulfur and nitrogen-containing heterocyclic compound (2-isobutylthiazole) (Table 1). The result was similar to the previous report in which 50 aroma volatile compounds were detected by GC-olfactometry in "Tasti-Lee" tomatoes [26]. Aldehydes were the most abundant volatile compounds in the tomatoes regardless of treatment, and the average ratio in the total volatiles was 86.3%, followed by ketones, 9.1%, and alcohols, 3.7%, with the rest each contributing less than 1% (Table 1). For control fruit, aldehydes alone occupied 93.8%, 90.9%, 92.7%, 88.9%, 85.7%, and 93.1% in fruit harvested at red, light red, pink, turning, breaker, and mature green stages, respectively (Table 1 and Table S1). cis-3-Hexenal was the predominant aldehyde among 18 aldehydes in the fruit, constituting 85.4%, 86.6%, 78.4%, 77.4%, 74.8%, and 79.1% of all aldehydes in the tomatoes picked at red, light red, pink, turning, breaker, and mature green stages, respectively (Table 1). The data showed that fruit harvested at the breaker and turning stages had the lowest aldehyde content compared to the earlier (mature green) and later harvest maturities (pink to red); the concentration being higher in mature green fruit (Table 1 and Table S1). However, opposite trends were found for alcohols, ketones, and oxygen-containing heterocyclic compounds. Pink or later harvested fruit had low hydrocarbon concentrations, but earlier harvested fruit had relatively higher concentrations (Table 1 and Table S1). On the other hand, the only sulfur compound, dimethyl disulfide, was not found in the early harvested fruit, but was detected in the pink or later harvested fruit (Table 1 and Table S1).

Table 1. Concentration (mg L^{-1}) of volatile compounds determined in ripe tomatoes harvested at six maturity stages and treated with refrigeration (chilling, c), blanching (heating, h), or not-treated (control, n).

			Treatment (Combination of Harvest Maturity and Temperature Treatment)																		ANOVA (Harvest Maturity × Treatment)
	Compounds	RI ^{Z}	Rn ^{Y}	Rh	Rc	Ln	Lh	Lc	Pn	Ph	Pc	Tn	Th	Tc	Bn	Bh	Bc	Mn	Mh	Mc	
	ALDEHYDES																				
1	Methacrolein	584	0.020 $^{d\ x}$	0.027 ^{d}	0.100 ^{ab}	0.027 ^{d}	0.020 ^{d}	0.100 ^{a}	0.033 ^{d}	0.053 ^{cd}	0.087 ^{bc}	0.091 ^{bc}	0.085 ^{bc}	0.116 ^{ab}	0.091 ^{bc}	0.093 ^{bc}	0.103 ^{ab}	0.117 ^{ab}	0.082 ^{bc}	0.117 ^{ab}	**
2	Butanal	590	0.009 ^{g}	0.009 ^{g}	0.024 ^{ab}	0.008 ^{g}	0.009 ^{g}	0.019 $^{b–d}$	0.015 $^{c–e}$	0.016 $^{c–e}$	0.02 ^{bc}	0.018 ^{cd}	0.018 ^{cd}	0.015 $^{d–f}$	0.016 $^{c–e}$	0.017 $^{c–e}$	0.019 $^{b–d}$	0.01 ^{fg}	0.012 $^{e–g}$	0.026 ^{a}	**
3	3-Methylbutanal #	638	0.113 ^{hi}	0.198 ^{fg}	0.091 ^{i}	0.138 $^{g–i}$	0.164 ^{gh}	0.176 ^{gh}	0.295 ^{de}	0.329 ^{de}	0.18 ^{gh}	0.408 ^{bc}	0.558 ^{a}	0.203 ^{fg}	0.602 ^{a}	0.625 ^{a}	0.267 ^{ef}	0.460 ^{b}	0.619 ^{a}	0.344 ^{cd}	**
4	2-Methylbutanal #	646	0.08 ^{gh}	0.146 ^{fg}	0.152 ^{fg}	0.038 ^{h}	0.106 ^{gh}	0.264 ^{de}	0.391 ^{c}	0.447 ^{bc}	0.305 ^{d}	0.483 ^{ab}	0.539 ^{a}	0.212 ^{ef}	0.534 ^{a}	0.495 ^{ab}	0.304 ^{d}	0.395 ^{c}	0.421 ^{bc}	0.405 ^{c}	**
5	2-Methyl-2-butenal	719	0.028 ^{de}	0.041 ^{d}	- ^{w}	0.010 ^{e}	0.036 ^{d}	-	0.076 ^{c}	0.089 ^{bc}	-	0.104 ^{b}	0.089 ^{bc}	-	0.125 ^{a}	0.130 ^{a}	-	0.084 ^{bc}	0.099 ^{b}	-	**
6	Tiglic aldehyde	1101	0.111 ^{e}	0.164 ^{e}	-	0.041 ^{f}	0.143 ^{e}	-	0.303 ^{d}	0.357 ^{cd}	-	0.414 ^{b}	0.357 ^{cd}	-	0.499 ^{a}	0.520 ^{a}	-	0.336 ^{d}	0.394 ^{bc}	-	**
7	*trans*-Penten-1-al	1131	0.136 ^{c}	0.107 ^{cd}	0.023 ^{f}	0.210 ^{b}	0.140 ^{c}	0.092 ^{d}	0.137 ^{c}	0.081 ^{de}	0.084 ^{de}	0.221 ^{b}	0.211 ^{b}	0.230 ^{b}	0.242 ^{b}	0.098 ^{cd}	0.079 ^{de}	0.312 ^{a}	0.100 ^{cd}	0.042 ^{ef}	**
8	*cis*-3-Hexenal #	771	23.833 ^{bc}	12.681 ^{f}	4.485 ^{g}	28.839 ^{b}	16.613 $^{d–f}$	11.931 ^{f}	27.966 ^{b}	16.269 $^{d–f}$	13.428 ^{ef}	28.782 ^{b}	21.748 ^{cd}	26.528 ^{bc}	26.017 ^{bc}	17.851 ^{de}	17.993 ^{de}	37.889 ^{a}	18.179 ^{de}	11.427 ^{f}	**
9	Hexanal #	774	1.113 ^{d}	0.784 ^{ef}	0.892 $^{d–f}$	1.045 ^{de}	0.813 ^{ef}	1.075 ^{de}	1.711 ^{bc}	1.017 ^{de}	1.013 ^{de}	1.973 ^{ab}	1.553 ^{c}	1.079 ^{de}	1.896 ^{ab}	0.965 $^{d–f}$	0.774 ^{ef}	2.045 ^{a}	0.964 $^{d–f}$	0.670 ^{f}	**
10	*trans*-2-Hexenal #	828	2.056 ^{de}	1.339 ^{gh}	0.503 ^{i}	2.446 ^{cd}	1.588 $^{e–h}$	1.540 $^{e–h}$	3.340 ^{b}	1.853 $^{e–g}$	1.766 $^{e–g}$	3.535 ^{b}	2.691 ^{c}	2.625 ^{c}	3.852 ^{b}	1.999 $^{d–f}$	1.450 $^{f–h}$	4.413 ^{a}	1.936 $^{d–f}$	1.142 ^{h}	**
11	Heptanal	875	0.008 ^{c}	0.009 ^{c}	0.011 ^{c}	0.007 ^{cd}	0.003 ^{d}	0.015 ^{bc}	0.014 ^{bc}	0.010 ^{c}	0.010 ^{c}	0.017 ^{b}	0.014 ^{bc}	0.014 ^{bc}	0.016 ^{b}	0.008 ^{c}	0.010 ^{c}	0.025 ^{a}	0.010 ^{c}	0.009 ^{c}	**
12	*trans*, *trans*-2, 4-Hexadienal	887	-	-	-	-	-	-	-	-	-	-	0.613 ^{a}	-	-	-	-	-	-	-	**
13	Benzaldehyde	945	0.010 ^{bc}	0.006 ^{de}	0.002 ^{ef}	0.010 ^{bc}	0.003 $^{d–f}$	0.007 ^{cd}	0.011 ^{b}	0.004 $^{d–f}$	0.002 ^{ef}	0.015 ^{a}	0.01 ^{bc}	0.006 ^{de}	0.011 ^{b}	0.001 ^{f}	0.002 ^{ef}	0.018 ^{a}	0.003 $^{d–f}$	0.002 ^{ef}	**
14	Octanal	969	0.001 ^{b}	0.002 ^{b}	0.001 ^{b}	0.001 ^{b}	-	0.002 ^{b}	-	0.001 ^{b}	0.004 ^{a}	-	0.001 ^{b}	0.001 ^{b}	-	-	0.001 ^{b}	0.001 ^{b}	-	-	**
15	2-phenylacetaldehyde	1049	0.366 ^{i}	0.375 ^{i}	0.887 $^{c–h}$	0.444 $^{g–i}$	0.296 ^{i}	0.93 $^{b–f}$	1.334 $^{a–c}$	0.674 $^{e–i}$	0.499 $^{f–i}$	1.125 $^{b–e}$	1.31 ^{ab}	1.148 $^{b–d}$	0.859 $^{c–h}$	0.504 $^{f–i}$	0.865 $^{c–g}$	1.742 ^{a}	0.844 $^{d–h}$	0.437 ^{hi}	**
16	2-Octenal	1074	0.032 ^{ab}	0.038 ^{a}	-	0.037 ^{a}	0.003 ^{c}	-	0.034 ^{a}	-	-	-	0.014 ^{bc}	-	-	-	-	0.034 ^{a}	-	-	**
17	Nonanal	1059	-	-	-	0.001 ^{a}	-	-	-	-	-	-	-	-	-	-	-	-	-	-	**
18	Neral #	1180	0.006	-	0.004	-	0.005	-	0.009	0.005	-	-	-	-	-	-	-	0.010	-	-	NS
	HYDROCARBONS																				
1	α-Pinene	910	-	0.001	-	-	-	-	-	-	0.001	-	-	0.001	-	-	0.001	-	-	0.001	NS
2	Cymene	994	-	-	-	-	-	-	-	-	-	-	-	0.001	-	-	0.001	-	-	0.001	NS
3	Limonene	998	0.024 $^{e–g}$	0.005 ^{g}	0.068 ^{d}	0.009 ^{fg}	0.012 ^{fg}	0.077 ^{d}	0.007 ^{fg}	0.011 ^{fg}	0.06 ^{de}	0.050 $^{d–f}$	0.116 ^{c}	0.342 ^{a}	0.062 ^{de}	0.044 $^{d–g}$	0.330 ^{a}	0.084 ^{cd}	0.123 ^{c}	0.281 ^{b}	**
4	Terpinolene	1048	-	-	-	-	-	-	-	-	-	-	-	0.001 ^{a}	-	-	0.001 ^{a}	-	-	-	**
	ALCOHOLS																				
1	2-Methylpropanol	612	0.006 ^{f}	0.018 ^{de}	-	-	0.014 ^{ef}	0.019 $^{c–e}$	0.017 ^{de}	0.022 $^{c–e}$	0.067 ^{a}	0.033 $^{b–d}$	0.020 $^{c–e}$	0.009 ^{ef}	0.035 ^{bc}	0.048 ^{b}	0.014 ^{ef}	0.005 ^{ef}	0.062 ^{a}	0.076 ^{a}	**
2	1-Penten-3-ol	1157	0.055 ^{d}	0.031 ^{e}	0.028 ^{e}	0.085 ^{b}	0.067 $^{b–d}$	0.067 $^{b–d}$	0.034 ^{e}	0.030 ^{e}	0.054 ^{d}	0.085 ^{b}	0.073 $^{b–d}$	0.083 ^{bc}	0.086 ^{b}	0.057 ^{d}	-	0.109 ^{a}	0.065 ^{cd}	-	**
3	3-Methylbutanol #	707	0.365 ^{de}	0.656 ^{cd}	0.068 ^{g}	0.329 $^{d–f}$	0.379 ^{de}	0.222 ^{ef}	0.315 $^{d–f}$	0.443 ^{d}	0.183 ^{fg}	0.604 ^{d}	0.767 ^{c}	0.296 $^{d–f}$	0.927 ^{b}	1.189 ^{a}	0.368 ^{de}	0.770 ^{c}	1.106 ^{a}	0.350 ^{de}	**
4	2-Methylbutanol #	711	0.190 $^{g–i}$	0.270 ^{gh}	0.067 ^{i}	0.144 ^{hi}	0.244 ^{gh}	0.417 ^{ef}	0.292 ^{fg}	0.424 ^{ef}	0.613 ^{bc}	0.508 $^{c–e}$	0.489 $^{c–e}$	0.313 ^{fg}	0.599 $^{b–d}$	0.703 ^{ab}	0.567 $^{b–d}$	0.464 ^{de}	0.606 $^{b–d}$	0.761 ^{a}	**
5	1-Pentanol	775	0.066 ^{b}	0.091 ^{a}	0.021 ^{ef}	0.049 ^{bc}	0.051 ^{bc}	0.026 $^{d–f}$	0.009 ^{fg}	0.044 ^{cd}	0.025 ^{def}	0.052 ^{bc}	0.042 $^{c–e}$	0.026 $^{d–f}$	0.057 ^{bc}	0.041 $^{c–e}$	0.016 ^{f}	0.052 ^{bc}	0.022 ^{ef}	-	**
6	4-Methylpentanol	809	-	0.003 ^{b}	0.005 ^{b}	0.002 ^{b}	0.003 ^{b}	-	-	-	-	-	-	-	0.009 ^{b}	0.039 ^{a}	-	0.009 ^{b}	0.031 ^{a}	-	**
7	3-Methylpentanol	817	0.015 $^{d–f}$	0.013 ^{f}	0.011 ^{f}	0.015 $^{d–f}$	0.013 ^{ef}	0.023 ^{bc}	0.014 ^{ef}	0.015 $^{d–f}$	0.023 ^{bc}	0.018 $^{c–e}$	0.020 $^{b–d}$	0.013 ^{ef}	0.030 ^{a}	0.023 ^{bc}	0.013 ^{ef}	0.025 ^{ab}	0.019 $^{c–e}$	0.019 $^{c–e}$	**
	KETONES																				
1	Acetone	533	0.423 ^{f}	0.453 ^{f}	3.090 ^{ab}	2.004 ^{d}	1.363 ^{e}	2.939 $^{a–c}$	1.096 ^{e}	0.727 ^{ef}	2.335 ^{cd}	1.998 ^{d}	1.988 ^{d}	3.106 ^{a}	2.569 $^{a–d}$	2.389 ^{cd}	2.45 $^{b–d}$	2.447 $^{a–d}$	1.285 ^{e}	2.171 ^{d}	**
2	2-Butanone	591	0.002 ^{b}	0.013 ^{a}	-	-	-	0.003 ^{b}	-	-	-	-	-	-	-	-	-	-	-	-	**
3	1-Penten-3-one #	665	0.197 $^{b–d}$	0.124 ^{ef}	0.071 ^{f}	0.253 ^{ab}	0.232 $^{a–c}$	0.172 $^{c–e}$	0.139 ^{de}	0.147 ^{de}	0.129 ^{ef}	0.244 ^{ab}	0.203 $^{a–d}$	0.230 $^{a–c}$	0.265 ^{a}	0.175 $^{c–e}$	0.008 ^{g}	0.255 ^{ab}	0.146 ^{de}	0.005 ^{g}	**
4	6-Methyl-5-hepten-2-one #	950	0.084 ^{i}	0.078 ^{i}	0.279 $^{a–e}$	0.107 ^{hi}	0.082 ^{i}	0.329 ^{a}	0.204 $^{e–g}$	0.161 ^{gh}	0.248 $^{c–f}$	0.254 $^{b–f}$	0.205 $^{e–g}$	0.304 $^{a–c}$	0.216 $^{d–g}$	0.197 ^{fg}	0.287 $^{a–d}$	0.317 ^{ab}	0.190 ^{fg}	0.247 $^{c–f}$	**
5	Geranyl acetone #	1367	0.345 ^{ef}	0.084 ^{h}	0.295 $^{e–g}$	0.233 ^{fg}	0.203 ^{g}	0.332 ^{ef}	0.553 ^{c}	0.201 ^{g}	0.307 $^{e–g}$	0.676 ^{ab}	0.401 ^{de}	0.491 ^{cd}	0.753 ^{a}	0.271 ^{fg}	0.408 ^{de}	0.592 ^{bc}	0.232 ^{fg}	0.227 ^{fg}	**

Table 1. *Cont.*

			Treatment (Combination of Harvest Maturity and Temperature Treatment)																		ANOVA (Harvest Maturity × Treatment)
OXYGEN-CONTAINING HETEROCYCLIC COMPOUNDS																					
1	2-Methylfuran	602	0.006 ^{f}	0.011 ^{ef}	0.062 ^{a}	0.005 ^{f}	0.009 ^{ef}	0.064 ^{a}	0.010 ^{ef}	0.014 $^{d-f}$	0.030 ^{b}	0.013 $^{d-f}$	0.016 ^{de}	0.024 ^{bc}	0.008 ^{ef}	0.011 ^{ef}	0.024 ^{bc}	0.014 $^{d-f}$	0.007 ^{f}	0.020 ^{cd}	**
2	2-Ethyl furan	676	0.014 $^{b-d}$	0.010 $^{d-f}$	0.010 $^{c-f}$	0.016 ^{b}	0.011 $^{b-f}$	0.015 $^{b-d}$	0.014 $^{b-e}$	0.008 ^{f}	0.010 $^{c-f}$	0.023 ^{a}	0.021 ^{a}	0.015 ^{bc}	0.023 ^{a}	0.014 $^{b-e}$	0.009 ^{ef}	0.027 ^{a}	0.014 $^{b-e}$	0.006 ^{f}	**
3	2-Pentyl-furan	996	0.054 $^{d-f}$	0.056 $^{d-f}$	0.267 ^{b}	0.057 $^{d-f}$	0.034 ^{ef}	0.256 ^{b}	0.088 ^{d}	0.080 ^{d}	0.074 ^{de}	0.064 ^{de}	0.045 $^{d-f}$	0.298 ^{b}	0.145 ^{c}	0.016 ^{f}	0.395 ^{a}	0.133 ^{c}	0.080 ^{d}	0.401 ^{a}	**
ESTERS																					
1	Butyl acetate	774	0.001 ^{c}	0.003 ^{bc}	0.004 ^{ab}	0.002 ^{c}	0.002 ^{c}	0.003 ^{c}	0.005 ^{a}	0.004 ^{ab}	0.001 ^{c}	0.003 ^{bc}	0.005 ^{ab}	0.001 ^{a}	0.003 ^{bc}	0.003 ^{bc}	-	0.003 ^{bc}	0.004 ^{ab}	0.001 ^{c}	**
2	2-Methylbutyl acetate	847	-	-	-	-	-	-	-	0.001 ^{a}	-	-	0.001 ^{a}	-	-	0.001 ^{a}	-	0.001 ^{a}	0.001 ^{a}	-	**
3	Methyl salicylate #	1156	0.003	0.009	-	0.011	0.013	-	0.004	0.001 ^{c}	0.002	0.005	0.003	-	-	-	-	-	-	-	NS
SULFUR- AND NITROGEN-CONTAINING HETEROCYCLIC COMPOUNDS																					
1	2-Isobutylthiazole #	1002	0.002 ^{de}	0.003 $^{b-e}$	-	0.001 ^{e}	0.001 ^{e}	0.002 $^{c-e}$	0.004 $^{a-c}$	0.004 $^{a-c}$	0.003 $^{b-e}$	0.004 $^{a-c}$	0.006 ^{a}	0.005 ^{a}	0.004 $^{a-c}$	0.002 ^{de}	0.004 $^{a-c}$	0.003 $^{b-e}$	0.001 ^{e}	0.003 $^{b-e}$	**
SULFUR COMPOUNDS																					
1	Disulfide, dimethyl	752	0.001 ^{b}	0.001 ^{b}	0.001 ^{b}	0.001 ^{b}	0.001 ^{b}	0.001 ^{b}	0.001 ^{b}	-	-	-	-	0.004 ^{a}	-	-	-	-	-	-	**
	TOTAL VOLATILES		29.78 ^{ef}	17.86 ^{j}	11.52 ^{k}	36.63 ^{cd}	22.68 ^{hi}	21.12 ^{ij}	38.48 $^{b-d}$	23.54 ^{ghi}	21.56 ^{ij}	41.82 ^{b}	34.23 ^{de}	37.74 $^{b-d}$	40.55 ^{bc}	28.53 ^{fg}	26.76 $^{f-h}$	51.44 ^{a}	29.42 $^{f-g}$	19.19 ^{ij}	**

^{z}: RI = retention index on a HP-5 column (Agilent). ^{y}: Abbreviations of combination of harvest maturity (R—red; L—light red; P—pink; T—turning; B—breaker; and M—mature green) and temperature treatment (h—heating; c—chilling; and n—non-treated control). ^{x}: Means followed by different superscripts in the same row indicate significant differences using the Duncan's multiple range test ($p < 0.05$). ^{w}: Not detectable. ** $p < 0.001$. NS: not significant. $^{\#}$ 13 important volatiles which are deemed key tomato volatiles [11,27–29] and were selected for PCA analysis.

Overall flavor quality is generally believed to be better when tomatoes are harvested during later maturity stages, and the best quality fruit are those that are vine-ripened [8,12]. However, different results were observed in this study: the earlier harvested fruit developed more abundant volatiles at the red ripe stage than later-harvested fruit (Table 1 and Table S1). One explanation could be because fruit at turning stage reached the full flavor quality potential, however, the later-harvested fruit had longer exposure to open-field weather stress (Figure 2). Seven and eight days before harvest, fruit experienced two days with low temperatures, 11 °C, which is below chilling injury temperatures in stored fruit [30] (Figure 2). During most of the fruiting season, the minimal air temperature was higher than 18 °C, except on December 19 and 20 when it dropped to 11 °C. This in addition to solar radiation resulted in high temperatures on the fruit surface (Figure 2) [31]. Thus, the pre-harvest low and high temperature stress may negatively affect flavor metabolism, especially for fruit close to being physiologically ripe.

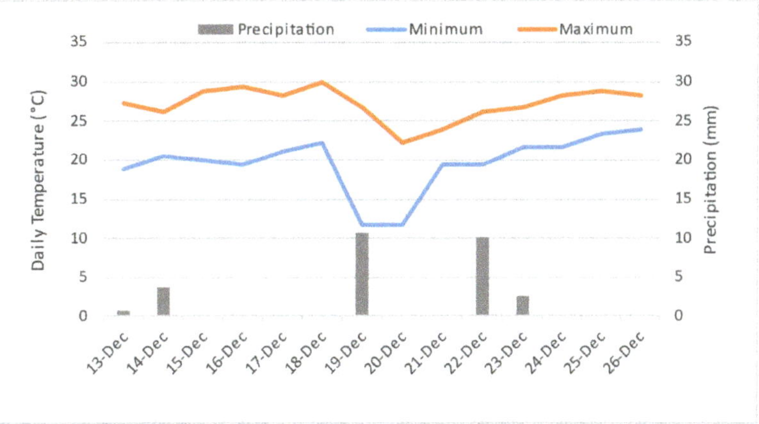

Figure 2. Daily maximum and minimum temperature and precipitation for two weeks before harvest in 2015. Recordings from the U.S. Climate Data (Station of Ft. Pierce Florida, 10 km from the farm) (https://www.usclimatedata.com/climate/fort-pierce/florida/united-states/usfl0156, accessed on 10 May 2021).

3.2. *Refrigeration and Blanching Treatment Reduce Volatile Compounds*

3.2.1. Refrigeration Treatment

Ripe tomatoes are often stored in a 5–10 °C refrigerator by consumers to extend their shelf life regardless of negative reports that refrigeration suppresses volatile production [18], a form of chilling injury (CI). Although tomatoes are chilling sensitive, full ripe fruit show a lower response to CI [18]. In this experiment, the tomatoes did not express any visual CI symptoms after storage at 5 °C for four days. However, an overall decrease in flavor volatiles was observed in the chilled fruit regardless of harvest maturities (Table 1 and Figure 3). The average volatile loss caused by chilling treatment over all harvest maturities was 42.46% (Table 1, Figure 3, and Table S1). Among those, the highest reduction occurred in the fruit harvested at the mature green stage with 63.82% loss, followed by red (61.20%), pink (43.95%), light red (42.25%), breaker (33.78%), and turning (9.78%) stages (Table 1 and Figure 3), suggesting that the turning and breaker fruits are less sensitive to chilling treatment, and that CI increased along with earlier or later harvests (Figure 3). Among the volatiles, aldehydes, especially cis-3-hexenal, hexanal, and trans-2-hexenal, decreased the most due to chilling treatment (Table 1 and Table S1, and Figure S1). Similar to aldehydes, chilling caused a decrease in esters in all fruit, and alcohols in early harvested fruit and fruit harvested at the red stage (Table 1 and Table S1 and Figure S1). These results are consistent with previous reports that volatile compounds of "FL 47" tomatoes decreased after chilling treatment applied to tomato fruit at the mature green stage, unlike in our study in which chilling treatment was performed when the fruit reached the red-ripe stage [17]. However, it was observed that chilling treatment stimulated some volatile compounds, such as hydrocarbons and oxygen-containing heterocyclic compounds in all materials regardless harvest maturities, and ketones in late harvested fruit (Table 1 and Table S1, and Figure S1). Nevertheless, the concentrations of these compounds were trace, and contributed very little to tomato flavor quality (Figure S1).

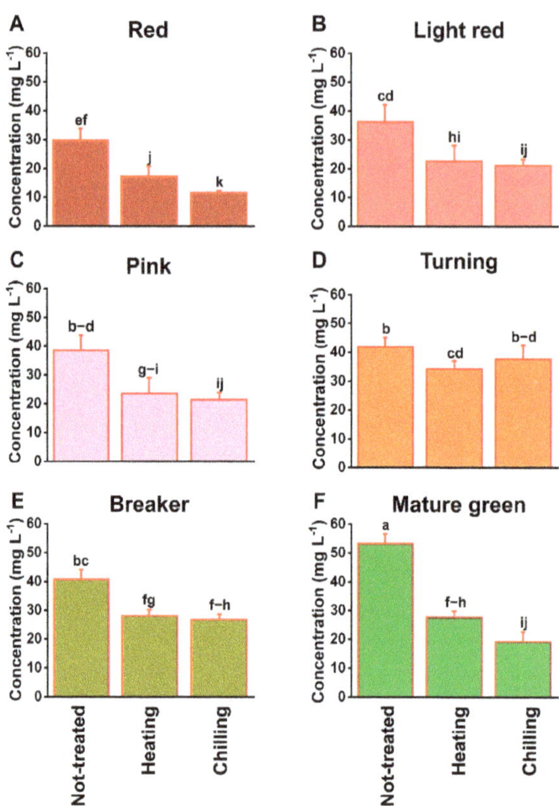

Figure 3. Effect of postharvest temperature treatments on the concentration of flavor compounds (mg L^{-1}) in ripe tomatoes harvested at six maturities: (**A**) red, (**B**) light red, (**C**) pink, (**D**) turning, (**E**) breaker, and (**F**) mature green. Each value is the mean of five replicates of three fruits each. Vertical bars represent the standard deviation of the means. The effect of "maturity × treatment" interaction was tested by two-way ANOVA ($p < 0.05$). Different letters on the top of columns represent significant differences between maturity × treatments combined using the Duncan's multiple range test based on the interaction ($p < 0.05$).

3.2.2. Blanching Treatment

Similar to refrigeration, the blanching-treated samples also showed a relative decrease in volatile compounds compared to the control group, although they reduced the volatile concentration through different mechanisms. Refrigeration (chilling) reduced the volatile concentration by slowing and disturbing plant physiological processes, but blanching (heating) reduced volatile concentration mainly by enhanced volatile evaporation/partitioning to the air [18]. The concentration of volatile compounds in blanching-treated tomatoes decreased by 35.4% on average for all harvest maturities (Table 1 and Table S1, and Figure S1). Loss of total volatile compounds was the lowest when the fruits were harvested at the turning stage (Figure 3D), and total volatiles decreased when fruits were harvested at mature/ripe when compared to the turning stage; fruits harvested at mature green lost the most volatiles, 48.00% (Table 1 and Figure 3F). Most of the aldehydes and ketones were reduced by blanching treatment, and some of them, such as cis-3-hexenal, hexanal, trans-2-hexenal, and geranyl acetone, were reduced by approximately 50% (Table 1 and Table S1). This means that even 1 min of blanching, which raised the core temperature of the fruit by about 1 °C on average in this experiment, caused substantial loss of these compounds from surface tissues. However, hydrocarbons, alcohols, and other compounds did not decrease, and some of them even increased (Figure S1B,C,F,G). The high loss ratios of aldehydes and ketones, partially due to low Henry's law coefficients for these compounds, means that they tend to escape from tissue/solution [32]. This could explain a loss of "freshness" when tomatoes are cooked. A longer cooking time and higher temperature showed significant changes in volatiles, including aldehydes [33].

3.2.3. Comparison of Volatile Profiles between Chilling and Heating Treated Tomatoes

To compare the volatile profiles of chilling and heating treated tomatoes, a PCA analysis was conducted for all samples harvested at various maturity stages based on all volatile compounds. Generally, heat treatment overlapped with the non-treated control at each harvest maturity, but showed differences to chilling treatment (Figure 4). There were fewer differences in samples harvested at turning and breaker stages, but more for samples harvested at earlier or later maturity (Figure 4). An additional PCA was performed based on 13 selected volatiles (Table 1, marked with #) which are key tomato flavor contributors [11,27–29] and represented volatiles from four major pathways: the amino acid pathway, the fatty acid pathway, the phenylalanine pathway, and the carotenoid pathway (Figure S2). Vector loadings showed that high levels of cis-3-hexenal, trans-2-hexenal, hexanal, 1-penten-3-one, and geranyl acetone, abundant acetaldehydes and ketones, were generally associated with the non-treated control. Only one or two compounds, such as 2-methylbutanol, were frequently associated with chilling treatment. All other compounds were often associated with heating and/or the non-treated control (Figure S2).

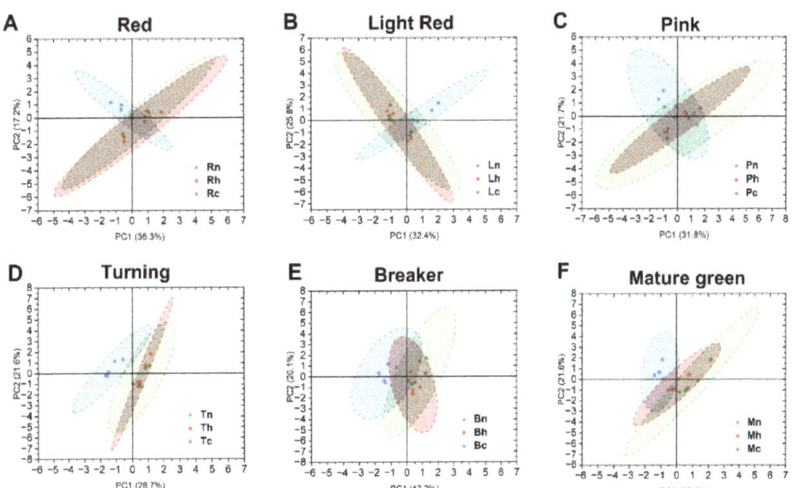

Figure 4. Principal component analysis (PCA) results based on all volatile compounds in "Tasti-Lee" tomatoes with different temperature treatments at six harvest maturities: (**A**) red, (**B**) light red, (**C**) pink, (**D**) turning, (**E**) breaker, and (**F**) mature green. Abbreviations represent combinations of harvest maturity (R—red; L—light red; P—pink; T—turning; B—breaker; and M—mature green) and temperature treatment (h—heating; c—chilling; n—non-treated control).

3.3. Response of the Flavor Synthesis Pathways to Refrigeration or Blanching Treatments

3.3.1. Fatty Acid Pathway

The formation of tomato volatile compounds can be categorized into four pathways, including the amino acid pathway, the fatty acid pathway, the phenylalanine pathway, and the carotenoid pathway [5]. Chemicals derived from the fatty acid pathway dominated the volatile compounds regardless of harvest maturity and temperature treatment (Table 1 and Figure S3). The concentrations of fatty acid derivatives substantially decreased after chilling treatment in fruit harvested at all maturities, except at the turning stage in which chilling only caused a small change (Figure 5A). The concentration of three typical C6 aldehydes, cis-3-hexenal, hexanal, and trans-2-hexenal, decreased on average by 50.05%, 38.27%, and 53.65%, respectively, due to chilling treatment (Table 1). At the same time, the concentrations of cis-3-hexenal, hexanal, and trans-2-hexenal in the blanching-treated tomatoes were reduced as well, although by much less in comparison with the chilling treatment (Table 1). Bai et al. [18] reported that in the oxylipin pathway, chilling treatment reduced the above three C6 aldehydes largely due to the downregulation of the gene expression of Tomlox C and HPL, and the activity of hydroperoxide lyase (HPL). HPL is a key enzyme in the fatty acid synthesis pathway and it is easily affected by temperature [34].

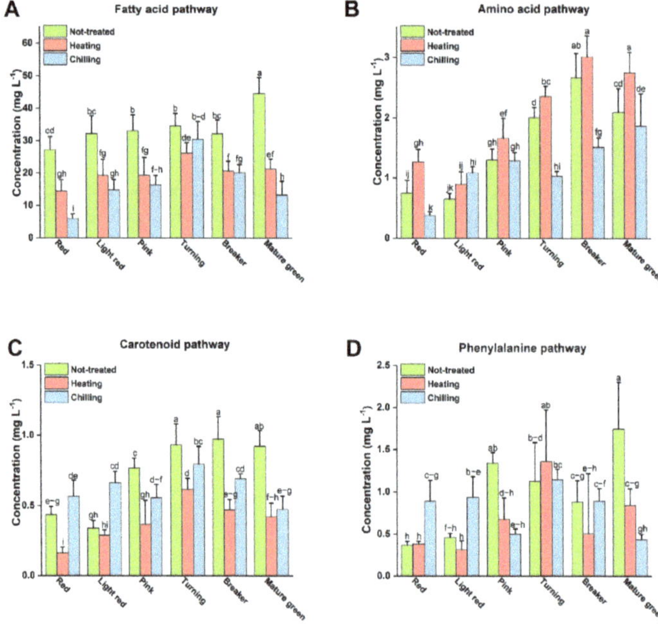

Figure 5. Effects of treatment combinations of harvest maturity and temperature on volatile concentration in tomatoes for aroma volatiles according to their synthesis pathway: (**A**) fatty acid pathway, (**B**) amino acid pathway, (**C**) carotenoid pathway, and (**D**) phenylalanine pathway. Each value is the mean of five replicates. Vertical bars represent standard deviation of the mean. Effect of "maturity × treatment" interaction was tested by two-way ANOVA ($p < 0.05$). Different letters on the top of columns represent significantly differences between treatments using the Duncan's multiple range test based on the interaction ($p < 0.05$).

3.3.2. Amino Acid Pathway

In tomato fruits, it has been speculated that amino acid derivatives are catalyzed by branched chain aminotransferases (BCATs) to remove amino groups from amino acids [5]. Subsequently, corresponding alcohols are formed after decarboxylation and reduction [35,36]. Isoleucine was a precursor of 2-methylbutanal, while 3-methylbutanol was derived from leucine [5].

Chilling treatment significantly reduced the concentrations of volatile compounds from the amino acid synthesis pathway in most harvest maturities: 11.00%, 43.38%, 48.75%, and 49.58% at mature green, breaker, turning, and red stages, respectively (Table 1, Figure 5B). However, fruit amino acid pathway volatiles, when harvested at the pink stage, were not affected by chilling treatment, and fruit harvested at the light red stage, in contrast, showed amino acid pathway volatiles that increased due to chilling treatment (Table 1, Figure 5B). The result was consistent with the work of Renard et al. [37], which proved that the production of amino acid derivatives decreased after chilling treatment. The reason for chilling-induced increase of amino acid derivatives at the light red stage was not clear.

With high expression of BCATs, there were also high yields of branched volatiles in banana fruits [38], indicating that the reduction of volatile compounds by chilling treatment in this study might be attributed to the decrease of BCATs enzyme activity.

In contrast to chilling treatment, blanching enhanced volatiles from the amino acid pathway (Table 1, Figure 5B). Blanching greatly increased the concentrations of 2-methylbutanal, 3-methylbutanal, 3-methylbutanol, 2-methylbutanol, and 2-isobutylthiazole (Table 1). The concentrations of amino acid pathway volatiles with blanching treatment were increased by 31.55%, 13.03%, 17.50%, 26.91%, 37.86%, and 69.49% at the mature green, breaker, turning, pink, light red, and red stages, respectively (Table 1).

3.3.3. Carotenoid Pathway

In this study, three carotenoid derivatives were detected: neral, 6-methyl-5-hepten-2-one, and geranyl acetone. Among these, the concentration of neral was very low. The amount of two important carotenoid derivatives, 6-methyl-5-hepten-2-one and geranyl acetone, decreased after blanching (Table 1). However, the concentration of 6-methyl-5-hepten-2-one increased after chilling treatment except for fruit harvested at the mature green stage (Table 1), contrary to Klee's result [36]. On the other hand, the reason for this result may be due to genetic differences between different varieties of tomatoes—both chilling and blanching reduced geranyl acetone concentration (Table 1). Simkin et al. [39] reported that the carotenoid derivatives in tomatoes were associated with genes potentially encoding carotenoid cleavage dioxygenases, LeCCD1A and LeCCD1B. Our results also showed that the carotenoid derivatives were reduced more by chilling or blanching treatment when harvested at early stages (Table 1, Figure 5C).

3.3.4. Phenylalanine Pathway

Two phenylalanine-derived volatiles (methyl salicylate and 2-phenylacetaldehyde) have been detected in this study (Table 1). In this study, no significant changes of methyl salicylate were found, and this might be due to the minimal concentration detected (Table 1). 2-phenylacetaldehyde is reported to be regulated by aromatic amino acid decarboxylases (AADCs), encoded by LeAADC1A, LeAADC1B, and LeAADC2 [40]. Previously, Zou et al. [41] found that the expression of LeAADC1A and LeAADC1B was downregulated during storage at 4 °C. In agreement with Wang et al. [25], the present study also found decreased concentrations of 2-phenylacetaldehyde in chilling injured tomatoes picked at the mature green stage (Table 1, Figure 5D). Intriguingly, here tomato picked at the red and light red stages contained more 2-phenylacetaldehyde when chilling was administered compared to controls, while no significant changes were found in the tomato picked at the breaker or turning stages (Table 1, Figure 5D).

3.4. Active Odorants

Based on quantitation, the calculation of odor activity values (OAVs) enables a more reliable evaluation of important odorants in foods [42]. Volatile compounds are considered to contribute to the overall flavor of the tomato when their concentration is greater than detection thresholds [43]. As shown in Table 2, a total of 19 aroma compounds, with their OAVs > 1, were selected. cis-3-Hexenal was the most active compound, which presented "green", "leafy" notes (Table 2). As shown in the Figure 6, tomatoes mainly emitted the aroma of "green" and "leafy", confirming that the fatty acid pathway was the important biosynthesis pathway of aroma for "Tasti-Lee" tomatoes. Tomatoes harvested at the mature green stage displayed the strongest "green" and "leafy" notes, as is consistent with many tomato flavor research reports [5,8]. 2-Methylbutanal and 3-methylbutanal were also major contributors to the aroma, which are the derivatives of amino acids and were described as "malt" notes (Table 2) [36]. The aroma of "malt" plays an important part in tomato flavor (Figure 6) [36]. After chilling or blanching treatment, the flavor of tomatoes generally decreased depending on harvest maturity and temperature treatment combinations (Figure 6). Fruit harvested at the turning stage had the second most abundant volatiles after ripening, only less than that in the mature green harvested fruit, and had very limited odor-active volatile loss after chilling and blanching treatments (Table 2, Figure 6).

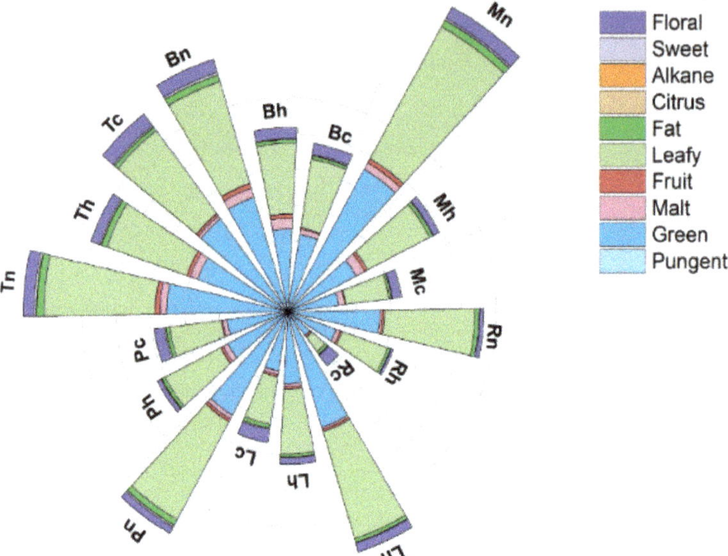

Figure 6. Effects of different treatment combinations of harvest maturity and temperature on aromatic volatiles by reported flavor sensory property (see Table 2). Abbreviations represent combinations of harvest maturity (R—red; L—light red; P—pink; T—turning; B—breaker; and M—mature green) and temperature treatment (h—heating; c—chilling; n—non-treated control).

Table 2. Odor activity values (OAVs) of active volatile compounds determined in the different treated tomatoes harvested at six maturity stages and treated by refrigeration (chilling, c), blanching (heating, h), or non-treated (control, n).

	Compounds	Odor Threshold in Water (mg L^{-1}) [Z]	Odor Description [Y]	Treatment (Combination of Harvest Maturity and Temperature Treatment)																	
				Rn [x]	Rh	Rc	Ln	Lh	Lc	Pn	Ph	Pc	Tn	Th	Tc	Bn	Bh	Bc	Mn	Mh	Mc
	ALDEHYDES																				
1	Butanal	0.0090	Pungent, green	1.0	1.0	2.7	0.9	1.0	2.1	1.7	1.8	2.2	2.0	2.0	1.7	1.8	1.9	2.1	1.1	1.3	2.9
2	3-Methylbutanal	0.0002	Malt	753.3	1320.0	606.7	920.0	1093.3	1173.3	1966.7	2193.3	1200.0	2720.0	3720.0	1353.3	4013.3	4166.7	1780.0	3066.7	4126.7	2293.3
3	2-Methylbutanal	0.0030	Cocoa, almond, malt	27.0	48.7	50.7	12.7	35.3	88.0	130.3	149.0	101.7	161.0	179.7	70.7	178.0	165.0	101.3	131.7	140.3	135.0
4	trans-Pentenal-1-al	0.0892	Strawberry, fruit, tomato	1.5	1.2	0.3	2.4	1.6	1.0	1.5	0.9	0.9	2.5	2.4	2.6	2.7	1.1	0.9	3.5	1.1	0.5
5	cis-3-Hexenal	0.0003	Leafy, green	95,332.0	50,724.0	17,940.0	115,356.0	66,452.0	47,724.0	111,864.0	65,076.0	53,712.0	115,128.0	86,992.0	106,112.0	104,068.0	71,404.0	71,972.0	151,556.0	72,716.0	45,708.0
6	Hexanal	0.0045	Grass, tallow, fat	247.3	174.2	198.2	232.2	180.7	238.9	380.2	226.0	225.1	438.4	345.1	239.8	421.3	214.4	172.0	454.4	214.2	148.9
7	trans-2-Hexenal	0.0170	Green, leafy	120.9	78.8	29.6	143.9	93.4	90.6	196.5	109.0	103.9	207.9	158.3	154.4	226.6	117.6	85.3	259.6	113.9	67.2
8	Heptanal	0.0030	Fat, citrus, rancid	2.7	3.0	3.7	2.3	1.0	5.0	4.7	3.3	3.3	5.7	4.7	4.7	5.3	2.7	3.3	8.3	3.3	3.0
9	Octanal	0.0007	Fat, soap, lemon, green	1.4	2.9	1.4	1.4	0.0	2.9	0.0	1.4	5.7	0.0	1.4	1.4	0.0	#VALUE!	1.4	1.4	0.0	0.0
10	2-phenylacetaldehyde	0.0063	Hawthorne, sweet, honey	58.1	59.5	140.8	70.5	47.0	147.6	211.7	107.0	79.2	178.6	207.9	182.2	136.3	80.0	137.3	276.5	134.0	69.4
11	2-Octenal	0.0030	Green, nut, fat	10.7	12.7	0.0	12.3	1.0	0.0	11.3	0.0	0.0	0.0	4.7	0.0	0.0	0.0	0.0	11.3	0.0	0.0
	HYDROCARBONS																				
1	Limonene	0.0100	Lemon, orange	2.4	0.5	6.8	0.9	1.2	7.7	3.4	0.0	0.0	5.0	11.6	34.2	6.2	4.4	33.0	8.4	12.3	28.1
	ALCOHOLS																				
1	3-Methylbutanol	0.2500	Whiskey, malt, burnt	1.5	2.6	0.3	1.3	1.5	0.9	0.1	0.0	0.0	2.4	3.1	1.2	3.7	4.8	1.5	3.1	4.4	1.4
2	2-Methylbutanol	0.2500	Malt, wine, onion	0.8	1.1	0.3	0.6	1.0	1.7	0.1	0.0	0.0	2.0	2.0	1.3	2.4	2.8	2.3	1.9	2.4	3.0
	KETONES																				
1	1-Penten-3-one	0.0015	Fruity, floral, green	131.3	82.7	47.3	168.7	154.7	114.7	22.7	0.0	0.0	162.7	135.3	153.3	176.7	116.7	5.3	170.0	97.3	3.3
2	6-Methyl-5-hepten-2-one	0.0500	Fruity, floral	1.7	1.6	5.6	2.1	1.6	6.6	0.7	0.0	0.0	5.1	4.1	6.1	4.3	3.9	5.7	6.3	3.8	4.9
3	Geranyl acetone	0.0600	Sweet, floral, estery	5.8	1.4	4.9	3.9	3.4	5.5	0.6	0.0	0.0	11.3	6.7	8.2	12.6	4.5	6.8	9.9	3.9	3.8
	OXYGEN-CONTAINING HETEROCYCLIC COMPOUNDS																				
1	2-Pentyl- furan	0.0058		9.3	9.7	46.0	9.8	5.9	44.1	5.9	0.0	0.0	11.0	7.8	51.4	25.0	2.8	68.1	22.9	13.8	69.1
	SULFUR- AND NITROGEN-CONTAINING HETEROCYCLIC COMPOUNDS																				
1	2-Isobutylthiazole	0.0035	Tomato leafy, green	0.6	0.9	0.0	0.3	0.3	0.6	9.7	0.0	0.0	1.1	1.7	1.4	1.1	0.6	1.1	0.9	0.3	0.9

[z]: Odor description of 6-methyl-5-hepten-2-one adapted from Klee [36], while others from Acree and Arn [44]. [y]: Odor threshold values (detection) in water adapted from Van Gemert [45]. When there was more than one value, those from the USDA, ARS laboratory in Albany (CA) were chosen for consistency of the methodology. [x]: Abbreviations of combination of harvest maturity (R—red; L—light red; P—pink; T—turning; B—breaker; and M—mature green) and temperature treatment (h—heating; c—chilling; n—non-treated control).

4. Conclusions

The concentration of volatile compounds and OAVs decreased along with advanced maturity in ripened "Tasti-Lee" tomatoes harvested after the turning or later harvest stages. The phenomenon contradicts the traditional understanding that the closer to full ripe at harvest, the better the fruit flavor quality and generally the more abundant the volatile concentrations. Both refrigeration and blanching applied when fruit reached full ripeness (red stage), an approach often used by consumers, substantially reduced the fatty acid derivatives, the dominant volatiles, and the total volatiles, and chilling resulted in greater suppression than blanching. However, the volatiles derived from the amino acid, carotenoid, and phenylalanine pathways showed variable changes related to harvest maturities and chilling/heating treatments. The responses of fruit to refrigeration were based on relatively long-term (4 days) physiological mechanisms and CI, however, responses to blanching were immediate and primarily from the evaporation/partition of the compounds with low Henry's law coefficient to the atmosphere, such as aldehydes and ketones. Fruit harvested at the turning stage had the highest tolerance to chilling and blanching treatments. This evidence provided in this study offers valuable insight into optimum harvest maturity. Mature green fruit were not able to develop maximum flavor quality based on aroma volatiles. Despite mature green fruit have the most volatile concentration, most of it was "leafy" and "green" cis-3-hexenal. On the other hand, the turning stage, the more advanced maturity, not only developed the best flavor quality (based on volatile profiles), but was also firm enough to resist compression and tough enough to tolerate environmental stress. The assumption that "vine-ripe fruit has the best quality" may mislead growers and consumers, because it is not only fragile to postharvest handling, but also does not have the best flavor quality. Post-ripening refrigeration and blanching both substantially suppress tomato flavor quality at the expense of extending holding time, removing skin and providing sanitation.

Supplementary Materials: The following are available online at https://www.mdpi.com/article/10.3390/foods10081727/s1, Figure S1: Effects of treatment combinations of harvest maturity and temperature on volatile concentration in tomatoes by chemical class: (A) aldehydes, (B) hydrocarbons, (C) alcohols, (D) ketones, (E) oxygen-containing heterocyclic compounds, (F) esters, (G) sulfur- and nitrogen-containing heterocyclic compounds, and (H) sulfur compounds. Each value is the mean of five replicates. Vertical bars represent standard deviation of the mean. The effect of the "maturity × treatment" interaction was tested by two-way ANOVA ($p < 0.05$). Different letters on the top of columns represent significantly differences between treatments using the Duncan's multiple range test ($p < 0.05$). Figure S2: Principal component analysis (PCA) results based on the 13 key volatile substances in "Tasti-Lee" tomatoes with different temperature treatments by six harvest maturities: (A) red, (B) light red, (C) pink, (D) turning, (E) breaker, and (F) mature green. Abbreviations represent combinations of harvest maturity (R—red; L—light red; P—pink; T—turning; B—breaker; and M—mature green) and temperature treatment (h—heating; c—chilling; n—non-treated control). Figure S3: Total volatile concentrations of all treatment combinations of harvest maturity and temperature treatment of tomatoes by synthesis pathways. Abbreviations represent combinations of harvest maturity (R—red; L—light red; P—pink; T—turning; B—breaker; and M—mature green) and temperature treatment (h—heating; c—chilling; n—non-treated control). Table S1: Relative concentration (%) of each chemical class determined in tomatoes harvested at six maturity stages and treated by refrigeration (chilling, c), blanching (heating, h), or non-treated control.

Author Contributions: Conceptualization, J.L. and J.B.; methodology, J.B.; software, Y.X.; validation, Q.L. and J.Y.; formal analysis, J.B.; investigation, J.L. and J.Y.; resources, E.R., J.C.H., and J.B.; data curation: J.L. and Y.X.; writing—original draft preparation, J.L. and Y.X.; writing—review and editing, E.B., A.P., E.R., J.Z., and J.B.; visualization, Y.X.; supervision, J.L. and J.B.; project administration, J.L. and J.B.; funding acquisition, J.L. All authors have read and agreed to the published version of the manuscript.

Funding: This research was supported by the National Natural Science Foundation of China (31772038).

Data Availability Statement: The data presented in this study are available in article and Supplementary Materials.

Conflicts of Interest: The authors declare no conflict of interest.

References

1. Davis, W.; Lucier, G. *Vegetable and Pulses Outlook: April 2021*; VGS-366; United States Dpartment of Agriculture, Economic Research Service: Washington, DC, USA, 2021. Available online: https://www.ers.usda.gov/webdocs/outlooks/100969/vgs-366.pdf?v=6983.5 (accessed on 2 June 2021).
2. Salehi, B.; Sharifi-Rad, R.; Sharopov, F.; Namiesnik, J.; Roointan, A.; Kamle, M.; Kumar, P.; Martins, N.; Sharifi-Rad, J. Beneficial effects and potential risks of tomato consumption for human health: An overview. *Nutrition* **2019**, *62*, 201–208. [CrossRef] [PubMed]
3. Buttery, R. Quantitative and sensory aspects of flavor of tomato and other vegertables and fruits. In *Flavor Science: Sensible Principles and Techniques*; American Chemical Society: Washington, DC, USA, 1993; pp. 259–286.
4. Rambla, J.L.; Tikunov, Y.M.; Monforte, A.J.; Bovy, A.G.; Granell, A. The expanded tomato fruit volatile landscape. *J. Exp. Bot.* **2014**, *65*, 4613–4623. [CrossRef]
5. Wang, L.; Baldwin, E.A.; Bai, J. Recent Advance in Aromatic Volatile Research in Tomato Fruit: The Metabolisms and Regulations. *Food Bioprocess Technol.* **2015**, *9*, 203–216. [CrossRef]
6. Sanz, C.; Olias, J.M.; Perez, A. Aroma biochemistry of fruits and vegetables. In *Proceedings-Phytochemical Society of Europe*; oxford University Press Inc.: New York, NY, USA, 1996; pp. 125–156.
7. Tieman, D.; Zhu, G.; Resende, M.F.; Lin, T.; Nguyen, C.; Bies, D.; Rambla, J.L.; Beltran, K.S.O.; Taylor, M.; Zhang, B. A chemical genetic roadmap to improved tomato flavor. *Science* **2017**, *355*, 391–394. [CrossRef]
8. Baldwin, E.; Plotto, A.; Narciso, J.; Bai, J. Effect of 1-methylcyclopropene on tomato flavour components, shelf life and decay as influenced by harvest maturity and storage temperature. *J. Sci. Food Agric.* **2011**, *91*, 969–980. [CrossRef] [PubMed]
9. Spricigo, P.C.; Freitas, T.P.; Purgatto, E.; Ferreira, M.D.; Correa, D.S.; Bai, J.; Brecht, J.K. Visually imperceptible mechanical damage of harvested tomatoes changes ethylene production, color, enzyme activity, and volatile compounds profile. *Postharvest Biol. Technol.* **2021**, *176*, 111503. [CrossRef]
10. Baldwin, E.; Nisperos-Carriedo, M.; Moshonas, M. Quantitative analysis of flavor and other volatiles and for certain constituents of two tomato cultivars during ripening. *J. Am. Soc. Hortic. Sci.* **1991**, *116*, 265–269. [CrossRef]
11. Klee, H.J.; Giovannoni, J.J. Genetics and Control of Tomato Fruit Ripening and Quality Attributes. *Annu. Rev. Genet.* **2011**, *45*, 41–59. [CrossRef] [PubMed]
12. Algazi, M. Effect of fruit ripeness when picked on flavor and composition in fresh market tomatoes. *J. Am. Soc. Hort. Sci.* **1977**, *102*, 724–731.
13. Xu, S.; Li, J.; Baldwin, E.A.; Plotto, A.; Rosskopf, E.; Hong, J.C.; Bai, J. Electronic tongue discrimination of four tomato cultivars harvested at six maturities and exposed to blanching and refrigeration treatments. *Postharvest Biol. Technol.* **2018**, *136*, 42–49. [CrossRef]
14. ScoTT, J. Interrelationship of sensory descriptors and chemical composition as affected by harvest maturity and season on fresh tomato flavor. *Proc. Fla. Statg Hort. Soc. I* **2000**, *13*, 289–294.
15. Baldwin, E.A.; Scott, J.W.; Bai, J. Sensory and chemical flavor analyses of tomato genotypes grown in Florida during three different growing seasons in multiple years. *J. Am. Soc. Hortic. Sci.* **2015**, *140*, 490–503. [CrossRef]
16. Loayza, F.E.; Brecht, J.K.; Simonne, A.H.; Plotto, A.; Baldwin, E.A.; Bai, J.; Lon-Kan, E. A brief hot-water treatment alleviates chilling injury symptoms in fresh tomatoes. *J. Sci. Food Agric.* **2021**, *101*, 54–64. [CrossRef]
17. Wang, L.; Baldwin, E.A.; Zhao, W.; Plotto, A.; Sun, X.; Wang, Z.; Brecht, J.K.; Bai, J.; Yu, Z. Suppression of volatile production in tomato fruit exposed to chilling temperature and alleviation of chilling injury by a pre-chilling heat treatment. *LWT Food Sci. Technol.* **2015**, *62*, 115–121. [CrossRef]
18. Bai, J.; Baldwin, E.A.; Imahori, Y.; Kostenyuk, I.; Burns, J.; Brecht, J.K. Chilling and heating may regulate C6 volatile aroma production by different mechanisms in tomato (*Solanum lycopersicum*) fruit. *Postharvest Biol. Technol.* **2011**, *60*, 111–120. [CrossRef]
19. Castro, S.M.; Saraiva, J.A.; Lopes-da-Silva, J.A.; Delgadillo, I.; Van Loey, A.; Smout, C.; Hendrickx, M. Effect of thermal blanching and of high pressure treatments on sweet green and red bell pepper fruits (*Capsicum annuum* L.). *Food Chem.* **2008**, *107*, 1436–1449. [CrossRef]
20. Maxin, P.; Weber, R.; Lindhard Pedersen, H.; Williams, M. Hot-water dipping of apples to control Penicillium expansum, Neonectria galligena and Botrytis cinerea: Effects of temperature on spore germination and fruit rots. *Eur. J. Hortic. Sci.* **2012**, *77*, 1.
21. Spadoni, A.; Guidarelli, M.; Phillips, J.; Mari, M.; Wisniewski, M. Transcriptional profiling of apple fruit in response to heat treatment: Involvement of a defense response during Penicillium expansum infection. *Postharvest Biol. Technol.* **2015**, *101*, 37–48. [CrossRef]
22. Bai, J.; Baldwin, E.A.; Fortuny, R.C.S.; Mattheis, J.P.; Stanley, R.; Perera, C.; Brecht, J.K. Effect of Pretreatment of IntactGala'Apple with Ethanol Vapor, Heat, or 1-Methylcyclopropene on Quality and Shelf Life of Fresh-cut Slices. *J. Am. Soc. Hortic. Sci.* **2004**, *129*, 583–593. [CrossRef]
23. Lurie, S. Postharvest heat treatments. *Postharvest Biol. Technol.* **1998**, *14*, 257–269. [CrossRef]

24. Buttery, R.G.; Teranishi, R.; Ling, L.C.; Flath, R.A.; Stern, D.J. Quantitative studies on origins of fresh tomato aroma volatiles. *J. Agric. Food Chem.* **1988**, *36*, 1247–1250. [CrossRef]
25. Wang, L.; Baldwin, E.A.; Plotto, A.; Luo, W.; Raithore, S.; Yu, Z.; Bai, J. Effect of methyl salicylate and methyl jasmonate pre-treatment on the volatile profile in tomato fruit subjected to chilling temperature. *Postharvest Biol. Technol.* **2015**, *108*, 28–38. [CrossRef]
26. Du, X.; Song, M.; Baldwin, E.; Rouseff, R. Identification of sulphur volatiles and GC-olfactometry aroma profiling in two fresh tomato cultivars. *Food Chem.* **2015**, *171*, 306–314. [CrossRef] [PubMed]
27. Petro-Turza, M. Flavor of tomato and tomato products. *Food Rev. Int.* **1986**, *2*, 309–351. [CrossRef]
28. Baldwin, E.A.; Scott, J.W.; Shewmaker, C.K.; Schuch, W. Flavor trivia and tomato aroma: Biochemistry and possible mechanisms for control of important aroma components. *HortScience* **2000**, *35*, 1013–1022. [CrossRef]
29. Beaulieu, J.C.; Grimm, C.C. Identification of volatile compounds in cantaloupe at various developmental stages using solid phase microextraction. *J. Agric. Food Chem.* **2001**, *49*, 1345–1352. [CrossRef] [PubMed]
30. Thorne, S.; Alvarez, J.S.S. The effect of irregular storage temperatures on firmness and surface colour in tomatoes. *J. Sci. Food Agric.* **1982**, *33*, 671–676. [CrossRef]
31. Schrader, L.E. Scientific basis of a unique formulation for reducing sunburn of fruits. *HortScience* **2011**, *46*, 6–11. [CrossRef]
32. Bai, J.; Hagenmaier, R.D.; Baldwin, E.A. Volatile response of four apple varieties with different coatings during marketing at room temperature. *J. Agric. Food Chem.* **2002**, *50*, 7660–7668. [CrossRef] [PubMed]
33. Buttery, R.G.; Teranishi, R.; Ling, L.C.; Turnbaugh, J.G. Quantitative and sensory studies on tomato paste volatiles. *J. Agric. Food Chem.* **1990**, *38*, 336–340. [CrossRef]
34. Canoles, M.; Soto, M.; Beaudry, R. Hydroperoxide lyase activity necessary for normal aroma volatile biosynthesis of tomato fruit, impacting sensory perception and preference. *HortScience* **2005**, *40*, 1130E–1131E. [CrossRef]
35. Mathieu, S.; Cin, V.D.; Fei, Z.; Li, H.; Bliss, P.; Taylor, M.G.; Klee, H.J.; Tieman, D.M. Flavour compounds in tomato fruits: Identification of loci and potential pathways affecting volatile composition. *J. Exp. Bot.* **2009**, *60*, 325–337. [CrossRef] [PubMed]
36. Klee, H.J. Improving the flavor of fresh fruits: Genomics, biochemistry, and biotechnology. *New Phytol.* **2010**, *187*, 44–56. [CrossRef]
37. Renard, C.M.; Ginies, C.; Gouble, B.; Bureau, S.; Causse, M. Home conservation strategies for tomato (*Solanum lycopersicum*): Storage temperature vs. duration–Is there a compromise for better aroma preservation? *Food Chem.* **2013**, *139*, 825–836. [CrossRef]
38. Yang, X.; Song, J.; Fillmore, S.; Pang, X.; Zhang, Z. Effect of high temperature on color, chlorophyll fluorescence and volatile biosynthesis in green-ripe banana fruit. *Postharvest Biol. Technol.* **2011**, *62*, 246–257. [CrossRef]
39. Simkin, A.J.; Schwartz, S.H.; Auldridge, M.; Taylor, M.G.; Klee, H.J. The tomato carotenoid cleavage dioxygenase 1 genes contribute to the formation of the flavor volatiles β-ionone, pseudoionone, and geranylacetone. *Plant J.* **2004**, *40*, 882–892. [CrossRef] [PubMed]
40. Tieman, D.; Taylor, M.; Schauer, N.; Fernie, A.R.; Hanson, A.D.; Klee, H.J. Tomato aromatic amino acid decarboxylases participate in synthesis of the flavor volatiles 2-phenylethanol and 2-phenylacetaldehyde. *Proc. Natl. Acad. Sci. USA* **2006**, *103*, 8287–8292. [CrossRef]
41. Zou, J.; Chen, J.; Tang, N.; Gao, Y.; Hong, M.; Wei, W.; Cao, H.; Jian, W.; Li, N.; Deng, W. Transcriptome analysis of aroma volatile metabolism change in tomato (*Solanum lycopersicum*) fruit under different storage temperatures and 1-MCP treatment. *Postharvest Biol. Technol.* **2018**, *135*, 57–67. [CrossRef]
42. Grosch, W. Evaluation of the key odorants of foods by dilution experiments, aroma models and omission. *Chem. Senses* **2001**, *26*, 533–545. [CrossRef]
43. Genovese, A.; Lamorte, S.A.; Gambuti, A.; Moio, L. Aroma of Aglianico and Uva di Troia grapes by aromatic series. *Food Res. Int.* **2013**, *53*, 15–23. [CrossRef]
44. Acree, T.; Heinrich, A. Flavornet and Human Odor Space, Gas Chromatography-Olfactometry (GCO) of Natural Products. Accessed at the website. Cornell University. 2004. Available online: http://www.flavornet.org/flavornet.html (accessed on 14 April 2021).
45. Van Gemert, L. *Odour Thresholds-Compilations of Odour Thresholds in Air, Water and Other Media*; Oliemans Punter & Partners BV: Utrecht, The Netherlands, 2003.

Article

Low Oxygen Storage Improves Tomato Postharvest Cold Tolerance, Especially for Tomatoes Cultivated with Far-Red LED Light

Fahrizal Yusuf Affandi [1,2,*], Jan A. Verschoor [3], Maxence J. M. Paillart [3], Julian C. Verdonk [1], Ernst J. Woltering [1,3] and Rob E. Schouten [1]

1 Horticulture and Product Physiology, Wageningen University and Research, P.O. Box 16, 6700 AA Wageningen, The Netherlands; julian.verdonk@wur.nl (J.C.V.); ernst.woltering@wur.nl (E.J.W.); Rob.schouten@wur.nl (R.E.S.)
2 Bioresource Technology and Veterinary Department, Vocational College, Universitas Gadjah Mada, Yogyakarta 55281, Indonesia
3 Food & Biobased Research, Wageningen University and Research, P.O. Box 17, 6700 AA Wageningen, The Netherlands; jan.verschoor@wur.nl (J.A.V.); maxence.paillart@wur.nl (M.J.M.P.)
* Correspondence: fahrizal.affandi@wur.nl

Abstract: We investigated the effects of low oxygen storage on chilling injury development, colour development, respiration and H_2O_2 levels of 'Merlice' tomatoes cultivated with and without far red (FR) LED lighting during 20 days of shelf-life. Mature green (MG) and red (R) tomatoes were stored at 2 °C in combination with 0.5, 2.5, 5 and 21 kPa O_2 for 15 days (experiment 1). MG tomatoes cultivated under either white LED or white LED light with FR LED light were stored at 2 °C in combination with 1, 5 and 21 O_2 kPa for 14 days (experiment 2). Chilled MG and R tomatoes from experiment 1 showed decay, firmness loss and higher weight loss during shelf-life which were reduced under low oxygen conditions. FR during cultivation improved chilling tolerance of MG tomatoes. Fastest colour development and lowest respiration rate during shelf-life were observed for MG fruit cultivated with FR lighting prior to storage at 1 kPa O_2/0 kPa CO_2. H_2O_2 levels during the shelf-life were not affected during cold storage. The improved cold tolerance of MG tomatoes cultivated with FR lighting is likely due to lower oxygen uptake that led to both higher lycopene synthesis and less softening.

Keywords: chilling injury; controlled atmosphere; far-red

Citation: Affandi, F.Y.; Verschoor, J.A.; Paillart, M.J.M.; Verdonk, J.C.; Woltering, E.J.; Schouten, R.E. Low Oxygen Storage Improves Tomato Postharvest Cold Tolerance, Especially for Tomatoes Cultivated with Far-Red LED Light. *Foods* **2021**, *10*, 1699. https://doi.org/10.3390/foods10081699

Academic Editors: Eleni Tsantili and Jinhe Bai

Received: 14 June 2021
Accepted: 13 July 2021
Published: 22 July 2021

1. Introduction

Tomato (*Solanum lycopersicum*) is a chilling sensitive fruit that will develop a disorder called chilling injury (CI) when exposed to low, but above freezing temperatures [1]. Chilling stress disrupts metabolic processes and causes alterations in membrane fluidity, followed by an increase in reactive oxygen species (ROS) production. In addition, low enzymatic activity causes reduced ROS scavenging, which promotes development of CI symptoms [2–4]. CI symptoms in tomatoes include surface pitting, interrupted pigment (lycopene) synthesis, rapid softening, loss of aroma and production of off-flavours, as well as increased susceptibility to fungal infection [5,6]. CI symptoms usually become visible during a shelf-life period after fruits have been exposure to chilling temperatures [5–7].

Controlled atmosphere (CA) storage and Modified Atmosphere Packaging (MAP) have been shown to reduce CI in mango, Japanese plum, guava, avocado and persimmon [8–13]. Low oxygen reduces respiration rate, and in addition, it may decrease ethylene production and ethylene sensitivity. CA storage downregulated the expression of ACC-synthase and ACC-oxidase genes, responsible for ethylene synthesis [14]. It may also limit ROS production, which could alleviate chilling injury symptoms [10,15,16]. CA storage induced activation of antioxidant scavenger enzymes such as catalase (CAT), superoxide dismutase

(SOD), ascorbate peroxidase (APX) and glutathione reductase (GR) in Japanese plum, apple and litchi [10,17,18], reducing ROS, often represented by lower hydrogen peroxide (H_2O_2) levels. H_2O_2 is both a toxic metabolite and signaling molecule [19,20]. Storage under CA slowed down the activities of cell wall degrading enzymes involved in lignification and softening [21,22]. In addition, low oxygen storage stabilised group VII of ethylene response factors (ERFVIIs) and transported these to the nucleus which induced expression of hypoxia-responsive genes. Hypoxia-responsive genes encode enzymes involved in sucrose catabolism (β-amylase, sucrose synthase and phosphofructokinase), fermentative metabolism (pyruvate decarboxylase, lactate dehydrogenase and alcohol dehydrogenase) and ROS scavenging (SOD, APX and CAT) [23–25].

The severity of CI symptoms depends on the ripening stage of the fruits; mature green (MG) tomatoes are more sensitive to CI than red (R) tomatoes [2]. Comparing the responses of R and MG fruit to chilling stress is expected to provide insights into the mechanism of how low oxygen alleviates CI in sensitive tomatoes [26–28]. We showed that addition of far-red (FR) lighting during cultivation alleviated CI in tomato. In MG fruit, additional FR lighting reduced weight loss, pitting and enhanced red colour development during shelf-life after prior cold storage. R fruit cultivated with additional far-red light were firmer at harvest and demonstrated reduced weight loss and less decay during shelf-life after prior cold storage [29]. In the current study we investigated the effect of varying low oxygen levels on CI occurrence in mature green (MG) and red (R) tomatoes during postharvest storage. In addition, we investigated the effect of FR lighting during cultivation on CI tolerance after prior low oxygen storage.

2. Materials and Methods

We carried out two experiments. In experiment 1, mature green (MG) and red (R) tomatoes were stored for 15 days at 2 °C either under regular atmosphere (21 kPa O_2, RA) or under 0.5, 2.5 and 5 kPa O_2, followed by a shelf-life period of 15 days at 20 °C. In experiment 2, MG tomatoes cultivated with or without FR were harvested and stored either under RA or under 1 and 5 kPa O_2 followed by a shelf-life period of 15–20 days at 20 °C. In both experiments, decay index, colour and firmness, respiration rate and hydrogen peroxide (H_2O_2) level were determined at harvest, during cold storage and during subsequent shelf-life.

2.1. *Plant Material and Growth Conditions*

For the first experiment, mature green (MG) and red (R) 'Merlice' tomatoes were harvested from a commercial greenhouse in Bleiswijk, the Netherlands in November 2016. The colour stage of the fruit was assessed using the NAI index (see Section 2.5). MG tomatoes were defined as tomatoes with a NAI value between −0.77 and −0.6 at harvest. R tomatoes were defined as having NAI values between 0.25 and 0.55 at harvest. For the second experiment, MG 'Merlice' tomatoes were harvested from a greenhouse at Wageningen University in May 2019 of plants grown under white LED lighting (WL) or WL with 8.3 μmol m^{-2} s^{-1} FR lighting, with a peak at 730 nm. For the FR treatment, 6% of the photons in the red region were replaced with FR. This resulted in 13 μmol m^{-2} s^{-1} FR in the FR treatment and hence this treatment was called WL + 13FR and the photon flux density was kept constant at 215 μmol m^{-2} s^{-1}. The greenhouse compartment was divided into four plots. The light intensity was 215 μmol m^{-2} s^{-1} at the top of the canopy. In this experiment, VYPRx PLUS modules (Fluence, TX, USA) were used as top lighting. For each of the plots there were six modules installed. Overhead lamps were switched on 16 h before sunset and switched off at sunset. Additionally, LED lighting was automatically switched off when the incoming sunlight exceeded 300 μmol m^{-2} s^{-1}. The spectral composition of the light treatments is shown in Figure S1. Light treatments were separated by double sided, non-transparent, white reflective plastic sheets. At harvest, uniform MG fruits were selected with a NAI value between −0.77 and −0.6. Further greenhouse management (fertigation, pollination) was conducted according to standard commercial practice.

2.2. Experimental Setup

In experiment 1, MG and R tomatoes were randomly assigned into five tomatoes per maturity per CA treatment at harvest, at the end of CA storage for 15 days at 2 °C and during subsequent shelf-life at 5, 10 and 15 days. This amounts to 125 MG and 125 R tomatoes. At harvest, colour and firmness was measured for all tomatoes. At each sampling point, colour, firmness and CI indices measurements were carried out. In experiment 2, the effect of far red illumination at harvest was characterised by randomly selecting 40 MG tomatoes per light treatment. Eight tomatoes per light treatment per CA treatment were assigned as a replicate of four tomatoes for repeated non-destructive measurement at harvest, after 7 and 14 days of CA storage, and after 4, 7, 10, 14 and 21 days of subsequent shelf-life. Prior to sampling during at 7 days CA storage, the CA was stop and tomatoes were taken out to be analysed. Eight randomly assigned tomatoes per light treatment and per CA treatment were taken for destructive analysis at 7 and 14 days of CA storage and after 7, 14 and 21 days of shelf-life. In total 240 FR and 240 non-FR cultivated MG tomatoes were selected for this experiment.

Tomatoes were individually marked on three positions on the equator for repeated colour and firmness measurements during shelf-life. In addition, fresh weight and three chilling indices were assessed approximately every 3 days during shelf-life. Individual fruits, assigned for destructive measurements, were cut into small pieces and quickly frozen in liquid nitrogen and later ground into a fine powder for H_2O_2 measurements.

2.3. CA Storage Conditions

Tomatoes were stored at 2 °C and 95% relative humidity (RH) under low oxygen conditions followed by subsequent shelf-life at 20 °C in darkness. Desired oxygen conditions were achieved by flushing humidified gas mixtures at a flow rate of 500 mL min^{-1} through 70 L stainless steel containers filled with tomatoes with an average weight of 5.15 $\pm$ 0.25 kg per container. In both experiments, tomatoes stored at RA and 2 °C served as low oxygen control whereas tomatoes stored at 12 °C and 95% RH under RA served as temperature control. All control treatment were carried out in identical containers and flow rate with the low oxygen treatments.

In experiment 1, MG and R tomatoes were subjected to low oxygen conditions of 0.5 kPa, 2.5 and 5 kPa O_2 combined with 0 kPa CO_2 (completed with balanced N_2) for 15 days. Following cold storage, fruit were transferred to shelf-life conditions at 20 °C and 85% RH for 15 days. In experiment 2, MG tomatoes were subjected to low oxygen storage at 1 and 5 kPa O_2 with 0 kPa CO_2 (completed with balanced N_2). During CA storage, respiration measurements were conducted. After 14 days of cold storage, tomatoes were exposed to shelf-life condition at 20 °C and 95% RH for 21 days.

2.4. Respiration Measurements

In experiment 2, respiration measurements were carried out according to method previously described by our group [30]. Analysis was carried out using an Interscience Compact GC system (Interscience, Breda, NL, USA) equipped with an RT-QBond column for detecting CO_2 at the back channel and a MolSieve 5A coupled with a back pressure column type RT-QBond for the detection of O_2 at the front channel. Helium with a constant pressure of 60 and 80 kPa was used as carrier gas for the back and front channel, respectively. Each column was connected to a Thermal Conductivity Detector (TCD) set at 110 °C. CGCeditor software (v1.5.5, 2008) was used to control the setting of the CompactGC. GC was continuously connected to the samples via tubing connected to a VICI valve (model EMTMA-CE). Valve and CompactGC were coordinated by EZChrom Elite software (v3.32 SP2).

Gas measurement were conducted directly from the container. Before measurement took place, the flow through the container was stopped to allow accumulation of CO_2 and depletion of O_2 and the first GC measurement was carried out. The second measurement was carried out at the end of the incubation period. The accumulation period was approximately 5 h. The difference in gas partial pressure between the first and second GC measurements was converted into consumption and production rates according to ideal gas law methods [31]. The measurement was carried out at day 4, 6, 10 and 12 during CA storage.

2.5. Colour and Firmness Measurement

Colour was assessed non-destructively by a hand-held photodiode array spectrophotometer (Pigment Analyzer PA1101, CP, Ibbenbüren, Germany). Remittance was assessed at 570 (R570) and 780 (R780) nm by calculating the normalised different vegetative index (*NDVI*, Equation (1)) and normalised anthocyanin index (*NAI*, Equation (2)) which are normalised value between −1 and 1 [32].

$$NDVI = \frac{R_{780} - R_{660}}{R_{780} + R_{660}} \quad (1)$$

$$NAI = \frac{R_{780} - R_{570}}{R_{780} + R_{570}} \quad (2)$$

Firmness was measured non-destructively using a commercial acoustic firmness tester (AFS, AWETA, Nootdorp, the Netherlands) with the tick power of the plunger set to 15. The AFS combines the single tomato resonant frequency (f in Hz) and mass (m, in kg), measured by an inbuild balance, into a *FI* (firmness index) [33] (Equation (3)).

$$FI = \frac{f^2 m^{2/3}}{10^4} \quad (3)$$

2.6. Disorder Index and Weight Loss

CI was assessed by three indices, a pitting index and uneven ripening for MG fruit, and a decay index for R tomatoes according to the previously described method [29]. All indices were visually assessed with the percentage of the tomato surface assigned to five classes (0 = no injury, 1 = <10%, 2 = 11–25%, 3 = 26–40%, 4 = >40% affected area). The average score of pitting and uneven ripening index for MG, and decay index tomatoes were termed general disorder index. Tomato weight loss over time was expressed as the percentage weight loss (*WL*, in %) with W_0 the initial weight (in g) and W_t the weight (in g) according to Equation (4).

$$WL = \frac{W_0 - W_t}{W_0} \times 100 \quad (4)$$

2.7. Hydrogen Peroxide (H_2O_2) Measurement

H_2O_2 was quantified via a colorimetric method [34]. Briefly, a 300 mg sample of frozen and ground tissue per tomato was extracted in a solution containing of 0.75 mL 0.1% (w/v) trichloroacetic acid (TCA), 0.75 mL 10 mM phosphate buffer (pH 7) and 1.5 mL 1 M KI. The homogenate was centrifuged (15,000× g, 4 °C, 15 min) and the supernatant transferred to a new tube and allowed to sit at RT for 20 min before obtaining the absorbance at 390 nm using a Varian CARY 4000 spectrophotometer (Agilent, Santa Clara, CA, USA). Measured absorbances were converted into H_2O_2 concentrations using a calibration curve constructed with a commercial H_2O_2 solution (Sigma Aldrich, St. Louis, MO, USA).

2.8. Statistical Analysis

Data obtained during shelf-life were subjected to mixed ANOVA, applying SPSS ver.21 (SPSS, Chicago, IL, USA) at $p < 0.05$. Data from the first experiment were analysed by mixed ANOVA with oxygen level and maturity as between subject factors and days in storage as within subject factor. For the second experiment, mixed ANOVA was carried out with oxygen level and FR as between subject factor and days in storage as within subject factor. Normality of the variables was tested applying the Shapiro-Wilk test. Mauchly's test of sphericity was carried out to test whether variances of the differences between all possible pairs of within-subject conditions were equal. If the sphericity assumption was not fulfilled, Greenhouse-Geisser's correction was applied to calculate the degrees of freedom.

In case of a significant interaction, a pairwise comparison was carried out for each shelf-life day with LSD (Least Significant Difference) values estimated.

3. Results

3.1. Experiment 1: Effects of Low Oxygen Conditions on CI Indices, Weight- and Firmness Loss

In the first experiment, typical CI symptoms such as pitting, uneven colouring and decay were observed for both MG and R tomatoes during low oxygen storage and shelf-life. Storage at 0.5 kPa oxygen resulted in necrosis, fungal infection and rotting and were therefore omitted from this study. In MG tomatoes there were generally no visible CI symptoms observed during cold storage, except for tomatoes stored at 5 kPa O_2 (Figure 1A). During the shelf-life, fruit (MG and R), prior stored at 2.5 kPa O_2, showed the lowest, and RA the highest disorder (Figure 1). MG tomatoes from the temperature control (12 °C) also developed some pitting, comparable to the tomatoes stored at 2.5 kPa O_2. R tomatoes stored at 12 °C (temperature control) developed the least decay. At 2 °C, the R tomatoes stored at 2.5 kPa showed the least decay while the fruit stored at RA developed the highest disorder after 20 days of shelf-life which prevented further measurements. On the other hand, R tomatoes from the temperature control (21 kPa at 12 °C) developed the lowest decay ($p < 0.0001$). This indicated that the storage at 12 °C also resulted in a small amount of CI symptoms. In general, MG tomatoes developed slower pitting than R tomatoes, indicating that R tomatoes were surprisingly more sensitive to cold storage than MG tomatoes.

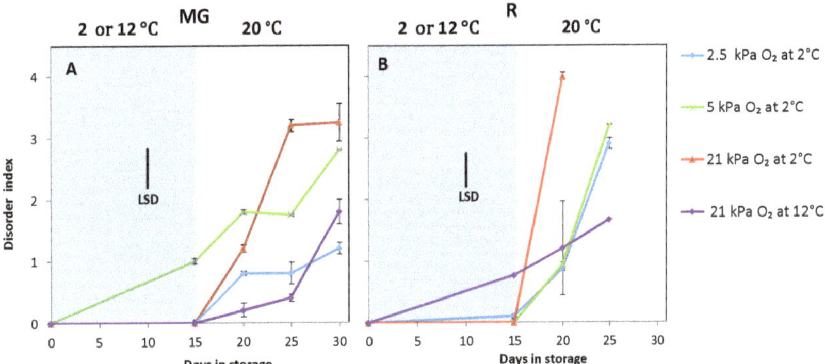

Figure 1. Chilling injury symptoms as indicated by the disorder index of MG (**A**) and R (**B**) tomatoes during cold storage at 2 or 12 °C (blue area) and subsequent shelf-life at 20 °C (white area). Blue, green, red and purple lines and symbols indicate 2.5, 5, 21 kPa O_2 (low oxygen control) applied during storage at 2 °C and 21 kPa O_2 at 12 °C (temperature control), respectively. The average decay index with indicated standard error is shown for five tomatoes. LSD values ($p < 0.05$) are indicated per panel. Disorder in MG fruit was determined by averaging the values of the pitting and uneven ripening index; disorder in R fruit was determined by the average decay incidence.

Weight loss was higher for MG compared to R tomatoes (Figure 2). Fruit stored at 12 °C showed highest weight loss. The lowest weight loss for both MG and R tomatoes was observed in fruit that had been stored at 2 °C and 2.5 kPa O_2 ($p < 0.005$). Fruit stored at 12 °C and stored at 2.5 or 5 kPa O_2 at 2 °C showed less softening compared to fruit stored at 2 °C and 21 kPa O_2 (Figure 3).

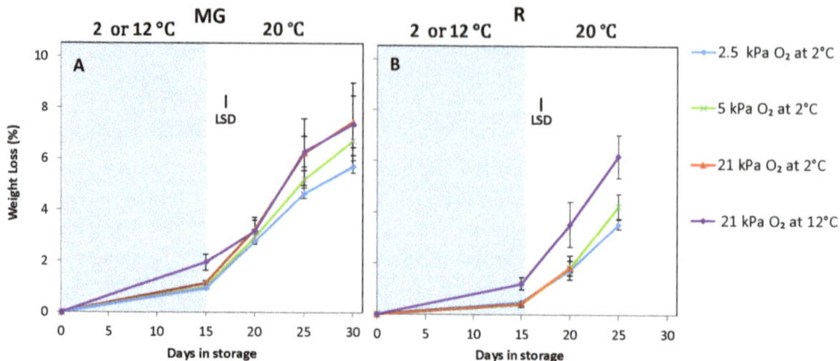

Figure 2. Weight loss of MG (**A**) and R (**B**) tomatoes during cold storage at 2 or 12 °C (blue area) and subsequent shelf-life at 20 °C (white area). Blue, green, red and purple lines and symbols indicate 2.5, 5, 21 kPa O_2 (low oxygen control) applied during storage at 2 °C and 21 kPa O_2 at 12 °C (temperature control), respectively. The average weight loss with indicated standard error is shown for five tomatoes. LSD values ($p < 0.05$) are indicated per panel.

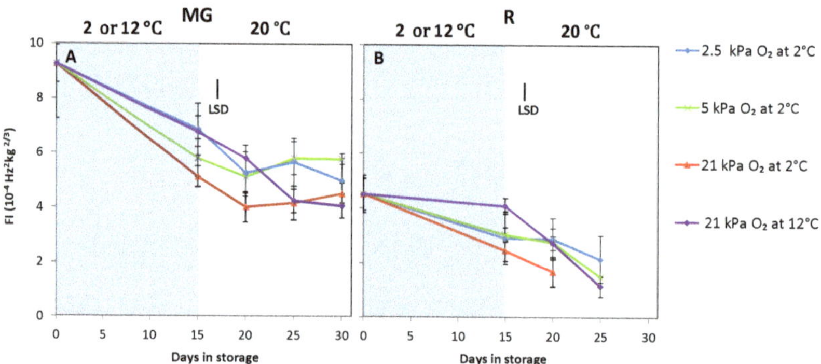

Figure 3. Firmness index of MG (**A**) and R (**B**) tomatoes during cold storage at 2 or 12 °C (blue area) and subsequent shelf-life at 20 °C (white area). Blue, green, red and purple lines and symbols indicate 2.5, 5, 21 kPa O_2 (low oxygen control) applied during storage at 2 °C and 21 kPa O_2 at 12 °C (temperature control), respectively. The average firmness index with indicated standard error is shown for five tomatoes. LSD values ($p < 0.05$) are indicated per panel.

3.2. Experiment 1: Effects of Low Oxygen Conditions on Tomato Colour Development

Red colour formation for MG fruit, as indicated by NAI values, was limited for all fruit that had been stored at 2 °C, independent of the oxygen level. Fruit stored at 12 °C showed colouration during subsequent shelf life at 20 °C. (Figure 4A). During low oxygen storage, more chlorophyll breakdown was observed with increasing oxygen levels. In R tomatoes, all treatments, except for the tomatoes in the temperature control, showed a reduction in the NAI values during cold storage. During shelf-life, fruit from all treatments showed increasing NAI values, except for the RA control (Figure 4B).

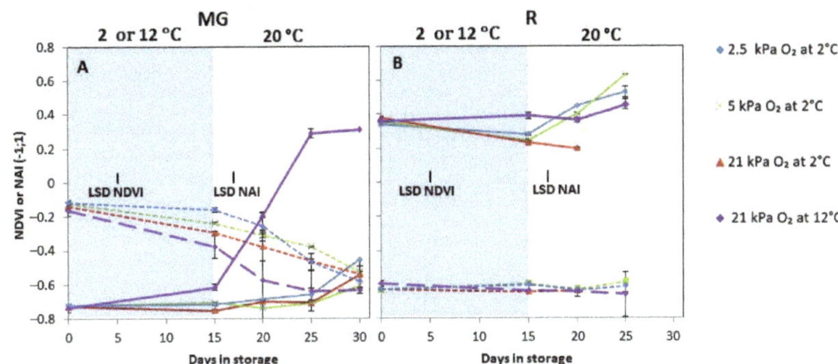

Figure 4. Colour as indicated by NDVI (dotted lines) and NAI (full lines) index of MG (**A**) and R (**B**) tomatoes during cold storage at 2 or 12 °C (blue area) and subsequent shelf-life at 20 °C (white area). Blue, green, red and purple lines and symbols indicate 2.5, 5, 21 kPa O_2 (low oxygen control) applied during storage at 2 °C and 21 kPa O_2 at 12 °C (temperature control), respectively. The NDVI or NAI with indicated standard error is shown for five individual tomatoes (repeated measure over times). LSD values ($p < 0.05$) are indicated per panel.

3.3. Experiment 2: Effects of Low Oxygen Storage of Mature Green Tomatoes Cultivated with and without Far Red Lighting on CI Symptoms, Weight- and Firmness Loss

Tomatoes cultivated without far red lighting during cultivation showed CI symptoms during shelf-life. The lowest pitting index was observed for MG tomatoes stored at 1 kPa O_2, the highest for the low oxygen control (Figure 5A). MG tomatoes cultivated with far-red lighting demonstrated reduced CI compared with tomatoes grown without FR lighting. In fact, no CI symptoms were observed for all low oxygen treatments, even after 3 weeks of shelf-life (Figure 5B). There were no chilling symptoms in fruit stored at 12 °C, and no differences were observed in terms of weight loss (Figure S2).

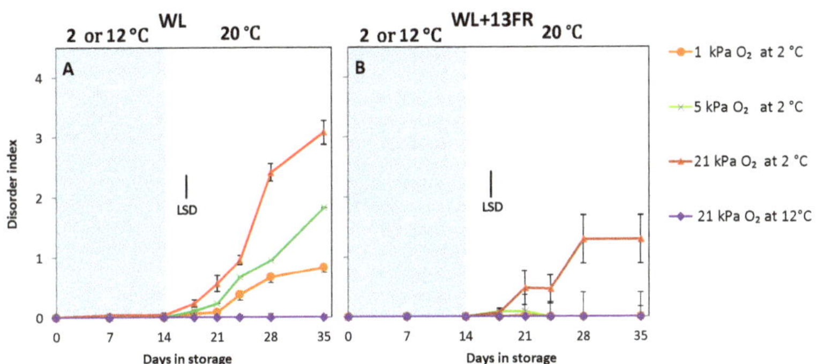

Figure 5. Chilling injury symptoms as indicated by the disorder index of MG tomatoes cultivated under white LED light (WL) without far red lighting (**A**) or with far red lighting during cultivation (**B**) during storage (blue area) at 2 or 12 °C and shelf-life at 20 °C (white area). Orange, green, red and purple lines and symbols indicate 1, 5 and 21 kPa O_2 (low oxygen control) applied during storage at 2 °C, and 21 kPa O_2 at 12 °C (temperature control), respectively. The average disorder index with indicated standard error is shown for two replicates of four tomatoes ($n = 2$); (repeated measure over times). LSD values ($p < 0.05$) are indicated per panel.

Firmness at harvest was similar for MG tomatoes cultivated with or without FR lighting ($p > 0.05$). Softening during storage at 2 °C for was faster for MG tomatoes that were cultivated without- compared to those with FR lighting ($p < 0.05$) (Figure 6). Softening of tomatoes cultivated without FR was similar during storage and shelf-life independent of the storage oxygen concentration. Tomatoes cultivated without FR from the temperature control treatment showed no softening during storage (Figure 6A), but tomatoes cultivated with FR showed similar softening for all treatments (Figure 6B).

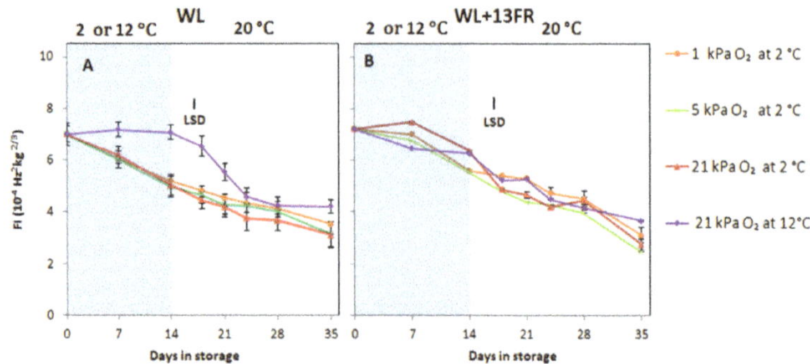

Figure 6. Firmness as indicated by firmness index (FI) of MG tomatoes cultivated under white LED light (WL) without far red lighting (**A**) or with far red lighting during cultivation (**B**) during cold storage (blue area) at 2 or 12 °C and subsequent shelf-life at 20 °C (white area). Orange, green, red and purple lines and symbols indicate 1, 5 and 21 kPa O_2 (low oxygen control) applied during storage at 2 °C, and 21 kPa O_2 at 12 °C (temperature control), respectively. The average firmness index with indicated standard error is shown for two replicates of four tomatoes (n = 2); (repeated measure over times). LSD values (p< 0.05) are indicated per panel.

3.4. *Effects of Low Oxygen Conditions on Colour Development of Mature Green Tomatoes Cultivated with and without Far Red Lighting*

During cold storage red colour development was blocked, as indicated by the constant NAI values, irrespective of low oxygen treatments for both MG fruit cultivated with and without FR lighting. Colour development for the temperature control tomatoes started immediately, although faster for the MG tomatoes cultivated with FR lighting (Figure 7). During shelf-life, colour development was similar for the different low temperature oxygen storage treatments in fruit without FR lighting. Fruit cultivated with FR lighting reached higher NAI values in fruit prior stored at the low oxygen concentrations ($p < 0.001$) (Figure 7B). NDVI values were not significantly affected by oxygen level nor FR treatment.

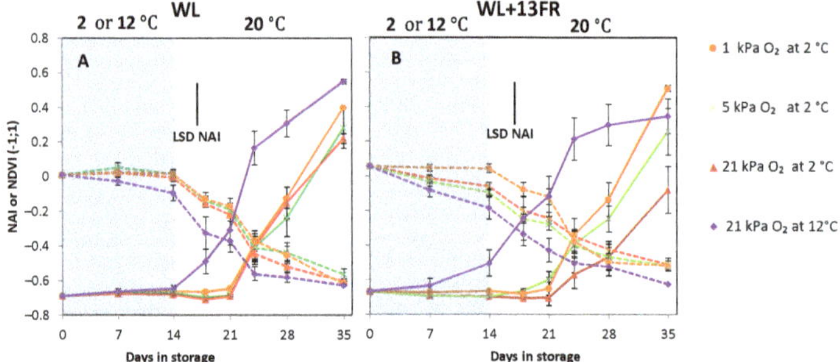

Figure 7. Colour indicated by NDVI (dotted lines) or NAI (full lines) values of MG tomatoes cultivated under white LED light (WL) without far red lighting (**A**) or with far red lighting during cultivation (**B**) during cold storage (blue area) at 2 or 12 °C and subsequent shelf-life at 20 °C (white area). Orange, green, red and purple lines and symbols indicate 1, 5 and 21 kPa O_2 (low oxygen control) applied during storage at 2 °C, and 21 kPa O_2 at 12 °C (temperature control), respectively. The average NDVI and NAI values with indicated standard error are shown for two replicates of four tomatoes (n = 2). LSD values (p < 0.05) are indicated per panel.

3.5. Effects of Low Oxygen Conditions on Respiration and H_2O_2 Production of Mature Green Tomatoes Cultivated with and without Far Red Lighting

Respiration rate measurements were carried out from the fourth day onwards to allow time to achieve the set low oxygen conditions. The O_2 consumption rate at 2 °C was observed to be lower for MG tomatoes stored at lower oxygen levels (Figure 8A,B). At 12 °C, both CO_2 production and O_2 consumption was higher than at 2 °C. The CO_2 production rate, however, was similar at the low oxygen levels (Figure 8C,D). The oxygen consumption rate over time was lower for MG fruits cultivated with FR lighting and stored at 1 kPa O_2.

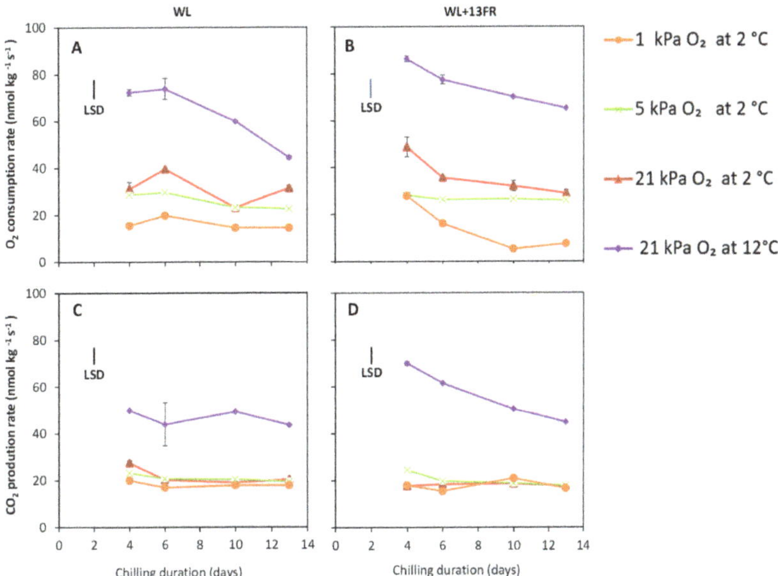

Figure 8. Respiration MG tomatoes cultivated under white LED light (WL) without- (**A**,**C**) or with far red lighting during cultivation (**B**,**D**) indicated as oxygen consumption (**A**,**B**) or CO_2 production (**C**,**D**) measured during cold storage at 2 or 12 °C. Orange, green, red and purple lines and symbols indicate 1, 5 and 21 kPa O_2 (low oxygen control), and 21 kPa O_2 at 12 °C (temperature control), respectively. The average O_2 or CO_2 production rates with indicated standard error are shown for two replicates from each respective container ($n = 2$) during cold storage at 2 °C or 12 °C. LSD values ($p < 0.05$) are indicated per panel.

H_2O_2 levels were stable during cold storage and steadily increased in all treatments during subsequent shelf-life ($p < 0.0001$, Figure S3). Varying oxygen levels during cold storage showed similar patterns of H_2O_2 production during subsequent shelf-life.

4. Discussion

4.1. Low Oxygen Storage Alleviated CI in Tomato Which Might Be Related to Lower Oxygen Uptake and Improved Lycopene Synthesis

When low temperature was combined with reduced oxygen concentrations, lower decay and lower weight loss was observed during shelf-life for both MG and R tomatoes (Figures 1 and 2). Our results showed that O_2 consumption decreased with lower oxygen levels while CO_2 production rates were similar (Figure 8). Low oxygen storage is reported to suppress respiration and ethylene production [15,35]. Low oxygen uptake might reduce O_2 availability for ROS production, such as singlet oxygen ($^1O_2$) and superoxide anions ($O_2^{\cdot-}$) [36]. O_2^- is dismutated into H_2O_2 by the action of SOD [4,36,37]. Lower levels of O_2^- are expected to yield lower levels of H_2O_2. However, we did not observe a lower level of H_2O_2 in the low oxygen stored fruit (Figure S3), perhaps indicating that low oxygen did not suppress oxidative stress initiated by the presence of O_2^-. As tomato stored under low oxygen showed further red colouration close to or even higher than the non-chilled control (Figure 4B) and faster red colouration (Figure 7) after transfer to 20 °C, we hypothesise that lycopene acted directly to quench $^1O_2$. Carotenoids are able to quench $^1O_2$ due to its high number of conjugated double bonds, whereas lycopene and its precursors, are the most effective $^1O_2$ quencher [38–41]. Quenching of $^1O_2$ by lycopene or its precursors might have resulted in delayed lycopene synthesis or lycopene degradation [31,42]. Therefore, uninterrupted colour synthesis might indicate that low oxygen prevents lycopene degradation as well as preserving the lycopene biosynthetic machinery during cold storage allowing new lycopene synthesis during shelf life [43–45].

The lowest oxygen concentration to delay or prevent CI symptoms was 1 kPa (Figure 7). A lower oxygen level (0.5 kPa) resulted in necrosis and fungal infection (data not shown), probably because of excessive fermentation. It was reported that MG 'Bermuda' tomatoes stored at 22 °C under 0.5 kPa O_2 developed identical symptoms after three days of storage [35].

4.2. *Low oxygen Storage Alleviated CI in Tomato Which Might Be Related to Lower Oxygen Uptake and Improved Lycopene Synthesis*

Tomatoes cultivated with FR during cultivation and kept at 1 kPa O_2 during cold storage were shown to completely alleviate CI symptoms (Figure 5B) in MG fruit. This confirmed our previous findings that FR addition during cultivation suppressed CI incidence [29]. It was observed that MG tomatoes cultivated with FR initiated colour development at higher firmness [29]. It means that tomato cultivated with FR, although they had the same firmness as those cultivated without FR at harvest, maintained higher firmness during cold storage (Figure 6). Excessive firmness loss during cold storage and during shelf-life is often regarded as one of the main symptoms of CI in tomato with firmness retention associated with lower decay and higher membrane integrity [46,47]. Improved cold tolerance of FR cultivated tomatoes might also be attributed to thicker cuticle wax layers [48] which in turn might lower the oxygen consumption rate (Figure 8). On contrary, no significant difference was on weight loss (Figure S3). This might be attributed to comparably high relative humidity during the shelf-life (>95% RH) which suppress weight loos induced-transpiration from the fruit [49].

Our findings suggests that when low oxygen storage is applied to accompany long cold storage or transport, higher CI tolerance will result in shelf-life extension when tomatoes are grown with FR in greenhouses or grown in the field characterised by a low red to far-red ratio.

5. Conclusions

This study assessed the application of low oxygen either alone or in combination with far-red cultivated tomatoes on CI development. Results obtained showed the efficacy of low oxygen in minimising CI in tomato. CI tolerance is improved when low oxygen storage of MG tomatoes is combined with FR lighting during cultivation, especially when stored at 2 °C. This is likely due to lower oxygen uptake that allowed for to uninterrupted lycopene production and less softening during shelf-life for prior cold stored MG tomatoes kept at 1 kPa O_2 and 0 kPa CO_2.

Supplementary Materials: The following are available online at https://www.mdpi.com/article/10.3390/foods10081699/s1, Figure S1: Spectra of 'PhysioSpec' Greenhouse white LED light and far-red light and the spectra of the 'PhysioSpec' Greenhouse lamp with only white LED light; Figure S2: Weight loss percentage of MG tomatoes during cold storage at 2 °C accompanied by 1 kPa, 5 kPa or 21 kPa O_2 (low oxygen control) and 21 kPa O_2 at 12 °C (temperature control), and subsequent shelf-life at 20 °C; Figure S3: H_2O_2 levels (nmol g^{-1} FW^{-1}) of MG tomatoes during cold storage at 2 °C accompanied by 1 kPa, 5 kPa or 21 kPa O_2 (low oxygen control) and 21 kPa O_2 at 12 °C (temperature control), and subsequent shelf-life at 20 °C.

Author Contributions: Conceptualisation, R.E.S. and E.J.W., methodology, F.Y.A. and M.J.M.P., validation, R.E.S., J.A.V. and E.J.W., formal analysis, F.Y.A.; investigation, F.Y.A., writing—original draft preparation, F.Y.A.; writing—review and editing, F.Y.A. and J.C.V.; supervision, J.C.V., E.J.W. and R.E.S. All authors have read and agreed to the published version of the manuscript.

Funding: This work is supported by the Indonesian Endowment Fund for Education (LPDP) (Grant numbers: PRJ-2348/LPDP/2015).

Institutional Review Board Statement: Not applicable.

Informed Consent Statement: Not applicable.

Data Availability Statement: The raw data will be made available upon request.

Acknowledgments: The authors acknowledge the Indonesian Endowment Fund for Education (LPDP) for financial support for F.Y.A.

Conflicts of Interest: The authors declared no conflict of interest.

References

1. Albornoz, K.; Cantwell, M.I.; Zhang, L.; Beckles, D.M. Integrative analysis of postharvest chilling injury in cherry tomato fruit reveals contrapuntal spatio-temporal responses to ripening and cold stress. *Sci. Rep.* **2019**, *9*, 1–14. [CrossRef]
2. Hodges, D.M.; Lester, G.E.; Munro, K.D.; Toivonen, P. Oxidative Stress: Importance for Postharvest Quality. *HortScience* **2004**, *39*, 924–929. [CrossRef]
3. Aghdam, M.S.; Bodbodak, S. Postharvest Heat Treatment for Mitigation of Chilling Injury in Fruits and Vegetables. *Food Bioprocess Technol.* **2013**, *7*, 37–53. [CrossRef]
4. Imahori, Y.; Bai, J.; Baldwin, E. Antioxidative responses of ripe tomato fruit to postharvest chilling and heating treatments. *Sci. Hortic.* **2016**, *198*, 398–406. [CrossRef]
5. Maul, F.; Sargent, S.; Sims, C.; Baldwin, E.; Balaban, M.; Huber, D. Tomato Flavor and Aroma Quality as Affected by Storage Temperature. *J. Food Sci.* **2000**, *65*, 1228–1237. [CrossRef]
6. Bai, J.; Baldwin, E.A.; Imahori, Y.; Kostenyuk, I.; Burns, J.; Brecht, J.K. Chilling and heating may regulate C6 volatile aroma production by different mechanisms in tomato (*Solanum lycopersicum*) fruit. *Postharvest Biol. Technol.* **2011**, *60*, 111–120. [CrossRef]
7. Zhang, X.; Shen, L.; Li, F.; Meng, D.; Sheng, J. Arginase induction by heat treatment contributes to amelioration of chilling injury and activation of antioxidant enzymes in tomato fruit. *Postharvest Biol. Technol.* **2013**, *79*, 1–8. [CrossRef]
8. Pesis, E.; Aharoni, D.; Aharon, Z.; Ben-Arie, R.; Aharoni, N.; Fuchs, Y. Modified atmosphere and modified humidity packaging alleviates chilling injury symptoms in mango fruit. *Postharvest Biol. Technol.* **2000**, *19*, 93–101. [CrossRef]
9. Singh, S.; Pal, R. Controlled atmosphere storage of guava (*Psidium guajava* L.) fruit. *Postharvest Biol. Technol.* **2008**, *47*, 296–306. [CrossRef]
10. Singh, S.P.; Singh, Z. Controlled and modified atmospheres influence chilling injury, fruit quality and antioxidative system of Japanese plums (*Prunus salicina* Lindell). *Int. J. Food Sci. Technol.* **2012**, *48*, 363–374. [CrossRef]
11. Alamar, M.C.; Collings, E.; Cools, K.; Terry, L.A. Impact of controlled atmosphere scheduling on strawberry and imported avocado fruit. *Postharvest Biol. Technol.* **2017**, *134*, 76–86. [CrossRef]
12. de Almeida Teixeira, G.; Santos, L.; Cunha Júnior, L.; Durigan, J. Effect of carbon dioxide (CO_2) and oxygen (O_2) levels on quality of 'Palmer' mangoes under controlled atmosphere storage. *J. Food Sci. Technol.* **2018**, *55*, 145–156. [CrossRef]
13. Zhao, Q.; Jin, M.; Guo, L.; Pei, H.; Nan, Y.; Rao, J. Modified atmosphere packaging and 1-methylcyclopropene alleviate chilling injury of 'Youhou' sweet persimmon during cold storage. *Food Packag. Shelf Life* **2020**, *24*, 100479. [CrossRef]
14. Esanhueza, D.; Evizoso, P.; Ebalic, I.; Campos-Vargas, R.; Emeneses, C. Transcriptomic analysis of fruit stored under cold conditions using controlled atmosphere in Prunus persica cv. "Red Pearl". *Front. Plant Sci.* **2015**, *6*, 788. [CrossRef]
15. Beaudry, R.M. Responses of Horticultural Commodities to Low Oxygen: Limits to the Expanded Use of Modified Atmosphere Packaging. *HortTechnology* **2000**, *10*, 491–500. [CrossRef]
16. Hodges, D.M.; Forney, C.F. The effects of ethylene, depressed oxygen and elevated carbon dioxide on antioxidant profiles of senescing spinach leaves. *J. Exp. Bot.* **2000**, *51*, 645–655. [CrossRef]
17. Sabban-Amin, R.; Feygenberg, O.; Belausov, E.; Pesis, E. Low oxygen and 1-MCP pretreatments delay superficial scald development by reducing reactive oxygen species (ROS) accumulation in stored 'Granny Smith' apples. *Postharvest Biol. Technol.* **2011**, *62*, 295–304. [CrossRef]
18. Ali, S.; Khan, A.S.; Malik, A.U.; Shahid, M. Effect of controlled atmosphere storage on pericarp browning, bioactive compounds and antioxidant enzymes of litchi fruits. *Food Chem.* **2016**, *206*, 18–29. [CrossRef]
19. Suzuki, N.; Mittler, R. Reactive oxygen species and temperature stresses: A delicate balance between signaling and destruction. *Physiol. Plant.* **2006**, *126*, 45–51. [CrossRef]
20. Gough, D.R.; Cotter, T.G. Hydrogen peroxide: A Jekyll and Hyde signalling molecule. *Cell Death Dis.* **2011**, *2*, e213. [CrossRef] [PubMed]

21. Gonzalez-Aguilar, G.A.; Villa-Rodriguez, J.A.; Zavala, J.F.A.; Yahia, E.M. Improvement of the antioxidant status of tropical fruits as a secondary response to some postharvest treatments. *Trends Food Sci. Technol.* **2010**, *21*, 475–482. [CrossRef]
22. Mditshwa, A.; Fawole, O.; Vries, F.; van der Merwe, K.; Crouch, E.; Opara, U.L. Impact of dynamic controlled atmospheres on reactive oxygen species, antioxidant capacity and phytochemical properties of apple peel (cv. Granny Smith). *Sci. Hortic.* **2017**, *216*, 169–176. [CrossRef]
23. Pucciariello, C.; Perata, P. New insights into reactive oxygen species and nitric oxide signalling under low oxygen in plants. *Plant Cell Environ.* **2017**, *40*, 473–482. [CrossRef] [PubMed]
24. Cukrov, D.; Brizzolara, S.; Tonutti, P. Physiological and biochemical effects of controlled and modified atmospheres. In *Postharvest Physiology and Biochemistry of Fruits and Vegetables*; Woodhead Publishing: Sawston, UK, 2019; pp. 425–441.
25. Cukrov, D.; Zermiani, M.; Brizzolara, S.; Cestaro, A.; Licausi, F.; Luchinat, C.; Santucci, C.; Tenori, L.; Van Veen, H.; Zuccolo, A.; et al. Extreme Hypoxic Conditions Induce Selective Molecular Responses and Metabolic Reset in Detached Apple Fruit. *Front. Plant Sci.* **2016**, *7*, 146. [CrossRef]
26. Whitaker, B.D. Changes in lipids of tomato fruit stored at chilling and non-chilling temperatures. *Phytochemistry* **1991**, *30*, 757–761. [CrossRef]
27. Lurie, S.; Sabehat, A. Prestorage temperature manipulations to reduce chilling injury in tomatoes. *Postharvest Biol. Technol.* **1997**, *11*, 57–62. [CrossRef]
28. Biswas, P.; East, A.R.; Brecht, J.K.; Hewett, E.W.; Heyes, J. Intermittent warming during low temperature storage reduces tomato chilling injury. *Postharvest Biol. Technol.* **2012**, *74*, 71–78. [CrossRef]
29. Affandi, F.Y.; Verdonk, J.C.; Ouzounis, T.; Ji, Y.; Woltering, E.J.; Schouten, R.E. Far-red light during cultivation induces postharvest cold tolerance in tomato fruit. *Postharvest Biol. Technol.* **2020**, *159*, 1–10. [CrossRef]
30. Bulens, I.; Van De Poel, B.; Hertog, M.L.; De Proft, M.P.; Geeraerd, A.H.; Nicolaï, B.M. Protocol: An updated integrated methodology for analysis of metabolites and enzyme activities of ethylene biosynthesis. *Plant Methods* **2011**, *7*, 17. [CrossRef]
31. Schouten, R.E.; Zhang, X.; Verschoor, J.A.; Otma, E.C.; Tijskens, P.; Van Kooten, O. Development of colour of broccoli heads as affected by controlled atmosphere storage and temperature. *Postharvest Biol. Technol.* **2009**, *51*, 27–35. [CrossRef]
32. Schouten, R.E.; Farneti, B.; Tijskens, P.; Alarcón, A.A.; Woltering, E.J. Quantifying lycopene synthesis and chlorophyll breakdown in tomato fruit using remittance VIS spectroscopy. *Postharvest Biol. Technol.* **2014**, *96*, 53–63. [CrossRef]
33. Schouten, R.E.; Fan, S.; Verdonk, J.C.; Wang, Y.; Kasim, N.F.M.; Woltering, E.J.; Tijskens, P. Mango Firmness Modeling as Affected by Transport and Ethylene Treatments. *Front. Plant Sci.* **2018**, *9*, 1647. [CrossRef] [PubMed]
34. Junglee, S.; Urban, L.; Sallanon, H.; Lopez-Lauri, F. Optimized Assay for Hydrogen Peroxide Determination in Plant Tissue Using Potassium Iodide. *Am. J. Anal. Chem.* **2014**, *05*, 730–736. [CrossRef]
35. Klieber, A.; Ratanachinakorn, B.; Simons, D. Effects of low oxygen and high carbon dioxide on tomato cultivar 'Bermuda' fruit physiology and composition. *Sci. Hortic.* **1996**, *65*, 251–261. [CrossRef]
36. Fahmy, K.; Nakano, K. The Individual and Combined Influences of Low Oxygen and High Carbon Dioxide on Chilling-injury Alleviation in Cucumber Fruit. *Environ. Control. Biol.* **2014**, *52*, 149–153. [CrossRef]
37. Malacrida, C.; Valle, E.M.; Boggio, S.B. Postharvest chilling induces oxidative stress response in the dwarf tomato cultivar Micro-Tom. *Physiol. Plant.* **2006**, *127*, 10–18. [CrossRef]
38. Di Mascio, P.; Kaiser, S.; Sies, H. Lycopene as the most efficient biological carotenoid singlet oxygen quencher. *Arch. Biochem. Biophys.* **1989**, *274*, 532–538. [CrossRef]
39. Min, D.B.; Boff, J.M. Chemistry and Reaction of Singlet Oxygen in Foods. *Compr. Rev. Food Sci. Food Saf.* **2002**, *1*, 58–72. [CrossRef]
40. Martinez, A.; Stinco, C.M.; Melendez-Martinez, A.J. Free Radical Scavenging Properties of Phytofluene and Phytoene Isomers as Compared to Lycopene: A Combined Experimental and Theoretical Study. *J. Phys. Chem. B* **2014**, *118*, 9819–9825. [CrossRef]
41. Lado, J.; Rodrigo, M.J.; López-Climent, M.; Gómez-Cadenas, A.; Zacarías, L. Implication of the antioxidant system in chilling injury tolerance in the red peel of grapefruit. *Postharvest Biol. Technol.* **2016**, *111*, 214–223. [CrossRef]
42. Farneti, B.; Schouten, R.E.; Woltering, E.J. Low temperature-induced lycopene degradation in red ripe tomato evaluated by remittance spectroscopy. *Postharvest Biol. Technol.* **2012**, *73*, 22–27. [CrossRef]
43. Garstka, M.; Venema, J.H.; Rumak, I.; Gieczewska, K.; Rosiak, M.; Koziol-Lipinska, J.; Kierdaszuk, B.; Vredenberg, W.J.; Mostowska, A. Contrasting effect of dark-chilling on chloroplast structure and arrangement of chlorophyll–protein complexes in pea and tomato: Plants with a different susceptibility to non-freezing temperature. *Planta* **2007**, *226*, 1165–1181. [CrossRef] [PubMed]
44. Yang, J.; Fu, M.-R.; Zhao, Y.-Y.; Mao, L.-C. Reduction of Chilling Injury and Ultrastructural Damage in Cherry Tomato Fruits after Hot Water Treatment. *Agric. Sci. China* **2009**, *8*, 304–310. [CrossRef]
45. Skupień, J.; Wójtowicz, J.; Łucja, K.; Mazur, R.; Garstka, M.; Gieczewska, K.; Mostowska, A. Dark-chilling induces substantial structural changes and modifies galactolipid and carotenoid composition during chloroplast biogenesis in cucumber (*Cucumis sativus* L.) cotyledons. *Plant Physiol. Biochem.* **2017**, *111*, 107–118. [CrossRef] [PubMed]
46. Mirdehghan, S.H.; Rahemi, M.; Martínez-Romero, D.; Guillen, F.; Valverde, J.; Zapata, P.J.; Serrano, M.; Valero, D. Reduction of pomegranate chilling injury during storage after heat treatment: Role of polyamines. *Postharvest Biol. Technol.* **2007**, *44*, 19–25. [CrossRef]
47. Rodoni, L.; Casadei, N.; Concellon, A.; Alicia, A.R.C.; Vicente, A. Effect of Short-Term Ozone Treatments on Tomato (*Solanum lycopersicum* L.) Fruit Quality and Cell Wall Degradation. *J. Agric. Food Chem.* **2010**, *58*, 594–599. [CrossRef] [PubMed]

48. Cozmuta, A.M.; Cozmuta, L.M.; Peter, A.; Nicula, C.; Vosgan, Z.; Giurgiulescu, L.; Vulpoi, A.; Baia, M. Effect of monochromatic Far-Red light on physical-nutritional-microbiological attributes of red tomatoes during storage. *Sci. Hortic.* **2016**, *211*, 220–230. [CrossRef]
49. Bhowmik, S.R.; Pan, J.C. Shelf Life of Mature Green Tomatoes Stored in Controlled Atmosphere and High Humidity. *J. Food Sci.* **1992**, *57*, 948–953. [CrossRef]

Article

The Influence of Internal Packaging (Liners) on Moisture Dynamics and Physical and Physiological Quality of Pomegranate Fruit during Cold Storage

Robert Lufu [1], Alemayehu Ambaw [1] and Umezuruike Linus Opara [1,2,*]

1 SARChI Postharvest Technology Research Laboratory, Africa Institute for Postharvest Technology, Faculty of AgriSciences, Stellenbosch University, Stellenbosch 7602, South Africa; lufurobert@gmail.com (R.L.); tsige@sun.ac.za (A.A.)
2 UNESCO International Centre for Biotechnology, Nsukka 410001, Nigeria
* Correspondence: opara@sun.ac.za; Tel.: +27-21-808-4064

Abstract: Weight loss and decay are common physiological disorders during postharvest handling and storage of pomegranates. The study focused on relating the ability of plastic liners as internal packaging to modify both gaseous and moisture atmosphere around the fruit to moisture dynamics and physical and physiological quality of pomegranate fruit (cv. Wonderful) during storage. Fruit were packed with no-liner, non-perforated 'Decco', non-perforated 'Zoe', micro-perforated Xtend®, 2 mm macro-perforated high density polyethylene (HDPE), and 4 mm macro-perforated HDPE plastic liners. After 84 days of storage at 5 °C and 90–95% relative humidity (RH), fruit packed with no-liner lost 15.6 ± 0.3% of initial weight. Non-perforated (Decco and Zoe) liners minimised losses to 0.79 and 0.82% compared to Xtend® micro-perforated (4.17%) and 2 mm HDPE (2.44%) and 4 mm macro-perforated HDPE (4.17%) liners, respectively. Clearly, micro- and macro-perforation of liners minimised moisture condensation, fruit decay, and shrivel severity. Micro-perforated Xtend® and macro-perforated 4 mm HDPE were the best treatments in minimising postharvest losses that are often associated with inadequate environment control inside packaging compared to the use of non-perforated liners.

Keywords: internal packaging; modified atmosphere packaging; storage quality; transpiration; water loss

Citation: Lufu, R.; Ambaw, A.; Opara, U.L. The Influence of Internal Packaging (Liners) on Moisture Dynamics and Physical and Physiological Quality of Pomegranate Fruit during Cold Storage. *Foods* **2021**, *10*, 1388. https://doi.org/10.3390/foods10061388

Academic Editors: Eleni Tsantili and Jinhe Bai

Received: 6 May 2021
Accepted: 10 June 2021
Published: 16 June 2021

1. Introduction

Production and consumption of pomegranate (*Punica granatum* L.) fruit is on the increase worldwide. The fruit has an edible portion of about 55–60% [1] and can be eaten fresh or processed into juice, wine, and jam [2–4]. Freshly harvested fruit is kept under cold storage awaiting export to distant markets. Fruit from South Africa takes about 42 d to reach Europe as the major export market destination and therefore a need to maintain good postharvest quality during prolonged storage and export conditions. Storing pomegranate (cv. Wonderful) for 3 months at 5 °C and above 92% relative humidity (RH) minimises physiological disorders and maintains internal and external quality attributes [5]. Chilling injury increases with storage duration and temperatures lower than 5 °C [6].

In postharvest fruit handling, weight loss and fruit decay are common physiological disorders, among others such as chilling injury and scalding, contributing to quantitative and qualitative loss [6,7]. Pomegranates are highly prone to moisture loss owing to the relatively high water permeability across the skin through minute openings, despite having a thick rind [6,8,9]. Fruit moisture loss if not well controlled results into shrinkage; shrivel; and quantitative loss in weight, taste, and overall acceptability of the fruit, and hence market loss [10].

Internal packaging techniques have been used in the fresh fruit industry to minimise moisture loss. Internal packaging refers to additional packaging materials applied around the fruit within the external package. Surface coating and waxing has been applied on apples, oranges, plums, and pomegranate to minimise moisture loss [8,11–15]. For pomegranates, heat shrinkable wraps on individual fruit in cartons have also been applied [8,16,17]. On the other hand, shrink wrapping, surface coating, or waxing can lead to anaerobic respiration by creating an oxygen deficit and yet promoting a high CO_2 atmosphere around the fruit. This results in the production of off flavors and a change in taste [18,19]. Plastic liners are a commonly applied internal packaging to minimise moisture loss for pomegranate fruit packaged in ventilated cartons [20–22]. Previous research has reported their ability to modify gaseous atmosphere around the fruit, preserving physical and physio-chemical quality [16,22]. However, non-perforated liners negatively affect the fruit cooling rate and increase energy usage during forced air-cooling operations [23,24]. The authors investigated the effect of individual carton design, stack orientation, and presence of internal packaging liners. The internal liner was identified as the most significant factor influencing the produce cooling rate, increasing the seven-eighth cooling time by more than twofold and the corresponding energy usages by up to threefold compared to stacks with no liners. In addition, non-perforated liners promote moisture condensation and consequential fruit decay. Mphahlele et al. [20] reported a higher decay incidence of 33.9% in fruit packed inside plastic liners compared to 29.2 and 16.7% for fruit packed in shrink-wraps and open cartons, respectively, by the end of three months of storage at 7 °C and 90% RH.

The use of perforations is identified as a solution to minimise moisture condensation on fruit surfaces and liner walls, thus minimising the consequential fruit decay as a result of the improved vapor transmission capabilities of the perforated liners [25,26]. However, there is still very limited literature on the effect of liner on the keeping quality of pomegranates. Specifically, perforated liners have not been properly studied and yet have the potential to reduce quality losses during prolonged fruit storage. Therefore, the role of liner perforation in counterbalancing between minimising excessive moisture loss and fruit decay will be demonstrated during prolonged storage. The present study focused on relating the ability of plastic liners (non-perforated, micro-perforated, and macro-perforated) as internal packaging to modify both gaseous and moisture atmosphere around the fruit, to moisture dynamics and physical and physiological quality of pomegranate fruit (cv. Wonderful) during storage.

2. Materials and Methods

2.1. Fruit Supply

Commercially mature, harvested pomegranate fruit (cv. Wonderful) of uniform diameter 81.8 ± 2.5 mm and mass 286 ± 15 g were procured from a farm in Bonnievale (33°58′12.02″ S, 20°09′21.03″ E), Western Cape, South Africa. Fruit were transported in refrigerated truck to Postharvest Technology Research Laboratory at Stellenbosch University.

2.2. Packaging and Storage

Fruit were portioned into six treatments: no-liner (control); non-perforated 'Decco' liner; non-perforated 'Zoe' liner (ZOEpac, South Africa); micro-perforated Xtend® liner; macro-perforated high density polyethylene (HDPE) liner with 54 holes of 2 mm diameter each (2 mm HDPE); macro-perforated HDPE liner with 36 holes of 4 mm diameter each (4 mm HDPE). For each treatment, 11 ventilated cartons each, loaded with 12 fruit were stored in cold rooms at 5 °C and 90–95% RH for 84 d. For each treatment, 12 fruit were randomly selected from the stack and assessed for quality after 28, 42, 56, and 84 d of cold storage.

2.3. Gas Analysis

A gas analyser (Checkmate 3, PBI Dansensor, Ringstead, Denmark) with a precision of ±0.5% was used to assess the gas atmosphere inside liners, across a rubber septum on the packaging film. The experiment was carried out in triplicate.

2.4. Water Vapor Transmission Rate (WVTR)

During storage of packaged fruit, moisture moves across films by diffusion force because of a concentration gradient created on opposite sides of the film. A modification of the dry cup technique (ASTM, 2005 method E96-95) was used to determine WVTR gravimetrically, as described by Opara et al. [27]. Film samples were cut out from areas of no-perforation on the liners in order to test for vapor transmission across the liner surface. Transmission across perforations was also assessed, and in this case film samples having one perforation at the centre were cut out from the macro-perforated liners (2 mm HDPE and 4 mm HDPE). The experiment was carried out in triplicate. Aluminium test cups were filled with 8.0 ± 0.5 g of anhydrous $CaCl_2$ salt. The cups were fitted with an O-ring and grease to provide proofing against moisture and air. A film sample was then laid on top and the cup tightly closed giving an active surface area of 25 cm^2. Cups were weighed and stored in sets under different conditions: 20 °C, 65% RH and 5 °C, 90% RH. The WVTR ($g\ m^{-2}\ d^{-1}$) of films were calculated on the basis of mass of water gained by $CaCl_2$ salt in the cup over time.

2.5. Condensation Assessment

Studies on moisture condensation with the five different liners were carried out in two set ups. The first set up was to determine how much visible condensate could be quantified inside the liner bags. Fruit were conditioned at ambient conditions of 17 ± 2 °C and 65 ± 5% RH for 12 h, weighed individually, packed and sealed in dozens in plastic liners, and placed inside ventilated cartons. Fruit were then stored on pallets in a cold room at 5 °C and 90 ± 5% RH for 24 h. Relative humidity and temperature of the room and inside individual carton liners was monitored using Tinytang sensors (Tinytag TV-4500, Hastings Data Loggers, Port Macquarie, Australia) at intervals of 600 s. Dry clean paper pads of known mass were used to sponge off the condensate water from the inside of the bag and on the fruit. The weight of wet pads was then immediately recorded. The amount of condensate was expressed in grams per day and as a percentage of the fruit mass. The experiment was repeated three times. The amount of condensate was also scored on a scale of 0–10 (where 0 = none; 1–2 = trace; 3–4 = slight; 5–6 = moderate; 7–8 = severe; 9–10 = extremely severe).

The second set up of the experiment was to determine the rate of change in the condensate within the bags over a period. In this case, fruit were conditioned at ambient temperatures while as the packaging material was conditioned at 5 °C in cold room for 12 h. The packed fruit were then weighed before storage at 5 °C and 90 ± 5% RH for 7 d. The condensate within the bags was scored on a 0 to 5 scale and the change in weight of the packed fruit were monitored per day. At the end of 7 d, the amount of remaining condensate in the liners was quantified as described in phase one above and the weight of fruit were also recorded. The rate of change in condensate was calculated in grams per day.

2.6. Weight and Size Loss Assessment

Twelve fruit were randomly selected, numbered, and monitored. The same individual fruit were monitored for weight, length, diameter, and circumference after 28, 42, 56, and 84 d of storage. Fruit weight was monitored using a digital scientific scale (Mettler Toledo, model ML3002E, Switzerland, 0.0001 g accuracy). Fruit circumference (C) was measured twice per sample fruit in the horizontal plane, using a fruit size (circumference) measurer strap band (GÜSS-FTA, Strand, South Africa). Fruit length (L) and diameter (D) were measured at two opposite longitudinal (excluding the fruit calyx) and equatorial fruit perimeters, respectively, using a digital Vernier calliper (Mitutoyo, Kawasaki, Japan, ±0.01 mm).

2.7. Shriveling and Decay

Incidence and severity of fruit physiological disorders of decay and shrivelling were assessed per treatment after 28, 42, 56, and 84 d of storage. The severity of each disorder was assessed subjectively using a hedonic scale of 0–5, where 0 = none; 1 = trace; 2 = slight; 3 = moderate; 4 = severe; 5 = extremely severe. Only severe injuries could be considered as commercially unacceptable [28]. Shrivel and decay indices were calculated by multiplying the scores of severity by the number of fruit affected, divided by the total number of fruit [28,29].

2.8. Respiration Rate

A closed system method [7] was applied to measure fruit respiration using five replicates per treatment. For each replicate, two fruit of known mass were placed inside a 3 L glass jar that was air-tight sealed with a lid with a rubber septum. The jars were incubated for 4 h at 5 °C and 90% RH. The accumulation of CO_2 inside each glass jar was monitored using an O_2/CO_2 gas analyser (Checkmate 3, PBI Dansensor, Denmark) and respiration rate presented as mean ± S.E. (mL CO_2 kg^{-1} h^{-1}).

2.9. Fruit Puncture Resistance

The ability of the fruit to resist a penetrating force was determined by a fruit puncture analyser (GÜSS-FTA, Strand, South Africa) with a 5 mm diameter probe, as described by Arendse et al. [5]. The probe was set to penetrate 8.9 mm into the fruit at 10 mm s^{-1}. The test was carried out on opposite sides of each of the 12 fruit per treatment, and the peak force (N) required to puncture the fruit was reported as puncture resistance mean ± standard error.

2.10. Aril Texture Analysis

Aril compression test was performed as described by Fawole and Opara [29]. Four arils were randomly chosen from each fruit segment to make a pool and then two arils selected from the pool, giving a total of 24 arils per treatment. A 35 mm diameter probe of the texture profile analyser TA. XT (Stable Micro System, Godalming, UK) was used to compress the aril at a test speed of 0.5 ms^{-1} and 0.20 N trigger force. Aril firmness was calculated as maximum force (N) required to completely break the aril. The means (±S.E.) of 24 determinations were reported per treatment.

2.11. Color Properties

A digital colorimeter (Minolta, model CR-400, Tokyo, Japan) was used. Fruit peel colour was monitored at two selected and ring-marked positions per fruit. Aril colour was monitored in a Petri dish at two random spots per sample. Values of L^*(lightness), a^* (redness), b^*(yellowness), and C^* (chroma) were measured. Where, a^* describes surface color in the range from green ($-a^*$) to red ($+a^*$), while b^* ranges from yellow ($+b^*$) to blue ($-b^*$). Monochromaticity L^* ranges from 0 (black) to 100 (white). C^* was calculated by equation 1 [30].Twelve replicates were considered per packaging treatment.

$$C^* = \left(a^{*2} + b^{*2}\right)^{\frac{1}{2}} \quad (1)$$

2.12. Statistical Analysis

Analysis of variance (ANOVA) was carried out using Statistica software (Statistica 13.0, StatSoft Inc., Tulsa, OK, USA). A two-way ANOVA was applied where applicable with packaging treatments and storage time being the major categories. Means were separated using Duncan's multiple range test, and significant difference between means was considered at $p < 0.05$. Results were presented as mean (±S.E.) of the studied variables. Relationship among selected parameters was determined by subjecting data to principal component analysis (PCA) using XLSTAT software version 2012.04.1 (Addinsoft, Paris, France).

3. Results and Discussion

3.1. Liner Properties

3.1.1. Gas Composition Inside Liners

There was a decrease in O_2 and an increase in CO_2 composition within non-perforated 'Decco' and 'Zoe' liners and to a slight extent inside micro-perforated Xtend® liners (Figure 1). Non-perorated liners provide the barrier that restricts movement of gases across packaging walls. However, there was no change in gas composition of the atmosphere inside the 2 mm macro-perforated and 4 mm macro-perforated HDPE liners. For fruit packed in non-perforated 'Decco' and 'Zoe' liners, O_2 composition inside the liners decreased from 21.4 to 15.9 and 15.6%, respectively, while CO_2 composition increased from 0.0 to 2.2 and 2.4%, respectively, after 5 d of cold storage. At 28 d of cold storage, CO_2 composition further increased to 3.1% and 4.0%, inside non-perforated 'Decco' and 'Zoe' liners, respectively. After 28 d, gas composition inside non-perforated liners remained more stable. Mphahlele et al. [20] observed a more stable O_2 concentration inside polyliners after a month of storing pomegranate (cv. Wonderful) at 7 °C. However, a steadier decrease in O_2 and increase in CO_2 concentrations inside different modified atmosphere packaging (MAP) liners was observed for other pomegranate cultivars ('Hicaznar' and 'Hicrannar') stored at 6 °C [21,22]. Quite similar to the current findings, Selcuk and Erkan [22] reported an increase in CO_2 from 0.0 to 3.9 and 2.5% for pomegranate packed in MAP1 and MAP2 liners, respectively, after 20 d of storage at 6 °C.

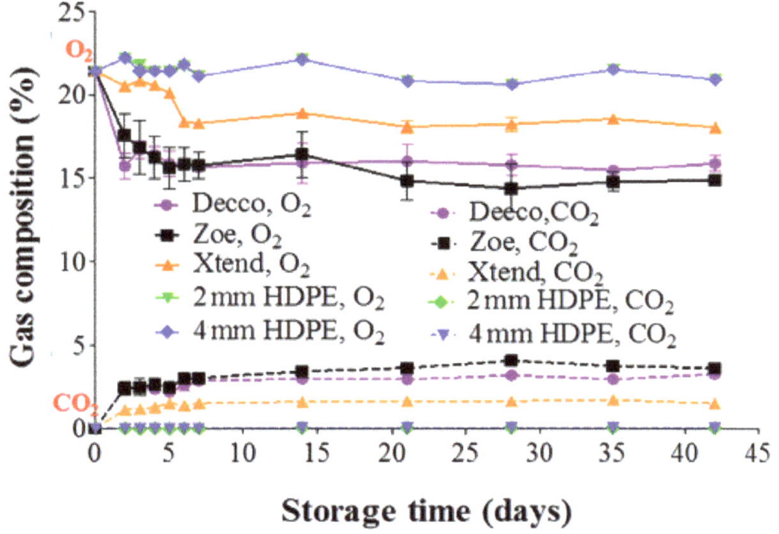

Figure 1. Gas composition inside plastic liners packed with pomegranate fruit (cv. Wonderful) stored at 5 °C and 90% relative humidity (RH). HDPE: high density polyethylene.

3.1.2. Water Vapor Transmission Rate (WVTR)

The rate at which a plastic liner is able to allow moisture across its walls is important in controlling humidity within the bag and around the fruit, and hence reducing condensation and associated risks of fruit decay during prolonged storage [31,32]. Water vapor transmission rate is dependent on liner permeability and prevailing storage temperatures and humidity differences inside and outside the plastic bags [27,33–35]. Generally, for all treatments WVTR decreased with time and then became more stable after about 15 d. Water vapor transmission rate was higher at 20 °C and 65 ± 5% RH than at 5 °C and 95% RH (Figures 2 and 3). The micro-perforated Xtend® liner exceptionally had a higher WVTR of 72.2 7 and 78.7 g m^{-2} d^{-1} at 5 °C and 20 °C, respectively. There was no difference in WVTR across all non-perforated films, irrespective of the type of plastic material and temperature of storage.

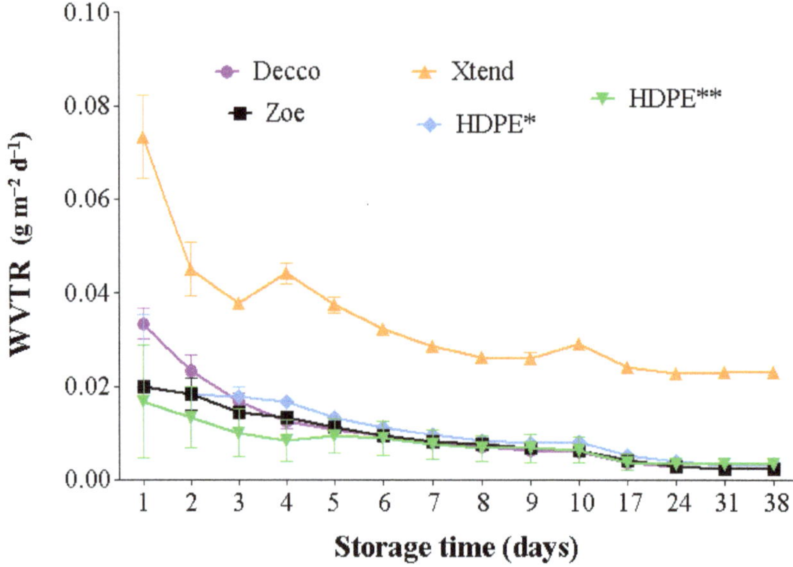

Figure 2. Water vapor transmission rate (WVTR) across plastic liner films under a controlled environment of 5 °C and 90% relative humidity (RH). The non-perforated film section of the 2 mm HDPE (*) and 4 mm HDPE (**) liners were used. HDPE: high density polyethylene.

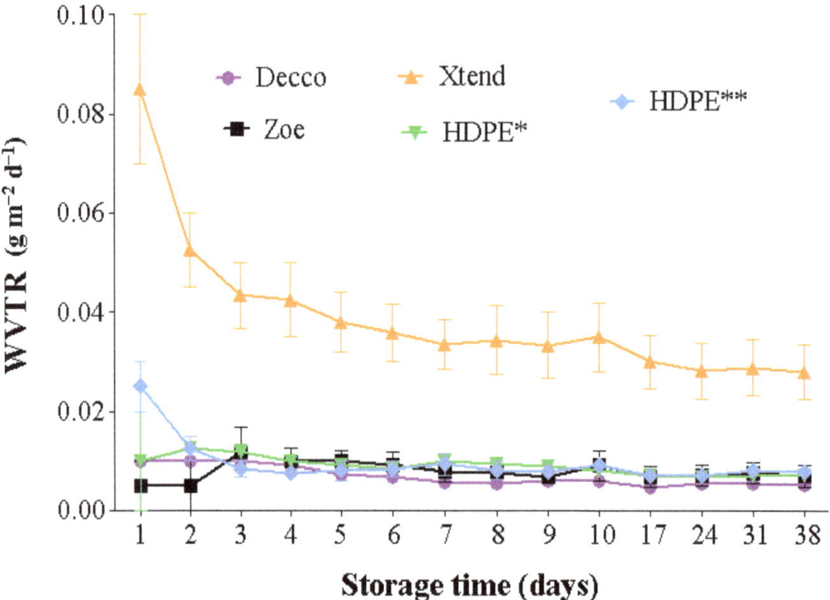

Figure 3. Water vapor transmission rate across plastic liner walls under a controlled environment of 20 °C and 65% relative humidity (RH). The non-perforated film section of the 2 mm HDPE (*) and 4 mm HDPE (**) liners were used. HDPE: high density polyethylene.

Perforations improved the WVTR of the HDPE films. The presence of one 4 mm diameter perforation improved ventilation area of the HDPE film by 2.56% compared to 0.64% by one 2 mm diameter perforation. At 20 °C, the HDPE film with one 4 mm diameter perforation had 66.6% and 44.6% faster WVTR compared to micro-perforated Xtend® film and HDPE film with one 2 mm diameter perforation, respectively (Figure 4). Therefore, the size of perforation plays a significant role in moisture transmission and controlling condensation within bags. Dirim et al. [33] reported a good relationship between film perforation area and WVTR at different temperature and RH conditions. Similar to our results, Opara et al. [27] observed increased WVTR with increased temperature, across biodegradable and synthetic polyfilms. The authors reported that increasing the number of perforations increased WVTR more than increasing storage temperature. Studies on water permeability across polypropyrene films showed increasing WVTR with increasing perforation diameter [34].

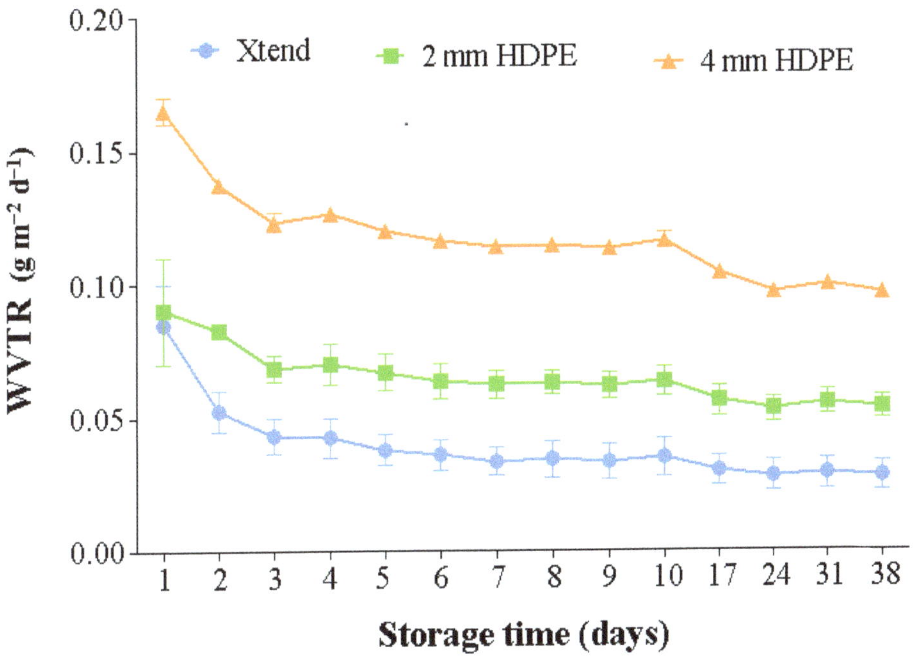

Figure 4. Effect of perforation on water vapor transmission rate (WVTR) under a controlled environment of 20 °C and 90% relative humidity (RH). HDPE: high density polyethylene.

3.1.3. Moisture Condensation Dynamics

One-Day Condensation Characteristics

The barrier effect of the liners permits them to retain a high RH around the fruit [36], resulting in moisture condensation. Generally, the rate of one-day condensate build-up was higher in non-perforated liner treatments than in perforated liner treatments (Table 1). Perforations improve vapor transmission capability of the liners, minimising vapor condensation inside MAP liners [25]. One-day condensate build-up was high in 2 mm macro-perforated HDPE liners, probably because of low perforation area (0.022%). However, one-day condensate build-up was lowest in micro-perforated Xtend® liners and in 4 mm macro-perforated HDPE liners because of their high moisture permeability. Similarly, a higher one-day condensation severity score was observed in non-perforated liners than in perforated liners. One-day condensation severity was such that non-perforated

'Decco' > non-perforated 'Zoe' > 2 mm macro-perforated HDPE > 4 mm macro-perforated HDPE > micro-perforated Xtend® liners (Table 2). A difference in the general characteristics (size and distribution) of condensate droplets formed within the different liner bags was observed (Table 2).

Table 1. Rate of moisture condensation and corresponding weight loss for 12 pomegranate fruit inside plastic liner bags, at 5 °C and 90% relative humidity (RH). HDPE: high density polyethylene.

Treatment	Weight Loss (g d^{-1})	Condensation Rate (g d^{-1})
Xtend	5.55	2.32
Decco	4.27	3.37
Zoe	3.46	2.85
2 mm HDPE	5.50	3.49
4 mm HDPE	6.81	2.48

Table 2. Condensate characterisation inside plastic liner bags for pomegranate fruit stored at 5 °C and 90% relative humidity (RH). HDPE: high density polyethylene.

Treatment	Condensation Score (0–10) [1]	Condensate Characteristics
Xtend	3.47	Large droplets. Condensate entirely on the inside-top wall of the liner. Droplets uniformly distributed on top wall. Very little condensate in the bottom corner. No condensation on fruit.
Decco	5.67	Medium droplets. Condensate on both the top and side walls within the liner. Droplets uniformly distributed on the walls. Visible condensate droplets on the fruit.
Zoe	5.33	Medium to large droplets. Condensate on both the top and side walls within the liner. Droplets non-uniformly distributed, creating a patch-like pattern Visible condensate droplets on the fruit.
2 mm HDPE	4.03	Very tinny/misty droplets on top of the bag. No condensation on the fruit and immediate area around perforations. Uniformly distributed.
4 mm HDPE	3.50	Very tinny/misty droplets on top of the bag. No condensation on the fruit and immediate area around perforations. Uniformly distributed.

[1] Condensation was scored using 0–10 score scale (where 0 = none; 1–2 = trace; 3–4 = slight; 5–6 = moderate; 7–8 = severe; 9–10 = extremely severe).

Condensation Behaviour over Prolonged Period

Condensate behaviour over time provides insight about water vapor transmission properties of the liners. A lower rate of condensation in the perforated liners suggests a faster moisture transmission rate across the walls of the liners, hence delayed build-up of humidity within the bags compared to non-perforated liner treatments. Severity of condensate within the bags decreased with time (Figure 5). The decrease in observable condensate was slowest in non-perforated liners compared to perforated liners. The rate at which condensate was decreasing was lowest in non-perforated 'Zoe' liners, followed by non-perforated 'Decco' liners. Condensate reduction rate was highest in micro-perforated Xtend® liners, followed by 4 mm macro-perforated HDPE and 2 mm macro-perforated HDPE liners. This can be attributed to a higher water vapor transmission rate across the micro-perforated liner compared to the rest of the liners. After 3 d of monitoring, condensation severity was in traces for micro-perforated Xtend® and 4 mm macro-perforated HDPE liners. By the end of 7 d of condensate monitoring, the micro-perforated Xtend® and macro-perforated 4 mm HDPE liners retained none of the condensate, while macro-perforated 2 mm HDPE and non-perforated 'Zoe' and 'Decco' liners retained 10.2, 33.7, and 29.8%, respectively (Figure 6). In another study, a particular MAP liner (Xtend®) was reported to eradicate vapor condensation in pomegranate fruit because of its high water vapor transmission compared to polypropylene bags, which showed progressive moisture accumulation [37].

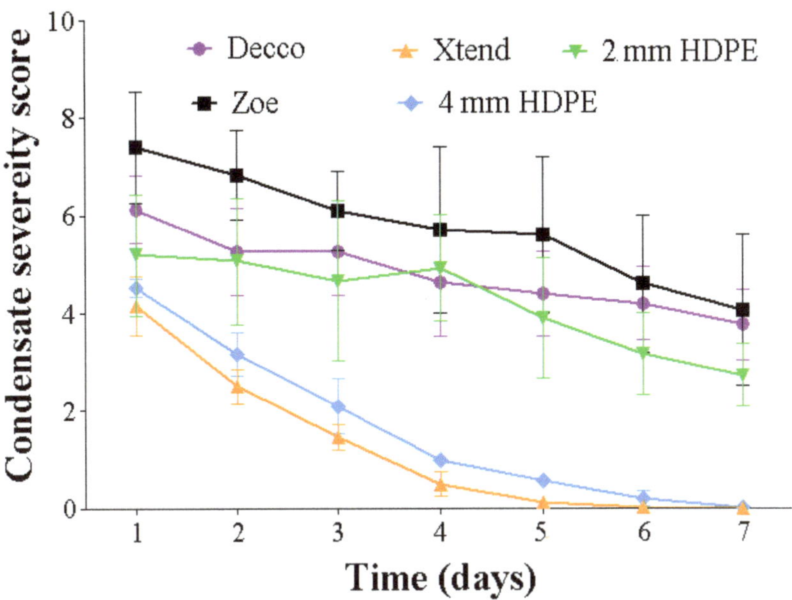

Figure 5. Variation of condensate inside liner bags, as depicted by 0–10 score scale (where 0 = none; 1–2 = trace; 3–4 = slight; 5–6 = moderate; 7–8 = severe; 9–10 = extremely severe). Pomegranate (cv. Wonderful) stored at 5 °C and 90% relative humidity (RH). HDPE: high density polyethylene.

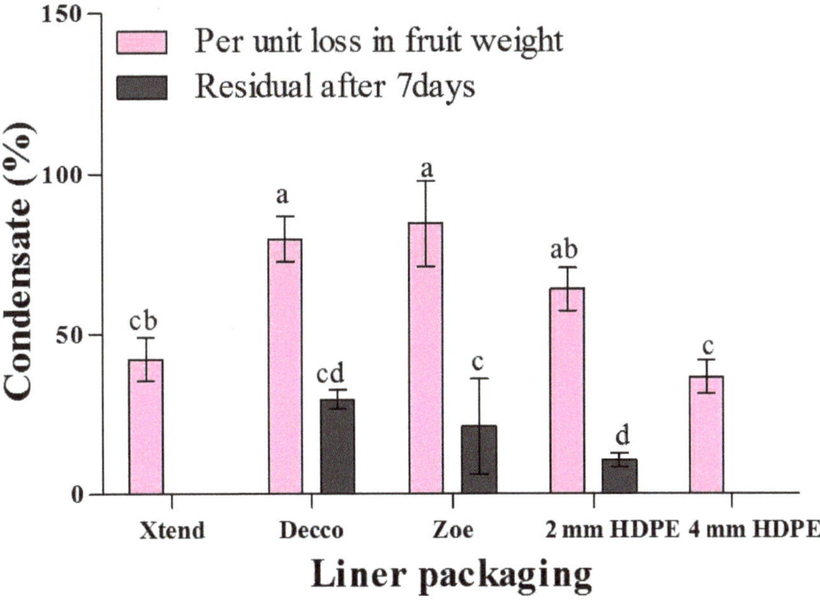

Figure 6. Condensate within liner bags with respect to weight lost for pomegranate (cv. Wonderful) stored at 5 °C and 90% RH for 1 d and condensate retained within plastic bags after a period of 7 d at 5 °C and 90% relative humidity (RH). Histograms columns with different letters are significantly different at $p < 0.05$ according to Duncan's multiple range test. Vertical bars represent S.E. HDPE: high density polyethylene.

Condensation and Fruit Mass Loss

The liner treatments with a lower rate of condensate reduction (high condensate retention) had a lower rate of fruit weight loss while treatments with a higher condensate reduction had a higher rate of fruit weight loss. Fruit in non-perforated liner treatments had a lower rate of weight loss than fruit in perforated liners during the 7 d of condensate monitoring (Figure 7). Fruit weight loss is commonly a result of moisture loss, while condensation results from the moisture lost by the fruit. In non-perforated 'Decco' and 'Zoe' liners, 79.6 and 84.4% of fruit moisture loss per day was retained as condensate compared to 42.1, 63.9, and 36.4% for micro-perforated Xtend®, 2 mm macro-perforated HDPE liners, and 4 mm macro-perforated HDPE liners, respectively (Figure 6).

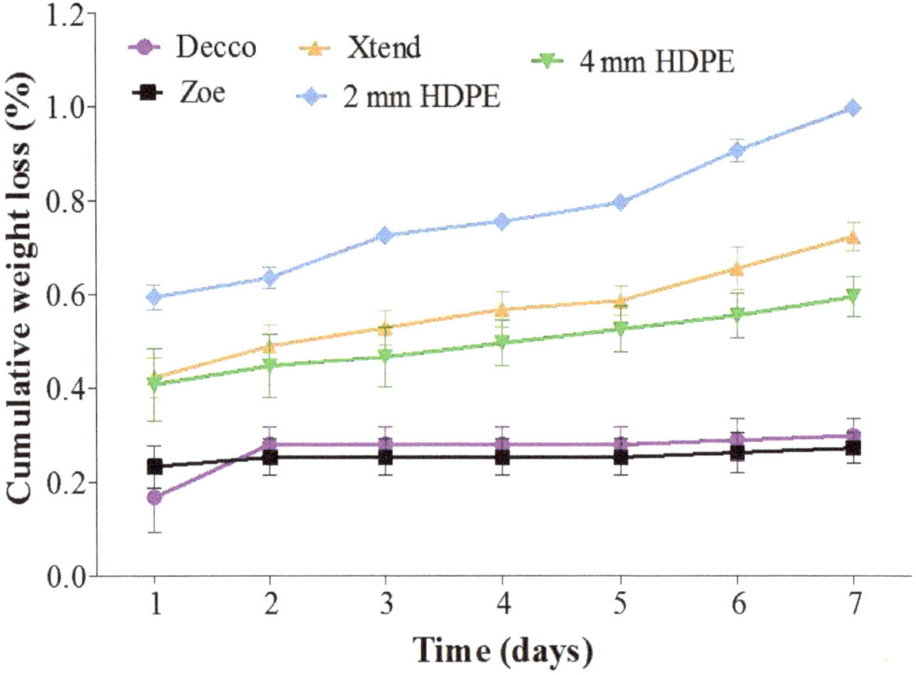

Figure 7. Cumulative percentage loss in weight during condensation variation within liner bags. Pomegranate (cv. Wonderful) stored at 5 °C and 90% relative humidity (RH). HDPE: high density polyethylene.

3.2. Weight Loss, Lineal Size Loss, and Shrivel

3.2.1. Fruit Weight Loss

Moisture loss is the major contributor to weight loss of harvested fruit during postharvest handling. Other physiological activities such as respiration can contribute to mass loss through the utilisation of fruit contents such as the carbohydrates in generating energy to support life processes of the fruit [38–40]. During storage, fruit packed with no-liner lost more weight than fruit packed in liners. At the end of 84 d of cold storage, the no-liner packed fruit lost 15.6 ± 0.3% of initial weight (Figure 8). However, fruit packed in non-perforated 'Decco' and 'Zoe' liners lost only 0.79 and 0.82%, respectively. Fruit packed in micro-perforated Xtend® liners lost 4.17%, compared to 2.44 and 4.17% by fruit packed in 2 mm macro-perforated HDPE and 4 mm macro-perforated HDPE liners, respectively. Non-perforated ('Decco' and 'Zoe') liners minimised fruit weight loss by 94.0% compared to micro-perforated Xtend® (73.2%), 2 mm macro-perforated HDPE (84.3%), and 4 mm macro-perforated HDPE (62.5%) liners. Weight loss increased with increasing ventilation area of the liners, as observed in kiwifruit [41]. The impact of liners on weight loss can be attributed to the fact that liners act as barriers to the moisture exchange between the immediate environment of the fruit inside liners and the outside environment. A high RH around the fruit minimises moisture loss from the fruit [42]. Liners maintain a high RH around the fruit, reducing the difference in vapor pressure inside the skin surface and immediate surrounding, hence reducing moisture diffusion [36]. Similar to our results, packing pomegranate (cv. Hicrannar) in MAP liners minimised fruit weight loss to 1.5 and 4.4% compared to 17.2% for fruit packed with no-liner, after 120 d of storage at 6 °C [22]. Al-Mughrabi et al. [43] observed 18.3% average weight loss for pomegranate fruit ('Taeifi', 'Banati', and 'Manfaloti' cultivars) packed in plastic crates only (without liners) for 42 d at 5 °C. Storing pomegranate (cv. Wonderful) fruit in MAP liners and shrink wraps maintained a weight loss less than 2% throughout storage period of 4 months, compared to 16.5% for fruit packed with no-liner after 90 days of storage at 7 °C [20]. Mukama et al. [44] reported that pomegranates packed in ventilated cartons without liners had a 17.5% more moisture loss than fruit packed in liners. Critical limits of weight loss in fresh fruit are scarce [45]. A weight loss of 5% can initiate shrivelling in pomegranates [38], which negatively affects fruit marketability. However, weight loss in pomegranates is majorly from the peel portion of the fruit [46] and therefore the arils (edible portion) remains largely preserved for consumption and use in juice processing.

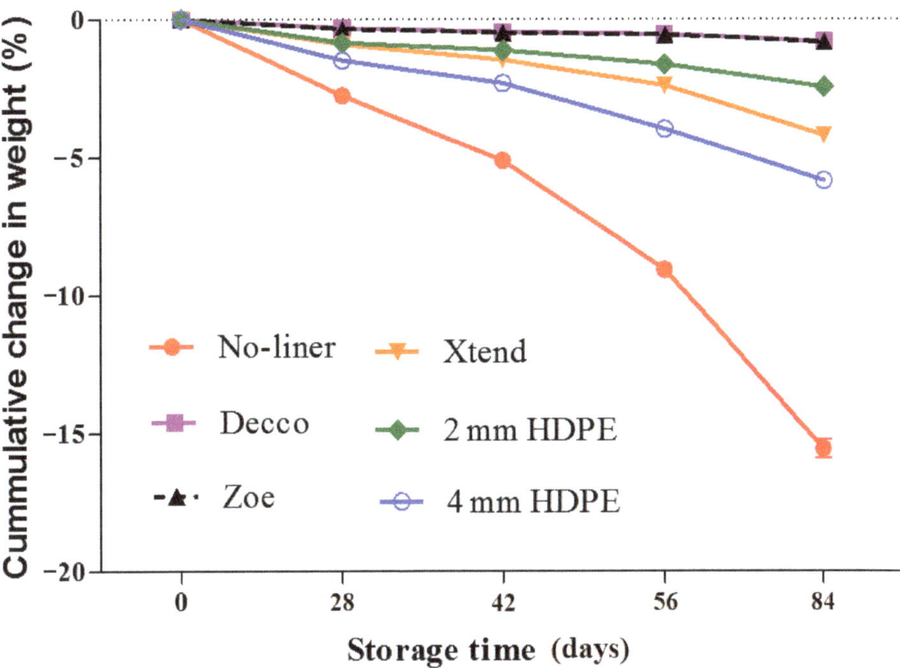

Figure 8. Cumulative change in weight during prolonged cold storage of pomegranate fruit (cv. Wonderful) at 5 °C and 90% relative humidity (RH). HDPE: high density polyethylene.

3.2.2. Fruit Lineal Size

The loss in moisture and weight leads to loss in fruit length, diameter, fruit circumference, and sphericity, which may lead to shrivelling, shrinkage, and loss in visual appeal. Generally, all liner treatments minimised loss in fruit length, diameter, and circumference compared to the no-liner treatment throughout the storage period. The non-perforated 'Decco' and 'Zoe' liners were significantly better in minimising loss in fruit length, diameter, and circumference compared to micro-perforated Xtend® and 2 and 4 mm macro-perforated HDPE liners (Table 3).

Table 3. Effect of plastic liner treatment on cumulative loss in fruit length, diameter, and circumference of pomegranate (cv. Wonderful) fruit stored at 5 °C and 90% relative humidity (RH).

		Cumulative Loss (%)		
Time (d)	**Treatment**	**Length**	**Diameter**	**Circumference**
28	No-liner	2.13 ± 0.07 ^{def}	1.67 ± 0.17 ^{def}	1.07 ± 0.18 ^{jg}
	Decco	0.64 ± 0.13 ^{ih}	0.34 ± 0.06 ^{jk}	0.38 ± 0.09 ^{l}
	Zoe	0.45 ± 0.07 ^{i}	0.23 ± 0.06 ^{k}	0.36 ± 0.07 ^{l}
	Xtend	1.40 ± 0.37 ^{geh}	0.76 ± 0.17 ^{jkh}	0.81 ± 0.07 ^{jl}
	2 mm HDPE	0.69 ± 0.18 ^{gi}	0.84 ± 0.07 ^{jh}	0.55 ± 0.09 ^{lk}
	4 mm HDPE	1.52 ± 0.21 ^{ge}	1.67 ± 0.16 ^{def}	0.86 ± 0.12 ^{jli}
42	No-liner	3.85 ± 0.13 ^{c}	3.01 ± 0.12 ^{c}	2.10 ± 0.08 ^{cd}
	Decco	1.11 ± 0.30 ^{gi4}	0.65 ± 0.18 ^{jki}	0.71 ± 0.11 ^{jl}
	Zoe	0.62 ± 0.07 ^{ih}	0.39 ± 0.09 ^{jk}	0.68 ± 0.14 ^{jl}
	Xtend	2.20 ± 0.64 ^{de}	1.19 ± 0.22 ^{gfhi}	1.42 ± 0.16 ^{fegh}
	2 mm HDPE	1.50 ± 0.08 ^{ge}	1.30 ± 0.10 ^{gfh}	0.95 ± 0.19 ^{jhk}
	4 mm HDPE	2.79 ± 0.12 ^{d}	2.04 ± 0.11 ^{de}	1.50 ± 0.12 ^{feg}
56	No-liner	5.39 ± 0.12 ^{b}	3.74 ± 0.18 ^{b}	2.93 ± 0.16 ^{b}
	Decco	1.34 ± 0.34 ^{gfh}	0.84 ± 0.15 ^{jh}	0.86 ± 0.11 ^{jli}
	Zoe	0.78 ± 0.06 ^{gi}	0.50 ± 0.17 ^{jk}	0.84 ± 0.16 ^{jli}
	Xtend	2.81 ± 0.15 ^{d}	1.58 ± 0.25 ^{dg}	1.88 ± 0.14 ^{cde}
	2 mm HDPE	2.06 ± 0.06 ^{def}	1.50 ± 0.12 ^{ge}	1.34 ± 0.24 ^{fghi}
	4 mm HDPE	3.56 ± 0.14 ^{c}	2.74 ± 0.10 ^{c}	2.07 ± 0.15 ^{cd}
84	No-liner	7.41 ± 0.14 ^{a}	5.27 ± 0.08 ^{a}	4.10 ± 0.17 ^{a}
	Decco	1.34 ± 0.32 ^{gfh}	1.09 ± 0.17 ^{ghi}	1.01 ± 0.10 ^{jhk}
	Zoe	0.98 ± 0.08 ^{gi}	0.82 ± 0.18 ^{jh}	0.89 ± 0.01 ^{jik}
	Xtend	3.87 ± 0.49 ^{c}	2.64 ± 0.47 ^{c}	2.34 ± 0.04 ^{c}
	2 mm HDPE	2.67 ± 0.10 ^{d}	2.09 ± 0.11 ^{d}	1.73 ± 0.13 ^{fd}
	4 mm HDPE	4.93 ± 0.24 ^{b}	3.71 ± 0.10 ^{b}	2.83 ± 0.03 ^{b}

Results presented as mean ± S.E. Different letter(s) on column per liner treatment indicate statistically significant differences ($p < 0.05$) according to Duncan's multiple range test. HDPE: high density polyethylene.

At the end of the 84 d of cold storage, fruit in no-liner lost 8.6% of the initial fruit length, while fruit packed in non-perforated 'Decco' and 'Zoe' liners lost 1.2% and 1.0% in fruit length, respectively. Fruit in micro-perforated Xtend® liners lost 2.7% of initial fruit length compared with 3.4 and 5.1% for fruit packed in 2 and 4 mm macro-perforated HDPE liners, respectively. A similar pattern was observed for loss in fruit diameter, where fruit packed with no-liner lost 5.4% compared to 1.1 and 0.8% for fruit packed with non-perforated 'Decco' and 'Zoe' liners, respectively. Micro-perforated Xtend®, 2 mm macro-perforated HDPE, and 4 mm macro-perforated HDPE liners minimised loss in fruit diameter to 2.2, 2.1, and 3.7%, respectively. A reduction in fruit circumference is a direct indicator of fruit shrinkage. After 84 d of cold storage, fruit packed with no-liner lost 4.1% of their initial circumference, compared to 1.0% and 0.8% for fruit packed in non-perforated 'Decco' and 'Zoe' liners, respectively. Perforated liners minimised the loss in fruit circumference to about half the loss in no-liner. Fruit packed with micro-perforated Xtend® liners lost 2.3% of their initial circumference compared to 1.8 and 2.8% for fruit packed in 2 and 4 mm macro-perforated HDPE liners, respectively.

Generally, the loss was more in fruit length than in fruit diameter. Shrivelling was more concentrated on the base of the fruit than on the sides. Quite similar results observed by Al-Mughrabi et al. [43] on different pomegranate cultivars conventionally stored in plastic boxes at storage temperatures of 5 °C, 10 °C, and ambient temperature for 56 d. The authors observed that the loss in fruit length and diameter is influenced by storage time, temperature, and cultivar. In their study, the cv. 'Manfaloti' with relatively lower fruit weight loss also registered lower loss in fruit diameter and length, as compared to cv. 'Banati'.

3.2.3. Peel Thickness

The dynamics of moisture loss of fruit may influence each of the fruit fractions differently. The porous nature and position of the pomegranate fruit skin makes it so prone to moisture loss because it comes into direct contact with the surrounding. Moisture loss in pomegranate fruit is primarily from the peel resulting in a reduction in peel thickness [5,44]. The greatest loss in peel thickness was observed in fruit packed with no-liner. Fruit packed in non-perforated liners retained more peel thickness than fruit packed in perforated liners (Figure 9). After 84 d of cold storage, fruit packed with no-liner lost 41.8% of the initial peel thickness. However, non-perforated 'Decco' and 'Zoe' liners minimised the loss in fruit peel thickness to 14.8 and 13.2%, respectively. Fruit lost 26.8% peel thickness when packed in micro-perforated Xtend® liners, and 22.0 and 26.7% in 2 mm macro-perforated and 4 mm macro-perforated HDPE liners, respectively. Similarly, Arendse et al. [5] reported a decrease in peel thickness with storage time of pomegranate (cv. Wonderful) packed in conventional corrugated boxes and stored at different temperatures (5, 7.5, 10, and 21 °C). The authors attributed the drastic decrease in peel thickness to low RH and high temperature (21 °C). The thicker peel of fruit packed with non-perforated liners can be attributed to higher RH inside bags compared to fruit packed with perforated liners.

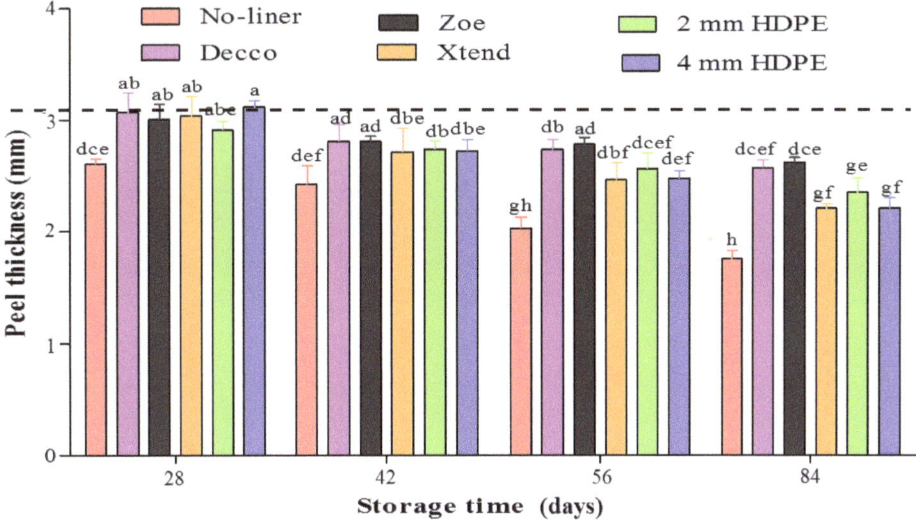

Figure 9. Peel thickness of pomegranate (cv. Wonderful) fruit stored at 5 °C and 90% relative humidity (RH). Mean values (vertical bars) with different letters are significantly different ($p < 0.05$) according to Duncan's multiple range test. HDPE: high density polyethylene.

3.2.4. Fruit Shrivelling

The effect of liner packaging on fruit shrivelling is summarised in Figure 10A,B. Fruit shrivelling results from moisture loss and subsequent loss in cell turgor pressure [47]. In pomegranates, shrivelling is expected after a 5% loss in fruit weight [38]. Fruit shrivelling was evident at 42 d of cold storage after 5.1% loss in weight for fruit packed with no-liner, with 86.1% incidences of shrivelling (Figure 10A). At 56 d of cold storage, shrivel incidence increased to 100% for fruit packed with no-liner. However, there was no incidences of fruit shrivelling observed for fruit packed with non-perforated 'Decco' and 'Zoe' liners throughout 84 d of storage. Slight shrivelling was observed especially at the crown area for fruit packed with micro-perforated Xtend® and 2 mm macro-perforated HDPE liners after 84 d of storage, with a shrivel incidence of 83.3 and 85.7%, respectively. However, shrivelling started at 56 d for fruit packed with 4 mm macro-perforated HDPE liners with an incidence of 72.7%.

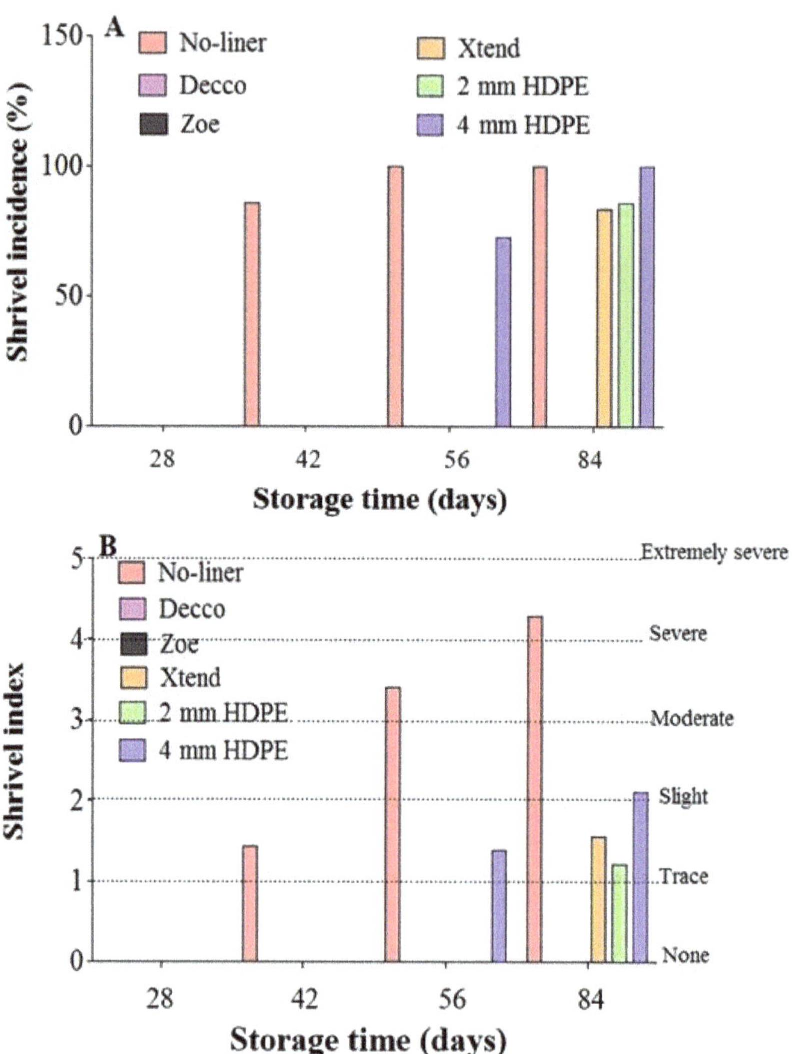

Figure 10. Percentage of total shrivelled fruit (shrivel incidence) (**A**) and shrivel index (incidence) (**B**) observed on pomegranate fruit stored for 84 d at 5 °C and 90% relative humidity (RH). HDPE: high density polyethylene.

The severity of fruit shrivelling (shrivel index) increased with storage time (Figure 10B). At 84 d of storage, fruit packed with no-liner were severely shrivelled with a shrivelling index of 4.3 (86.0%) compared to cases of extreme shrivelling with an index of 5 (100%). However, fruit packed with micro-perforated Xtend® and 2 mm macro-perforated HDPE liners were tracely shrivelled with shrivel index of 1.6 (31.1%) and 1.2 (24.3%), respectively. Fruit packed with 4 mm macro-perforated HDPE liners were slightly shrivelled, having a shrivel index of 2.1 (42.0%). The high shrivel incidence and index in fruit packed with no-liner is attributed to excessive moisture loss during storage. Plastic liners, due to their barrier ability, maintain high relative humidity around the fruit, minimising moisture loss and subsequent shrivelling. Wiley et al. [41] did not observe shrivelling in kiwifruit packed in non-perforated and macro-perforated liners, but reported shrivelling for fruit packed with micro-perforated liners, after 119 d storage at 0 °C.

3.3. Respiration Rate

Respiration rate (RR) of pomegranates (non-climacteric fruit) was generally low, and the decrease with storage period (Figure 11) may be attributed to senescence after harvest. Throughout the storage period, respiration rate was highest in fruit packed with no-liner, followed by fruit packed with perforated liners and lowest in fruit packed with non-perforated liners. Respiration rate for fruit packed in non-perforated liners decreased from 8.1 to about 3.3 mL CO_2 kg^{-1} h^{-1} within 42 d of cold storage and remained stable to the end of storage. Mphahlele et al. [20] reports quite similar trend for pomegranate (cv. Wonderful) packed in MAP liners, where RR decreased within 28 d and stayed stable throughout 84 d of storage at 7 °C. The authors observed higher RR in control fruit than fruit packed with MAP at the end of 3 months. The initial respiration rate of fruit before storage decreased by 28.4% at the end of 84 d of storage for fruit packed with no-liner, compared to 61.7 and 59.3% for fruit packed in non-perforated 'Decco' and 'Zoe' liners, respectively. Micro-perforated Xtend® and 4 mm macro-perforated HDPE liners reduced respiration rate of the fruit by 42.0% compared to 37.0% by 2 mm macro-perforated HDPE liners.

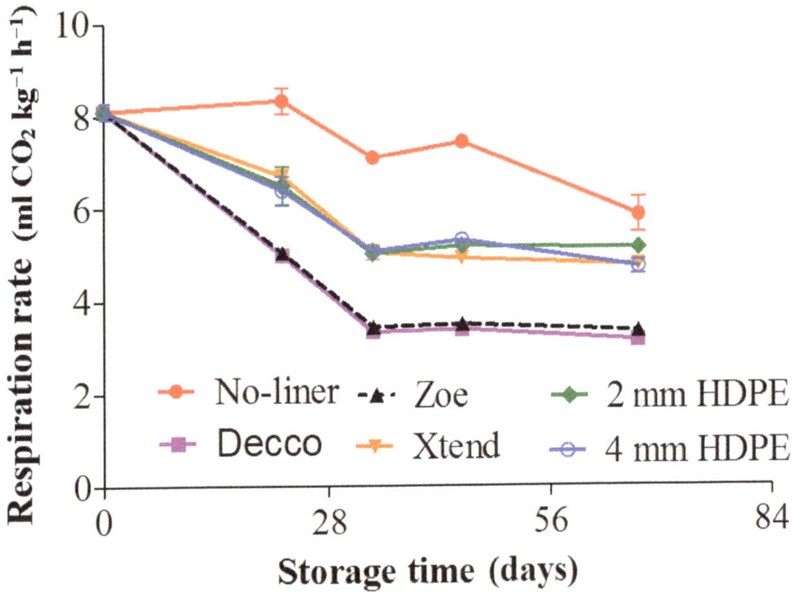

Figure 11. Respiration rate of pomegranate fruit determined by a closed system at 5 °C and 90% relative humidity (RH). HDPE: high density polyethylene.

Other researchers also reported a decline in respiration rate with storage time for pomegranate fruit [6,48]. Passive MAP achieved by non-perforated 'Decco' and 'Zoe' liners is probably responsible for the low respiration rate. Nanda et al. [8] reported that MAP inform of shrink-wrapping reduced respiration rate of pomegranate, attributing it to the ability of the films having a low permeability to gases. Furthermore, the lower RR in fruit packed with non-perforated and perforated liners compared to fruit packed with no-liner can be attributed to alleviation of water stress from around the fruit [49].

3.4. *Textural Properties*

3.4.1. Fruit Puncture Resistance

The ability of harvested fruit to resist a puncturing force provides information on the structural integrity. There was a decline in fruit puncture resistance with storage time for all treatments. The baseline (initial) fruit texture was best retained by fruit packed with non-perforated liners followed by fruit packed in perforated liners and no-liner packaging. At the end of 84 d of cold storage, fruit packed with no-liner lost 28.3% of the initial whole fruit firmness (116.1 ± 2.0 N). However, packing fruit in non-perforated 'Decco' and 'Zoe' liners reduced fruit firmness by 8.0% and 6.8%, respectively. Micro-perforated Xtend® liners minimised fruit firmness by 12.0% compared to 15.8% and 15.5% by 2 mm macro-perforated HDPE and 4 mm macro-perforated HDPE liners, respectively (Table 4). The general decline in texture with storage time can be attributed to fruit softening resulting from enzymatic disintegration of cell wall structure [50]. Similar results reported by Mansouri et al. [51], and Arendese et al. [5] reported declines in whole fruit firmness with storage time for different conventionally packed pomegranate fruit (cv. Hondos-e-Yalabad, Malas-e-Saveh, and Wonderful) in boxes. The higher respiration rate observed in fruit packed with no-liner and macro-perforated liners may have contributed to the higher loss in fruit texture, compared to fruit packed in passively modified atmosphere by non-perforated liners. Drake et al. [52] observed that 'Bartlett' pears at low temperatures of 1 °C packed in MAP liners retained more fruit firmness than pear packed under regular atmosphere. The authors reported that storing pears under controlled atmosphere retained fruit firmness throughout cold storage, irrespective of packaging treatment. Similar to our results, Kumar et al. [37] reported that pomegranate (cv. 'Baghwa') packed in Xtend® MAP liners retained better and desirable firmness compared to fruit packed with polypropylene liners and with no-liner stored at 4 °C for 120 d.

Table 4. Whole fruit puncture resistance and aril firmness for pomegranate fruit packed in different liner bags for 84 d at 5 °C and 90% relative humidity (RH).

Storage Time (d)	Treatment	Whole Fruit Puncture Resistance (N)	Aril Firmness (N)
0		116.11 ± 1.96 ^{ab}	143.91 ± 1.51 ^{dce}
28	No-liner	119.66 ± 2.99 ^{a}	146.43 ± 1.96 ^{db}
	Decco	114.54 ± 2.19 ^{abc}	143.11 ± 2.20 ^{de}
	Zoe	115.008 ± 1.90 ^{abc}	143.39 ± 3.93 ^{de}
	Xtend	114.51 ± 2.46 ^{abcd}	142.22 ± 2.11 ^{df}
	2 mm HDPE	113.79 ± 2.62 ^{abcd}	141.87 ± 1.69 ^{df}
	4 mm HDPE	114.36 ± 2.30 ^{abcd}	140.87 ± 2.01 ^{df}
42	No-liner	104.50 ± 2.06 ^{hf}	150.67 ± 2.02 ^{ab}
	Decco	114.88 ± 1.39 ^{abc}	142.84 ± 2.04 ^{df}
	Zoe	115.00 ± 1.71 ^{abc}	142.98 ± 1.59 ^{df}
	Xtend	111.74 ± 1.34 ^{eb}	138.26 ± 1.98 ^{fe}
	2 mm HDPE	109.86 ± 2.52 ^{ebf}	140.06 ± 2.11 ^{df}
	4 mm HDPE	106.85 ± 1.61 ^{eh}	138.86 ± 1.95 ^{fe}
56	No-liner	92.68 ± 1.65 ^{i}	153.93 ± 2.27 ^{a}
	Decco	113.05 ± 1.36 ^{eb}	141.30 ± 1.05 ^{df}
	Zoe	113.89 ± 2.17 ^{abcg}	142.76 ± 1.51 ^{df}
	Xtend	109.30 ± 1.05 ^{ecf}	139.84 ± 1.90 ^{df}
	2 mm HDPE	102.51 ± 1.97 ^{hgi}	138.74 ± 1.25 ^{fe}
	4 mm HDPE	101.15 ± 2.03 ^{hi}	140.29 ± 1.60 ^{df}
84	No-liner	83.30 ± 2.60 ^{j}	149.82 ± 1.30 ^{abc}
	Decco	106.82 ± 1.16 ^{eh}	141.69 ± 1.34 ^{df}
	Zoe	108.18 ± 1.50 ^{edfg}	141.01 ± 1.53 ^{df}
	Xtend	101.39 ± 1.80 ^{hi}	139.82 ± 1.16 ^{df}
	2 mm HDPE	97.74 ± 1.05 ^{ji}	135.98 ± 0.75 ^{f}
	4 mm HDPE	98.09 ± 1.43 ^{ji}	138.88 ± 1.24 ^{fe}

Mean values with different letters are significantly different ($p < 0.05$) according to Duncan's multiple range test. HDPE: high density polyethylene.

3.4.2. Aril Firmness

Generally, aril firmness increased in fruit packed with no-liner compared to decreasing aril firmness in fruit packed with liners (Table 4). The increase in aril firmness for fruit packed with no-liner could be attributed to moisture loss, leading to hardening of aril tissues. The decrease in aril firmness is often associated to quality deterioration and may be due to physiological activity such as respiration that brings about softening and disintegration of cell wall structure [50,53]. There was no significant difference in aril firmness for fruit packed with liners throughout the storage period. Fruit packed with non-perforated 'Decco' and 'Zoe' liners retained more aril firmness compared to fruit packed with perforated liners. At the end of 84 d of storage, fruit packed in either of the non-perforated ('Decco' and 'Zoe') liners lost 2.0% of initial aril firmness (143.9 ± 1.5 N), compared to 2.8, 5.5, and 3.5% for fruit in micro-perforated Xtend®, 2 mm macro-perforated HDPE liners, and 4 mm macro-perforated HDPE liners, respectively. Liners have been reported to maintain desirable firmness in pomegranate and table grape [36,37]. Similar results have been reported with the application of heat shrinkable films on pomegranate fruit [8,17].

3.5. Fruit Decay

The incidence of decayed fruit increased with storage time in all treatments. A similar trend was observed in pomegranate cultivars 'Mollar de Elche' and 'Wonderful' stored at 6 °C and 7 °C, respectively [20,54]. At the end of 84 d of cold storage, 35.4% of fruit packed with no-liner were lost to visible mould. However, packing fruit in non-perforated 'Decco' and 'Zoe' liners minimised decay incidence to 24.0 and 26.0%, respectively. Furthermore, packing fruit in micro-perforated Xtend® liners minimised fruit decay incidence to 17.7%, compared to 24.0 and 18.5% for fruit packed in 2 mm macro-perforated and 4 mm macro-perforated HDPE liners, respectively (Figure 12A). Selcuk and Erkan [22] reported similar results on 'Hicrannar' pomegranate stored at 6 °C for 120 d, where the no-liner control registered 40% decay compared to 13.3 and 26.7% for MAP liner treatments. On the contrary, Laribi et al. [54] and Mphahlele et al. [20] reported higher decay incidence in pomegranate (cv. 'Mollar de Elche' and 'Wonderful') packed with MAP liners than with no-liners at the end of 84 and 140 d of cold storage, respectively. However, no significant difference in decay incidence between shrink-wrapped and non-wrapped pomegranate (cv. 'Primosole) at 70 d of cold storage was reported by D'Aquino et al. [17]. The higher decay incidence of fruit packed in non-perforated liners could be attributed to higher moisture condensation within liner bags and lower WVTR across film, resulting into accelerated fruit moulding compared to fruit packed in perforated liners (Figure 12A).

Fruit decay severity provides insight into the extent of the decay on a particular fruit. The influence of packaging treatments on fruit decay severity was different from their influence on decay incidence. Fruit packed with no-liner had the highest decay severity index than fruit packed in liners. The severity (index) of decay was higher in fruit packed with perforated liners compared to fruit packed in non-perforated liners (Figure 12B). This could be attributed to a lower respiration rate observed in fruit packed with non-perforated 'Decco' and 'Zoe' liners compared to fruit packed in perforated liners. Selcuk and Erkan [22] reported a no significant difference in decay index for control treatment and MAP liner treatments for pomegranate stored at 6 °C for 120 d.

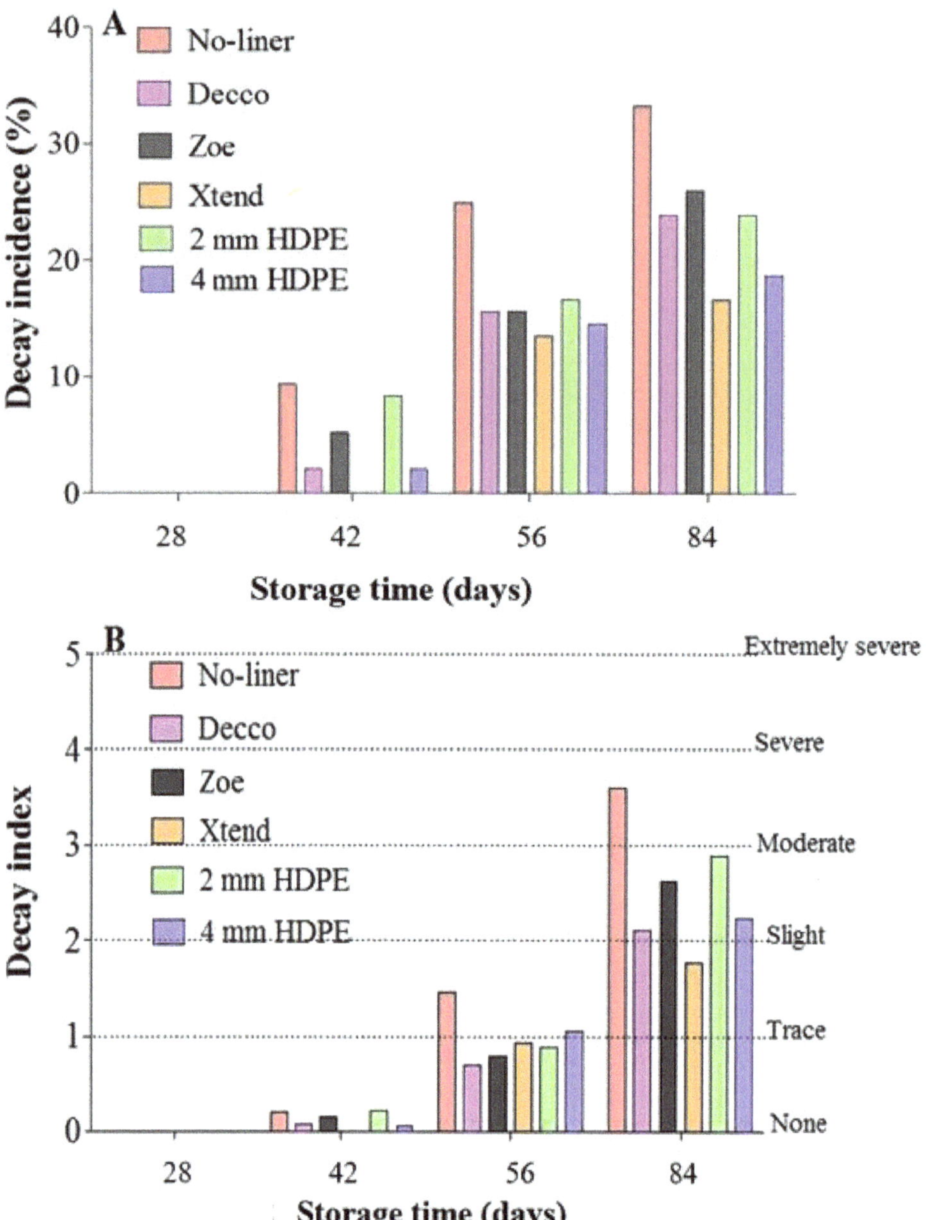

Figure 12. (**A**) Percentage cumulative decay incidence. (**B**) Cumulative decay index (severity). Pomegranate fruit stored for 84 d at 5 °C and 90% relative humidity (RH). HDPE: high density polyethylene.

3.6. Colour Attributes

3.6.1. Fruit Peel Colour

Fruit peel colour is an important contributor to visual appeal and acceptance of pomegranate fruit by consumers. Generally, there was a progressive decline in the lightness (L^*) values with storage time (Table 5). Fruit skin lightness was 51.7 ± 2.4 before storage, after 84 d of storage; fruit packed with no-liner lost 30.0% of the lightness. This can be attributed to excessive moisture loss causing the peel to become pale. The darkening ($L^* \rightarrow 0$) of the peel is expected as the fruit ages during storage. However, packing fruit with non-perforated 'Decco' and 'Zoe' liners significantly ($p < 0.05$) minimised the loss in skin colour lightness to 5.7% and 3.6%, respectively. Fruit packed with micro-perforated Xtend® lost 14.1% compared to 15.0% and 15.7% for fruit packed with 2 mm macro-perforated and 4 mm macro-perforated HDPE liners, respectively. The difference in results can be attributed to differences in the ability of liners to minimise moisture loss and respiration rate. Similarly, Selcuk and Erkan [22] reported higher skin colour lightness for pomegranate (cv. Hicrannar) fruit stored under MAP liners with the fruit looking brighter and fresher compared to the no liner control fruit at the end of 4 months of cold storage and additional 3 d of shelf life.

Table 5. Impact of liners on pomegranate fruit peel colour parameters. Fruit were stored at 5 °C and 90% relative humidity (RH).

Time (d)	Treatment	L^* [1]	a^* [2]	C^* [3]
0		51.70 ± 2.36 [a]	29.74 ± 0.34 [a]	40.72 ± 0.51 [ab]
28	No-liner	51.39 ± 2.10 [a]	29.15 ± 0.66 [ab]	40.56 ± 0.73 [ab]
	Xtend	50.11 ± 2.36 [ab]	29.73 ± 1.56 [a]	40.74 ± 0.98 [ab]
	Decco	51.32 ± 1.67 [a]	29.85 ± 1.40 [a]	40.36 ± 0.78 [ab]
	Zoe	51.21 ± 2.00 [a]	29.68 ± 1.80 [a]	40.91 ± 0.94 [a]
	2 mm HDPE	50.83 ± 2.78 [a]	29.71 ± 1.76 [a]	40.16 ± 1.71 [ab]
	4 mm HDPE	51.40 ± 1.52 [a]	29.56 ± 1.08 [ab]	41.16 ± 0.70 [a]
42	No-liner	46.22 ± 1.08 [ad]	27.10 ± 0.66 [ad]	38.05 ± 0.59 [db]
	Xtend	48.56 ± 1.72 [abc]	29.21 ± 1.61 [ab]	39.98 ± 0.95 [ab]
	Decco	49.15 ± 1.80 [abc]	28.33 ± 1.68 [abc]	39.97 ± 1.02 [ab]
	Zoe	51.07 ± 2.19 [a]	28.72 ± 0.42 [ab]	40.05 ± 0.78 [ab]
	2 mm HDPE	48.99 ± 2.78 [abc]	28.95 ± 2.01 [ab]	39.74 ± 1.37 [ab]
	4 mm HDPE	50.11 ± 1.10 [ab]	27.17 ± 0.52 [ad]	39.67 ± 0.50 [ab]
56	No-liner	41.68 ± 0.54 [d]	26.75 ± 0.76 [ad]	36.45 ± 0.56 [dec]
	Xtend	46.39 ± 0.99 [ad]	26.75 ± 0.65 [ad]	39.41 ± 0.44 [ab]
	Decco	49.13 ± 1.22 [abc]	26.33 ± 0.48 [ad]	39.45 ± 0.48 [ab]
	Zoe	49.95 ± 2.46 [ab]	27.02 ± 0.36 [ad]	39.91 ± 0.49 [ab]
	2 mm HDPE	46.24 ± 1.09 [ad]	26.81 ± 0.20 [ad]	38.01 ± 1.20 [db]
	4 mm HDPE	46.27 ± 0.85 [ad]	25.83 ± 0.51 [db]	38.92 ± 0.73 [abc]
84	No-liner	36.20 ± 0.85 [e]	24.23 ± 0.48 [d]	32.90 ± 0.67 [e]
	Xtend	44.44 ± 0.58 [db]	26.43 ± 0.59 [ad]	38.54 ± 0.56 [ab]
	Decco	48.75 ± 1.19 [abc]	26.15 ± 0.48 [ad]	38.80 ± 0.47 [abc]
	Zoe	49.84 ± 0.70 [ab]	26.90 ± 0.68 [ad]	38.06 ± 0.90 [abc]
	2 mm HDPE	43.95 ± 0.72 [dc]	24.91 ± 0.42 [dc]	35.43 ± 0.66 [d]
	4 mm HDPE	43.58 ± 0.77 [dc]	24.47 ± 0.47 [d]	35.99 ± 0.40 [d]

[1] Lightnesss index describes surface color in the range from 0 (black) to 100 (white). [2] Redness index describes surface color in the range from green ($-a^*$) to red ($+a^*$). [3] Chroma. Mean values with different letters are significantly different ($p < 0.05$) according to Duncan's multiple range test. HDPE: high density polyethylene.

There was no difference in peel redness colour (a^*) among treatments throughout the study; however at 84 d, fruit packed with micro-perforated Xtend® and non-perforated 'Decco' and 'Zoe' liners retained the initial skin redness colour (a^*) before storage. However, Drake [52] reported that packing pears in MAP liners preserved more of the green colour at 90 d of cold storage than did the pears under regular atmosphere at 30 d of storage. This could be attributed to the ability of liners modifying the atmosphere around the fruit, thereby minimising break down of colour pigments.

The effect of storage time on chroma (C^*) was only significant on fruit packed with no-liner and macro-perforated HDPE liners. At 84 d of cold storage, fruit packed with non-perforated 'Decco', non-perforated 'Zoe', and micro-perforated Xtend® liners significantly retained the initial skin C^* (colour saturation) compared with fruit in other treatments. Furthermore, fruit packed with macro-perforated HDPE liners significantly retained higher

C^* than fruit packed with no-liner. Selcuk and Erkan [22] reported a no significant impact of liner packaging on the chroma (C^*) for 'Hicrannar' pomegranate stored for 120 d at 6 °C. A decrease in skin colour parameters L^* and C^* was observed with minimal changes for wrapped fruit compared to un-wrapped pomegranate (cv. Primosole) stored at 8 °C for 84 d storage [17].

3.6.2. Aril Colour

The colour of arils is very important, especially in terms of the consumption of fresh pomegranate fruit. There was a significant effect of storage time on lightness (L^*), redness (a^*), and chroma (C^*) colour attributes of arils for all treatments (Table 6). Fruit packed with liners significantly retained higher L^* and a^* aril colour attributes in comparison with fruit packed with no-liner at 84 d of cold storage. Fruit packed with no-liner retained 55.7% of the initial aril L^* colour attribute, compared to 82.4 and 76.9% for fruit packed with non-perforated 'Decco' and 'Zoe' liners, respectively. Fruit packed in micro-perforated Xtend® liner retained 67.3% of aril L^* colour attribute, with no significant difference compared to 68.4% and 70.4% for fruit packed with 2 mm macro-perforated and 4 mm macro-perforated HDPE liners, respectively. These results could be attributed to the influence of liner packaging on fruit weight loss and respiration rate. Excessive loss of moisture and high respiration rate by the no-liner packed fruit could have resulted in the loss of aril colour lightness and redness due to water stress and degradation of colour pigments.

Table 6. Impact of liner treatment on pomegranate aril colour parameters. Fruit were stored for 84 d at 5 °C and 90% relative humidity (RH).

Time (d)	Treatment	L^* [1]	a^* [2]	C^* [3]
0		20.86 ± 0.38 [abc]	19.19 ± 0.41 [a]	21.09 ± 0.66 [a]
28	No-liner	20.80 ± 0.54 [abc]	18.20 ± 1.55 [abc]	19.78 ± 1.71 [abcd]
	Xtend	21.90 ± 0.68 [a]	18.08 ± 0.46 [abc]	20.30 ± 0.55 [abc]
	Decco	21.50 ± 0.51 [ab]	18.83 ± 0.52 [ab]	20.33 ± 0.59 [abc]
	Zoe	21.59 ± 1.06 [ab]	18.60 ± 0.62 [ab]	20.66 ± 0.91 [ab]
	2 mm HDPE	21.09 ± 0.42 [abc]	18.25 ± 0.83 [abc]	19.75 ± 1.02 [abcd]
	4 mm HDPE	21.90 ± 1.44 [a]	18.30 ± 0.52 [abc]	19.97 ± 0.39 [abcd]
42	No-liner	16.56 ± 1.06 [ei]	15.31 ± 1.08 [fd]	16.45 ± 1.21 [ge]
	Xtend	19.04 ± 1.25 [eb]	17.01 ± 1.31 [abcde]	18.38 ± 1.42 [abcdef]
	Decco	19.65 ± 0.39 [abcd]	15.10 ± 0.91 [fd]	16.32 ± 1.04 [ge]
	Zoe	19.35 ± 0.88 [abcd]	16.43 ± 1.01 [fb]	17.67 ± 1.16 [gb]
	2 mm HDPE	18.77 ± 1.06 [ecf]	14.49 ± 0.67 [feg]	15.74 ± 0.83 [geh]
	4 mm HDPE	18.15 ± 0.90 [edfg]	15.86 ± 0.66 [fc]	17.06 ± 0.82 [gd]
56	No-liner	14.02 ± 0.69 [i]	12.26 ± 0.75 [hg]	13.47 ± 0.89 [ih]
	Xtend	16.35 ± 0.40 [if]	15.80 ± 0.66 [fc]	17.82 ± 0.62 [gb]
	Decco	17.42 ± 0.42 [edfg]	17.16 ± 0.44 [abcd]	18.30 ± 0.57 [abcdef]
	Zoe	18.05 ± 0.35 [edfg]	17.37 ± 0.59 [abcd]	18.58 ± 0.74 [abcde]
	2 mm HDPE	17.29 ± 1.46 [edfg]	16.03 ± 0.83 [fc]	17.48 ± 0.91 [fgc]
	4 mm HDPE	15.60 ± 0.63 [ig]	15.48 ± 0.51 [fd]	16.95 ± 0.57 [gd]
84	No-liner	11.62 ± 0.68 [j]	11.34 ± 0.43 [h]	12.94 ± 0.62 [i]
	Xtend	14.14 ± 0.76 [i]	14.04 ± 0.29 [fg]	15.24 ± 0.31 [gi]
	Decco	17.19 ± 0.67 [edfgh]	16.04 ± 1.06 [fc]	17.22 ± 1.23 [gd]
	Zoe	16.04 ± 0.56 [ig]	16.01 ± 0.72 [fc]	18.03 ± 0.94 [gb]
	2 mm HDPE	14.26 ± 0.59 [i]	14.50 ± 0.58 [feg]	15.39 ± 0.64 [gif]
	4 mm HDPE	14.69 ± 0.58 [ih]	14.46 ± 0.58 [feg]	16.68 ± 0.83 [ge]

[1] Lightnesss index, describes surface color in the range from 0 (black) to 100 (white). [2] Redness index, describes surface color in the range from green ($-a^*$) to red ($+a^*$). [3] Chroma. Mean values with different letters are significantly different ($p < 0.05$) according to Duncan's multiple range test. HDPE: high density polyethylene.

There was no significant difference in a^* and C^* aril colour attributes among all fruit packed with liners (Table 6). Therefore, perforation of liners did not have an effect on the redness and tone saturation colour attributes of the arils. Similarly, Arendse et al. [5] observed significant decrease of aril colour parameters L^*, a^*, and C^* with storage time for pomegranate (cv. 'Wonderful') fruit packed in boxes and stored at different temperature conditions.

3.7. Principal Component Analysis

The averages of quality attributes of pomegranate fruit packed with no-liner, non-perforated 'Decco' and 'Zoe', micro-perforated Xtend®, and macro-perforated 2 and 4 mm HDPE liners are shown in Figures 13 and 14. Total variability was explained by five principal factors. Shipping fruit takes 42 d across the Atlantic Ocean from South Africa to Europe, which is the main export market. After 42 d of storage, the first two principal factors (F1 and F2) explained 85.8% of the total variability. Along F1 (explaining 70.9% of total variability), packaging fruit with no-liner was associated with higher weight loss, shrivelling, high respiration rate, and aril hardening by 42 d of storage. The same attributes associated with no-liner packaging had high negative values along F1 (Table 7). On the other hand, packing fruit with both non-perforated and perforated liners was associated with retaining fruit puncture resistance and peel colour attributes of L^*, C^*, and a^*. The same attributes associated with liner packaging had high positive values along F1 (Table 7). Along F2 (explaining 14.9% of total variability), packing fruit with no-liner, non-perforated 'Decco' and 'Zoe', and macro-perforated 2 mm HDPE were associated to facilitating fruit decay (incidence and index). Variables of decay incidence and index had high positive values along F2 (Table 7).

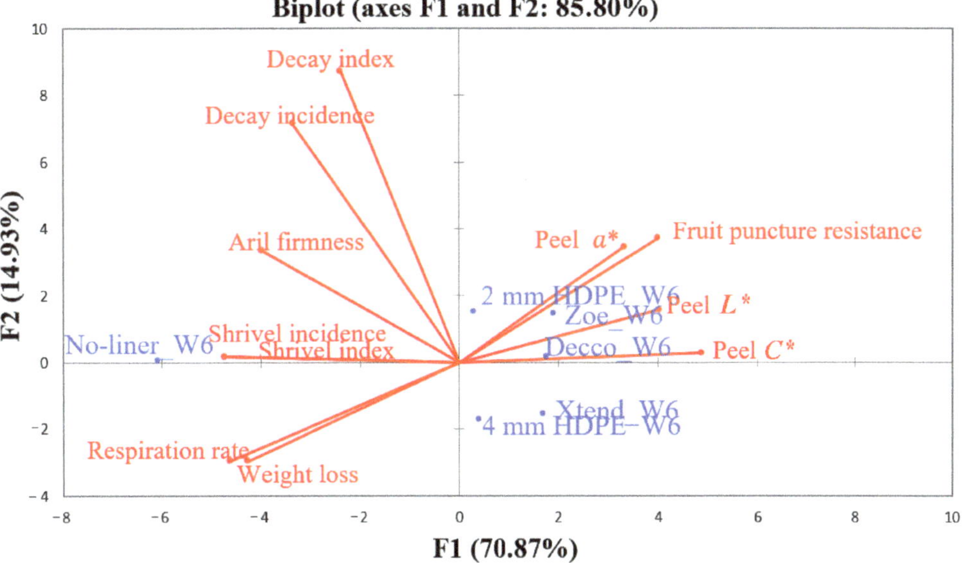

Figure 13. Principal component analysis of the first two factors (F1 and F2) due to physical and physiological attributes of pomegranate (cv. Wonderful) after 42 d of storage at 5 °C and 95% relative humidity (RH). W6 = 42 d. L^*: lightness index describes surface color in the range from 0 (black) to 100 (white); a^*: redness index describes surface color in the range from green ($-a^*$) to red ($+a^*$); C^*: chroma; HDPE: high density polyethylene.

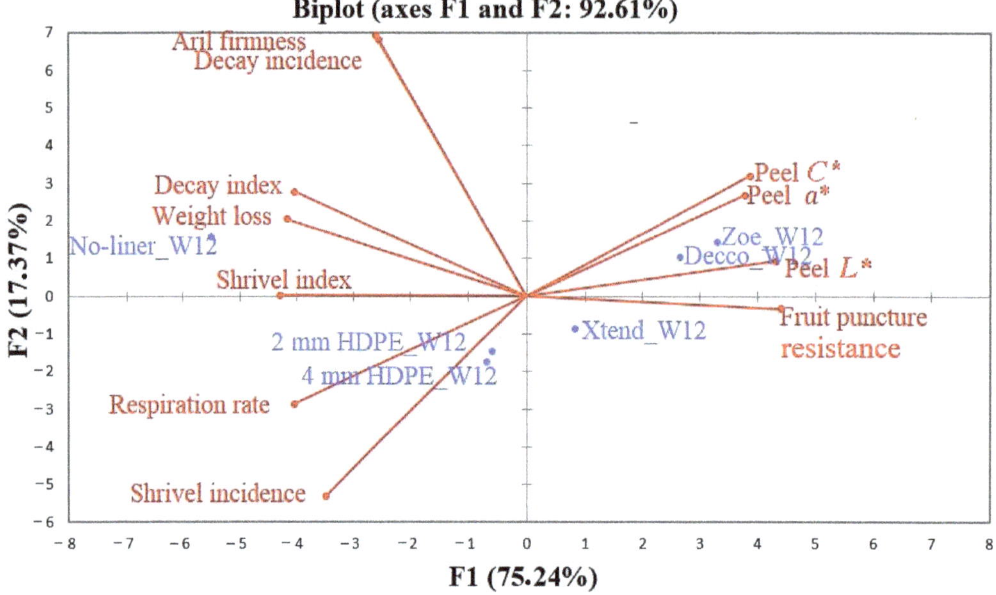

Figure 14. Principal component analysis of the first two factors (F1 and F2) due to physical and physiological attributes of pomegranate (cv. Wonderful) after 84 d of storage at 5 °C and 95% relative humidity (RH). W12 = 84 d. L^*: lightness index describes surface color in the range from 0 (black) to 100 (white); a^*: redness index describes surface color in the range from green ($-a^*$) to red ($+a^*$); C^*: chroma; HDPE: high density polyethylene.

Table 7. Factor loadings and scores for the first two principal factors (F1 and F2) for pomegranate packed in no-liner and with different liners. HDPE: high density polyethylene.

	42 d of Storage		84 d of Storage	
	F1	F2	F1	F2
Factor Loadings				
Weight loss	−0.955	−0.278	−0.945	0.224
Fruit firmness	0.818	0.349	0.999	−0.036
Aril firmness	−0.824	0.317	−0.588	0.755
Decay incidence	−0.695	0.674	−0.579	0.741
Decay index	−0.496	0.825	−0.913	0.302
Shrivel incidence	−0.972	0.017	−0.791	−0.578
Shrivel index	−0.972	0.017	−0.973	0.003
Respiration rate	−0.879	−0.279	−0.913	−0.312
Peel L^*	0.826	0.150	0.979	0.102
Peel a^*	0.682	0.328	0.858	0.292
Peel C^*	0.998	0.029	0.878	0.348
Factor Scores				
No-liner_W6	−6.070	0.050	−5.506	1.587
Decco_W6	1.782	0.166	2.648	1.042
Zoe_W6	1.898	1.486	3.296	1.419
Xtend_W6	1.709	−1.537	0.821	−0.869
2 mm HDPE_W6	0.300	1.536	−0.584	−1.453
4 mm HDPE_W6	0.380	−1.701	−0.676	−1.727

After 84 d of storage, a clearer separation between packaging treatments was observed (Figure 14). In this case, the first two component factors (F1 and F2) explained 92.6% of the total variability, with F1 and F2 accounting for 75.2 and 17.4%, respectively. Along F1, packaging fruit with no-liner and macro-perforated 2 and 4 mm HDPE liners was associated with facilitating fruit weight loss, shrivelling (incidence and index), respiration rate, and decay index. On the other hand, packing fruit with non-perforated 'Decco' and 'Zoe' and micro-perforated Xtend® liners was associated with retaining fruit puncture resistance and peel colour attributes of L^*, C^*, and a^*. Along F2, packing fruit with no-liner and non-perforated 'Decco' and 'Zoe' liners was associated with decay incidence and aril firmness (or hardness as applicable to no-liner packed fruit).

4. Conclusions

High incidence of postharvest fruit losses and waste is a major problem facing the pomegranate industry worldwide [55,56]. The use of plastic liners as internal packages in the multi-scale packaging of pomegranate fruit plays a major role in minimising quantitative and qualitative losses during prolonged cold storage. Packaging pomegranate (cv. Wonderful) in non-perforated liners greatly minimises mass loss and maintains fruit colour and textural quality during cold storage for 84 d at 5 °C. Micro-perforated Xtend® and 4 mm macro-perforated HDPE liners were able to minimise moisture condensation within the bags and reduced decay incidence, which are some of the challenges of packing fruit in non-perforated liners. Packing fruit with perforated liners also greatly minimised fruit mass and size loss and retained acceptable quality during prolonged storage compared to packing fruit with no-liner. Using micro-perforated Xtend® and macro-perforated 4 mm HDPE can be considered to minimise postharvest losses often associated with inadequate environment control inside packaging compared to the use of non-perforated liners (note that 4 mm HDPE liners are over three times cheaper than micro-perforated Xtend® liners).

Author Contributions: Conceptualisation, U.L.O.; methodology, R.L.; validation, A.A.; formal analysis, A.A., R.L.; investigation, A.A., R.L.; writing—original draft preparation, R.L.; writing—review and editing, A.A., U.L.O.; visualisation, A.A., R.L.; supervision, U.L.O., A.A.; project administration, U.L.O.; funding acquisition, U.L.O. All authors have read and agreed to the published version of the manuscript.

Funding: This work is based on the research supported wholly/in part by the National Research Foundation of South Africa (grant numbers: 64813). The opinions, findings, and conclusions or recommendations expressed are those of the author(s) alone, and the NRF accepts no liability whatsoever in this regard. Research reported in this publication was supported in part by the Foundation for Food and Agriculture Research under award number 434—grant ID: DFs-18-0000000008.

Institutional Review Board Statement: Not applicable.

Informed Consent Statement: Not applicable.

Data Availability Statement: Data sharing is contained within the article.

Acknowledgments: We acknowledge the award of postgraduate scholarship to R. Lufu by the Intra-ACP Sharing Capacity to Build Capacity for Quality Graduate Training in Agriculture in African Universities (SHARE), and the Regional Universities' Forum for Capacity Building in Agriculture (RUFORUM) for support.

Conflicts of Interest: The authors declare no conflict of interest.

References

1. Fawole, O.A.; Opara, U.L. Developmental changes in maturity indices of pomegranate fruit: A descriptive review. *Sci. Hortic.* **2013**, *159*, 152–161. [CrossRef]
2. Kader, A.A. Postharvest Biology and Technology of Pomegranates. In *Pomegranates: Ancient Roots to Modern Medicine*; Seeram, N.P., Schulman, R.N., Heber, D., Eds.; CRC/Taylor & Francis: Boca Raton, FL, USA, 2006; pp. 211–220, ISBN 9788578110796.
3. Opara, L.U.; Al-Ani, M.R.; Al-Shuaibi, Y.S. Physico-chemical Properties, Vitamin C Content, and Antimicrobial Properties of Pomegranate Fruit (*Punica granatum* L.). *Food Bioprocess Technol.* **2009**, *2*, 315–321. [CrossRef]
4. Wetzstein, H.Y.; Zhang, Z.; Ravid, N.; Wetzstein, M.E. Characterization of Attributes Related to Fruit Size in Pomegranate. *HortScience* **2011**, *46*, 908–912. [CrossRef]

5. Arendse, E.; Fawole, O.A.; Opara, U.L. Influence of storage temperature and duration on postharvest physico-chemical and mechanical properties of pomegranate fruit and arils. *CyTA J. Food* **2014**, *12*, 389–398. [CrossRef]
6. Elyatem, S.M.; Kader, A.A. Post-harvest physiology and storage behaviour of pomegranate fruits. *Sci. Hortic.* **1984**, *24*, 287–298. [CrossRef]
7. Caleb, O.J.; Opara, U.L.; Witthuhn, C.R. Modified Atmosphere Packaging of Pomegranate Fruit and Arils: A Review. *Food Bioprocess Technol.* **2011**, *5*, 15–30. [CrossRef]
8. Nanda, S.; Rao, D.S.; Krishnamurthy, S. Effects of shrink film wrapping and storage temperature on the shelf life and quality of pomegranate fruits cv. Ganesh. *Postharvest Biol. Technol.* **2001**, *22*, 61–69. [CrossRef]
9. Opara, L.U.; Studman, C.J.; Banks, N.H.; Opara, U.L. Fruit Skin Splitting and Cracking. *Hortic. Rev.* **2010**, *19*, 217–262. [CrossRef]
10. Vigneault, C.; Thompson, J.; Wu, S. Designing Container for Handling Fresh Horticultural Produce. In *Postharvest Technologies for Horticultural Crops*; Benkeblia, N., Ed.; Research Signpost: Kerala, India, 2009; Volume 2, pp. 25–47, ISBN 9788130803562.
11. Fawole, O.A.; Riva, S.C.; Opara, U.L. Efficacy of Edible Coatings in Alleviating Shrivel and Maintaining Quality of Japanese Plum (*Prunus salicina* Lindl.) during Export and Shelf Life Conditions. *Agronomy* **2020**, *10*, 1023. [CrossRef]
12. Safizadeh, M.R. The Effect of Various Film Packaging, Wax Coating and Storage Conditions on the Shelf Life and Quality of Pomegranate Fruits. *J. Hortic. Res.* **2019**, *27*, 47–54. [CrossRef]
13. Motamedi, E.; Nasiri, J.; Malidarreh, T.R.; Kalantari, S.; Naghavi, M.R.; Safari, M. Performance of carnauba wax-nanoclay emulsion coatings on postharvest quality of 'Valencia' orange fruit. *Sci. Hortic.* **2018**, *240*, 170–178. [CrossRef]
14. Cofelice, M.; Lopez, F.; Cuomo, F. Quality Control of Fresh-Cut Apples after Coating Application. *Foods* **2019**, *8*, 189. [CrossRef]
15. Bai, J.; Hagenmaier, R.D.; Baldwin, E.A. Coating selection for 'Delicious' and other apples. *Postharvest Biol. Technol.* **2003**, *28*, 381–390. [CrossRef]
16. Artés, F.; Villaescusa, R.; Tudela, A.J. Modified atmosphere packaging of pomegranate. *J. Food Sci.* **2000**, *65*, 1112–1116. [CrossRef]
17. D'Aquino, S.; Palma, A.; Schirra, M.; Continella, A.; Tribulato, E.; La Malfa, S. Influence of film wrapping and fludioxonil application on quality of pomegranate fruit. *Postharvest Biol. Technol.* **2010**, *55*, 121–128. [CrossRef]
18. Gil, M.I.; Sánchez, R.; Marin, J.G.; Artés, F. Quality changes in pomegranates during ripening and cold storage. *Eur. Food Res. Technol.* **1996**, *202*, 481–485. [CrossRef]
19. Cantín, M.C.; Crisosto, H.C.; Day, R.K. Evaluation of the effect of different modified atmosphere packaging box liners on the quality and shelf life of 'Friar' plums. *Horttechnology* **2008**, *18*, 261–265. [CrossRef]
20. Mphahlele, R.R.; Fawole, O.A.; Opara, U.L. Influence of packaging system and long term storage on physiological attributes, biochemical quality, volatile composition and antioxidant properties of pomegranate fruit. *Sci. Hortic.* **2016**, *211*, 140–151. [CrossRef]
21. Selcuk, N.; Erkan, M. Changes in phenolic compounds and antioxidant activity of sour–sweet pomegranates cv. 'Hicaznar' during long-term storage under modified atmosphere packaging. *Postharvest Biol. Technol.* **2015**, *109*, 30–39. [CrossRef]
22. Selcuk, N.; Erkan, M. Changes in antioxidant activity and postharvest quality of sweet pomegranates cv. Hicrannar under modified atmosphere packaging. *Postharvest Biol. Technol.* **2014**, *92*, 29–36. [CrossRef]
23. Ambaw, A.; Mukama, M.; Opara, U. Analysis of the effects of package design on the rate and uniformity of cooling of stacked pomegranates: Numerical and experimental studies. *Comput. Electron. Agric.* **2017**, *136*, 13–24. [CrossRef]
24. Mukama, M.; Ambaw, A.; Berry, T.M.; Opara, U.L. Energy usage of forced air precooling of pomegranate fruit inside ventilated cartons. *J. Food Eng.* **2017**, *215*, 126–133. [CrossRef]
25. Ben-Yehoshua, S.; Rodov, V.; Fishman, S.; Peretz, J. Modified-atmosphere packaging of fruits and vegetables: Reducing condensation of water in bell peppers and mangoes. *Acta Hortic.* **1998**, 387–392. [CrossRef]
26. Opara, U.L.; Atukuri, J.; Fawole, O.A. Application of physical and chemical postharvest treatments to enhance storage and shelf life of pomegranate fruit—A review. *Sci. Hortic.* **2015**, *197*, 41–49. [CrossRef]
27. Opara, U.L.; Hussein, Z.; Caleb, O.J.; Mahajan, P. Investigating the effects of perforation and storage temperature on water vapour transmission rate of packaging films: Experimental and modelling approaches. *Wulfenia J.* **2015**, *9*, 498–509.
28. Artés, F.; Tudela, J.A.; Gil, M.I. Improving the keeping quality of pomegranate fruit by intermittent warming. *Z. Lebensm. Forsch. A* **1998**, *207*, 316–321. [CrossRef]
29. Fawole, O.A.; Opara, U.L. Harvest Discrimination of Pomegranate Fruit: Postharvest Quality Changes and Relationships between Instrumental and Sensory Attributes during Shelf Life. *J. Food Sci.* **2013**, *78*, S1264–S1272. [CrossRef]
30. Pathare, P.; Opara, U.L.; Vigneault, C.; Delele, M.A.; Al-Said, F.A.-J. Design of Packaging Vents for Cooling Fresh Horticultural Produce. *Food Bioprocess Technol.* **2012**, *5*, 2031–2045. [CrossRef]
31. Almenar, E.; Del-Valle, V.; Hernández-Muñoz, P.; Lagarón, J.M.; Catalá, R.; Gavara, R. Equilibrium modified atmosphere packaging of wild strawberries. *J. Sci. Food Agric.* **2007**, *87*, 1931–1939. [CrossRef]
32. Mistriotis, A.; Giannoulis, A.; Giannopoulos, D.; Briassoulis, D. Analysis of the effect of perforation on the permeability of biodegradable non-barrier films. *Procedia Food Sci.* **2011**, *1*, 32–38. [CrossRef]
33. Dirim, S.; Özden, H.; Bayındırlı, A.; Esin, A. Modification of water vapour transfer rate of low density polyethylene films for food packaging. *J. Food Eng.* **2004**, *63*, 9–13. [CrossRef]
34. Mastromatteo, M.; Lucera, A.; Lampignano, V.; Del Nobile, M.A. A new approach to predict the mass transport properties of micro-perforated films intended for food packaging applications. *J. Food Eng.* **2012**, *113*, 41–46. [CrossRef]
35. Hussein, Z.; Caleb, O.J.; Opara, U.L. Perforation-mediated modified atmosphere packaging of fresh and minimally processed produce—A review. *Food Packag. Shelf Life* **2015**, *6*, 7–20. [CrossRef]

36. Ngcobo, M.E.; Delele, M.A.; Opara, U.L.; Meyer, C.J. Performance of multi-packaging for table grapes based on airflow, cooling rates and fruit quality. *J. Food Eng.* **2013**, *116*, 613–621. [CrossRef]
37. Kumar, A.; Babu, J.; Bhagwan, A.; Kumar, M.R. Effect of modified atmosphere packaging on shelf life and quality of 'bhagwa' pomegranate in cold storage. *Acta Hortic.* **2013**, 963–969. [CrossRef]
38. Kader, A.A.; Chordas, A.; Elyatem, S. Response of pomegranate to ethylene treatment and storage temperature. *Calif. Agric.* **1984**, *38*, 14–15.
39. Xanthopoulos, G.T.; Templalexis, C.G.; Aleiferis, N.P.; Lentzou, D.I. The contribution of transpiration and respiration in water loss of perishable agricultural products: The case of pears. *Biosyst. Eng.* **2017**, *158*, 76–85. [CrossRef]
40. Lufu, R.; Ambaw, A.; Opara, U.L. The contribution of transpiration and respiration processes in the mass loss of pomegranate fruit (cv. Wonderful). *Postharvest Biol. Technol.* **2019**, *157*, 110982. [CrossRef]
41. Wiley, P.; Crisosto, C.H.; Mitchell, F.G. Adapting perforated box liners to the california kiwifruit industry. *Acta Hortic.* **1999**, *498*, 299–306. [CrossRef]
42. Thompson, J.F.; Mitchell, F.G.; Rumsey, T.R.; Kasmire, R.F.; Crisosto, C.H. *Commercial Cooling of Fruits, Vegetables, and Flowers*, Revised ed.; Division of Agriculture and Natural Resources, University of California: Oakland, CA, USA, 2008; ISBN 9781601076199.
43. Al-mughrabi, M.A.; Bacha, M.A.; Abdelrahman, A.O. Effect of storage temperature and duration on fruit quality of three pomegranate cultivars. *J. King Saudi Univ.* **1995**, *7*, 239–248.
44. Mukama, M.; Ambaw, A.; Berry, T.M.; Opara, U.L. Analysing the dynamics of quality loss during precooling and ambient storage of pomegranate fruit. *J. Food Eng.* **2019**, *245*, 166–173. [CrossRef]
45. Lufu, R.; Ambaw, A.; Opara, U.L. Water loss of fresh fruit: Influencing pre-harvest, harvest and postharvest factors. *Sci. Hortic.* **2020**, *272*, 109519. [CrossRef]
46. Lufu, R.; Tsige, A.A.; Opara, U.L. *Characterising Water Loss in Pomegranate Fruit Cultivars ('Acco', 'Herskawitz' & 'Wonderful') under Cold and Shelf Storage Conditions*; Research Square: Durham, NC, USA, 2020.
47. Paull, R. Effect of temperature and relative humidity on fresh commodity quality. *Postharvest Biol. Technol.* **1999**, *15*, 263–277. [CrossRef]
48. Artés, F.; Marín, J.G.; Martínez, J.A. Controlled atmosphere storage of pomegranate. *Eur. Food Res. Technol.* **1996**, *203*, 33–37. [CrossRef]
49. Dhall, R.K.; Sharma, S.R.; Mahajan, B.V.C. Effect of shrink wrap packaging for maintaining quality of cucumber during storage. *J. Food Sci. Technol.* **2012**, *49*, 495–499. [CrossRef]
50. Martin-Cabrejas, M.A.; Waldron, K.W.; Selvendran, R.R.; Parker, M.L.; Moates, G.K. Ripening-related changes in the cell walls of Spanish pear (*Pyrus communis*). *Physiol. Plant.* **1994**, *91*, 671–679. [CrossRef]
51. Mansouri, Y.; Khazaei, J.; Hassan-Beygi, S.; Mohtasebi, S.S. Post harvest characteristics of pomegranate (*Punica granatum* L.) fruit. *Cercet. Agron. Mold.* **2011**, *44*, 5–16. [CrossRef]
52. Drake, S.; Elfving, D.; Visser, D.; Drake, S. Quality of modified atmosphere packaged "bartlett" pears as influenced by time and type of storage. *J. Food Process. Preserv.* **2005**, *28*, 348–358. [CrossRef]
53. Ekrami-Rad, N.; Khazaei, J.; Khoshtaghaza, M.-H. Selected Mechanical Properties of Pomegranate Peel and Fruit. *Int. J. Food Prop.* **2011**, *14*, 570–582. [CrossRef]
54. Laribi, A.I.; Palou, L.; Taberner, V.; Pérez-gago, M.B. Modified atmosphere packaging to extend cold storage of pomegranate cv. 'Mollar de Elche'. In *Post Harvest, Food and Process Engineering, Proceedings of the International Conference of Agricultural Engineering—CIGR-AgEng 2012: Agriculture and Engineering for a Healthier Life, Valencia, Spain, 8–12 July 2012*; CIGR-EurAgEng: Valencia, Spain, 1992; p. 6.
55. Opara, I.K.; Fawole, O.A.; Kelly, C.; Opara, U.L. Quantification of On-Farm Pomegranate Fruit Postharvest Losses and Waste, and Implications on Sustainability Indicators: South African Case Study. *Sustainability* **2021**, *13*, 5168. [CrossRef]
56. Opara, I.K.; Fawole, O.A.; Opara, U.L. Postharvest Losses of Pomegranate Fruit at the Packhouse and Implications for Sustainability Indicators. *Sustainability* **2021**, *13*, 5187. [CrossRef]

MDPI

Article

Extending the Shelf-Life of Fresh-Cut Green Bean Pods by Ethanol, Ascorbic Acid, and Essential Oils

Asmaa H. R. Awad [1], Aditya Parmar [2,*], Marwa R. Ali [3], Mohamed M. El-Mogy [1,*] and Karima F. Abdelgawad [1]

[1] Vegetable Crops Department, Faculty of Agriculture, Cairo University, 12613 Giza, Egypt; soma_h_97@yahoo.com (A.H.R.A.); karima.abdelgawad@agr.cu.edu.eg (K.F.A.)
[2] Natural Resources Institute, University of Greenwich, Central Avenue, Chatham Maritime, Kent ME4 4TB, UK
[3] Food Science Department, Faculty of Agriculture, Cairo University, 12613 Giza, Egypt; marwa3mrf@agr.cu.edu.eg
* Correspondence: a.parmar@gre.ac.uk (A.P.); elmogy@agr.cu.edu.eg (M.M.E.-M.)

Abstract: Green beans are a perishable crop, which deteriorate rapidly after harvest, particularly when minimally processed into ready-to-eat fresh-cut green beans. This study investigated the effectiveness of ethanol, ascorbic acid (AsA), tea tree essential oil (TTO), and peppermint essential oil (PMO) on the quality and storability of fresh-cut green bean pods samples stored at 5 °C for 15 days. Our results indicated that samples treated with ethanol, AsA, TTO, and PMO preserved appearance, firmness (except ethanol), chlorophyll content, and moisture compared with the samples without any treatment (control). Additionally, higher vitamin C, total soluble solids (TSS), total sugars, and total phenolic compounds (TPC) were observed in samples treated with ethanol, AsA, TTO, and PMO compared with the control. The most effective treatments for controlling microbial growth were ethanol followed by either TTO or PMO. All the treatments had positive effects on shelf life, maintained quality, and reducing microbial growth during 15 days of cold storage. A particular treatment can be selected based on the economic feasibility and critical control point in the value chain.

Keywords: *Phaseolus vulgaris*; peppermint; tea tree; storability; minimal processed; ready to eat

Citation: Awad, A.H.R.; Parmar, A.; Ali, M.R.; El-Mogy, M.M.; Abdelgawad, K.F. Extending the Shelf-Life of Fresh-Cut Green Bean Pods by Ethanol, Ascorbic Acid, and Essential Oils. *Foods* **2021**, *10*, 1103. https://doi.org/10.3390/foods10051103

Academic Editors: Eleni Tsantili and Jinhe Bai

Received: 21 April 2021
Accepted: 14 May 2021
Published: 17 May 2021

1. Introduction

Green bean (*Phaseolus vulgaris*) belongs to the family of *Fabaceae* and is considered one of the most important legume crops worldwide. The pods of green bean can be harvested at an immature stage (as a fresh vegetable) or mature stage (for dried seeds). Green bean is a rich source of minerals, vitamins, and dietary fibre that play a significant role in the human diet and wellbeing [1]. However, green bean pods are highly perishable with limited shelf-life due to their high respiration rate. During postharvest, green beans are susceptible to mechanical damage, shriveling, chlorophyll pigment degradation, and increased fibre content [2,3]. These biochemical changes reduce the quality and consumption of green bean pods and decrease their economic and nutritional values.

Due to rapid urbanisation in developing countries, demand for fresh minimally processed refrigerated fruit and vegetable has increased significantly. Minimal processing includes trimming, peeling, coring, cutting, and packing. These unit operations result in some undesirable morphological and physiological changes such as browning, pigmentations, and microbial growth. Additionally, the moisture loss and respiration rate of minimally processed vegetables are much higher during refrigerated storage compared to non-processed vegetables.

The effects of several bioactive compounds in essential oils (EOs) and plant extracts as anti-microbial and shelf life enhancing agents in horticultural crops are well known [4,5]. The United States Food and Drug Administration (FDA) classified many EOs as safe, nontoxic, antioxidant, and anticancer compounds including tea-tree essential oil (TTO) and peppermint essential oil (PMO) [6,7]. TTO and PMO are obtained from *Melaleuca alternifolia* and *Mentha piperita* by hydrodistillation, respectively. Several previous studies have reported the positive effect of TTO for controlling postharvest diseases of fresh fruit and vegetables such as strawberry [8] and lettuce [9]. It has been reported that calcium chloride + chitosan + TTO emulsion treatment preserved the firmness and overall quality of fresh-cut red bell pepper for 18 d at 4 °C [10]. New investigations on the effect of PMO as a postharvest treatment showed preserved quality and storability of fresh fruit such as table grapes [11] and dragon fruit [12].

Ascorbic acid (AsA) plays an important role in plant antioxidant systems and human health [13,14]. Positive effects of AsA for controlling enzymatic browning in fruit and vegetables such as plums [15], mung bean sprouts [16], and fresh-cut artichoke [17] has been reported previously. Ethanol is another natural compound that is used in various postharvest treatments. Ethanol is considered safe as per the Generally Recognized As Safe (GRAS) guidelines [18]. Previous reports mentioned that postharvest ethanol treatments (dips or vapour) extend the storage duration of several fresh horticultural products. For example, ethanol has been shown to reduce postharvest fungal diseases of table grapes [19] and Chinese berries [20], delay yellowing of broccoli florets [21], retard senescence in vegetables [22], inhibit the ethylene pathway biosynthesis of melons [23], and suppress the ripening of tomatoes [24].

To the best of the author's knowledge, this is a novel investigation. There were no previous reports on the effect of ethanol, AsA, TTO, and PMO on postharvest behaviour and quality of green beans. The objective of the current study was to assess the effect of these natural compounds on the quality parameters and shelf life of fresh-cut green beans stored for 15 d at 5 °C.

2. Materials and Methods

2.1. Preparation of Plant Material and Treatments

Green bean pods (*Phaseolus vulgaris* L., variety Hama) were harvested from a local private farm and transferred under cooling condition (4 °C) within two hours to the postharvest laboratory. Green bean pods free from defects and damage, with uniform diameter and length, were prepared by cutting the two ends of the pod with a sterile sharp knife. The fresh-cut green bean pods were immersed in four different treatment solutions:

A. Ethanol (0.5%),
B. Ascorbic acid (AsA) (0.5%),
C. Tea tree oil (TTO) (0.5%), and
D. Peppermint oil (PMO) (0.5%).

The control group left without any treatment. TTO and PMO concentrations were prepared by dissolving the required amounts of oils in 50 mL of 0.05 mL Tween-20 and then sterile distilled water was added to obtain 1000 mL of desire concentrations. The treated samples were then dried in a laminar airflow hood for 2 h, and then packed and sealed in micro-perforated (five perforations/cm^2, perforation diameter 0.7 mm) polypropylene bags by using autoclaved forceps. The dimensions of the bags were as follows: length 22 cm, breadth 16.5 cm, and 1 mm thickness; the bags contained 250 g of samples each. Samples were stored at 5 °C and 90% RH for 15 d. Each treatment was carried out in triplicate and the whole experiment was repeated. For each treatment, samples were divided into two groups. One group was used to determine weight loss, decay, and general appearance throughout full storage time and the other was used to determine pod quality parameters (firmness and TSS), chemical compounds (vitamin C, TPC, and chlorophyll content), mould, yeast, and total microbial count. All parameters were measured at time intervals of 0, 3, 6, 9, 12, 15 d after treatments.

2.2. Gas Chromatography-Mass Spectrometry Analysis (GC-MS)

To analyse the chemical compositions of the TTO and PMO, the GC-MS system (Agilent Technologies) equipped with gas chromatography (7890B) and mass spectrometer detector (5977A) (Shimadzu, Kyoto, Japan) was used. Samples were diluted with hexane (1:19, v/v). The GC was equipped with an HP-5MS column (30 m × 0.25 mm internal diameter and 0.25 μm film thickness). Analyses were carried out using helium as the carrier gas at a flow rate of 1.0 mL/min at a split ratio of 1:10, injection volume of 1 μL and the following temperature program: 40 °C for 1 min; rising at 4 °C min^{-1} to 150 °C and held for 6 min; rising at 4 °C min^{-1} to 210 °C and held for 5 min. The injector and detector were held at 280 °C and 220 °C, respectively. Mass spectra were obtained by electron ionization (EI) at 70 eV; using a spectral range of m/z 40 to 550 and solvent delay of 3 min. The identification of different constituents was determined by comparing the spectrum fragmentation pattern with those stored in Wiley and NIST Mass Spectral Library data.

2.3. Free Radical Scavenging Using DPPH

The free radical scavenging activity of the TTO and PMO against 2,2-diphenyl-1-picrylhydrazyl (DPPH) was measured according to the method described in [25]. First, EOs samples were diluted with methanol (50 μL mL^{-1}); then, the sample was mixed with 4 mL DPPH solution (0.24 mg L^{-1} in methanol). The control sample was prepared by mixing methanol with DPPH solution at the same volume. The mixture was mixed well and incubated in the dark at room temperature for 30 min; then, the absorbance was measured at 517 nm using a spectrophotometer (Unico UV-2000, UNICO company, Fairfield, NJ, USA). The free radical scavenging activity of each essential oil was calculated according to Equation (1).

$$Free\ radical\ scavenging\ (\%) = [(A_{control} - A_{sample})/A_{control}] \times 100 \quad (1)$$

where: $A_{control}$ = *absorbance of control sample*, and A_{sample} = *absorbance of essential oil sample.*

2.4. Determination of Weight Loss, General Appearance, Firmness, and Total Soluble Solids

Green bean samples were weighed immediately after (drying in a laminar airflow hood for 2 h) and at every sampling time to measure weight loss by using a digital laboratory scale. Another set of the green bean samples of 250 g each in triplicates) were used for further chemical analysis. To determine general appearance, the following scale was used: 9 = excellent, 7 = very good, 5 = good, 3 = poor, and 1 = unacceptable. The appearance score was assessed by a group of three trained laboratory panelists.

The firmness values of each pod were determined at three different points by using the digital penetrometer (PCE-PTR 200, PCE Americas Inc., Jupiter, FL, USA) with 6 mm diameter probe (range 0 to 1 kg*) [26]. The firmness values were expressed in Newton (N). To determine total soluble solids (TSS), a digital refractometer (model PR101, Atago [0–45%] Co. Ltd., Tokyo, Japan) was used at room temperature [27].

2.5. Determination of Total Sugar, Vitamin C, Total Phenolic Compounds, and Chlorophyll Content

The total sugar content was determined by the anthrone method at 630 nm as described in [28]. Briefly, 200 mg fruit were extracted three times with ethanol (80%) at 80 °C. Then, the extracts were evaporated to dryness and redissolved in 2 mL distilled water. One millilitre of sample extracts was added to 1.5 mL of anthrone reagent (0.2% in H_2SO_4) and mixed thoroughly. The sample was brought to boil using a boiling water bath. The solution was cooled to room temperature and absorbance was measured. The formation of the blue-green complex indicates the presence of total sugars. Glucose was used as a standard. Vitamin C content was determined using the titrimetric method with 2,6-dichlorophenolindophenol described by the Association of Official Analytical Chemistry [29].

TPC was calculated by using the Folin–Ciocalteu reagent with some alteration by using gallic acid as a standard curve [17]. Five grams of the sample was diluted using 5 mL of methanol (80%). The solution was blended with 2.5 mL of Folin–Ciocalteu (10-fold with distilled water) and added to 2.5 mL of distilled water. Afterwards, 2 mL of aqueous sodium carbonate solution (7.5%, w/v) was added after incubation for 5 min. The final solution was mixed and incubated in the dark at room temperature for 1 h. The absorption was assessed at 765 nm using the spectrophotometer, and the results were expressed as milligrams of gallic acid equivalent (GAE) per 100 mg of fresh fruit weight.

Chlorophyll content was determined as described in [30]. In brief, 0.5 g of fresh sample were homogenized with 5 mL dimethyl formamide and kept in the dark in the refrigerator for 48 h. The absorbance was then measured at 470, 647 and 663 nm with a spectrophotometer (model UV-2401 PC, International Equipment Trading LTD. (IET), Milano, Italia).

2.6. Determination of Mould, Yeast and Total Microbial Count

Ten grams of each treatment were homogenized with 90 mL sterile saline for 2 min by a stomacher (Stomacher BW-400P, Turelab, Shanghai, China). Total count and mould and yeast count (MY) were enumerated on plate count agar and potato dextrose agar after incubation at 37 °C for 48 h and 28 °C for 5 d, respectively [31]. The results were expressed as log10 of colony-forming units per gram sample (CFU g^{-1}).

2.7. Statistical Analysis

Statistical analyses of the pooled data from the two experiments were performed with SPSS software. The data were analyzed by one-way analysis of variance (ANOVA). The least significant differences among the treatment means were determined at $p \leq 0.05$ by using the Tukey test. Moreover, a second one-way analysis was carried out to investigate the storage time (in the supplements).

3. Results

3.1. Chemical Composition of Essential Oils and Free Radical Scavenging

The chemical composition of the EOs is shown in Table 1. A total of nine and 13 compounds were identified in the PMO and TTO, respectively. Levomenthol was reported to be the predominant compound (36.27%) in PMO followed by Cyclohexanone, 5-methyl-2-(1-methyl ethyl)-, Cyclohexanone, 5-methyl-2-(1-methyl ethyl)-, trans-, and D-Limonene. In TTO, 4-Terpinenol was reported to be the predominant compound (42.56%) followed by Gamma-Terpinene, Alpha-Terpinene and Terpineol.

Table 1. Identified volatile compounds of peppermint and tea tree essential oils.

Compound	Rt (min) *	(%)
Peppermint oil (PMO)		
Bicyclo [3.1.0]hex-2-ene, 4-methyl-1-(1-methylethyl)-	8.17	1.3
Bicyclo [3.1.0]hexane, 4-methylene-1-(1-methylethyl)-	9.592	1.97
D-Limonene	11.43	11.19
Decanal	11.53	2.41
Isopulegol	15.63	1.67
Cyclohexanone, 5-methyl-2-(1-methylethyl)-, cis-	15.92	25.32
Cyclohexanone, 5-methyl-2-(1-methylethyl)-, trans-	16.3	14.68
Levomenthol	16.6	36.27
Cyclohexanol, 5-methyl-2-(1-methylethyl)-, acetate, (1.alpha.,2.alpha.,5.beta.)-	20.84	5.19
Tea tree oil (TTO)		
Alpha.-Pinene, (-)-	8.16	5.71
(-)-.beta.-Pinene	9.58	2.55
Alpha.-Terpinene	11.00	10.77
Benzene, 1-methyl-3-(1-methylethyl)-	11.29	6.36
D-Limonene	11.43	3.24
Eucalyptol	11.50	2.06

Table 1. *Cont.*

Compound	Rt (min) *	(%)
Gamma.-Terpinene	12.53	10.12
Alpha.-Terpinolene	13.59	3.60
trans-β-Terpineol	15.60	0.49
4-Terpinenol	16.76	42.56
Terpineol	17.23	7.91
γ-Terpineol	17.48	1.43
Caryophyllene	24.90	3.19

* Retention time (RT) (as minutes).

In this study, the free-radical scavenging activity of PMO and TTO were assessed through the DPPH radical scavenging test. The radical scavenging activity was observed in EOs at the concentration of 50 μL mL^{-1} for each PMO and TTO. PMO had lower radical scavenging activity (54%) compared to TTO (63.56%).

3.2. Appearance

The appearance score was gradually lowered during storage time (results presented are from one-way ANOVA analysis for each treatment compound, with storage time being the factor, Supplementary Table S1). At the end of storage (15 d), the control samples showed the lowest appearance score (Figure 1A), whereas the ethanol and AsA treatment showed the highest appearance score from 6 d storage until the end of the storage period in comparison to other treatments. For PMO and TTO, a gradual decrease in appearance was observed during the storage period.

3.3. Weight Loss

Weight loss in samples increased during the storage period for all treatments (Supplementary Table S1). The samples treated with ethanol showed lower levels of weight loss in comparison to control during the entire storage time, and the lowest on days 6 and 12 among treatments (Figure 1B). At the end of the storage, the control samples showed the highest level of weight loss. Moreover, AsA, TTO, and PMO treatments reduced the weight loss when compared to the control from 9 to 15 d.

3.4. Firmness

A reduction in firmness was observed for all the samples with increasing storage period (Supplementary Table S1). There was no difference in firmness between samples treated with ethanol and the control during storage periods except after 9 d of storage (Figure 1C). Moreover, treatments with AsA, PMO and TTO were equally effective in maintaining firmness at higher levels than controls from day 9 and thereafter. It was also shown that no difference was observed in firmness between treatments with TTO and PMO and the control only after three and six days of storage (Figure 1C).

3.5. Chlorophyll Content

Chlorophyll content decreased with an increase in storage time (Supplementary Table S1). These results indicated that there was no difference among treatments after three and six days of storage. However, all treated samples had higher chlorophyll content compared to the control after 9 d and 12 d, whereas ethanol resulted in the highest chlorophyll content after 15 d (Figure 1D). Samples treated with AsA, and TTO had higher chlorophyll content than the control 15 d of storage.

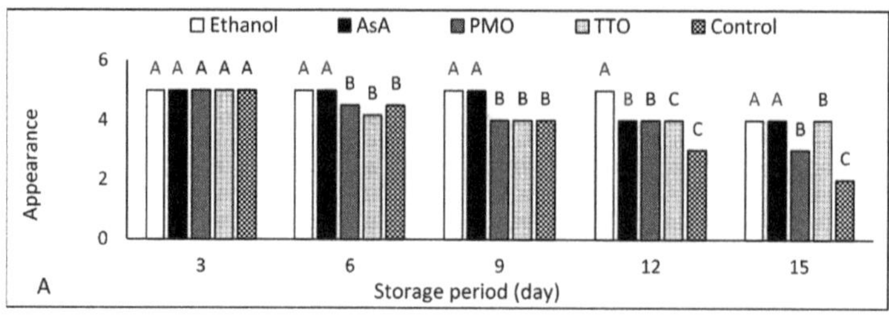

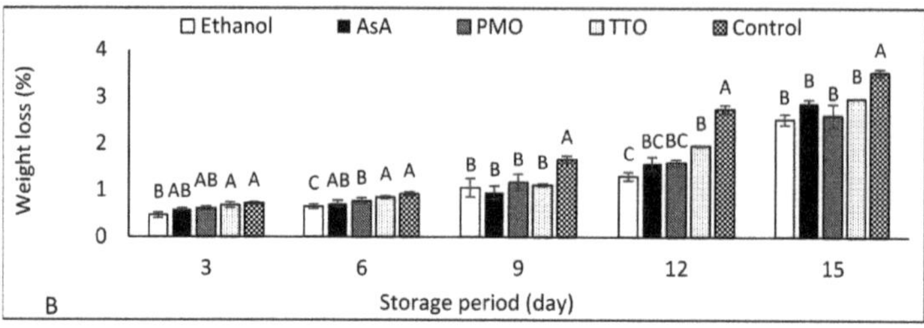

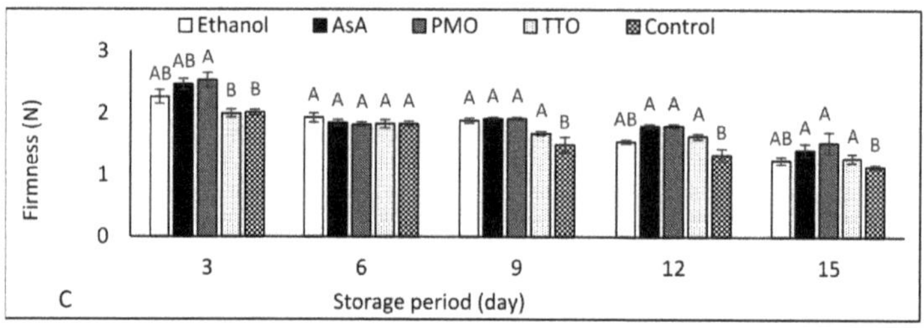

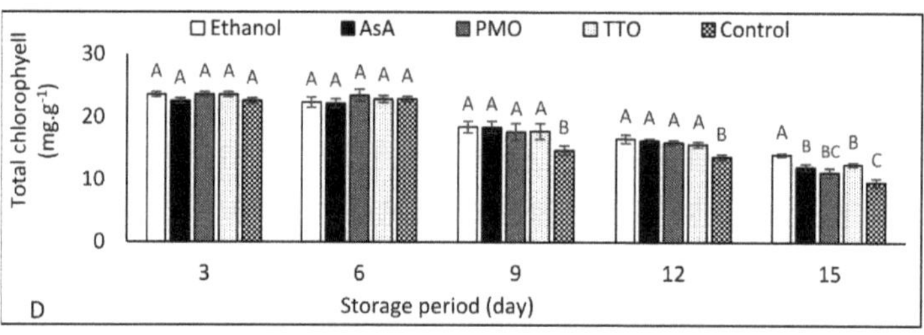

Figure 1. Effect of ethanol, ascorbic acid (AsA), peppermint oil (PMO), and tea tree oil (TTO) on (**A**) appearance, (**B**), weight loss, (**C**) firmness, and (**D**) total chlorophyll of fresh-cut green bean pods stored at 5 °C for 15 d. Different letters indicate significant differences ($p < 0.05$) using the Tukey test at every storage point. Data are means of three replicates. Vertical bars represent standard error.

3.6. Vitamin C

The vitamin C content of fresh green beans decreased with increasing storage time (Supplementary Table S1). The difference among treatments from 3 up to 9 d of storage was not noticeable, while all treatments showed higher vitamin C content compared to the control from 12 to 15 d of storage (Figure 2A).

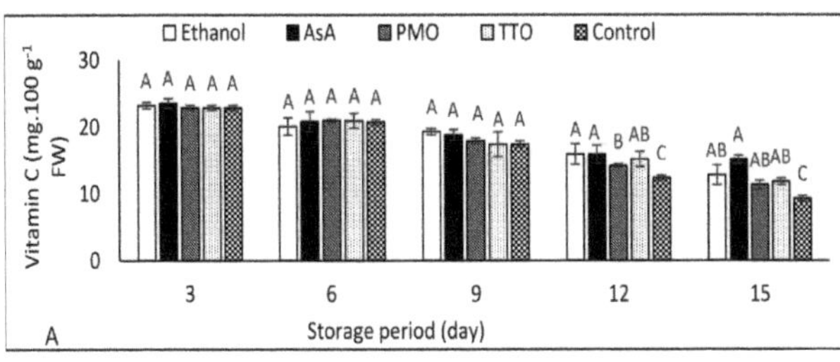

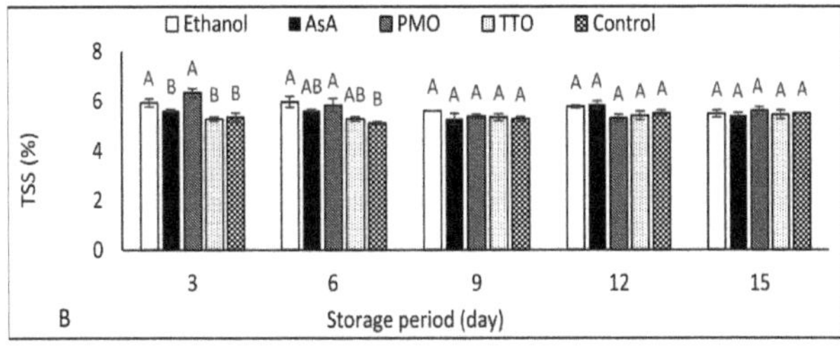

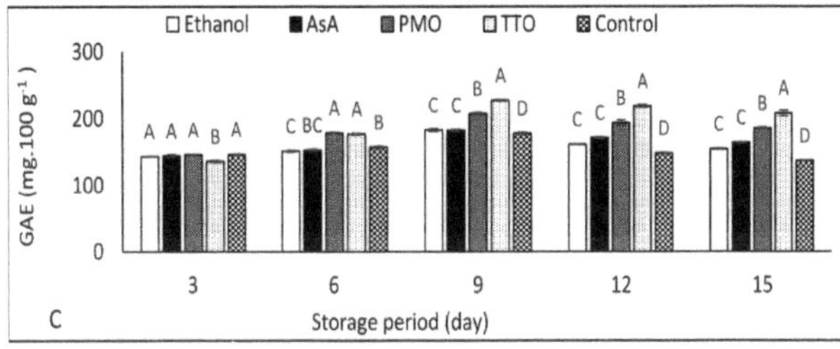

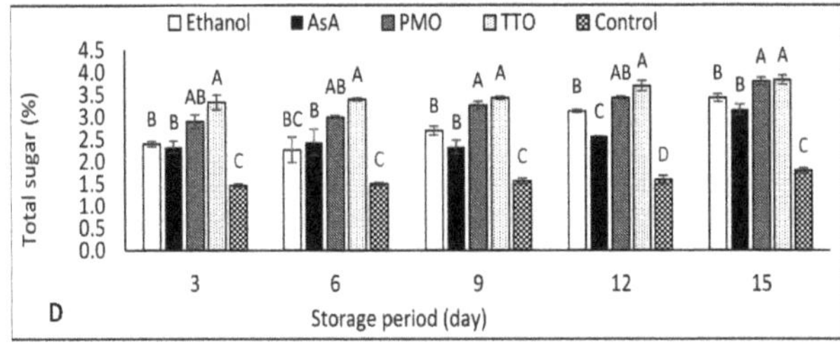

Figure 2. Effect of ethanol, ascorbic acid (AsA), peppermint oil (PMO), and tea tree oil (TTO) on (**A**) vitamin C, (**B**) total soluble solids (TSS), (**C**) total phenolic compound (TPC), and (**D**) total sugars of fresh-cut green bean pods stored at 5 °C for 15 d. Different letters indicate significant differences ($p < 0.05$) using the Tukey test at every storage point. Data are means of three replicates. Vertical bars represent standard error.

3.7. Total Soluble Solids (TSS)

TSS of green beans samples decreased at the beginning of storage in all samples and then slightly increased in AsA, PMO and controls at the end of storage (Supplementary Table S2). The results in Figure 2B indicated that samples treated with ethanol and PMO had a higher TSS than the control after 3 d of storage. However, for the last three storage periods (9, 12, and 15 d), there was no major difference among all treatments.

3.8. Total Phenolic Compounds (TPC)

TPC increased up to 9 d of storage, then slightly decreased in all treatments and control (Supplementary Table S2). TPC was higher in all the treatments when compared to control from 9 d until the end of storage (Figure 2C). TTO treatment exhibited the highest TPC levels, followed by PMO from day 9 and afterwards.

3.9. Total Sugars

The total sugar content of samples treated with ethanol, PMO, TTP, and control slightly increased during the storage time, while the concentration in samples treated with AsA did not change substantially (Supplementary Table S2). As shown in Figure 2D, samples treated with either ethanol or AsA maintained a higher total sugar than the control, but lower than the remaining treatments from day 9 and thereafter.

Samples treated with either PMO or TTO exhibited a higher total sugar content compared to the control and other treatments without a significant difference between them (Figure 2D).

3.10. Mould, Yeast and Total Microbial Count

Mould and yeast (MY) and total microbial count increased with increasing storage time (Supplementary Table S2). In this study, the samples treated with ethanol suppressed the MY and total microbial count until 12 d of storage (Table 2). Our study showed that samples treated with either PMO or TTO suppressed MY and total microbial count during storage until 9 d of storage.

Table 2. Effect of ethanol, ascorbic acid (AsA), peppermint oil (PMO), and tea tree oil (TTO) on mould and yeast and total count (log CFU/g) of fresh-cut green bean pods stored at 5 °C for 15 d. Data are mean of three replicates ± standard errors. Different letters indicate significant differences (Tukey test, $p < 0.05\%$).

	Mould and Yeast (CFU/g)				
	3 d	**6 d**	**9 d**	**12 d**	**15 d**
Ethanol	ND *	ND	ND	ND	1.35 ± 0.03 c
AsA	ND	1.28 ± 0.12 b	1.37 ± 0.03 b	1.48 ± 0.09 ab	1.32 ± 0.02 c
PMO	ND	ND	ND	1.44 ± 0.03 b	1.49 ± 0.01 b
TTO	ND	ND	ND	1.41 ± 0.03 b	1.37 ± 0.02 c
Control	ND	1.54 ± 0.04 a	1.66 ± 0.01 a	1.68 ± 0.01 a	1.79 ± 0.03 a
	Total count (CFU/g)				
	3 d	**6 d**	**9 d**	**12 d**	**15 d**
Ethanol	ND	ND	ND	ND	0.53 ± 0.03 d
AsA	ND	ND	0.63 ± 0.06 b	0.90 ± 0.06 c	1.30 ± 0.05 c
PMO	ND	ND	ND	1.47 ± 0.01 b	1.49 ± 0.01 b
TTO	ND	ND	ND	1.55 ± 0.02 b	1.56 ± 0.04 b
Control	ND	1.71 ± 0.01 a	1.76 ± 0.03 a	1.93 ± 0.02 a	2.05 ± 0.03 a

* ND: not detected (there is no fungal growth found).

4. Discussion

4.1. Chemical Composition of Essential Oils and Free Radical Scavenging

The chemical composition of EOs agreed with Wu et al. [25], who stated that PMO mostly composed of monoterpenes and their derivatives. Vasile et al. [32] showed previously that the TTO contained both light monoterpenes and numerous sesquiterpenes, which are represented as the main component in 4-terpineol, followed by Alpha-Terpinene and Gamma-Terpinene. These were some of the important constituents of PMO and TTO in our results.

Our results indicated that PMO had higher radical scavenging activity than the results reported by others [25] who found that IC50 was recorded at 500 $\mu L\ mL^{-1}$, which is a very high concentration compared to the concentration used in this study (50 μL mL^{-1}). Variations in the antioxidant potential of EOs have been reported, primarily due to the presence of conjugated double bond compounds, which serve as hydrogen/electron donors [33]. PMO and TTO are also able to scavenge free radicals, which are harmful to the body because of their antioxidant function. TTO is also known for its therapeutic properties such as anti-inflammatory, antibacterial, and anticancer activity [34].

4.2. Appearance

Appearance is the most important quality parameter for the consumer's acceptance. Figure 3 shows the appearance differences among treatments after 3 and 15 d from storage of green bean samples. It has been reported that the quality and acceptability of stored shredded carrots was maintained by exogenous AsA application [35], which agrees with our result (Figure 1A). This could be due to the inhibitory effects of AsA on chlorophyll degradation, ripening, and senescence. A sensory evaluation of the overall acceptability of the products, over time, and according to the proposed treatments, would be desirable. Our results shown in Figure 1A agree with a number of previous studies [8,36] that reported that TTO treatment maintained the overall acceptability and appearance of strawberry fruit stored at 20 °C for 3 d. Exogenous application of PMO has been shown to retard ripening, maintain appearance and suppress the decay of Mangosteen fruit [37], primarily due to the preservative effect of EOs as an antioxidant as its major components (Table 1). However, the synergistic or antagonistic influence of one compound in a small proportion of the mixture must be recognized [38].

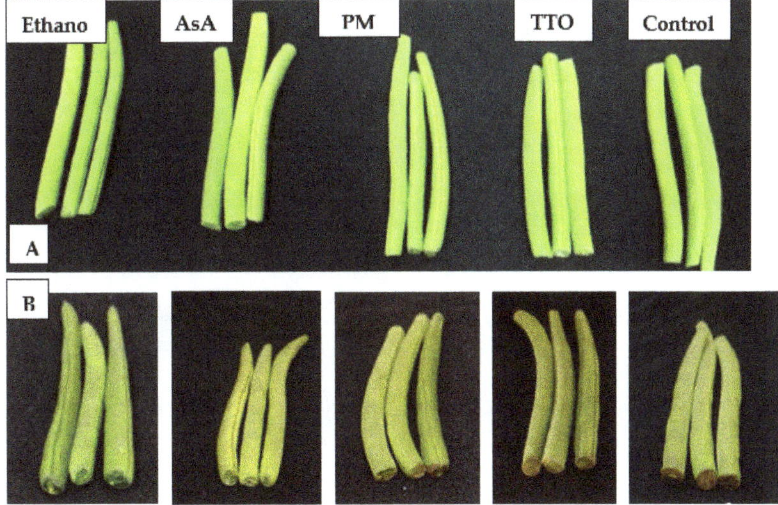

Figure 3. Green bean samples treated with ethanol, AsA, PMO, TTO, and control after (**A**) 3 d and (**B**) 15 d of storage at 5 °C.

4.3. Weight Loss

Weight loss of fresh fruit and vegetables during cold storage is affected by factors such as the structure of the cuticle [39], transpiration, and respiration [17]. Our results agree with Jin et al. [23], who reported a gradual decrease of water loss in sweet melon treated with vapour ethanol stored at 23 °C for 19 d. The reduced weight loss of the green bean samples by the 1% ethanol treatment could be attributed to the inhibitory effects on ethylene biosynthesis resulting in a decrease respiration process [22]. Previously, Ali et al. [40] reported a considerable decrease in the weight loss of litchi fruit treated by AsA compared to untreated samples. They suggested that AsA treatment reduced senescence and metabolic activity, which leads to a reduction of weight loss.

Regarding EOs treatments, our results shown in Figure 1B agree with previous works, where weight loss was decreased in strawberry fruit treated with vapour TTO stored at 20 °C for 3 d [36]. Results in Figure 1B support our hypothesis that PMO could reduce the weight loss of green bean samples. The exogenous application of PMO on dragon fruit reduced water loss during storage at 25 °C for 21 d [12]. Water loss reduction by the application of EOs is primarily due to EOs acting as a semi-permeable layer, resulting in a decrease in the respiratory rate, evaporation, and transpiration [4].

4.4. Firmness

The results in Figure 1C agree with the findings of other researchers [41] who reported that the firmness of Chinese bayberry is not affected by ethanol treatment. In previous studies, the ethanol-treated loquat fruit showed lower firmness values as compared to the non-treated fruit [18]. It has been reported that ethanol vapour treatment maintained the firmness of sweet melon during storage [23]. Therefore, more research is required to investigate the influence of ethanol treatment on the firmness of fresh fruit and vegetables.

AsA treatment retained the firmness of the green bean pods (Figure 1C); similar results were shown in previous works, where sweet pepper and plum fruit treated with AsA had higher firmness than the control due to its antioxidant properties [15,42]. The higher firmness of green bean samples treated with AsA could be explained by higher scavenging of ROS of cells leading to a decreased respiration rate [43]. It is well known that slowing oxidation preserves freshness and colour in fruits and vegetables.

EOs maintained the firmness of treated samples with either TTO or PMO could be due to its properties to inhibit the pectin degradation on the surface of green beans [44]. Additionally, EOs can act as a coating agent, which has a positive effect on respiration rate, leading to a reduction in loss of firmness [12].

4.5. Chlorophyll Content

It has been confirmed that ethanol retards chlorophyll degradation [45], repining and senescence [46]. The maintained chlorophyll content of the samples treated with ethanol is due to its inhibited effects on the activities and gene expression of chlorophyll enzymes [21]. It has been reported that AsA reduced the degradation of chlorophyll, which is correlated to photosystem and therefore resulted in higher chlorophyll content [47]. Our results shown in Figure 1d are in concurrence with Barzegar et al. [42], who found that AsA treatment maintained chlorophyll content in stored sweet pepper and green chilies, respectively.

Our results indicated that PMO and TTO treatments had a positive role in maintaining the chlorophyll content of fresh-cut green beans. This effect can be explained by a lower breakdown of chlorophyll pigments induced by PMO treatment, which acts as a coating agent [48]. Moreover, the thin coating layer around samples formed by TTO treatment could protect them from oxidation and colour changes [49].

4.6. Vitamin C

Vitamin C is classified as a natural antioxidant compound and has a potential role in reducing the risk of cancer by scavenging ROS in the human body [50]. However, vitamin C is rapidly degraded in fresh fruit and vegetables by several factors including storage [17]. Therefore, there is a need to minimize the decrease in vitamin C during refrigerated storage. Our results showed that the applied treatments of essential oil, ethanol and AsA presented a higher vitamin C value only towards the end of the storage period (9 and 12 d) The same results have been observed in previous studies for treated loquat fruit [18] with ethanol and litchi fruit with AsA [40].

The effect of postharvest treatment of TTO on vitamin C content of fruit and vegetable is not identified well due to limited research on this application. For example, a higher concentration of vitamin C was observed in lettuce plants treated before harvest with TTO and stored for 20 d at −2 °C [51]; however, it was mentioned that no significant difference was observed after 3 d of storage. The difference in our study could be due to the difference in time of application and different concentration of TTO, also due to the difference in plant spices.

Our results shown in Figure 2A are in concurrence with previous work by Naeem et al. [48], who found that application of PMO acting as a semi-permeable coating around the samples' surfaces resulting in a reduction of vitamin C loss. The reduction in vitamin C loss by EOs application could be explained by the antioxidant properties of EOs resulting in a reduction in oxygen diffusion and respiration rate [52].

4.7. Total Soluble Solids (TSS)

Treatment with either ethanol or PMO resulted in a higher TSS in green bean samples compared to control (Figure 2B). Previously, Wang et al. [18] observed a higher TSS content in loquat fruit treated with ethanol compared with the control. Additionally, our results agree with early work reporting a gradual increase in TSS in mangosteen fruit treated [37] with PMO. Higher retention of TSS in samples treated with PMO can be related to the reduction of evaporation transpiration and respiration rate leading to conserving TSS [53].

4.8. Total Phenolic Compounds (TPC)

It is well known that TPC are considered the most important antioxidant compounds, responsible for scavenging free radicals, resulting in higher antioxidant defense [54]. Our results shown in Figure 2C agree with Wang et al. [18], who reported that TPC in loquat fruit increased at the beginning of storage and then decreased during the end of storage. In this study, ethanol and AsA treatments conserved the TPC of samples during storage. Additionally, it has been reported that that TPC in mung bean sprouts and litchi fruit was increased by AsA treatment, respectively [16,40]. Higher TPC in white bottom mushrooms treated with PMO compared to the control was recorded, consistent with the results seen in this study (Figure 2C) [55]. One of the explanations for higher TPC is that PMO have phenolic compounds that accumulate and result in higher TPC. Moreover, EOs also affect delaying the senescence process, resulting in maintaining TPC [37].

Samples treated with TTO showed higher TPC in this study; this agrees with previous work that observed higher TPC content during refrigerated storage of lettuce heads treated with TTO [51]. They also mentioned that this result could be ascribed to the induction of PAL by TTO, resulting in higher biosynthesis of phenolic compounds. Additionally, the potential of EOs antioxidant activity could reduce the oxidation of phenolic compound [56].

4.9. Total Sugars

In this study, either ethanol or AsA slowed the reduction of total sugar during cool storage (Figure 2D). Previously [57], a higher total sugar content in stored bitter melon was recorded when treated with ethanol plus melatonin. Another study indicated that treatment with AsA inhibited the reduction of the sugars content in strawberry fruit during cold storage [58]. This result could be due to the reduction of PAL activity by AsA treatment resulting in a lower loss of sugars during storage [59]. Total sugar preservation in samples treated with either PMO or TTO could be explained by the role of EOs in reducing the activity of enzymes that convert starch to sugar in stored crops [53]. Our results agree with previous work, where a reduction in total sugar loss was recorded in mushrooms treated after harvest with PMO compared to control [55].

4.10. Mould & Yeast (MY) and Total Microbial Count

Controlling postharvest diseases by ethanol application has been tested by various previous reports; e.g., *Botrytis cinerea* was controlled in table grape [60] and anthracnose in loquat fruit [18]. The reduced fungal growth in samples treated by ethanol is due to its striking lethal interactions with fungal spore that occurred in the mitochondrial membrane [61]. In this study and previous study on rocket [62], AsA treatment reduced

MY and total microbial count. Regarding AsA application, it reduces the pH; a lower pH value is unsuitable for microbial growth [35].

Previous works demonstrated the inhibited role of EOs and their several compounds for controlling human and plant pathogens [4]. Our results in Table 2 show that both PMO and TTO had an inhibitory effect on microbial growth on the samples surface. The same trend was confirmed by previous work on dragon fruit [12], where PMO inhibited the mould growth during storage for 21 d. It was found that TTO treatment suppressed the growth of the pathogens on the surface of strawberry fruit during refrigerated storage [36], mostly due to inhibitory effects of the main chemical compositions of PMO such as Levomenthol [63] and TTO such as 4-Terpinenol [64] against microbial growth (Table 1).

5. Conclusions

Our study suggested that ethanol, AsA, PMO, and TTO can extend the life of fresh-cut green beans pods under refrigerated conditions by reducing weight loss and maintaining appearance, chlorophyll content, firmness, vitamin C, TSS, total sugar, and TPC. Moreover, a reduction in the mould and yeast count and total microbial count during storage was observed. Ethanol was the most effective application, followed by PMO and TTO, for preserving quality during refrigerated storage at 5 °C up to 15 d. Although all the treatments had positive effects, the best treatment shall be selected based on particular requirement and economic feasibility. For example, if there is a problem of higher microbial load along the value chain, we recommend the use of ethanol as it was the most effective, followed by PMP, TTO. Moreover, further research is required to enhance the storage ability of fresh-cut green beans and other legume crops including peas and board bean.

Supplementary Materials: The following are available online at https://www.mdpi.com/article/10.3390/foods10051103/s1, Table S1: Effect of ethanol, ascorbic acid (AsA), peppermint oil (PMO), and tea tree oil (TTO) on appearance, weight loss, firmness, total chlorophyll, and vitamin C of fresh-cut green bean pods storied at 5 °C for 15 d, Table S2: Effect of ethanol, ascorbic acid (AsA), peppermint oil (PMO), and tea tree oil (TTO) on total soluble solids (TSS), and total phenolic compound (TPC), mold and yeast (CFU/g), and total count (CFU/g) of fresh-cut green bean pods storied at 5 °C for 15 d.

Author Contributions: Conceptualization, A.H.R.A. and K.F.A.; methodology, M.R.A., A.P., K.F.A., and M.M.E.-M.; software, M.M.E.-M., K.F.A.; validation, A.H.R.A. and A.P.; investigation, A.H.R.A., K.F.A. and M.R.A.; resources, M.M.E.-M.; data curation, A.H.R.A.; writing—original draft preparation, A.H.R.A.; writing—review and editing, A.P. and K.F.A.; visualization, M.R.A., A.P. and M.M.E.-M.; supervision, M.M.E.-M. and M.R.A.; project administration, K.F.A.; funding acquisition, M.M.E.-M. All authors have read and agreed to the published version of the manuscript.

Funding: This research received no external funding.

Institutional Review Board Statement: The study did not require ethical approval.

Informed Consent Statement: Not applicable.

Data Availability Statement: Not applicable.

Acknowledgments: We express our thanks to Said Abdullah Shehata, Cairo University, Faculty of Agriculture, Giza, Egypt for providing facilities via his running project "Increase export competition for vegetable crops".

Conflicts of Interest: The authors declare no conflict of interest.

References

1. Brigide, P.; Canniatt-Brazaca, S.G.; Silva, M.O. Nutritional characteristics of biofortified common beans. *Food Sci. Technol.* **2014**, *34*, 493–500. [CrossRef]
2. Xie, G.F.; Zhang, M.S. Research advances in the postharvest storage and preservation techniques of fresh common bean (*Phaseolus vulgaris* L.). *Sci. Technol. Food Ind.* **2019**, *40*, 326–330.
3. Miao, Y.; Tian, W.N.; Hao, C.M.; Rao, L.; Cao, J.K.; Jiang, W.B. Study on pods fibrosis delaying of postharvest common bean by chitosan treatment. *J. China Agr. Univ.* **2012**, *17*, 132–137.

4. El-Mogy, M.M.; Alsanius, B.W. Cassia oil for controlling plant and human pathogens on fresh strawberries. *Food Control* **2012**, *28*, 157–162. [CrossRef]
5. Bakkali, F.; Averbeck, S.; Idaomar, M. Biological effects of essential oils–A review. *Food Chem. Toxicol.* **2008**, *46*, 446–475. [CrossRef]
6. Falleh, H.; Ben Jemaa, M.; Saada, M.; Ksouri, R. Essential oils: A promising eco-friendly food preservative. *Food Chem.* **2020**, *330*, 127268. [CrossRef]
7. Xu, Y.; Wei, J.; Wei, Y.; Han, P.; Dai, K.; Zou, X.; Jiang, S.; Xu, F.; Wang, H.; Sun, J.; et al. Tea tree oil controls brown rot in peaches by damaging the cell membrane of Monilinia fructicola. *Postharvest Biol. Technol.* **2021**, *175*, 111474. [CrossRef]
8. Wei, Y.; Wei, Y.; Xu, F.; Shao, X. The combined effects of tea tree oil and hot air treatment on the quality and sensory characteristics and decay of strawberry. *Postharvest Biol. Technol.* **2018**, *136*, 139–144. [CrossRef]
9. Goñi, M.; Tomadoni, B.; Moreira, M.; Roura, S. Application of tea tree and clove essential oil on late development stages of Butterhead lettuce: Impact on microbiological quality. *LWT Food Sci. Tech.* **2013**, *54*, 107–113. [CrossRef]
10. Sathiyaseelan, A.; Saravanakumar, K.; Mariadoss, A.V.A.; Ramachandran, C.; Hu, X.; Oh, D.-H.; Wang, M.-H. Chitosan-tea tree oil nanoemulsion and calcium chloride tailored edible coating increase the shelf life of fresh cut red bell pepper. *Prog. Org. Coat.* **2021**, *151*, 106010. [CrossRef]
11. Servili, A.; Feliziani, E.; Romanazzi, G. Exposure to volatiles of essential oils alone or under hypobaric treatment to control postharvest gray mold of table grapes. *Postharvest Biol. Technol.* **2017**, *133*, 36–40. [CrossRef]
12. Chaemsanit, S.; Matan, N.; Matan, N. Effect of peppermint oil on the shelf-life of dragon fruit during storage. *Food Control* **2018**, *90*, 172–179. [CrossRef]
13. El-Mogy, M.M.; Garchery, C.; Stevens, R. Irrigation with salt water affects growth, yield, fruit quality, storability and marker-gene expression in cherry tomato. *Acta Agric. Scand. Sect. B Plant Soil Sci.* **2018**, *68*, 727–737. [CrossRef]
14. Abdelgawad, K.F.; El-Mogy, M.M.; Mohamed, M.I.A.; Garchery, C.; Stevens, R.G. Increasing Ascorbic Acid Content and Salinity Tolerance of Cherry Tomato Plants by Suppressed Expression of the Ascorbate Oxidase Gene. *Agronomy* **2019**, *9*, 51. [CrossRef]
15. Liu, K.; Yuan, C.; Chen, Y.; Li, H.; Liu, J. Combined effects of ascorbic acid and chitosan on the quality maintenance and shelf life of plums. *Sci. Hortic.* **2014**, *176*, 45–53. [CrossRef]
16. Sikora, M.; Świeca, M. Effect of ascorbic acid postharvest treatment on enzymatic browning, phenolics and antioxidant capacity of stored mung bean sprouts. *Food Chem.* **2018**, *239*, 1160–1166. [CrossRef]
17. El-Mogy, M.M.; Parmar, A.; Ali, M.R.; Abdel-Aziz, M.E.; Abdeldaym, E.A. Improving postharvest storage of fresh artichoke bottoms by an edible coating of Cordia myxa gum. *Postharvest Biol. Technol.* **2020**, *163*, 111143. [CrossRef]
18. Wang, K.; Cao, S.; Di, Y.; Liao, Y.; Zheng, Y. Effect of ethanol treatment on disease resistance against anthracnose rot in postharvest loquat fruit. *Sci. Hortic.* **2015**, *188*, 115–121. [CrossRef]
19. Lichter, A.; Zutkhy, Y.; Sonego, L.; Dvir, O.; Kaplunov, T.; Sarig, P.; Ben-Arie, R. Ethanol controls postharvest decay of table grapes. *Postharvest Biol. Technol.* **2002**, *24*, 301–308. [CrossRef]
20. Wang, K.; Jin, P.; Tang, S.; Shang, H.; Rui, H.; Di, H.; Cai, Y.; Zheng, Y. Improved control of postharvest decay in Chinese bayberries by a combination treatment of ethanol vapor with hot air. *Food Control* **2011**, *22*, 82–87. [CrossRef]
21. Fukasawa, A.; Suzuki, Y.; Terai, H.; Yamauchi, N. Effects of postharvest ethanol vapor treatment on activities and gene expression of chlorophyll catabolic enzymes in broccoli florets. *Postharvest Biol. Technol.* **2010**, *55*, 97–102. [CrossRef]
22. Asoda, T.; Terai, H.; Kato, M.; Suzuki, Y. Effects of postharvest ethanol vapor treatment on ethylene responsiveness in broccoli. *Postharvest Biol. Technol.* **2009**, *52*, 216–220. [CrossRef]
23. Jin, Y.Z.; Lv, D.Q.; Liu, W.W.; Qi, H.Y.; Bai, X.H. Ethanol vapor treatment maintains postharvest storage quality and inhibits internal ethylene biosynthesis during storage of oriental sweet melons. *Postharvest Biol. Technol.* **2013**, *86*, 372–380. [CrossRef]
24. Suzuki, Y.; Nagata, Y. Postharvest ethanol vapor treatment of tomato fruit stimulates gene expression of ethylene biosynthetic enzymes and ripening related transcription factors, although it suppresses ripening. *Postharvest Biol. Technol.* **2019**, *152*, 118–126. [CrossRef]
25. Wu, Z.; Tan, B.; Liu, Y.; Dunn, J.; Guerola, P.M.; Tortajada, M.; Cao, Z.; Ji, P. Chemical Composition and Antioxidant Properties of Essential Oils from Peppermint, Native Spearmint and Scotch Spearmint. *Molecules* **2019**, *24*, 2825. [CrossRef]
26. El-Mogy, M.M.; Ali, M.R.; Darwish, O.S.; Rogers, H.J. Impact of salicylic acid, abscisic acid, and methyl jasmonate on postharvest quality and bioactive compounds of cultivated strawberry fruit. *J. Berry Res.* **2019**, *9*, 333–348. [CrossRef]
27. El-Mogy, M.M.; Ludlow, R.A.; Roberts, C.; Müller, C.T.; Rogers, H.J. Postharvest exogenous melatonin treatment of strawberry reduces postharvest spoilage but affects components of the aroma profile. *J. Berry Res.* **2019**, *9*, 297–307. [CrossRef]
28. Mostafa, H.S.; Ali, M.R.; Mohamed, R.M. Production of a novel probiotic date juice with anti-proliferative activity against Hep-2 cancer cells. *Food Sci. Technol.* **2020**. [CrossRef]
29. Association of Official Analytical Chemistry (AOAC). *Official Methods of Analysis of AOAC International*, 17th ed.; Association of Official Analytical Chemistry (AOAC): Gaithersburg, MD, USA, 2000.
30. Shehata, S.A.; Elmogy, M.; Mohamed, H.F.Y. Postharvest quality and nutrient contents of long sweet pepper enhanced by supplementary potassium foliar application. *Int. J. Veg. Sci.* **2018**, *25*, 196–209. [CrossRef]
31. Shahi, N.; Min, B.; Bonsi, E.A. Microbial decontamination of fresh produce (*Strawberry*) using washing solutions. *J. Food Res.* **2015**, *4*, 128. [CrossRef]

32. Vasile, C.; Sivertsvik, M.; Mitelut, A.C.; Brebu, M.A.; Stoleru, E.; Rosnes, J.T.; Tănase, E.E.; Khan, W.; Pamfil, D.; Cornea, C.P.; et al. Comparative Analysis of the Composition and Active Property Evaluation of Certain Essential Oils to Assess their Potential Applications in Active Food Packaging. *Materials* **2017**, *10*, 45. [CrossRef]
33. Kokina, M.; Kalušević, A.; Nedović, V.; Nikšić, M.; Shamtsyan, M.; Šavikin, K.; Pljevljakušić, D.; Pantić, M.; Lević, S.; Salević, A. Characterization, Antioxidant and Antibacterial Activity of Essential Oils and Their Encapsulation into Biodegradable Material Followed by Freeze Drying. *Food Technol. Biotechnol.* **2019**, *57*, 282–289. [CrossRef]
34. Loose, M.; Pilger, E.; Wagenlehner, F. Anti-Bacterial Effects of Essential Oils Against Uropathogenic Bacteria. *Antibiotics* **2020**, *9*, 358. [CrossRef] [PubMed]
35. Xylia, P.; Clark, A.; Chrysargyris, A.; Romanazzi, G.; Tzortzakis, N. Quality and safety attributes on shredded carrots by using Origanum majorana and ascorbic acid. *Postharvest Biol. Technol.* **2019**, *155*, 120–129. [CrossRef]
36. Shao, X.; Wang, H.; Xu, F.; Cheng, S. Effects and possible mechanisms of tea tree oil vapor treatment on the main disease in postharvest strawberry fruit. *Postharvest Biol. Technol.* **2013**, *77*, 94–101. [CrossRef]
37. Owolabi, I.O.; Songsamoe, S.; Matan, N. Combined impact of peppermint oil and lime oil on Mangosteen (*Garcinia Mangostana*) fruit ripening and mold growth using closed system. *Postharvest Biol. Technol.* **2021**, *175*, 111488. [CrossRef]
38. Misharina, T.A.; Samusenko, A.L. Antioxidant properties of essential oils from lemon, grapefruit, coriander, clove, and their mixtures. *Appl. Biochem. Microbiol.* **2008**, *44*, 438–442. [CrossRef]
39. Hernández-López, G.; Ventura-Aguilar, R.I.; Correa-Pacheco, Z.N.; Bautista-Baños, S.; Barrera-Necha, L.L. Nanostructured chitosan edible coating loaded with α-pinene for the preservation of the postharvest quality of *Capsicum annuum* L. and Alternaria alternata control. *Int. J. Biol. Macromol.* **2020**, *165*, 1881–1888. [CrossRef] [PubMed]
40. Ali, S.; Khan, A.S.; Malik, A.U.; Anwar, R.; Anjum, M.A.; Nawaz, A.; Shafique, M.; Naz, S. Combined application of ascorbic and oxalic acids delays postharvest browning of litchi fruits under controlled atmosphere conditions. *Food Chem.* **2021**, *350*, 129277. [CrossRef]
41. Zhang, W.; Li, X.; Wang, X.; Wang, G.; Zheng, J.; Abeysinghe, D.C.; Ferguson, I.B.; Chen, K. Ethanol vapour treatment alleviates postharvest decay and maintains fruit quality in Chinese bayberry. *Postharvest Biol. Technol.* **2007**, *46*, 195–198. [CrossRef]
42. Barzegar, T.; Fateh, M.; Razavi, F. Enhancement of postharvest sensory quality and antioxidant capacity of sweet pepper fruits by foliar applying calcium lactate and ascorbic acid. *Sci. Hortic.* **2018**, *241*, 293–303. [CrossRef]
43. Lin, L.; Li, Q.P.; Wang, B.G.; Cao, J.K.; Jiang, W.B. Inhibition of core browning in 'Yali' pear fruit by post-harvest treatment with ascorbic acid. *J. Hortic. Sci. Biotechnol.* **2007**, *82*, 397–402. [CrossRef]
44. Aminifard, M.H.; Mohammadi, S. Essential oils to control Botrytis cinerea in vitro and in vivo on plum fruits. *J. Sci. Food Agric.* **2012**, *93*, 348–353. [CrossRef]
45. Suzuki, Y.; Uji, T.; Terai, H. Inhibition of senescence in broccoli florets with ethanol vapor from alcohol powder. *Postharvest Biol. Technol.* **2004**, *31*, 177–182. [CrossRef]
46. Saltveit, M.E.; Mencarelli, F. Inhibition of ethylene synthesis and action in ripening tomato fruit by ethanol vapors. *J. Am. Soc. Hortic. Sci.* **1988**, *113*, 572–576.
47. Rosenqvist, E.; Van Kooten, O. Chlorophyll fluorescence: A general description and nomenclature. In *Practical Applications of Chlorophyll Fluorescence in Plant Biology*; DeEll, J.R., Toivonen, P.M.A., Eds.; Springer: Boston, MA, USA, 2003; pp. 31–77.
48. Naeem, A.; Abbas, T.; Ali, T.M.; Hasnain, A. Effect of guar gum coatings containing essential oils on shelf life and nutritional quality of green-unripe mangoes during low temperature storage. *Int. J. Biol. Macromol.* **2018**, *113*, 403–410. [CrossRef] [PubMed]
49. Chaudhary, S.; Kumar, S.; Kumar, V.; Sharma, R. Chitosan nanoemulsions as advanced edible coatings for fruits and vegetables: Composition, fabrication and developments in last decade. *Int. J. Biol. Macromol.* **2020**, *152*, 154–170. [CrossRef] [PubMed]
50. Carr, A.; Frei, B. Does vitamin C act as a pro-oxidant under physiological conditions? *FASEB J.* **1999**, *13*, 1007–1024. [CrossRef]
51. Viacava, G.E.; Goyeneche, R.; Goñi, M.G.; Roura, S.I.; Agüero, M.V. Natural elicitors as preharvest treatments to improve postharvest quality of Butterhead lettuce. *Sci. Hortic.* **2018**, *228*, 145–152. [CrossRef]
52. Shehata, S.A.; Abdeldaym, E.A.; Ali, M.R.; Mohamed, R.M.; Bob, R.I.; AbdelGawad, K.F. Effect of Some Citrus Essential Oils on Post-Harvest Shelf Life and Physicochemical Quality of Strawberries during Cold Storage. *Agronomy* **2020**, *10*, 1466. [CrossRef]
53. Yin, C.; Huang, C.; Wang, J.; Liu, Y.; Lu, P.; Huang, L. Effect of Chitosan- and Alginate-Based Coatings Enriched with Cinnamon Essential Oil Microcapsules to Improve the Postharvest Quality of Mangoes. *Materials* **2019**, *12*, 2039. [CrossRef] [PubMed]
54. Lu, Y.; Zhang, J.; Wang, X.; Lin, Q.; Liu, W.; Xie, X.; Wang, Z.; Guan, W. Effects of UV-C irradiation on the physiological and antioxidant responses of button mushrooms (*Agaricus bisporus*) during storage. *Int. J. Food Sci. Technol.* **2016**, *51*, 1502–1508. [CrossRef]
55. Qu, T.; Li, B.; Huang, X.; Li, X.; Ding, Y.; Chen, J.; Tang, X. Effect of Peppermint Oil on the Storage Quality of White Button Mushrooms (*Agaricus bisporus*). *Food Bioprocess Technol.* **2020**, *13*, 404–418. [CrossRef]
56. Yang, F.; Li, H.; Li, F.; Xin, Z.; Zhao, L.; Zheng, Y.; Hu, Q. Effect of Nano-Packing on Preservation Quality of Fresh Strawberry (*Fragaria ananassa Duch.* cv Fengxiang) during Storage at 4 °C. *J. Food Sci.* **2010**, *75*, C236–C240. [CrossRef]
57. Lin, X.; Wang, L.; Hou, Y.; Zheng, Y.; Jin, P. A Combination of Melatonin and Ethanol Treatment Improves Postharvest Quality in Bitter Melon Fruit. *Foods* **2020**, *9*, 1376. [CrossRef] [PubMed]
58. Zhao, H.; Liu, S.; Chen, M.; Li, J.; Huang, D.; Zhu, S. Synergistic effects of ascorbic acid and plant-derived ceramide to enhance storability and boost antioxidant systems of postharvest strawberries. *J. Sci. Food Agric.* **2019**, *99*, 6562–6571. [CrossRef]

59. Piscopo, A.; Zappia, A.; Princi, M.P.; De Bruno, A.; Araniti, F.; Antonio, L.; Abenavoli, M.R.; Poiana, M. Quality of shredded carrots minimally processed by different dipping solutions. *J. Food Sci. Technol.* **2019**, *56*, 2584–2593. [CrossRef]
60. Lurie, S.; Pesis, E.; Gadiyeva, O.; Feygenberg, O.; Ben-Arie, R.; Kaplunov, T.; Zutahy, Y.; Lichter, A. Modified ethanol atmosphere to control decay of table grapes during storage. *Postharvest Biol. Technol.* **2006**, *42*, 222–227. [CrossRef]
61. Gabler, F.M.; Mansour, M.F.; Smilanick, J.L.; Mackey, B.E. Survival of spores of Rhizopus stolonifer, Aspergillus niger, Botrytis cinerea and Alternaria alternata after exposure to ethanol solutions at various temperatures. *J. Appl. Microbiol.* **2004**, *96*, 1354–1360. [CrossRef]
62. Zappia, A.; De Bruno, A.; Piscopo, A.; Poiana, M. Physico-chemical and microbiological quality of ready-to-eat rocket (*Eruca vesicaria* L. Cav.) treated with organic acids during storage in dark and light conditions. *Food Sci. Biotechnol.* **2019**, *28*, 965–973. [CrossRef]
63. Hashem, M.; Alamri, S.A.; Alqahtani, M.S.; AlShehri, S.R. A multiple volatile oil blend prolongs the shelf life of peach fruit and suppresses postharvest spoilage. *Sci. Hortic.* **2019**, *251*, 48–58. [CrossRef]
64. Kong, Q.; Zhang, L.; An, P.; Qi, J.; Yu, X.; Lu, J.; Ren, X. Antifungal mechanisms of α-terpineol and terpene-4-alcohol as the critical components of Melaleuca alternifolia oil in the inhibition of rot disease caused by Aspergillus ochraceus in postharvest grapes. *J. Appl. Microbiol.* **2019**, *126*, 1161–1174. [CrossRef] [PubMed]

MDPI

Article

Evaluation of Quality and Storability of "Italia" Table Grapes Kept on the Vine in Comparison to Cold Storage Techniques

Francesca Piazzolla, Maria Luisa Amodio, Sandra Pati and Giancarlo Colelli *

Dipartimento di Scienze Agrarie, Alimenti, Risorse Naturali e Ingegneria (DAFNE), Università di Foggia, Via Napoli 25, 71122 Foggia, Italy; francesca.piazzolla@unifg.it (F.P.); marialuisa.amodio@unifg.it (M.L.A.); sandra.pati@unifg.it (S.P.)
* Correspondence: giancarlo.colelli@unifg.it

Abstract: The aim of the study was to compare the quality of table grapes (cv. Italia) held on the vine compared to grapes stored in cold rooms with or without modified-atmosphere packaging (MAP). The grapes were harvested from 12 plants in 2 vineyards in the same area, differing for the age of the plant. Four- and a fourteen-year-old vines were cultivated with the "Apulia tendone" system. After the first harvest, grapes were divided into small clusters and used for storage treatments in air and in MAP. Samples of 400 g were packaged in polypropylene (PP) trays sealed with a polypropylene/polyamide (PP/PA) film with 20% CO_2 in air. MAP and control samples were then stored in the same cold room at 0 °C. Initially and after 8, 21, and 28 days, grapes stored in air and MAP were compared to fresh harvested grapes, stored on the plants. Quality attributes included color, texture, maturity index, phenols, antioxidant activity, sugars, organic acids, sensory parameters, and volatile compounds. The results obtained demonstrated that grapes held on the plant and in MAP showed better quality in terms of appearance scores compared to grapes stored in air. In particular, the application of high CO_2 contributed to reduce the deterioration rate of the clusters, minimizing weight loss, and delaying degradation processes, and this particularly for grapes from the 14-year-old vine, where grapes held on the plant degraded faster than grapes in the younger vines. Most volatile compounds did not change their concentration with the storage treatment, except for ethyl acetate and ethanol, which increased in MAP at the end of storage, and to some compound responsible for green odor. In conclusion, keeping the grapes on the plant can be considered a good agronomic practice to preserve the quality, whereas MAP can be applied to better maintain postharvest quality of the product throughout storage and distribution.

Keywords: modified atmosphere; carbon dioxide; phenols; antioxidant; ethanol; acetaldehyde

Citation: Piazzolla, F.; Amodio, M.L.; Pati, S.; Colelli, G. Evaluation of Quality and Storability of "Italia" Table Grapes Kept on the Vine in Comparison to Cold Storage Techniques. *Foods* **2021**, *10*, 943. https://doi.org/10.3390/foods10050943

Academic Editors: Eleni Tsantili and Jinhe Bai

Received: 9 March 2021
Accepted: 21 April 2021
Published: 26 April 2021

1. Introduction

Table grapes are one of the most consumed fruits in the world, and a valuable source of phytonutrients [1,2].

It is a non-climacteric fruit, and for this reason, unlike many other fruit crops, the ripening process does not continue off the vine. Sugars produced by photosynthesis are translocated from leaves to the grapes only until they are on the vine or until the maximum Brix degree is reached. Also, there is no conversion of starch into sugars, as the amount of starch in berries is very low [3]. Though the concentration of berry solutes can change after harvest, and some other quality-related compounds can evolve or be degraded, in general, grapes should be harvested only when the target quality parameters have been reached.

A common practice used in southern Italy is to appositely cover canopies with plastic film (i.e., low-density polyethylene) during August to delay the harvest times of table grapes from October to the late November or early December [4]. This "late" forcing, conversely to the "early" one, aimed to anticipate the shoot sprouting, is executed to delay the harvest as much as possible, [5] while protecting fruit form autumnal rains. In addition, forced "storage" has the main advantages of maintaining green and hydrated stalks, and, above all, to allow the use of fungicides, which cannot be applied during storage. Some authors [5,6] reported the benefits of this agronomic technique to meet the demand for high-quality table grapes during Christmas time, resulting in a higher profitability for the producers.

Piazzolla et al. [6], in fact, demonstrated that "late forcing" on the vine is a feasible approach to preserve and even improve table grape quality related to sensorial aspects, even if the authors reported that extreme delay could reduce the quality of the fruit. Nonetheless, nothing is known about quality and storability of "on-vine-stored" grapes and on the potential of this method compared to conventional postharvest storage techniques.

Table grapes are not chilling-sensitive [7]; the respiration rate and the rate of ethylene production for grapes are very low (2 mg $CO_2 \cdot kg^{-1} \cdot h^{-1}$ at 0 °C and less than 0.1 $\mu L \cdot kg \cdot h^{-1}$ at 20 °C, respectively). Table grape quality is reduced by the occurrence of stem browning and *Botrytis cinerea* infections [8]. Recently, the main physical, chemical and bio-based treatments in postharvest for the control of *Botrytis cinerea* on table grapes have been reviewed by De Simone et al. [9]. Ideally, grapes are stored in cold rooms operating at −1 to 0 °C with 95% RH, and a very low airflow; in these conditions, grapes may be stored up to 4 weeks [10], but generally SO_2 fumigation are also applied to better control the mold growth [11,12]. Standard commercial practices include initial sulfur dioxide (SO_2) fumigation during pre-cooling, followed by weekly fumigations with similar doses during cold storage, or most commonly, SO_2 generator pads inside boxes are widely used for table grape storage and transport [13]. As alternative, chitosan, and aloe vera gel treatments have been successfully proposed to maintain table grape quality and extend their shelf-life [14–16].

Other methods used to extend the storability of table grapes include controlled (CA) or modified atmospheres (MA), and active packaging.

As for CA, the addition of CO_2 (10 to 15 KPa in air) can be effective in controlling grey mold (*Botrytis cinerea*) for 2 to 4 weeks depending on cultivars [7].

Crisosto et al. [17] concluded that the CA treatment with 10 kPa CO_2 combined with O_2 levels from 3 to 12 kPa limited grey mold infection on "Red globe" grapes during 12-weeks of cold storage, whereas early harvested "Red globe" could be stored only for 4 weeks in 10 kPa CO_2 + 6 kPa O_2. In addition, CA treatment [18] and low temperature storage combined with SO_2 slow-release generators [19] may be effective as insecticidal control.

As for MAP, Artés-Hernández et al. [20], reported that "Autumn seedless" grapes stored for 2 months in MAP with 15 kPa O_2 and 10 kPa CO_2 at 0 °C followed by 1 week at 15 °C in air, helped to prevent rachis browning and flavor losses.

Costa et al. [21] selected the best packaging material to achieve at the equilibrium the desired gas conditions (10–15% O_2 and 4% CO_2) that could control the respiration, the water loss, and the rachis color changes of the grapes. Among the tested polymeric material, oriented polypropylene (80 µm of thickness) could allow a shelf-life longer than 70 days.

Silva-Sanzana et al. [22] reported that modified-atmosphere packaging controlled the green color losses on stalks of "Red Globe" grapes stored for 90 days at 0 °C, compared with a conventional storage even after a shelf-life period, but no comparison is available with grapes "stored on the vine".

Cefola et al. [23] reported that the storage of "Italia" table grapes at high CO_2 atmosphere (20 KPa) induced the shift to anaerobic metabolism, reporting an increase in respiration rate, and acetaldehyde and ethanol production, and a lower evaluation at the sensorial test, than grapes stored with lower CO_2 concentrations (up to 10%) or in air. Finally, the use of an active packaging based on PET coated with a layered double Hydroxide (LDH) hosting 2-acetoxybenzoic anion (salicylate) as antimicrobial molecule was shown to cause a significant reduction in total mesophilic aerobic count and mold and yeast population with respect to control [24].

This work aimed to assess for the first time the quality and storability of table grapes stored on the vine, also considering quality evolution of grapes from the same plants stored in cold rooms (with or without modified-atmosphere packaging).

2. Material and Methods

2.1. Plant Material and Sample Preparation

"Italia" table grapes were grown in Foggia (Puglia, Italy, 41°28′39.2″ N), in 2 contiguous vineyards of the same grower, but with plants of 2 different ages, 4 and 14 years. The vegetative system consisted of the "tendone" which was covered with a plastic net and a plastic film of LDPE. Grapes were harvested randomly from 12 marked plants in October, when according to the grower procedure, the commercial maturity was reached. Grapes were rapidly transported to the Postharvest Laboratory of the University of Foggia where clusters were divided into smaller clusters. Three replicates of 400 g were used for initial quality determinations while 18 samples of the same size were prepared for storage in air in macroperforated bags (AIR) or in modified-atmosphere packaging (MAP). For MAP, 9 samples were packaged in polypropylene (PP) trays sealed with a PP/PA film (polypropylene/polyamide, 30 μm, with CO_2 transmission rate of 48 mL·m^{-2} day and O_2 transmission rate of 135 mL·m^{-2} day) in active modified-atmosphere (20% CO_2 in air) using a semi-automatic tabletop tray sealer (ILPRA termosaldatrici, FoodPack, Vigevano, PV, Italy). For AIR treatment the trays were wrapped in macroperforated bags containing wet paper pads to keep a high-RH environment around the product. Three replicate trays were prepared for each treatment and sampling time, with a total of 18 trays. All samples were stored at 0 °C for 28 days. After 8, 21, and 28 days of storage, 3 trays for each treatment (MAP and AIR) were used for quality determinations together with samples freshly harvested from the 12 marked plants (PLANT).

2.2. Quality Determination

The following quality parameters were analyzed initially, and on each sampling date.

Weight loss, berry firmness, gas concentration (for MAP samples only), visual quality (appearance score), color parameters (Hue Angle, Chroma) and sensorial quality (aroma, flavor, crunchiness, sweetness, acidity, fizzy taste, resistance to berry detachment, and overall evaluation) were monitored on fresh samples. About 100 g of berries were then frozen and stored at −80 °C until analysis with TSS, pH, TA, volatile compounds, organic acid, and sugar composition measured on the squeezed juice, whereas total phenol content and antioxidant activity were extracted from the skin. Additionally, at each sampling date ethanol and acetaldehyde extracts were prepared and frozen at −20 °C.

2.2.1. Headspace Gas Determination and Weight Loss

Before opening the packages, the gas composition was determined using a gas analyzer with an accuracy of 0.5% for both O_2 and CO_2 (Witt, Gascontrol 100, MAPY4.0, Witten, Germany), equipped with an aspiration pump and a needle. A gas volume of 0.5 mL was automatically withdrawn and used for the gas analysis.

Weight loss was calculated as percentage of difference from initial weight.

2.2.2. Color Analysis

Color indexes were extracted from hyperspectral images acquired with a Spectral scanner (DV SRL, Padova, Italy), in the VisNir range (400–1000 nm, resolution 5 nm) as described in Piazzolla et al. [6]. From the primary L*, a*, and b*, Hue angle, and Chroma were calculated.

2.2.3. Sensory Evaluation

Sensory attributes of the berries were scored by a 5-judges trained panel. The judges evaluated 3 berries with pedicel from each sampling replicate for each harvest time. Judges evaluated the aroma, flavor, crunchiness, sweetness, sourness, fizzy taste, resistance to berry detachment and the overall evaluation using a scale of 5 to 1, where 5 = very high; 3 = fair; 1 = very low.

2.2.4. Firmness

Firmness was evaluated on 45 berries per replicate by a compression test, performed with an Instron Universal Testing Machine (model 3343, Norwood, MA, US), at a speed of 50 mm·min^{-1}. The force (N) required for a 3 mm compression between two parallel plates (diameter of 10 cm) was recorded.

2.2.5. Total Soluble Solids, pH, and Titatrable Acidity

Initially and for each sampling date, total soluble solids (TSS), pH and titratable acidity (TA) were assessed using 4 g of juice sample from 15 berries, for each replicate. TSS were measured using a digital refractometer (PR-32 Palette, Atago, Tokyo, Japan), while pH and TA were assessed with an automatic titrator (TitroMatic 1S, Crison, Toledo, Spain), titrating to pH 8.1; the value was expressed as percentage of tartaric acid.

2.2.6. Total Phenolic Content and Antioxidant Activity

Total phenol content and antioxidant activity were determined on frozen samples. One gram of skin was added of 3 mL g^{-1} of methanol plus 3% formic acid and was homogenized with an Ultraturrax (IKA T18 basic, Wilmington, NC, USA) [25]. The extract was then centrifuged at 5 °C and 9000 rpm for 10 min. Total phenols were determined according to the method of Singleton and Rossi [26] and expressed as grams of gallic acid per kilogram of fresh weight (g GA·kg^{-1}). Antioxidant assay was performed following the procedure described by Brand-Williams et al. [27] and reported in grams of Trolox equivalents per kilogram of fresh weight (g TE·kg^{-1}).

2.2.7. Simultaneous Analysis of Organic Acids and Sugars

All samples were thawed and then the juice from each sample was filtered with a C_{18} Sep-Pak cartridge (Grace Pure ^{TM}, New York, NY, USA) and then with a 0.2 µm filter. After dilution (1:1) with ultrapure water, 10 µL-samples were injected into an HPLC system (Agilent 1200 series) equipped with an UV detector, set at 210 nm, and a refractive index detector. Compounds were separated on a Rezex ROA-Organic Acid H + (8%) column (300 × 7.80 mm) (Phenomenex, Torrance, CA, USA), using an aqueous mobile phase of 0.1% phosphoric acid, (flow rate of 0.5 mL·min^{-1}) and an oven temperature of 30 °C. The different organic acids and sugars were quantified by chromatographic comparison with analytical standards. Resulted were expressed as g·kg^{-1}.

2.2.8. Determination of Ethanol and Acetaldehyde

For ethanol and acetaldehyde, 5 g of fresh table grape tissues were homogenized with 10 mL of water and then centrifuged at 5 °C and 9000 rpm for 10 min. Five mL of supernatant were placed into 20 mL glass vials and stored at −20 °C until analysis according to the method of Mateos et al. [28]. After thawing for 1 h in a water bath at 65 °C, 1 mL headspace gas sample was withdrawn and injected into a gas chromatograph (Shimadzu GC-14A, Tokyo, Japan) equipped with a FID detector (150 °C). Ethanol and acetaldehyde were identified by co-chromatography with standards and quantified by a calibration curve.

2.2.9. Headspace Solid-Phase Microextraction (HS-SPME) and Gas-Chromatography Mass Spectrometry (GC-MS) Analysis

The extraction of volatile compounds was carried out by HS-SPME using an 85 µm Carboxen/Polydimethylsiloxane fibre (Supelco, Bellefonte, PA, USA) and a GC-MS instrument, according to Piazzolla et al. [6].

After thawing the fruit, and detaching seeds and pedicels, 100 g of fruit tissue were added with 2 g of $CaCl_2$, 20 g of NaCl, 100 µL of internal standard solution (100 ppm

2-methyl pentanol methanolic solution) and homogenized using a commercial blender. The homogenized (8 g) was placed into a 15 mL capped SPME vial and stirred for 20 min, at 40 °C. Then, the fibre was exposed for 30 min to the capped vial headspace, manually injected into the GC (splitless mode) and kept for 4 min to allow for desorption of volatile compounds. The separation was achieved on a DB-WAX capillary column (60 m × 250 µm × 0.25 µm, J&W Scientific Inc., Folsom, CA, USA) and the identification by comparison of retention time and mass spectra with pure compounds when available, or with data system library (NIST 02, $p > 80$). All compound concentrations were expressed as µg of 2-methyl pentanol equivalent g^{-1}.

2.2.10. Statistical Analysis

For each vineyard, a 2-way factorial design for treatment (Air, AM, and PLANT) and time of storage (8, 21 and 28 days) was conducted. At each storage time, a 1-way ANOVA for the treatment was performed. Mean values were separated with Tukey test ($p < 0.05$).

The data were analyzed with StatGraphics Centurion software (ver.16.1.11, StatPoint Technologies, Inc., The Plains, VA, USA).

3. Results and Discussion

In Table 1 is reported, for each vineyard, the effect of the treatment and time of storage on quality attributes. Data of both vineyards were enough in agreement, and as will be better explained with data discussion, main differences were due to the higher quality of grapes from 4-year old vines, which degraded much slower, in comparison to the 14-year old grapes, particularly when stored in air, or kept on the plant. Treatment and time of storage affected more parameters for grapes from 14-year-old vines, than in the case of grapes from the 4-year-old vines. For grapes from 14-year-old vines, treatment influenced firmness, weight loss, hue angle, chroma, titratable acidity, phenol content, antioxidant activity, acetaldehyde, ethanol, citric acid and all sensorial parameters (except for sweetness). The time of storage affected most of parameters except for weight loss, phenols, antioxidant activity, fumaric acid, and fizzy taste score. On the other side, for grapes of 4-year-old vines, hue angle, acidity, phenols, were not affected by treatment and much less parameters were affected by time of storage. Since interaction between treatment and time of storage, was often significant when treatment was significant, and mostly for sensorial score, the simple effect of treatment for each quality attribute was evaluated at each sampling time.

As for gas evolution within packages, CO_2 concentration was reduced for both experiment to 10% after 20 days of storage and remained constant until the end storage, while O_2 stayed up to the atmospheric level (18–20%). After 28 days of storage, the CO_2 and O_2 concentrations were approximately 10.5 and 20%, respectively.

In Figure 1 is shown the effect of treatment on firmness during the storage; in particular firmness of berries held in MAP remained practically unchanged until the end of storage for both experiments, when grapes hold on the PLANT showed higher firmness values, compared to grapes stored in air and in MAP. For grapes of 4-year-old vines, a singular increase of firmness values was observed on berries left on the plant for 28 days, suggesting a concentration of pectin and cellulose when there was less competition for nutrients between fruit. As for berry stored in AIR weight loss increased to 3.7 and 2.2% at the end of the storage, for grapes of 4- and 14-year-old vines, respectively.

Figure 2 shows the effect of treatment on cluster and stalk appearance scores during storage. Here the main difference due to the age of the plant can be observed. Although for 4-year-old vines, there was no difference over storage time for cluster appearance of grapes stored on the plant and in MAP, for grapes from 14-year-old vines, cluster score was best preserved in MAP up to 21 days of storage, with no difference after 28 days. In this case, grapes stored on the vine degraded much faster when hold on the plant or in the cold room, whereas the presence of CO_2 was effective on delaying ethylene effects [29,30] for grapes from both vineyards.

Table 1. Results of the 2-way ANOVA for treatment (TR) and time of storage (T) a on quality attributes of "Italia" table grapes from a 4-year and 14-year-old vineyard. Within each row, each factor and their interaction have a significant effect for $p \leq 0.05$ (*); $p \leq 0.01$ (**); $p \leq 0.001$ (***); $p \leq 0.0001$ (****), or not significant (ns).

Quality Attributes	4-Year Old Vine			14-Year Old Vine		
	Treatment (TR)	Time (T)	TR X T	Treatment (TR)	Time (T)	TR X T
Firmness (N)	****	ns	**	**	**	***
Weight loss (%) [1]	**	ns	ns	***	ns	ns
Hue angle (°)	ns	ns	ns	***	****	ns
Chroma	*	ns	*	****	*	*
TSS (%)	ns	ns	ns	ns	****	ns
pH-value	ns	ns	ns	ns	****	**
Titratable acidity (%)	ns	ns	ns	*	**	*
Phenols (g kg^{-1})	ns	ns	ns	*	ns	ns
Antioxidant activity (g kg^{-1})	ns	ns	ns	*	ns	ns
Acetaldheyde (nmole/g)	***	****	**	**	****	**
Ethanol (nmole/g)	*	****	ns	**	****	**
Tartaric acid (g kg^{-1})	ns	*	ns	ns	**	ns
Malic acid (g kg^{-1})	ns	ns	ns	ns	*	ns
Fumaric acid (g kg^{-1})	ns	ns	ns	ns	ns	ns
Citric acid (g kg^{-1})	ns	ns	ns	*	**	*
Succinic acid (g kg^{-1})	***	ns	ns	ns	*	ns
Sucrose (g kg^{-1})	ns	*	ns	ns	****	****
Glusose (g kg^{-1})	ns	ns	ns	ns	****	ns
Fructose (g kg^{-1})	ns	ns	ns	ns	*	ns
Sensorial Score (1 to 5)						
Cluster appearance score	****	****	****	****	****	****
Stalk appearance score	****	****	****	****	****	****
Berry appearance score	****	****	*	*	****	ns
Crunchiness	****	**	**	****	***	ns
Berry detachment	****	***	ns	****	****	****
Aroma	***	****	ns	*	****	**
Flavor	**	****	**	***	****	**
Sweetness	***	****	****	ns	****	**
Sourness	**	****	ns	**	****	ns
Fizzy taste	***	ns	ns	****	ns	ns
Overall evaluation	****	****	*	**	****	ns

[1] AIR and MAP treatment.

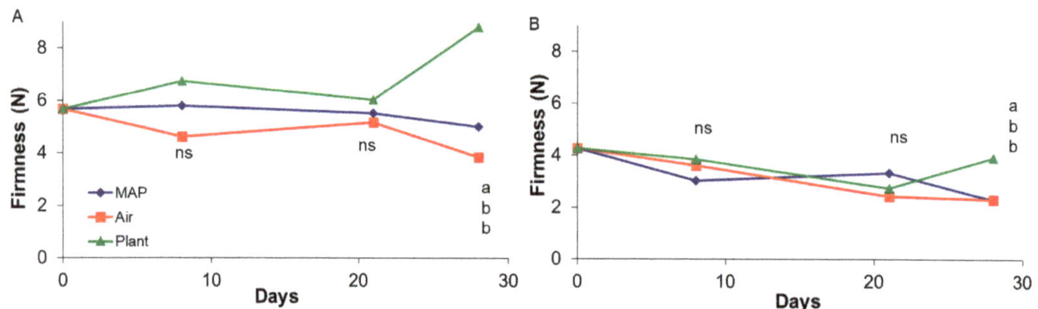

Figure 1. Effect of storage treatment on firmness of "Italia" table grape from a 4-year (**A**) and 14-year-old (**B**) vineyard during storage. At each sampling time different letters indicate mean values significantly different ($p < 0.05$ and Tukey test; ns: not significant).

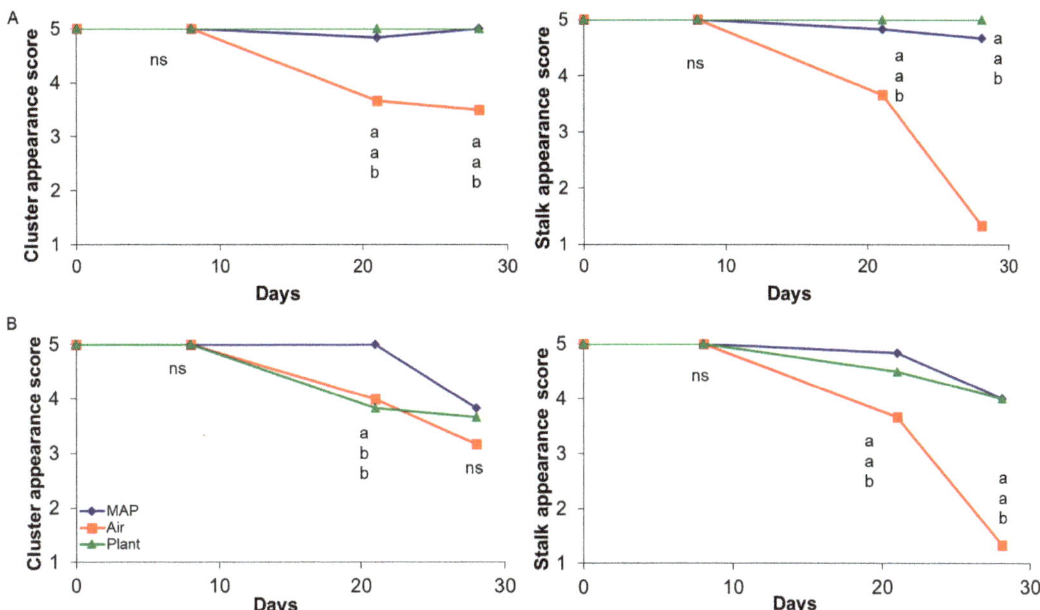

Figure 2. Effect of treatment on stalk and cluster appearance scores of "Italia" table grapes from a 4-year (**A**) and 14-year-old (**B**) vineyard during storage. At each sampling time different letters indicate mean values significantly different ($p < 0.05$ and Tukey test; ns: not significant).

In terms of visual appearance, the main effect of the different storage treatment was observed on the stalk, which showed dehydration and discoloration, for grapes stored in AIR, while less differences were observed for the berries, particularly for those from 14-year-old vines. Grape berries were less influenced by water loss, since they are well protected with waxy layers.

Stalk appearance showed a severe deterioration for grapes stored in AIR, whereas appearance scores remained almost unchanged for PLANT and MAP stored grapes. The score for AIR samples reached the value of 1 at 28 days, certainly for the observed dehydration, which occurred despite the protected conditions (macroperforated bag and humidified water pad). On the other hand, the high score values registered for samples in MAP are certainly related to low levels of dehydration but also to the effect of high-CO_2 atmospheres on slowing down chlorophyll degradation rate. The effect of atmospheres with 29 kPa CO_2 and 1 kPa O_2 on chlorophyll retention has been demonstrated by Pariasca et al. [29] on pea pods and by Cefola et al. [31] on broccoli raabs stored with 10% CO_2. Also, Silva-Sanzana et al. [22] reported that modified-atmosphere packaging helped to maintain green color of the stalk for "Red Globe" grapes stored for 90 days at 0 °C, with no negative effect on the quality of the berries. Similar results were found in "Autumn seedless" table grapes [20], indicating that clusters stored in air showed extreme browning of the stalk while clusters stored in CA (5 kPa O_2+ 15 kPa CO_2) and MAP (15 kPa O_2+ 10 kPa CO_2) had a good visual appearance at 60 days of storage at 0 °C and after additional 7 days in air at 15 °C.

Berry resistance to detachment decreased over time for all treatments; grapes stored in AIR showed a lower resistance to berry detachment than grapes hold on the PLANT or in MAP for grapes from 4-year old vines (Figure 3), and up to 21 days of storage for grapes from 14-year-old vine, where at 28 days PLANT showed highest value (3.7 N) and both air and MAP the same value of about 2.4 N. These results can be in part attributed to the different degree of water loss of stalk suffered by samples of different treatments. Dehydration stress, in fact stimulates ethylene production which in turn favors the formation of the abscission layer, which reduces the force required for berry detachment. Pariasca et al. [29] and Bailén [30], found that the CO_2 inhibits the abscission layer in MAP samples, because of the known competition of CO_2 with ethylene on binding sites. In this case, for grapes of 14-year-old vines CO_2 could had prevented ethylene effects, up to 21 days, but at the end of the storage in both MAP and AIR samples senescence processes were not inhibited, and no differences in berry detachment were observed.

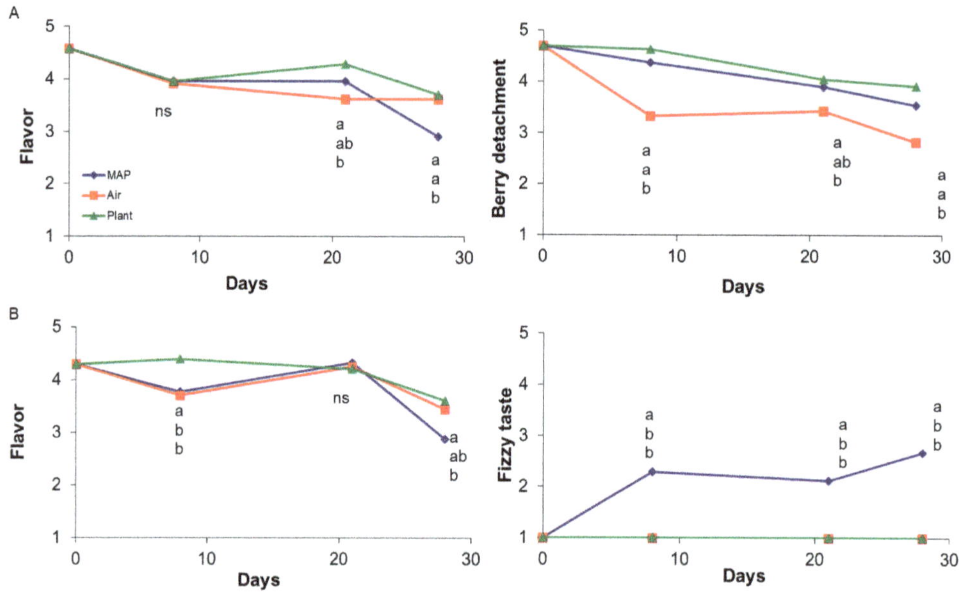

Figure 3. Effect of treatment on flavor and berry detachment over storage of "Italia" table grapes from a 4-year vineyard (**A**) and on flavor and fizzy taste on grapes from 14-year-old (**B**) vineyard. At each sampling time different letters indicate mean values significantly different ($p < 0.05$ and Tukey test; ns: not significant).

As for the difference in color observed particularly for grapes of the oldest vine, a slightly higher decrease of b*, Chroma (also for the 4-year-old-vine), and Hue angle values was observed for grapes stored on the plant, which also showed a higher increase of the a* value. This may be explained by a higher enzymatic activity for grapes stored on the PLANT, and on the same time by the reduction of photosynthetic activity. Nonetheless difference in color were very minimum and not perceived by eyes. Regarding chemical attributes: titratable acidity, phenols, citric acid, and sucrose were significantly affected by the treatment, showing the same trend for grapes from both vineyards (data for grapes of 14-year-old vines is shown in Figure 4).

Titratable acidity and citric acid content increased after 28 days of storage for grapes store on the PLANT, which presented a higher content than AIR, while intermediate values were observed for MAP. Probably, the reduction of the fruit load could have led to a stimulation of the vegetative activity of the plant, inducing the increase of the acidity in the fruit. No differences for the other individual organic acids were found; tartaric and malic acids were the most abundant acids, followed by citric and succinic. In addition, traces of fumaric acid were also detected, in agreement with what reported in the literature for other varieties [32,33].

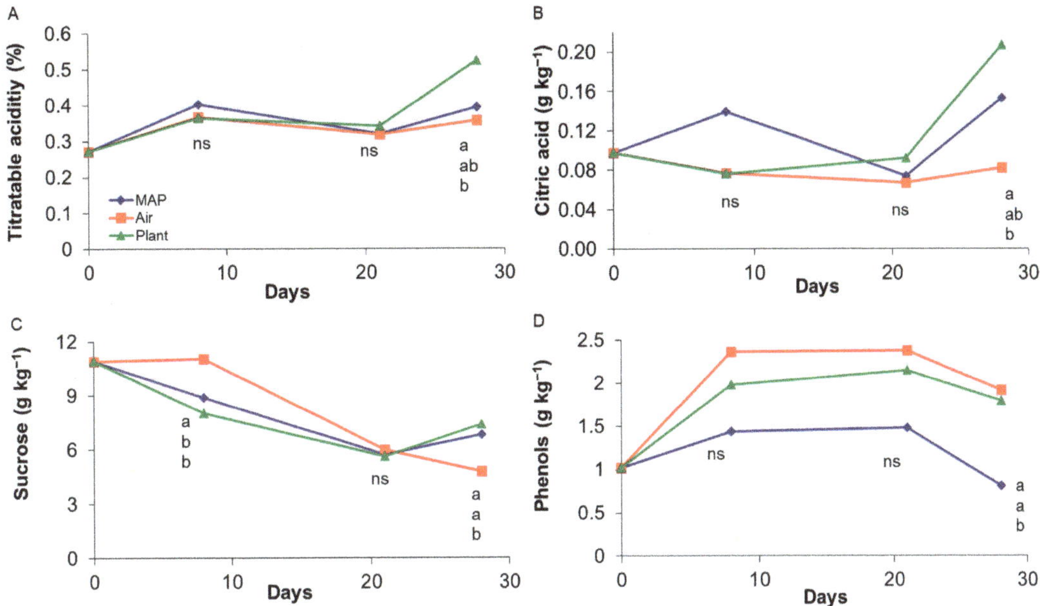

Figure 4. Effect of storage treatment on titratable acidity (**A**), citric acid (**B**), sucrose (**C**) and phenols (**D**) of "Italia" table grapes from a 14-year-old vineyard during storage. At each sampling time different letters indicate mean values significantly different ($p < 0.05$ and Tukey test; ns: not significant).

On the other hand, sucrose content decreased during storage; after 28 days, grapes stored in AIR showed a lower content (4.8 $g \cdot kg^{-1}$) than other treatments (about 7 $g \cdot kg^{-1}$ for both PLANT and MAP). As for glucose, at 28 days of storage was higher in MAP than in AIR, confirming a lower metabolism in MAP grapes probably associated with the presence of CO_2. These results are confirmed by the soluble solids content that report a slight decrease during the storage without significant difference between treatments.

For all treatments and in both vineyards, an increase of phenolic content was also observed (much lower for grapes stored in MAP) immediately after 8 days of storage and up 21 days, followed by a reduction at the end storage. In Figure 4, data refer to 14-year-old vines, where after 28 days, grapes stored in AIR and on PLANT showed a higher phenolic content (1.92 and 1.80 $g \cdot kg^{-1}$, respectively) than the grapes stored in MAP (0.81 $g \cdot kg^{-1}$), but the same trend was observed also in the youngest vineyard. Probably, these results are associated with the biosynthesis of phenols, which is inhibited during postharvest storage in fruit and vegetables treated with elevated CO_2 concentrations [34].

Regarding sensorial attributes, changes in flavor and presence of fizzy taste over time, are shown in Figure 3. It is important to notice that samples stored in MAP at the end of storage received the lowest flavor score, possibly related to the accumulation of CO_2 in the cell sap in the form of carbonic acid which is the cause of the increase of the fizzy taste score, reported in figure for grapes from 14-year-old vines. On the other hand, it is important to highlight that the fizzy taste can also be due to fermentative processes caused by excessive accumulation of carbon dioxide and oxygen depletion typical of MAP [23] and associated with the accumulation of volatile compounds (ethyl acetate and ethanol). The latter aspect is more critical since when due to carbonic acid; it disappear after a few hours, with the evaporation of CO_2.

With regard to aroma compounds a total of 21 volatiles, including six aldehydes (3-methylbutanal, pentanal, Z-2-butenal, hexanal, E-2-hexenal), six alcohols (ethanol, 1-hexanol, Z-3-hexen-1-ol, E-3-hexen-1-ol, 2-methyl-3-buten-2-ol, 1-pentanol), one ester (ethylacetate), four terpenes (D-limonene, cis-, and trans-linaloloxide, linalool), three ketones (3-penten-2-one, 1-penten-3-one, 6-methyl-5-hepten-2-one) one furan derivative (2-ethylfuran), one acid (acetic acid), were found in both grapes from 14-year old vines and 4-year old vines. Grapes from 14-year-old vines also showed the presence of E-2-butenal. For both grapes from 14-year-old vines and 4-year-old vines, most of the volatiles were not affected by the storage treatment except for ethyl acetate and ethanol accumulating in MAP and some typical compounds better preserved in plant. Particularly ethyl acetate and ethanol showed the same trend during the storage, with significant differences only at 28 days of storage, in which the highest concentrations were found in the treatment with MAP (Figure 5). In general, the increase in ethyl acetate and ethanol concentrations has been also reported in the headspace of apples and strawberries stored in high CO_2 due to the fermentative metabolism [35,36]. Also in this trial, the ethanol and ethyl acetate could be used as indicators to determine the grade of degradation of table grapes.

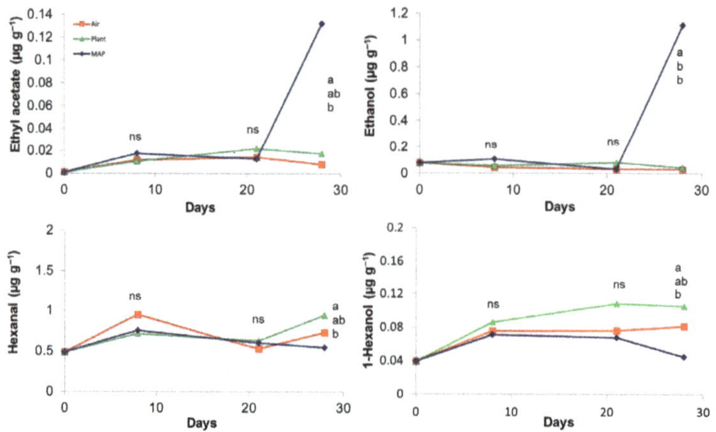

Figure 5. Effect of the storage treatment on the contents of ethyl acetate (**A**), ethanol (**B**), hexanal (**C**) (E-2-hexenal showed a similar trend), 1-hexanol (**D**), in grapes from 14-year-old vines (grapes from 4-year-old vines showed a similar trend). At each sampling different letters indicate significantly different mean values ($p < 0.05$ and Tukey test; ns: not significant).

On the other side, for both grapes from 14-year-old vines and 4-year-old vines, grapes kept on PLANT showed the highest concentration of E-2-hexenal, hexanal, and 1-hexanol, at the end of the storage (Figure 5), showing that grapes on the vine better maintained some typical compounds. The contents of all the other compounds, including e.g., the linalool, which is known to give floral notes, were not significantly different among the treatments.

Therefore, we can affirm that the MA packaging showed some advantages during the first 21 days of storage, as showed by the sensory evaluation of firmness, aroma, cluster, and stalk appearance, while at the end of storage the MA treatment suffered an accelerated process of fermentation likely characterized by the high content of ethanol and ethyl acetate, inducing the perception of fizzy taste.

4. Conclusions

Results of this experiment demonstrated that holding "Italia" table grapes on the PLANT, allowed a good preservation of the quality attributes (phenolic content, and flavor) compared to harvested product. Some additional benefits on cluster appearance score could be obtained by using MAP conditions, particularly if the grapes come from old vines, being more perishable than fruit from younger vine. Most volatile compounds did not change their concentration with the storage treatment, but ethyl acetate ant ethanol increased in MAP stored grapes at 28 days of storage, suggesting the occurrence of fermentation processes confirmed by a highest perception of fizzy taste in grape stored with MAP conditions. Nonetheless, 21 days are a very reasonable time for the commercialization of this product, considering that normally for packaged fresh produce a shelf-life of 7–12 days is normally accepted. Therefore, depending on the market and distribution needs and to the age of the vineyard, different storage strategies may be applied.

Author Contributions: Conceptualization, G.C., M.L.A. Methodology, formal analysis, and investigation, F.P., S.P., G.C., M.L.A. Data curation, writing—original draft preparation, F.P. Writing—review and editing, supervision, G.C., M.L.A., S.P. Funding acquisition, G.C., M.L.A. All authors have read and agreed to the published version of the manuscript.

Funding: This research received no external funding.

Acknowledgments: We thank the company Azienda Agricola F.lli Carpentiere Srl (Barletta, BT, Italy) for its kind cooperation in hosting this research in their vineyards.

Conflicts of Interest: The authors declare no conflict of interest.

References

1. Pezzuto, J.M. Grapes and Human Health: A Perspective. *J. Agric. Food Chem.* **2008**, *56*, 6777–6784. [CrossRef] [PubMed]
2. FAO OIV. Table and Dried Grapes: World Data Available. Available online: http://www.fao.org/documents/card/en/c/709ef071-6082-4434-91bf-4bc5b01380c6/ (accessed on 17 April 2021).
3. Amerine, M.A.; Root, G.A. Carbohydrate content of various parts of grape cluster. II. *Am. J. Enol. Vitic.* **1960**, *11*, 137–139.
4. Colapietra, M. La coltivazione dell'uva da tavola: Le cure colturali e la raccolta. *Vita Camp.* **2012**, *9*, 1–4.
5. de Palma, L.; Novello, V.; Tarricone, L.; Lopriore, G.; Tarantino, A. Semiforzatura precoce e tardiva del vigneto a uva da tavola. *Spec. Vitic. Tavola* **2005**, *13*, 9–16.
6. Piazzolla, F.; Pati, S.; Amodio, M.L.; Colelli, G. Effect of harvest time on table grape quality during on-vine storage. *J. Sci. Food Agric.* **2015**, *96*, 131–139. [CrossRef]
7. Crisosto, C.H.; Smilanick, J.L.; Dokoozlian, N.K.; Luvisi, D.A. Maintaining table grape postharvest quality for long distant markets. In Proceedings of the International Symposium on Table Grape Production, Anaheim, CA, USA, 28–29 June 1994; pp. 195–199.
8. Crisosto, H.C.; Smilanick, J.L. Table grape postharvest quality maintenance guidelines. Online publication. Available online: http://kare.ucanr.edu/files/123831.pdf (accessed on 24 April 2021).
9. De Simone, N.; Pace, B.; Grieco, F.; Chimienti, M.; Tyibilika, V.; Santoro, V.; Capozzi, V.; Colelli, G.; Spano, G.; Russo, P. Botrytis cinerea and Table Grapes: A Review of the Main Physical, Chemical, and Bio-Based Control Treatments in Post-Harvest. *Foods* **2020**, *9*, 1138. [CrossRef] [PubMed]
10. Ngcobo, M.E.K.; Delelea, M.A.; Chenc, L.; Opara, U.L. Investigating the potential of a humidification system to control moisture loss and quality of 'Crimson Seedless' table grape during cold storage. *Postharvest Biol. Technol.* **2013**, *86*, 201–211. [CrossRef]
11. Smilanick, J.L.; Henson, D.J. Minimum gaseous sulfur-dioxide concentrations and exposure periods to control Botrytis cinerea. *Crop Prot.* **1992**, *11*, 535–540. [CrossRef]
12. Palou, L.; Crisosto, C.H.; Garner, D.; Basinal, L.M.; Smilanick, J.L.; Zoffoli, J.P. Minimum constant sulfur dioxide emission rates to control gray mold of coldstored table grape. *Am. J. Enol. Vitic.* **2002**, *53*, 110–115.
13. Nelson, K.E.; Ahmedullah, M. Packaging and decay control system for storage and transit of table grape for export. *Am. J. Enol. Vitic.* **1976**, *24*, 74–79.
14. Chen, R.; Wu, P.; Cao, D.; Tian, H.; Chen, C.; Zhu, B. Edible coatings inhibit the postharvest berry abscission of table grapes caused by sulfur dioxide during storage. *Postharvest Biol. Technol.* **2019**, *152*, 1–8. [CrossRef]
15. Nia, A.E.; Taghipour, S.; Siahmansour, S. Pre-harvest application of chitosan and postharvest Aloe vera gel coating enhances quality of table grape (*Vitis vinifera* L. cv. 'Yaghouti') during postharvest period. *Food Chem.* **2021**, *347*, 129012. [CrossRef]
16. Liguori, G.; Sortino, G.; Gullo, G.; Inglese, P. Effects of Modified Atmosphere Packaging and Chitosan Treatment on Quality and Sensorial Parameters of Minimally Processed cv. 'Italia' Table Grapes. *Agronomy* **2021**, *11*, 328. [CrossRef]
17. Crisosto, C.H.; Garner, D.; Crisosto, G. Carbon dioxide-enriched atmosphere during cold storage limit losses from Botrytis but accelerate rachis browning of "Red globe" table grape. *Postharvest Biol. Technol.* **2002**, *26*, 181–189. [CrossRef]
18. Ahumada, M.H.; Mitcham, E.J.; Moore, D.G. Postharvest Quality of 'Thompson Seedless' Grapes after Insecticidal Controlled-atmosphere Treatments. *HortScience* **1996**, *31*, 833–836. [CrossRef]
19. Yokoyama, V.Y.; Miller, G.T.; Crisosto, C.H. Los temperature storage combined with sulfur dioxide slow release pads for quarantine control of omnivorous leafroller Platynota stultana (Lepidoptera: Tortricidae). *J. Econ. Entom.* **1999**, *92*, 235–238. [CrossRef]
20. Artés-Hernández, F.; Aguayo, E.; Artés, F. Alternative atmosphere treatments for keeping quality of 'Autumn seedless' table grapes during long-term cold storage. *Postharvest Biol. Technol.* **2004**, *31*, 59–67. [CrossRef]

21. Costa, C.; Lucera, A.; Conte, A.; Mastromatteo, M.; Speranza, B.; Antonacci, A.; Del Nobile, M. Effects of passive and active modified atmosphere packaging conditions on ready-to-eat table grape. *J. Food Eng.* **2011**, *102*, 115–121. [CrossRef]
22. Silva-Sanzana, C.; Balic, I.; Sepúlveda, P.; Olmedo, P.; León, G.; Defilippi, B.G.; Blanco-Herrera, F.; Campos-Vargas, R. Effect of modified atmosphere packaging (MAP) on rachis quality of 'Red Globe' table grape variety. *Postharvest Biol. Technol.* **2016**, *119*, 33–40. [CrossRef]
23. Cefola, M.; Damascelli, A.; Lippolis, V.; Cervellieri, S.; Linsalata, V.; Logrieco, A.F.; Pace, B. Relationships among volatile metabolites, quality and sensory parameters of 'Italia' table grapes assessed during cold storage in low or high CO_2 modified atmospheres. *Postharvest Biol. Technol.* **2018**, *142*, 124–134. [CrossRef]
24. Gorrasi, G.; Bugatti, V.; Vertuccio, L.; Vittoria, V.; Pace, B.; Cefola, M.; Quintieri, L.; Bernardo, P.; Clarizia, G. Active packaging for table grapes: Evaluation of antimicrobial performances of packaging for shelf life of the grapes under thermal stress. *Food Packag. Shelf Life* **2020**, *25*, 100545. [CrossRef]
25. Artés-Hernández, F.; Artés, F.; Tomás-Barberán, F.A. Quality and Enhancement of Bioactive Phenolics in Cv. Napoleon Table Grapes Exposed to Different Postharvest Gaseous Treatments. *J. Agric. Food Chem.* **2003**, *51*, 5290–5295. [CrossRef]
26. Singleton, V.L.; Rossi, J.A., Jr. Colorimetry of Total Phenolics with Phosphomolybdic-Phosphotungstic Acid Reagents. *Am. J. Enol. Vitic.* **1965**, *16*, 144–158.
27. Brand-Williams, W.; Cuvelier, M.; Berset, C. Use of a free radical method to evaluate antioxidant activity. *LWT* **1995**, *28*, 25–30. [CrossRef]
28. Mateos, M.; Ke, D.; Cantwell, M.; Kader, A.A. Phenolic metabolism and ethanolic fermentation of intact and cut lettuce exposed to CO_2-enriched atmospheres. *Postharvest Biol. Technol.* **1993**, *3*, 225–233. [CrossRef]
29. Pariasca, J.A.; Miyazaki, T.; Hisaka, H.; Nakagawa, H.; Sato, T. Effect of modified atmosphere packaging (MAP) and controlled atmosphere (CA) storage on the quality of snow pea pods (*Pisum sativum* L. var. saccharatum). *Postharvest Biol. Technol.* **2001**, *21*, 213–223. [CrossRef]
30. Bailén, G.; Guillen, F.; Castillo, S.; Serrano, M.; Valero, D.; Martínez-Romero, D. Use of Activated Carbon inside Modified Atmosphere Packages To Maintain Tomato Fruit Quality during Cold Storage. *J. Agric. Food Chem.* **2006**, *54*, 2229–2235. [CrossRef] [PubMed]
31. Cefola, M.; Amodio, M.L.; Cornacchia, R.; Rinaldi, R.; Vanadia, S.; Colelli, G. Effect of atmosphere composition on the quality of ready-to-use broccoli raab (*Brassica rapa* L.). *J. Sci. Food Agric.* **2010**, *90*, 789–797. [CrossRef]
32. Takeda, F.; Saunders, M.S.; Saunders, J.A. Physical and chemical changes in 'Muscadine' grape during postharvest storage. *Am. J. Enol. Vitic.* **1983**, *34*, 180–185.
33. Souci, S.W.; Fachmann, W.; Kraut, H. *Food Composition and Nutrition Tables*, 3rd ed.; Deutsche Forschungsanstalt für Lebensmittelchemie: Freising, Germany, 1986; p. 1032.
34. Kader, A.A. Effects on Nutritional Quality. In *Modified on Controlled Atmospheres for the Storage, Transportation, and Packaging of Horticultural Commodities*; Elhadi, M.Y., Ed.; CRC Press Taylor & Francis Group: Boca Raton, FL, USA, 2009; pp. 111–118.
35. Lumpkin, C.; Fellman, J.K.; Rudell, D.R.; Mattheis, J.P. 'Fuji' apple (*Malus domestica* Borkh.) volatile production during high pCO_2 controlled atmosphere storage. *Postharvest Biol. Technol.* **2015**, *100*, 234–243. [CrossRef]
36. Pelayo, C.; Ebeler, S.E.; Kader, A.A. Postharvest life and flavor quality of three strawberry cultivars kept at 5 °C in air or air +20 kPa CO_2. *Postharvest Biol. Technol.* **2003**, *27*, 171–183. [CrossRef]

Article

Post-Harvest Treatment with Methyl Jasmonate Impacts Lipid Metabolism in Tomato Pericarp (*Solanum lycopersicum* L. cv. Grape) at Different Ripening Stages

Silvia Leticia Rivero Meza [1], Eric de Castro Tobaruela [1], Grazieli Benedetti Pascoal [2], Isabel Louro Massaretto [1] and Eduardo Purgatto [1,*]

1 Department of Food Science and Experimental Nutrition/Food Research Center, Faculty of Pharmaceutical Sciences, University of São Paulo (USP), Av. Prof. Lineu Prestes 580, bl 14, Butantã, 05508-000 São Paulo, SP, Brazil; silviarm@usp.br (S.L.R.M.); erictobaruela@gmail.com (E.d.C.T.); isabelmassaretto@gmail.com (I.L.M.)

2 Faculty of Medicine, Federal University of Uberlândia (UFU), Av. Pará, 1720, bl 2U, Umuarama, 38405-320 Uberlândia, MG, Brazil; grazi.nutri13@gmail.com

* Correspondence: epurgatt@usp.br; Tel.: +55-113-0911-480

Citation: Rivero Meza, S.L.; de Castro Tobaruela, E.; Benedetti Pascoal, G.; Louro Massaretto, I.; Purgatto, E. Post-Harvest Treatment with Methyl Jasmonate Impacts Lipid Metabolism in Tomato Pericarp (*Solanum lycopersicum* L. cv. Grape) at Different Ripening Stages. *Foods* **2021**, *10*, 877. https://doi.org/10.3390/foods10040877

Academic Editors: Eleni Tsantili and Jinhe Bai

Received: 22 March 2021
Accepted: 13 April 2021
Published: 16 April 2021

Publisher's Note: MDPI stays neutral with regard to jurisdictional claims in published maps and institutional affiliations.

Abstract: The application of exogenous jasmonate can stimulate the production of ethylene, carotenoids, and aroma compounds and accelerate fruit ripening. These alterations improve fruit quality and make fruit desirable for human consumption. However, fruit over-ripening results in large losses of fruit crops. This problem is overcome by applying 1-methylcyclopropene to the fruits, due to its capacity to block the ethylene receptors, suppressing fruit ripening. In this study, treatments with only 1-methylcyclopropene and both 1-methylcyclopropene and methyl jasmonate were administered to observe whether exogenous methyl jasmonate can improve the metabolite levels in fruits with blocked ethylene receptors. Fruit pericarps were analyzed at 4, 10, and 21 days after harvest (DAH) and compared with untreated fruits. The post-harvest treatments affected primary metabolites (sugars, organic acids, amino acids, and fatty acids) and secondary metabolites (carotenoids, tocopherols, and phytosterols). However, the lipid metabolism of the tomatoes was most impacted by the exogenous jasmonate. Fatty acids, carotenoids, tocopherols, and phytosterols showed a delay in their production at 4 and 10 DAH. Conversely, at 21 DAH, these non-polar metabolites exhibited an important improvement in their accumulation.

Keywords: post-harvest treatment; jasmonate; metabolite profiling; lipid metabolism; *Solanum lycopersicum*; ethylene inhibition; fruit quality

1. Introduction

At the onset of tomato ripening, changes in primary metabolites are observed, such as the accumulation of glucose and fructose and the presence of citric and malic acids in ripe fruits [1]. Sugars and organic acids are critical to good flavor, contributing to sweetness and acid balance. Consequently, they are responsible for consumer acceptance [2].

Additionally, changes in secondary metabolites of tomato fruit are observed mainly for those related to health benefits as well as lycopene, β-carotene, α-tocopherol, β-tocopherol, and β-sitosterol. Carotenoids and tocopherols play an important role in human nutrition, mainly due to antioxidant properties and the visual perception of ripe fruits, while phytosterols are associated with reducing LDL cholesterol and total cholesterol [3–5]. Many of these ripening processes are regulated by plant hormones such as ethylene, methyl jasmonate, abscisic acid, and other phytohormones [6,7].

Methyl jasmonate can interact with other phytohormones, such as ethylene, in promoting biological activity, such as antibacterial and antifungal activities and signaling plant defenses [8]. Application of exogenous jasmonate stimulates ethylene production, the degradation of chlorophyll, accumulation of β-carotene, and production of aroma compounds, which can accelerate fruit ripening [9].

Although these changes can improve the fruit quality, making it desirable for consumption, fruit over-ripening can result in large losses of fruit crops. This problem can be overcome by exogenous application of 1-methylcyclopropene to tomato fruits, due to its ability to reduce the ethylene production and respiration rate of climacteric fruits [10]. This action prolongs the shelf life of tomato fruits by retaining their firmness and delaying lycopene production and consequently color development [11,12]. This study investigates the metabolic response to methyl jasmonate applied concomitantly with 1-methylcyclopropene to harvested tomato fruits during their ripening.

2. Materials and Methods

2.1. *Plant Material and Post-Harvest Treatment*

Tomatoes (*Solanum lycopersicum* cv. Grape) at the mature green stage (N = 1200) were collected from a standard commercial greenhouse in Ibiúna (23°39′21″ S; 47°13′22″ W), São Paulo, Brazil. Fruits were sterilized with 0.1% aqueous sodium hypochlorite solution for 15 min. Four biological replicates were applied in the experiment, each comprising 100 fruits. Tomatoes were randomly separated into 3 groups (N = 400 by group): (1) control group (CTRL), with no treatment; (2) treated with 1-methylcyclopropene (MCP); (3) treated with both 1-methylcyclopropene and methyl jasmonate (MCP+MeJA). Fruits were left to ripen spontaneously in a 323 L chamber at a constant temperature (20 ± 2 °C) and humidity (80% ± 5% RH) in a 16 hour-day/8-hour-night cycle. For the MCP treatment, the instructions of the manufacturer for "manual addition" were followed: 2.45 g of 1-methylcyclopropene (powder, 3.3% w/w active ingredient, SmartFresh post-harvest treatment; AgroFresh Solutions, Inc., Philadelphia, PA, USA) was weighed and transferred to a 500 mL Erlenmeyer flask capped with a rubber stopper. Using a syringe, 75 mL of de-ionized water was added to the flask, dissolving the powder, and releasing the 1-methylcyclopropene gas. The flask was placed in the chamber, the stopper was removed, and the chamber was closed immediately. A small fan was placed in the chamber, directed at the flask to aid in the dispersion of the gas. For the MCP+ MeJA group, methyl jasmonate (Sigma-Aldrich, Saint Louis, MO, USA) was applied to a filter paper left on the chamber wall for evaporation (100 ppm, final concentration in gas phase), and 1-methylcyclopropene treatment was made as described above. Both treatments were conducted for the second time 12 h after the first exposure to the hormone, totaling 24 h of treatment. Samples of 10 fruits from each replicate were randomly taken at 4, 10, and 21 days after harvest (DAH), considering the control group as a reference. Biological replicates were composed of pericarp tissues by removing the placenta and fruit seeds. The pericarp samples were frozen in liquid nitrogen and stored at −80°C for subsequent analyses.

2.2. *Ripening Parameters*

2.2.1. Ethylene Emission

Ethylene emission was analyzed by placing five tomato fruits in 600 mL airtight glass containers at 25 °C for 1 h. Five 1 mL samples of gas produced in the headspace were then collected with gastight syringes through a rubber septum. A gas chromatograph with a flame ionization detector (Flame Ionization Detector (FID) for GC; (Agilent Technologies, Santa Clara, CA, USA, model HP-6890) and HP-PLOT Q column (30 m × 0.53 mm × 40 μm) were used to evaluate ethylene emission. The injector and detector temperatures were equally established at 250 °C and the oven at 30 °C. The helium gas flow was set at 1 mL.min^{-1}, and the injections were performed using a pulsed splitless method.

2.2.2. Fruit Surface Color

Fruit surface color was measured using a HunterLab ColorQuest XE colorimeter (Hunter Associates Laboratory, Inc.) in terms of L*, a*, and b* space. The experimental data were treated to obtain values of the hue angle. Three measurements were made at the equatorial zone of six tomato fruits [13].

2.3. Analysis of Metabolite Profiling of Tomato Fruit Using GC-MS

2.3.1. Extraction and Derivatization of Polar Metabolites

The extraction and derivatization of polar metabolites were conducted as described in [14]. For the extraction process, 100 mg of frozen pericarp powder was mixed with 100% distilled methanol at −20 °C (1400 μL) and ribitol (200 μg.mL^{-1}, internal standard; 60 μL). The mixture was vortexed, incubated in a thermomixer at 950 rpm for 10 min at 70 °C, centrifuged at 11,000× *g* for 10 min, and the supernatant was collected. To the upper phase was added chloroform at −20 °C (750 μL) and Milli-Q water (1500 μL), followed by mixing and centrifugation at 2200× *g* for 15 min. The upper hydrophilic phase (150 μL) was collected and dried under nitrogen gas. Sample derivatization comprised adding 20 mg.mL^{-1} methoxyamine hydrochloride (Sigma-Aldrich, St. Louis, MO, USA; 40 μL) and pyridine with subsequent incubation in an orbital shaker at 1000 rpm and 37 °C for 2 h. Consecutively, N-methyl-N-(trimethylsilyl) trifluoroacetamide (MSTFA; 70 μL) was added to the sample, followed by incubation in an orbital shaker at 1000 g and 37 °C for 30 min. Finally, the derivatized samples were moved into glass vials and analyzed by GC-MS. A pool of polar metabolite external standards (1 mg.mL^{-1}, Sigma-Aldrich) was applied to certify the identified metabolites by mass spectral comparison: D-glucose; D-fructose; maltose; sucrose; D-galactose; myo-inositol; citric acid; L-alanine; L-serine; L-proline; L-aspartate; L-glutamate [15].

2.3.2. Extraction and Derivatization of Non-Polar Metabolites

For the extraction process, 1000 mg of frozen pericarp powder was mixed with chloroform (1250 μL), methanol (2500 μL), and n-tridecane (800 μg.mL^{-1}, internal standard; 20 μL), followed by vortexing for 10 s and incubation on ice for 30 min. Then, 1.5% sodium sulfate (1250 μL) and chloroform (1250 μL) were added to the mixture, incubated on ice for 5 min and centrifuged at 4 °C for 1000× *g* and 15 min. The upper polar phase was collected and dried under nitrogen gas. The sample was redissolved in hexane (1000 μL), toluene (200 μL), methanol (1500 μL), and 8% chloridric acid (300 μL), mixed for 10 s, and incubated for 1.5 h at 100 °C. Subsequently, hexane (1000 μL) and Milli-Q water were added to the sample and mixed [16–18]. The hexane phase was separated and dried under nitrogen gas. The sample was redissolved in hexane (80 μL) and pyridine (20 μL) and derivatized with MSTFA (40 μL). Finally, the derivatized samples were moved into glass vials and analyzed by GC-MS. A pool of fatty acid methyl ester (FAME) external standards (Sigma-Aldrich) was applied to certify the identified metabolites by mass spectral comparison: methyl laurate (C12:0, 0.8 mg.mL^{-1}); methyl tetradecanoate (C14:0, 0.8 mg.mL^{-1}); methyl palmitate (C16:0, 0.8 mg.mL^{-1}); methyl octadecanoate (C18:0, 0.4 mg.mL^{-1}); methyl arachidate (C20:0, 0.4 mg.mL^{-1}); methyl docosanoate (C22:0, 0.4 mg.mL^{-1}); methyl lignocerate (C24:0, 0.4 mg.mL^{-1}); methyl linoleate (C 18:2, 0.4 mg.mL^{-1}); (Z)-9-oleyl methyl ester (C 18:1, 0.4 mg.mL^{-1}); methyl linolenate (C 18:3, 0.4 mg.mL^{-1}); methyl palmitoleate (C 16:1, 0.8 mg.mL^{-1}) [15].

2.3.3. GC-MS Analysis

Derivatized samples were analyzed by GC-MS (Agilent 5977 Series GC/MSD, Agilent Technologies, Santa Clara, CA, USA) [15]. Trimethylsilyl derivatives (1 μL) were injected into an injector at 230 °C in splitless mode. The oven temperature ramp applied was 80 °C (initial temperature), held for 2 min, heated at 15 °C.min^{-1} to 330 °C, and held for 6 min. The electron impact ionization mass spectrometer was set at ionization voltage 70 eV; ion source temperature 250 °C; injection port temperature 250 °C; and mass scan range 70–600 m/z at 20 scans.s^{-1}. The column used was an HP5ms column (30 m × 0.25 m × 0.25 μm). The flow rate of helium gas was 2 mL.min^{-1}. Acquisition, deconvolution, and analyses of experimental data were processed by MassHunter Quantitative Analysis software (Agilent, CA, EUA). The NIST mass spectral library (NIST 2011, Gaithersburg, MD, USA) was used for retention index (RI) comparison and data validation. Some of the identified metabolites were also confirmed by mass spectral comparison with the authentic external standards previously described.

2.4. Analysis of Carotenoids by HPLC

Frozen pericarp powder (200 mg) was mixed with 100 μL of 30% NaCl (w:v) solution and 200 μL of dichloromethane to extract carotenoids. Hexane:ether (1:1; 500 μL) was added to the mixture and centrifuged at 13,000× g at 4 °C for 5 min. This protocol was repeated thrice, and the organic phases were pooled [19]. The upper phase was dried under nitrogen gas and dissolved in ethyl acetate. Samples were analyzed by HPLC (Analytical HPLC, 1260 Infinity II LC System; Agilent Technologies, Santa Clara, CA, USA), coupled to a diode array detector (DAD), and equipped with a YMC Carotenoid HPLC C30 (5 μm × 250 mm × 4.6 mm) column [20]. Lycopene, β-carotene, and lutein from Sigma-Aldrich were used as external standards.

2.5. Statistical Analysis

Experimental data were expressed as mean ± standard deviation (SD) of four biological replicates. Statistical analysis was performed by one-way analysis of variance (ANOVA), and Tukey's test was applied to establish significant differences among mean values at $p < 0.05$, using the Minitab 19.0 software package (State College, PA, USA). For multivariate analysis, raw data were normalized by the internal standard area, processed using log transformation (log 2), mean-centered, and divided by the square root of the deviation of each variable (Pareto scaling). Principal component analysis (PCA), heatmaps, and fold-change analysis were executed to evaluate differences between treated and untreated groups, using the MetaboAnalyst 4.0 server (https://www.metaboanalyst.ca accessed on 15 April 2021) [21].

3. Results and Discussion

3.1. Effect of Methyl Jasmonate on the Ethylene Emission and Fruit Surface Color of Tomatoes

In this study, the alterations of metabolites identified in tomato fruits under post-harvest treatments were observed. Therefore, one group of fruits with ethylene inhibited by 1-methylcyclopropene were exposed to methyl jasmonate hormone (MCP+MeJA), other fruit groups were treated only with 1-methylcyclopropene (MCP), and the untreated tomato fruits (CTRL) were used as a reference for the assays. The three groups of fruits are visualized in Figure 1A.

The CTRL-group fruits achieved the breaker stage at 4 DAH and the ripe stage at 10 DAH. Regarding treated fruits, the breaker and red stages were achieved with MCP at 13 and 21 DAH, respectively, and MCP + MeJA at 10 and 13 DAH, respectively. The ripening stages of the CTRL group were characterized by measuring the ethylene emission and surface color of the tomato fruits from the day of harvest to 21 DAH (Figure 1B,C). The metabolite profiling was analyzed at 4, 10, and 21 DAH, aiming to observe the effect of treatments compared to the CTRL.

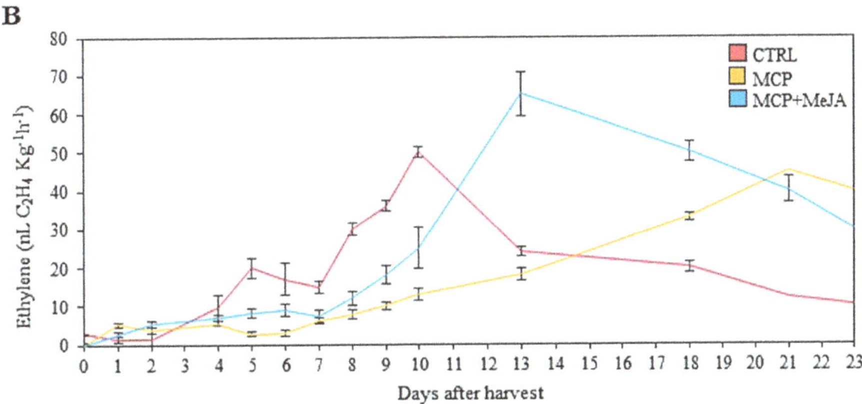

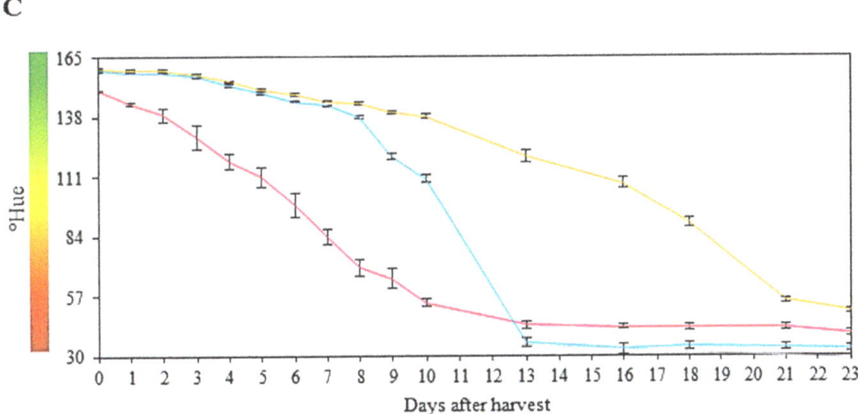

Figure 1. Characterization of tomato (*Solanum lycopersicum* L. cv. Grape) fruits treated with 1-methylcyclopropene (MCP) and both hormones, namely 1-methylcyclopropene and methyl jasmonate (MCP+MeJA), during ripening. (**A**) Representative images of tomatoes. Effects of MCP and MCP + MeJA on ethylene emission (**B**) and fruit color (**C**) compared to the control group (CTRL). Values are means ± standard error of four biological replicates of at least 10 fruits each.

Treatments with both 1-methylcyclopropene and methyl jasmonate, and only 1-methylcyclopropene, showed a delay in fruit ripening by reducing ethylene emission and fruit surface color, as compared to the CTRL group. A similar result was observed in a study where tomatoes were treated with 1-methylcyclopropene, which reported a reduction in ethylene emission and respiration rate [22]. Both groups, MCP and MCP + MeJA, presented the characteristic curves of ethylene emission of climacteric fruits. Fruits treated only with 1-methylcyclopropene showed the longest delay in fruit ripening, characterized by their ethylene peak and redness at 21 DAH. However, tomatoes treated with both 1-methylcyclopropene and methyl jasmonate showed an ethylene peak at 13 DAH when they acquired a reddish color.

It was observed that using exogenous methyl jasmonate hormone in fruits with ethylene receptors blocked by 1-methylcyclopropene stimulated the ripening process, as compared to fruits treated only with 1-methylcyclopropene. This behavior indicates that 1-methylcyclopropene efficiently blocks ethylene receptors and consequently may avoid the interaction of ethylene with other phytohormones related to ripening processes such as endogenous methyl jasmonate, delaying fruit ripening. However, when exogenous methyl jasmonate hormone was applied to these fruits, an acceleration in ripening was observed by the accumulation of pigments and anticipation of an ethylene peak from 21 to 13 DAH. Additionally, the highest peak of ethylene emission was observed for the MCP + MeJA group, which may be related to stimulation of ethylene biosynthesis in climacteric fruits by methyl jasmonate hormone. Thus, our results suggest that exogenous methyl jasmonate can act independently of ethylene, or the blockage of ethylene receptors was reversed after some time. Therefore, the synthesis of new receptors in tomato fruits after 1-methylcyclopropene treatment could be possible, as this occurs in several fruits [22,23]. This behavior may be responsible for the increased ethylene production after some time, as observed after 10 DAH.

3.2. *Primary Metabolite Profiling Affected by Post-Harvest Hormone Treatment*

Primary metabolites are important components related to fruit quality. Additionally, they are considered crucial for plant growth and development. Thus, understanding the fruit metabolism can support developing future approaches for its manipulation [24]. In this work, a total of 46 primary metabolites were identified by GC-MS metabolomics analysis: 10 sugars (glucose, fructose, sucrose, allose, gulose, glucaric acid, myo-inositol, mannose, ribose, and arabinofuranose); 9 organic acids (oxaloacetic, citric, succinic, aconitic, malic, citraconic, fumaric, propanoic, and butanoic acids); 12 amino acids (proline, serine, valine, threonine, aspartic acid, glutamic acid, glutamine, γ-aminobutyric acid (GABA), asparagine, tryptophan, phenylalanine, and tyrosine); 12 saturated fatty acids (capric, lauric, myristic, palmitic, stearic, eicosanoic, docosanoic, tricosanoic, lignoceric, hyenic, cerotic, and montanic acids); 3 unsaturated fatty acids (oleic, linoleic, and linolenic acids) at 4, 10, and 21 DAH (Table 1). Table 1 shows the effects of methyl jasmonate and 1-methylcyclopropene on the accumulation or reduction of each metabolite at the three different maturation stages, indicated by the area normalized by the internal standard.

Moreover, a global overview of the metabolic changes occurring in tomatoes during ripening was obtained to evaluate significant differences among accumulated metabolites in treated fruits compared with the control group (Figure 2).

Table 1. Primary metabolites in tomato pericarp (*Solanum lycopersicum* L. cv. Grape) exposed to 1-methylcyclopropene (MCP) and both 1-methylcyclopropene and methyl jasmonate (MCP+MeJA) treatments at 4, 10, and 21 days after harvest (DAH), detected by gas chromatography-mass spectrometry (GC-MS).

Metabolite	4 DAH			10 DAH			21 DAH		
	CTRL	MCP	MCP + MeJA	CTRL	MCP	MCP + MeJA	CTRL	MCP	MCP + MeJA
Sugars									
Glucose	1534.5 ± 76.0 ^{d}	70.6 ± 4.7 ^{g}	598.7 ± 37.9 ^{f}	1977.6 ± 11.4 ^{c}	156.5 ± 7.5 ^{g}	2977.0 ± 253.0 ^{b}	4352.0 ± 281.0 ^{a}	189.4 ± 15.2 ^{g}	1112.6 ± 140.1 ^{e}
Fructose	27474.0 ± 4039.0 ^{d}	3935.0 ± 489.0 ^{e}	37418.0 ± 5231.0 ^{c}	59266.0 ± 6310.0 ^{b}	4858.0 ± 544.0 ^{e}	26343.0 ± 3352.0 ^{d}	101194.0 ± 5662.0 ^{a}	25944.0 ± 1592.0 ^{d}	24324.0 ± 1808.0 ^{d}
Sucrose	38205.0 ± 569.0 ^{c}	5105.0 ± 559.0 ^{f}	20507.0 ± 1161.0 ^{d}	54654.0 ± 716.0 ^{b}	3661.0 ± 356.0 ^{f}	11961.0 ± 469.0 ^{e}	84839.0 ± 4545.0 ^{a}	10885.0 ± 358.0 ^{e}	11248.0 ± 284.0 ^{e}
Allose	1098.6 ± 44.2 ^{c}	172.6 ± 12.6 ^{ef}	115.1 ± 4.8 ^{f}	1563.4 ± 21.5 ^{b}	630.6 ± 39.0 ^{d}	687.0 ± 56.9 ^{d}	3309.0 ± 380.0 ^{a}	469.1 ± 49.6 ^{d,e}	584.1 ± 27.8 ^{d}
Gulose	221.5 ± 9.8 ^{d}	196.9 ± 15.1 ^{d}	229.8± 10.9 ^{d}	790.4 ± 38.3 ^{b}	92.6 ± 4.5 ^{e}	228.8 ± 10.4 ^{d}	1017.0 ± 60.4 ^{a}	183.5 ± 17.6 ^{d}	404.0 ± 12.9 ^{c}
Glucaric acid	42.2 ± 1.6 ^{d}	21.9 ± 2.1 ^{e}	23.1 ± 0.4 ^{e}	72.0 ± 1.30 ^{c}	93.4 ± 6.3 ^{b}	126.7 ± 8.5 ^{a}	124.6 ± 8.7 ^{a}	15.1 ± 1.3 ^{e}	21.2 ± 0.8 ^{e}
Myo-inositol	77.9 ± 2.9 ^{e}	33.1 ± 3.2 ^{f}	91.3 ± 4.5 ^{e}	169.6 ± 2.0 ^{d}	178.5 ± 8.1 ^{d}	156.6 ± 4.1 ^{d}	340.1 ± 20.6 ^{c}	412.4 ± 23.9 ^{b}	481.0 ± 9.7 ^{a}
Mannose	42.2 ± 3.1 ^{d}	3.4 ± 0.2 ^{f}	6.0 ± 0.3 ^{f}	100.0 ± 2.3 ^{b}	69.5 ± 3.5 ^{c}	148.3 ± 4.7 ^{a}	142.4 ± 11.0 ^{a}	28.9 ± 2.4 ^{e}	66.9 ± 9.7 ^{c}
Ribose	174.6 ± 7.4 ^{c}	5.9 ± 0.4 ^{g}	14.3 ± 0.6 ^{g}	249.4 ± 3.9 ^{b}	26.2 ± 1.3 ^{g}	128.5 ± 3.7 ^{d}	386.4 ± 27.1 ^{a}	60.7 ± 6.4 ^{f}	89.8 ± 3.5 ^{f}
Arabinofuranose	15.1 ± 0.7 ^{c}	2.1 ± 0.3 ^{e}	4.7 ± 0.1 ^{e}	25.5 ± 0.8 ^{b}	9.3 ± 0.6 ^{d}	8.1 ± 0.5 ^{d}	45.2 ± 3.3 ^{a}	17.2 ± 1.4 ^{c}	17.6 ± 1.6 ^{c}
Total	68885.0 ± 4082.0 ^{c}	9546.0± 771.0 ^{e}	59009.0 ± 5831.0 ^{c}	118868.0 ± 5993.0 ^{b}	9775.0 ± 904.0 ^{e}	42764.0 ± 3869.0 ^{d}	195750.0 ± 9973.0 ^{a}	38206.0 ± 1874.0 ^{d}	38348.0 ± 2196.0 ^{d}
Organic acids									
Oxaloacetic acid	573.3 ± 24.3 ^{e}	173.3± 18.6 ^{f}	214.5 ± 6.2 ^{f}	2380.4 ± 56.5 ^{a}	1234.0 ± 59.1 ^{c}	1482.7 ± 55.4 ^{b}	1241.9 ± 80.7 ^{c}	703.7 ± 72.1 ^{d}	823.0 ± 24.2 ^{d}
Citric acid	6517.0 ± 413.0 ^{c}	765.4 ± 85.1 ^{e}	998.9 ± 62.0 ^{e}	7878.0 ± 457.0 ^{c}	889.9 ± 79.5 ^{e}	4117.0 ± 591.0 ^{d}	18901.0 ± 1208.0 ^{a}	15494.0 ± 1393.0 ^{b}	15948.0 ± 747.0 ^{b}
Succinic acid	2646.0 ± 360.0 ^{c,d}	642.8 ± 49.6 ^{f}	1909.3 ± 81.4 ^{d,e}	12894.0 ± 485.0 ^{a}	2826.0 ± 650.0 ^{c}	3587.5 ± 102.0 ^{b}	2862.0 ± 254.0 ^{b,c}	1851.1 ± 135.1 ^{e}	2372.9 ± 71.7 ^{c,d,e}
Aconitic acid	61.1 ± 2.9 ^{c}	2.9 ± 0.3 ^{f}	52.9 ± 2.6 ^{d}	83.1 ± 0.9 ^{b}	4.2 ± 0.6 ^{f}	51.0 ± 1.1 ^{d}	101.4 ± 5.70 ^{a}	18.2 ± 1.6 ^{e}	67.1 ± 2.8 ^{c}
Malic acid	2537.5 ± 101.3 ^{d}	118.7 ± 5.3 ^{f}	156.1 ± 6.2 ^{f}	6653.7 ± 174.0 ^{c}	1208.0 ± 109.4 ^{e,f}	2076.1 ± 46.3 ^{d,e}	16800.0 ± 1014.0 ^{a}	13024.0 ± 1137.0 ^{b}	12285.0 ± 385.0 ^{b}
Citraconic acid	17.3 ± 0.7 ^{c}	4.1 ± 0.3 ^{d}	3.7 ± 0.2 ^{d}	104.5 ± 12.4 ^{a}	3.9 ± 0.2 ^{d}	2.7 ± 0.3 ^{d}	101.9 ± 6.1 ^{a}	20.9 ± 1.9 ^{c}	49.0 ± 2.2 ^{b}
Fumaric acid	167.9 ± 6.7 ^{c}	61.0 ± 5.3 ^{f,g}	73.4 ± 2.7 ^{f}	181.7 ± 1.2 ^{b}	47.7 ± 2.0 ^{h}	99.0 ± 3.0 ^{e}	237.8 ± 10.3 ^{a}	55.8 ± 5.2 ^{g,h}	128.6 ± 5.4 ^{d}
Propanoic acid	111.7 ± 6.8 ^{c}	14.6 ± 2.1 ^{e}	19.0 ± 1.2 ^{d,e}	145.6 ± 1.5 ^{c}	15.4 ± 0.8 ^{e}	15.4 ± 0.5 ^{e}	458.6 ± 79.1 ^{b}	90.7 ± 5.0 ^{c,d}	1318.0 ± 51.3 ^{a}
Butanoic acid	284.5 ± 18.9 ^{c,d}	221.3 ± 19.7 ^{d}	314.9 ± 8.6 ^{c,d}	393.4 ± 10.9 ^{c}	179.5 ± 8.6 ^{d}	274.4 ± 5.9 ^{c,d}	641.8 ± 96.9 ^{b}	412.7 ± 22.5 ^{c}	2404.7 ± 175.3 ^{a}
Total	12917.0 ± 804.0 ^{d}	2003.4 ± 173.2 ^{f}	3743.0±130.7 ^{e,f}	30715.0±907.0 ^{c}	6409.0±787.0 ^{e}	11706.0±691.0 ^{d}	41346.0±2535.0 ^{a}	31671.0±2748.0 ^{c}	35397.0±1239.0 ^{b}
Amino acids									
Proline	501.2 ± 14.8 ^{d}	798.4 ± 71.8 ^{b,c,d}	992.4 ± 55.3 ^{b,c}	1167.2 ± 26.8 ^{b}	693.5 ± 40.9 ^{c,d}	470.9 ± 38.8 ^{d}	2863.8 ± 483.0 ^{a}	2546.8 ± 166.2 ^{a}	2452.7 ± 44.8 ^{a}
Serine	78.0 ± 5.7 ^{b}	24.8 ± 1.4 ^{e}	58.5 ± 4.3 ^{c}	93.6 ± 2.3 ^{a}	27.7 ± 1.9 ^{e}	24.8 ± 1.0 ^{e}	49.0 ± 1.5 ^{d}	16.1 ± 1.9 ^{f}	26.9 ± 1.3 ^{e}
Valine	6.2 ± 0.2 ^{e}	1.9 ± 0.2 ^{f}	2.2 ± 0.1 ^{f}	17.6 ± 0.2 ^{a}	17.0 ± 0.8 ^{a}	14.9 ± 0.5 ^{b}	7.9 ± 0.4 ^{d}	6.6 ± 0.9 ^{e}	9.1 ± 0.3 ^{c}
Threonine	6.0 ± 0.45 ^{e}	4.3 ± 0.6 ^{e}	6.4 ± 0.1 ^{d,e}	25.9 ± 0.4 ^{b}	9.9 ± 0.5 ^{d}	14.9 ± 0.5 ^{c}	25.2 ± 0.8 ^{b}	26.4 ± 2.7 ^{b}	50.5 ± 3.5 ^{a}
Aspartic acid	1523.2 ± 54.2 ^{b}	54.7 ± 4.8 ^{e}	141.7 ± 3.9 ^{e}	2166.8 ± 58.7 ^{a}	823.2 ± 42.5 ^{d}	1163.1 ± 43.0 ^{c}	1540.8 ± 161.0 ^{b}	1148.6 ± 97.4 ^{c}	1116.9 ± 43.0 ^{c}
Glutamic acid	1744.3 ± 75.6 ^{d}	215.3 ± 17.3 ^{e}	228.7 ± 8.3 ^{e}	4906.7 ± 42.6 ^{a}	1614.2 ± 71.8 ^{d}	1530.1 ± 50.3 ^{d}	3957.0 ± 424.0 ^{b}	2729.0 ± 208.0 ^{c}	2551.6 ± 68.0 ^{c}
Glutamine	185.1 ± 7.7 ^{c}	77.8 ± 5.6 ^{e}	172.9 ± 4.8 ^{c}	519.3 ± 9.2 ^{a}	99.9 ± 5.3 ^{d,e}	148.3 ± 4.7 ^{c,d}	475.8 ± 38.7 ^{a}	351.8 ± 38.2 ^{b}	314.9 ± 24.4 ^{b}
GABA	207.6 ± 17.2 ^{b,c}	148.4 ± 13.2 ^{d,e}	173.9 ± 4.5 ^{c,d}	1060.8 ± 45.9 ^{a}	104.3 ± 4.6 ^{e,f}	76.9 ± 4.4 ^{f}	216.9 ± 20.6 ^{b,c}	221.9 ± 23.6 ^{b,c}	242.5 ± 12.0 ^{b}
Asparagine	139.7 ± 6.2 ^{c}	19.8 ± 1.7 ^{e}	22.6 ± 0.5 ^{e}	260.2 ± 6.9 ^{a}	117.6 ± 5.6 ^{d}	142.8 ± 4.8 ^{c}	235.6 ± 11.7 ^{b}	134.2 ± 10.9 ^{c,d}	148.5 ± 11.1 ^{c}
Tryptophan	173.7 ± 7.1 ^{b}	21.9 ± 1.9 ^{d}	22.8 ± 0.9 ^{d}	321.4 ± 8.6 ^{a}	154.6 ± 7.9 ^{c}	154.2 ± 4.9 ^{c}	27.1 ± 0.9 ^{d}	25.0 ± 1.8 ^{d}	31.9 ± 2.8 ^{d}
Phenylalanine	7.6 ± 0.5 ^{e}	76.2 ± 9.8 ^{a}	69.1 ± 2.1 ^{a}	26.4 ± 1.3 ^{d}	5.9 ± 0.3 ^{e}	8.3 ± 0.4 ^{e}	50.9 ± 2.7 ^{b}	36.3 ± 4.9 ^{c}	36.4 ± 1.6 ^{c}
Tyrosine	21.1 ± 0.8 ^{e}	47.8 ± 3.6 ^{c}	71.0 ± 3.9 ^{a}	27.8 ± 2.3 ^{d,e}	35.7 ± 2.0 ^{d}	27.7 ± 1.1 ^{d,e}	62.3 ± 3.7 ^{b}	51.9 ± 5.0 ^{c}	46.4 ± 4.3 ^{c}
Total	4593.8 ± 170.8 ^{d}	1491.4 ± 125.1 ^{f}	1962.3 ± 70.7 ^{f}	10593.8 ± 94.5 ^{a}	3703.5 ± 149.7 ^{e}	3777.0 ± 148.3 ^{d,e}	9512.0 ± 940.0 ^{b}	7294.0 ± 482.0 ^{c}	35397.0 ± 1239.0 ^{b}

Table 1. *Cont.*

Metabolite	4 DAH			10 DAH			21 DAH		
	CTRL	MCP	MCP + MeJA	CTRL	MCP	MCP + MeJA	CTRL	MCP	MCP + MeJA
Saturated fatty acids									
Capric acid	61.9 ± 1.9 ^{a}	26.6 ± 0.6 ^{c}	52.2 ± 0.7 ^{b}	15.0 ± 0.4 ^{d}	1.5 ± 0.1 ^{g}	2.3 ± 0.0 ^{g}	6.8 ± 0.0 ^{e}	4.0 ± 0.1 ^{f}	5.7 ± 0.1 ^{e,f}
Lauric acid	40.5 ± 2.1 ^{a}	18.9 ± 0.1 ^{c}	22.4 ± 0.3 ^{b}	19.7 ± 1.2 ^{c}	1.4 ± 0.1 ^{e}	2.1 ± 0.0 ^{e}	6.8 ± 0.0 ^{d}	5.6 ± 0.1 ^{d}	5.9 ± 0.1 ^{d}
Myristic acid	19.0 ± 0.8 ^{a}	0.8 ± 0.0 ^{f}	2.1 ± 0.0 ^{e}	15.0 ± 0.5 ^{b}	1.3 ± 0.1 ^{f}	2.6 ± 0.0 ^{e}	6.1 ± 0.1 ^{c}	4.9 ± 0.1 ^{d}	5.9 ± 0.1 ^{c}
Palmitic acid	436.7 ± 10.4 ^{a}	148.7 ± 5.3 ^{e}	285.9 ± 23.0 ^{d}	413.1 ± 10.9 ^{a,b}	23.9 ± 0.9 ^{f}	36.2 ± 0.3 ^{f}	159.6 ± 1.4 ^{e}	349.8 ± 6.2 ^{c}	400.5 ± 9.3 ^{b}
Stearic acid	16.5 ± 0.5 ^{a}	9.2 ± 0.5 ^{e,f}	8.6 ± 0.1 ^{f}	14.6 ± 0.3 ^{b}	1.0 ± 0.1 ^{h}	13.2 ± 0.1 ^{c}	9.9 ± 0.2 ^{d}	7.6 ± 0.2 ^{g}	9.3 ± 0.1 ^{d,e}
Eicosanoic acid	93.9 ± 1.6 ^{b}	84.1 ± 6.0 ^{c}	170.9 ± 2.8 ^{a}	10.2 ± 0.3 ^{d}	5.8 ± 0.3 ^{d,e}	8.7 ± 0.2 ^{d}	7.2 ± 0.2 ^{d}	7.1 ± 0.1 ^{d}	0.7 ± 0.0 ^{d,e}
Docosanoic acid	43.7 ± 1.0 ^{b}	23.6 ± 2.4 ^{c}	53.8 ± 1.6 ^{a}	41.3 ± 1.1 ^{b}	6.7 ± 0.3 ^{f}	10.9 ± 0.1 ^{e}	16.0 ± 0.1 ^{d}	8.2 ± 0.2 ^{f}	9.2 ± 0.4 ^{e,f}
Tricosanoic acid	27.6 ± 1.1 ^{b}	22.7 ± 1.3 ^{c}	45.2 ± 0.3 ^{a}	20.3 ± 0.3 ^{d}	5.5 ± 0.3 ^{g}	8.8 ± 0.2 ^{e}	7.1 ± 0.8 ^{f}	6.6 ± 0.1 ^{f,g}	7.7 ± 0.3 ^{e,f}
Lignoceric acid	22.1 ± 1.8 ^{c}	163.6 ± 5.1 ^{b}	618.9 ± 23.1 ^{a}	13.9 ± 0.2 ^{c}	5.3 ± 0.2 ^{c}	11.1 ± 0.5 ^{c}	7.2 ± 0.2 ^{c}	6.8 ± 0.1 ^{c}	7.8 ± 0.4 ^{c}
Hyenic acid	32.5 ± 1.0 ^{b}	25.8 ± 0.2 ^{c}	37.2 ± 1.7 ^{a}	8.8 ± 0.2 ^{d}	3.3 ± 0.1 ^{f}	6.9 ± 0.0 ^{e}	6.6 ± 0.1 ^{e}	5.6 ± 0.1 ^{e}	7.3 ± 0.1 ^{d,e}
Cerotic acid	10.0 ± 0.7 ^{b}	7.9 ± 0.4 ^{c,d}	26.9 ± 1.6 ^{a}	10.2 ± 0.3 ^{b}	5.8 ± 0.3 ^{e}	8.7 ± 0.2 ^{b,c}	7.3 ± 0.2 ^{c,d}	7.2 ± 0.1 ^{d,e}	7.5 ± 0.3 ^{c,d}
Montanic acid	34.4 ± 1.3 ^{a}	15.2 ± 0.4 ^{c}	17.7 ± 0.6 ^{b}	9.5 ± 0.2 ^{d}	5.4 ± 0.2 ^{f}	7.4 ± 0.1 ^{e}	9.8 ± 0.2 ^{d}	6.1 ± 0.1 ^{f}	9.1 ± 0.1 ^{d}
Total	838.8 ± 20.6 ^{b}	547.0 ± 6.9 ^{d}	1341.7 ± 15.0 ^{a}	591.6 ± 14.5 ^{c}	66.7 ± 2.7 ^{i}	119.0 ± 0.5 ^{h}	250.4 ± 2.0 ^{g}	419.7 ± 7.5 ^{f}	476.8 ± 11.0 ^{e}
Unsaturated fatty acids									
Oleic acid	610.5 ± 26.0 ^{b}	278.4 ± 5.4 ^{d}	866.7 ± 20.9 ^{a}	339.4 ± 7.6 ^{c}	19.9 ± 0.9 ^{g}	44.8 ± 0.4 ^{g}	213.0 ± 7.1 ^{e}	103.6 ± 10.5 ^{f}	340.2 ± 6.5 ^{c}
Linoleic acid	1079.5 ± 56.5 ^{c}	8.8 ± 0.5 ^{g}	46.9 ± 0.3 ^{g}	749.6 ± 21.4 ^{d}	285.0 ± 14.9 ^{f}	588.5 ± 16.7 ^{e}	593.1 ± 5.1 ^{e}	5065.0 ± 93.4 ^{b}	5880.9 ± 144.9 ^{b}
α-Linolenic acid	94.9 ± 4.4 ^{c}	354.6 ± 6.6 ^{a}	292.1 ± 1.5 ^{b}	24.0 ± 0.3 ^{e,f}	11.3 ± 0.5 ^{g}	33.2 ± 0.1 ^{d}	17.3 ± 2.3 ^{f,g}	14.1 ± 0.4 ^{g}	25.3 ± 1.1 ^{e}
Total	1784.9 ± 78.7 ^{c}	641.8 ± 10.3 ^{f}	1205.7 ± 22.1 ^{d}	1113.1 ± 29.0 ^{d}	316.2 ± 15.8 ^{g}	666.5 ± 16.4 ^{f}	823.4 ± 11.1 ^{e}	5183.7 ± 89.5 ^{b}	6246.4 ± 147.8 ^{a}

Values were presented as normalized area by ribitol or n-tridecane (internal polar and non-polar standards, respectively). CTRL: Control fruits. Different superscript letters (a–g) indicate statistical significance ($p < 0.05$) at the same line (mean ± standard deviation, n = 4). GABA, γ-aminobutyric acid.

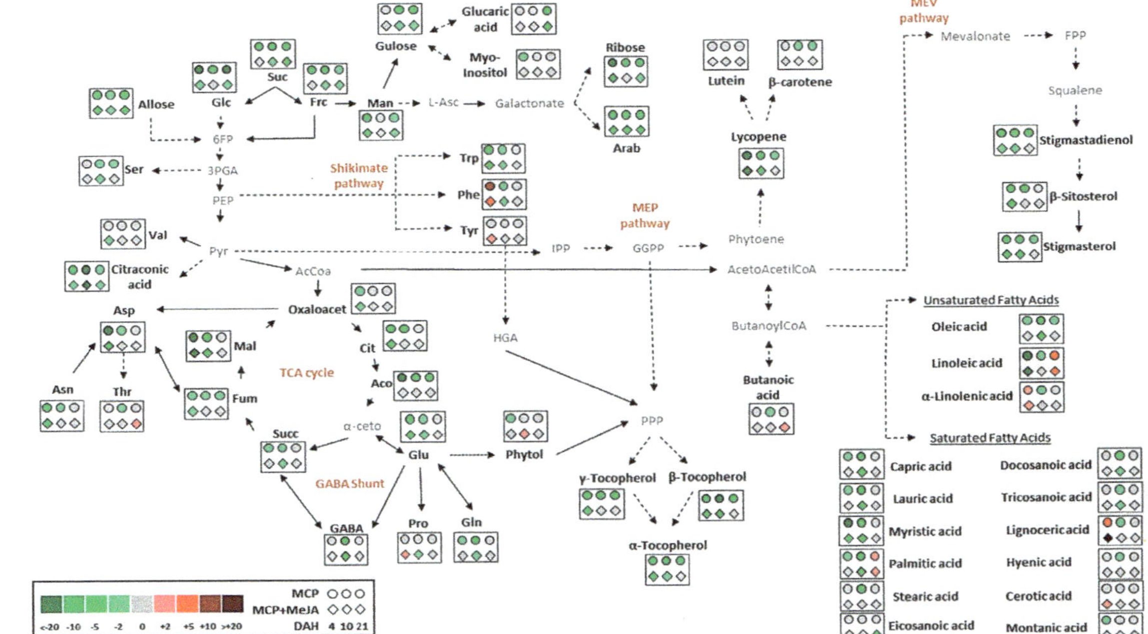

Figure 2. Global overview of metabolic changes occurring in tomato pericarp (*Solanum lycopersicum* L. cv. Grape) treated with 1-methylcyclopropene (MCP) and both hormones, 1-methylcyclopropene and methyl jasmonate (MCP + MeJA), compared to the control group (CTRL). Data were normalized to the CTRL. Metabolites showing up- or down-regulation in each treatment exceeding twofold compared to the CTRL are shown. The color scale displays the different amounts of metabolite in terms of proportional change relative to the level in the appropriate control. Suc, sucrose; Glc, glucose; Frc, fructose; Man, mannose; L-Asc, L-ascorbic acid; Arab, arabinofuranose; 6FP, fructose-6-phosphate; 3-GPA, glyceraldehyde-3-phosphate; Ser, serine; PEP, phosphoenolpyruvate; Trp, tryptophan; Phe, phenylalanine; Tyr, tyrosine; HGA, homogentisic acid; Pyr, pyruvic acid; IPP, isopentenyl pyrophosphate; GGPP, geranylgeranyl; PPP, phytyl pyrophosphate; Val, valine; AcCoA, acetyl-CoA; Oxaloacet, oxaloacetic acid; Cit, citric acid; Aco, aconitic acid; α-keto, α-etoglutaric acid; Succ, succinic acid; Fum, fumaric acid; Mal, malic acid; Glu, glutamic acid; GABA, γ-aminobutyric acid; Gln, glutamine; Pro, proline; Arg, arginine; Asp, aspartic acid; Thr, threonine; Asn, asparagine; FPP, farnesyl pyrophosphate.

A PCA was performed on primary metabolites at the 4th, 10th, and 21st ripening stages, confirming the high reproducibility among the four biological replicates and groups analyzed. Moreover, clear separation of the CTRL group and both treated groups was evidenced for the primary metabolites in the PCA score. Heatmap analysis was used to analyze the differences between treated and untreated groups regarding the metabolite changes on each day after harvest (Figures 3–5).

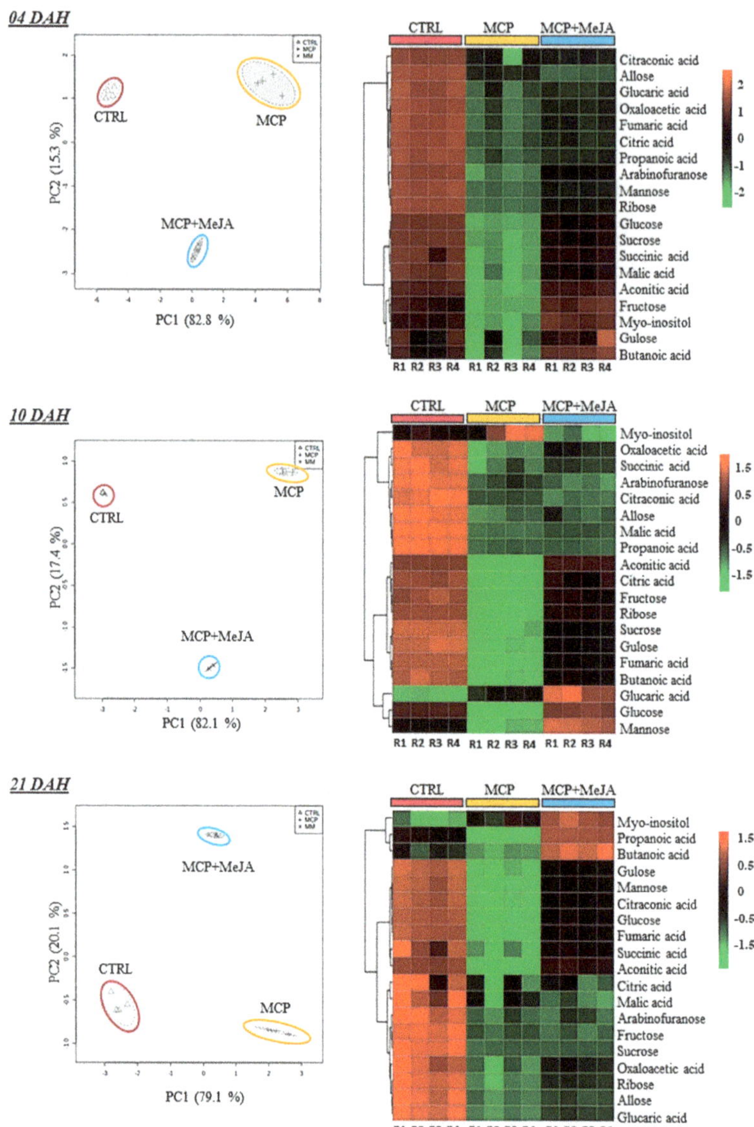

Figure 3. Relative contents of sugars and organic acids in tomato pericarp (*Solanum lycopersicum* L. cv. Grape) exposed to 1-methylcyclopropene (MCP) and both 1-methylcyclopropene and methyl jasmonate (MCP+MeJA) compared to the control group (CTRL). Unsupervised principal component analysis (PCA-score) and heatmap analysis represent the major sources of variability. Color scale represents the variation in the relative concentration of compounds, from low (green) to high (red) contents at 4, 10, and 21 days after harvest (DAH).

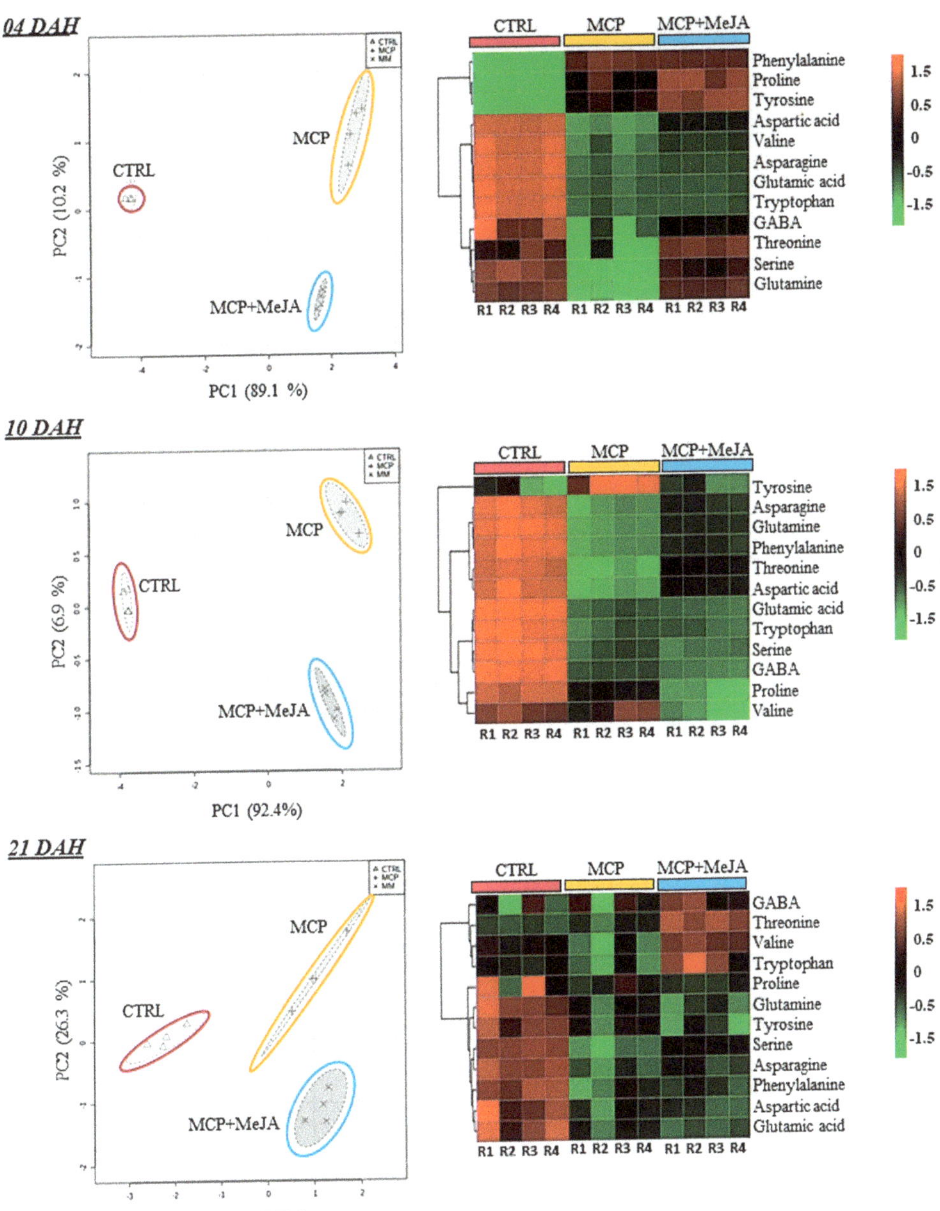

Figure 4. Relative contents of amino acids in tomato pericarp (*Solanum lycopersicum* L. cv. Grape) exposed to 1-methylcyclopropene (MCP) and both hormones, 1-methylcyclopropene and methyl jasmonate (MCP+MeJA), as compared to the control group (CTRL). Unsupervised principal component analysis (PCA-score) and heatmap analysis represent the major sources of variability. Color scale represents the variation in the relative concentration of compounds, from low (green) to high (red) contents at 4, 10, and 21 days after harvest (DAH).

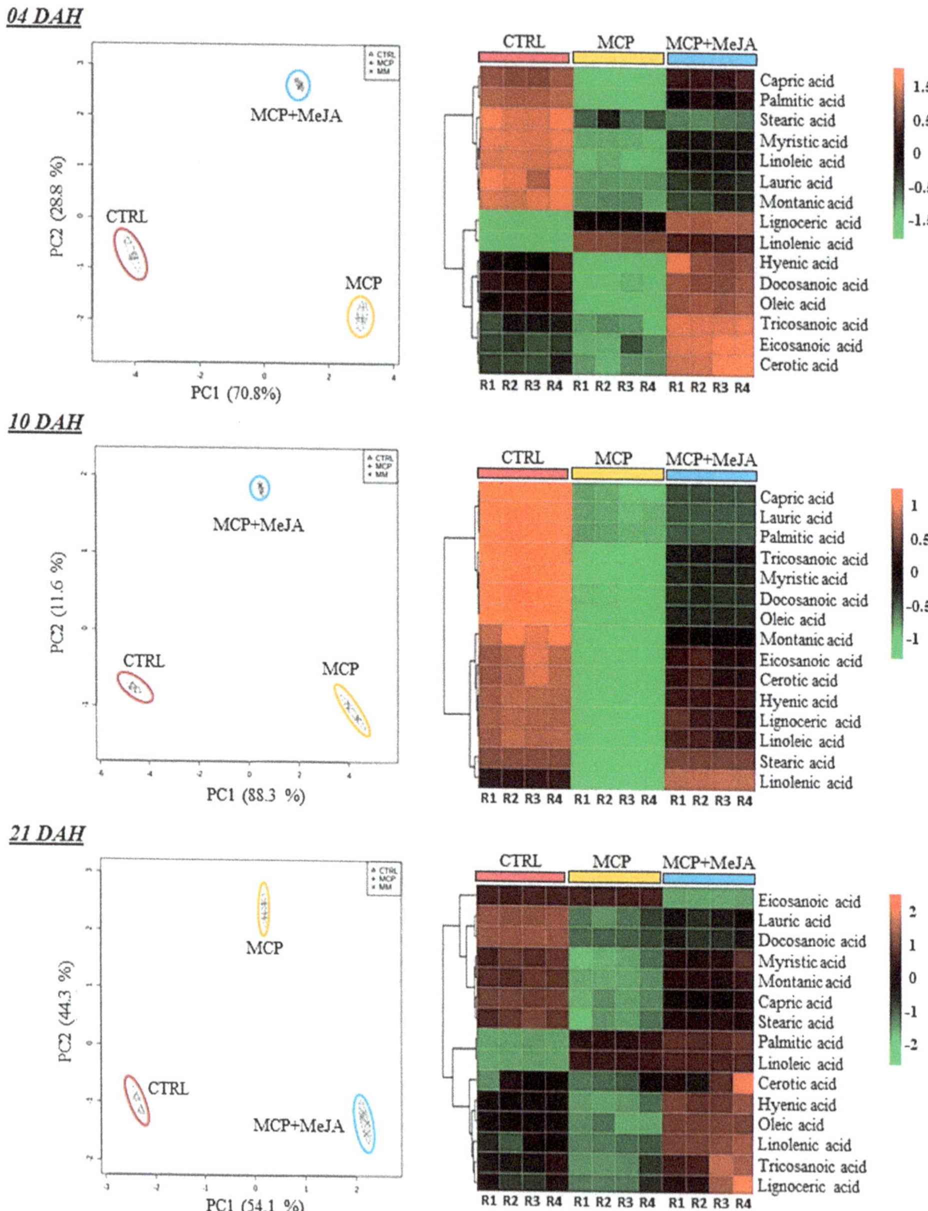

Figure 5. Relative contents of fatty acids in tomato pericarp (*Solanum lycopersicum* L. cv. Grape) exposed to 1-methylcyclopropene (MCP) and both hormones, 1-methylcyclopropene and methyl jasmonate (MCP+MeJA), as compared to the control group (CTRL). Unsupervised principal component analysis (PCA-score) and heatmap analysis represent the major sources of variability. Color scale represents the variation in the relative concentration of compounds, from low (green) to high (red) contents at 4, 10, and 21 days after harvest (DAH).

Primary metabolism is essential for fruit quality. Sugars, organic acids, and amino acids are responsible for the taste of tomato fruits, facilitating sensory perception. Amino acids and fatty acids play important roles as precursors of aroma compounds [7]. Treatment with 1-methylcyclopropene impacted sugar and organic acids, inhibiting their production during ripening. Fruits treated only with 1-methylcyclopropene were most affected, showing a greater delay in accumulating sugars and organic acids than fruits treated with both 1-methylcyclopropene and methyl jasmonate (Figure 3). For instance, glucose showed a 22-, 13-, and 23-fold reduction at 4-, 10-, and 21 DAH, respectively, in MCP, as compared to the CTRL. Mannose, ribose, and malic and aconitic acids exhibited a 14-, 30-, 21-, and 20-fold decrease in levels at 4 DAH, respectively. Conversely, fructose, sucrose, and citraconic acid showed a 12-, 15-, and 27-fold decrease in levels at 10 DAH when MCP was compared with the CTRL (Table 1). Reductions in the levels of these metabolites in fruits treated with 1-methylcyclopropene are shown in (Figure 2), based on the fold-change analysis of the treated fruits and control group.

Exceptionally, glucose, glucaric acid, and mannose levels showed an increase at 10 DAH in MCP + MeJA, as compared to the CTRL. Similar behavior was observed for myo-inositol, propanoic, and butanoic acids at 21 DAH (Table 1). As shown in Figure 3, heatmap analysis demonstrated a tendency of these metabolites to increase at 10 DAH. As observed by ethylene emission, the minor impact on the production of sugars and organic acids observed for MCP + MeJA may suggest that methyl jasmonate plays an important role in ripening. This may act independently of endogenous ethylene, stimulate the synthesis of new receptors, or reverse the blockage of ethylene receptors after some time.

Amino acid profiling was also affected by the action of 1-methylcyclopropene. Inhibition of the production of amino acids during ripening was observed for both MPC and MCP + MeJA compared with the control (Figure 4). The most affected amino acids were aspartic acid at 4 DAH and GABA at 10 DAH, showing a 28- and 10-fold reduction in their levels with MCP, respectively. However, MCP + MeJA showed 11- and 14-fold decreases, respectively, as compared to the CTRL, as shown in (Figure 2). Conversely, tyrosine and phenylalanine showed levels two- and ninefold higher for MCP and MCP + MeJA at 4 DAH, as compared to the CTRL (Table 1, Figure 2). It is important to highlight that phenylalanine and tyrosine are aromatic amino acids, which participate in the shikimate pathway and are responsible for the aroma development of fruit. Table 1 shows that the total amino acid level was represented mostly by proline, glutamic, and aspartic acids, which are important to fruit quality, as they provide sweetness and umami flavor.

Additionally, fatty acid profiling was also affected by the post-harvest treatments. The action of 1-methylcyclopropene showed a greater impact on fatty acids such as oleic, capric, lauric, palmitic, stearic, and myristic acids at 10 DAH, as shown in (Figure 5), decreasing their levels by 17-, 10-, 14-, 17-, 14-, and 14-fold in the MPC group, respectively, and 7-, 6-, 9-, 11-, 1-, and 7-fold in the MCP + MeJA group, respectively, as compared to the CTRL (Table 1). The reduction in fatty acids by 1-methylcyclopropene was evident when the fold-change analysis was applicable (Figure 2). The MCP + MeJA group also showed a reduction in fatty acid levels; however, this was less impactful than the MCP group (Figure 5). The most impacted were the linoleic and myristic acids at 4 DAH with a reduction of 119- and 26-fold in MCP, respectively, and a 23- and 9-fold decrease in MCP + MeJA, respectively, as compared to the CTRL (Table 1).

Conversely, a tendency of increased levels of some fatty acids was also detected, as well as in lignoceric, cerotic, and α-linolenic acids at 4 DAH, and palmitic and linoleic acids at 21 DAH for the MCP and MCP + MeJA groups (Figure 5). In the MCP group, an increase was detected in the levels of lignoceric and α-linolenic acids at 4 DAH by 7- and 4-fold, respectively, while in MCP + MeJA, the increases were 28 and 3-fold, respectively. Moreover, palmitic and linoleic acids were increased by 2- and 10-fold, respectively, in MCP, and 3- and 10-fold, respectively, in MCP + MeJA at 21 DAH (Table 1). Interestingly, the MCP + MeJA group was less impacted when reductions were observed and more impacted when increases were observed compared with the MCP group. This behavior may indicate that methyl jasmonate can act as a stimulator in fatty acid production. Palmitic and eicosanoic acids contributed crucially to the total saturated fatty acid level, and oleic and linoleic acids contributed to the total unsaturated fatty acid level, which is notable, as they play an important role in the fruit quality and nutritional value.

3.3. Secondary Metabolite Profiling Affected by Post-Harvest Hormone Treatment

The secondary metabolites identified in tomato fruits at 4, 10, and 21 DAH were lycopene, β-carotene, and lutein by HPLC analysis; α-tocopherol, β-tocopherol, γ-tocopherol, phytol, β-sitosterol, stigmasterol, and stigmastadienol were identified by GC-MS analysis.

Lycopene was the most affected by the action of 1-methylcyclopropene, reducing its level not only in MCP but also in MCP + MeJA by 29- and 25-fold, respectively, at 4 DAH. However, at 10 DAH, lycopene was reduced by eight- and sixfold, respectively, compared with the CTRL (Figure 6A); β-carotene and lutein showed a decrease less than threefold by 1-methylcyclopropene at the ripening stages (Table S1). These remarkable impacts on the synthesis of carotenoids are illustrated in (Figure 2), mainly at the onset of ripening.

However, the action of 1-methylcyclopropene had a lesser impact on lycopene accumulation at 21 days of hormone treatment, decreasing its production by 2.8-fold, and its action was completely reversed by the methyl jasmonate hormone. Fruits treated with methyl jasmonate showed an increase not only in lycopene production but also in β-carotene and lutein at 21 DAH, indicating the important role that methyl jasmonate plays in the synthesis of carotenoids (Figure 6A). Lycopene and β-carotene showed an increase of 10%, and lutein of 20%, as compared to the CTRL (Figure 6A, Table S1), which is considered relevant since these bioactive compounds have been associated with health benefits, leading to decreases in the occurrence of chronic non-communicable diseases [25]. The total carotenoid level was represented mainly by lycopene.

Tocopherol profiling showed similar behavior to carotenoids during ripening, decreasing its levels in both treated groups at 4 and 10 DAH (Figure 2). At 21 DAH, it presented a decrease, in the MCP group, and an increase, in MCP+MeJA, of tocopherols, as compared to the CTRL. The α-tocopherol levels showed a reduction in MCP and MCP + MeJA of 5- and 4-fold, respectively, at 4 DAH, while at 10 DAH, they decreased by 12 and 3-fold, respectively. The β-tocopherol levels showed a reduction of 14 and 12-fold at 4 DAH, and 23- and 9-fold at 10 DAH in MCP and MCP+MeJA, respectively. Additionally, γ-tocopherol was decreased by 6-fold at 4 and 10 DAH in both treatment groups, except for the MCP + MeJA at 10 DAH, which decreased 1.7 fold, as compared to the CTRL (Figure 6B, Table S2).

Conversely, at 21 DAH, tocopherol profiling was less affected by 1-methylcyclopropene and positively impacted by the concomitant treatment of 1-methylcyclopropene and methyl jasmonate, showing increases of 40% in the α-tocopherol and β-tocopherol levels and 21% in the γ-tocopherol levels, as compared to the CTRL (Figure 6B, Table S2). The total tocopherol level was characterized mainly by the α-tocopherol content. An acyclic diterpenoid identified was phytol, which presented a twofold reduction in MCP at 4 DAH and a twofold increase in MCP + MeJA at 10 DAH (Figure 6B, Table S2). The impact of these treatments at 4 and 10 DAH is also shown in (Figure 2).

Phytosterols were also affected by 1-methylcyclopropene, showing fivefold reductions in β-sitosterol levels in MCP at 4 and 10 DAH and threefold reductions in MCP + MeJA at 4 and 10 DAH, as compared to the CTRL. Stigmasterol exhibited four and sevenfold reductions in MCP at 4 and 10 DAH, respectively, while MCP + MeJA showed three and fivefold decreases at 4 and 10 DAH, respectively. Stigmastadienol was the phytosterol most affected by 1- methylcyclopropene, decreasing ninefold at 4 DAH (Figure 6C, Table S2); β-sitosterol and stigmasterol were the major sources of the total phytosterol level. Moreover, down-regulation exceeding twofold, as compared to the CTRL, is observed for the phytosterols in (Figure 2). Divergent behavior of phytosterols profiled at 4 and 10 DAH, β-sitosterol, stigmasterol, and stigmastadienol showed an increase in their levels by 42%, 34%, and 32%, respectively, in fruits treated with both 1-methylcyclopropene and methyl jasmonate at 21 DAH (Figure 6C, Table S2).

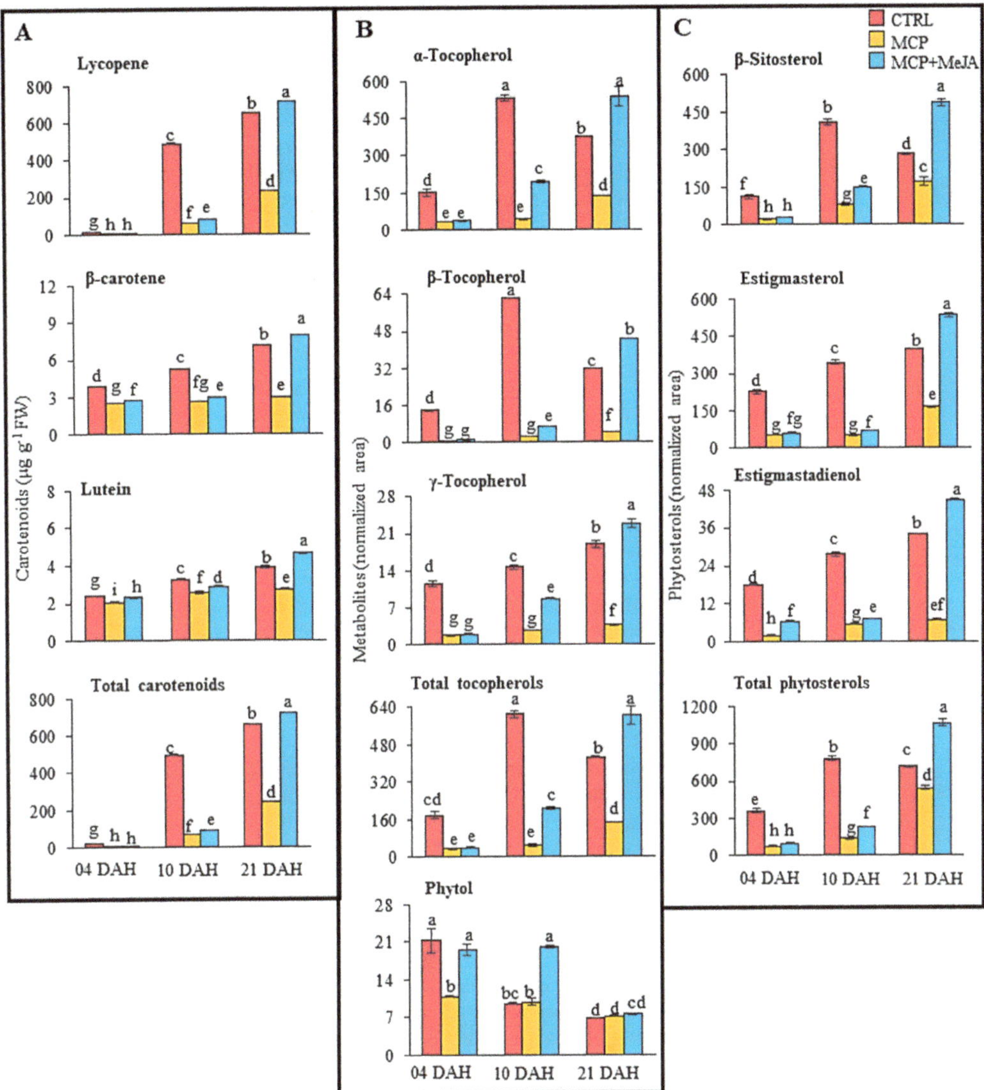

Figure 6. Secondary metabolites in tomato pericarp (*Solanum lycopersicum* L. cv. Grape) exposed to 1-methylcyclopropene (MCP) and both hormones, 1-methylcyclopropene and methyl jasmonate (MCP+MeJA), compared to the control group (CTRL) at 4, 10, and 21 days after harvest (DAH). Contents of carotenoids (**A**), normalized area of tocopherols and phytol (**B**), and phytosterols (**C**). Values are means ± SE of four biological replicates of 10 fruits each. Different letters indicate statistically significant differences ($p < 0.05$).

3.4. *Lipid Metabolism Affected by the Post-Harvest Jasmonate Treatment*

The metabolite profiling of the tomato fruit pericarp treated only with 1-methylcyclopropene and with both 1-methylcyclopropene and methyl jasmonate showed a significant impact on the fruit quality and, consequently, the ripening process. Although the profiles of sugars, organic acids, and amino acids were affected by the jasmonate treatment, the most remarkable difference observed was in the lipid metabolism.

Fruits treated with methyl jasmonate showed a positive impact on the accumulation of metabolites, mainly in the non-polar metabolites (fatty acids, carotenoids, tocopherols, and phytosterols). The application of exogenous methyl jasmonate in fruits with blocked ethylene receptors showed that methyl jasmonate could act independently of endogenous ethylene, suggesting that the blocking of ethylene receptors was reversed after 10 DAH, or new ethylene receptors were synthesized. Post-harvest treatment with jasmonate showed that it is possible to obtain an improved fruit quality with a prolonged shelf life.

Oleic, capric, lauric, palmitic, stearic, and myristic acids showed a 17-fold reduction when treated with only 1-methylcyclopropene and up to an 11-fold reduction when treated with both methyl jasmonate and 1-methylcyclopropene, as compared to the untreated fruits at 10 DAH. It is noteworthy that a drastic decrease was observed in the levels of linoleic and myristic acids at 4 DAH with both treatments (Table 1, Figure 2). Additionally, reductions in the levels of carotenoids, tocopherols, and phytosterols were also detected at the onset of ripening (Figure 6, Tables S1 and S2).

Conversely, an interesting increase in the levels of lignoceric, cerotic, and α-linolenic acids at 4 DAH and palmitic and linoleic acids at 21 DAH with both treatments was detected (Figures 2 and 5). Moreover, notable accumulations in the levels of secondary metabolites, such as lycopene, β-carotene, lutein, α-, β-, and γ-tocopherols, β-sitosterol, stigmasterol, and stigmastadienol, were detected at 21 DAH by the action of methyl jasmonate (Figure 6, Tables S1 and S2). However, it is noteworthy that the maturation stage and hormonal regulation may not be the only factors responsible for the improved lipid metabolism in tomato fruits; other factors such as genetic factors, cultural practices, cultivation, and environmental conditions should be considered [25]. Understanding the interactions between hormone treatment and environmental factors, genotype, and agronomic practices is essential to produce high-quality fruits by improving the synthesis of high-value nutrients.

The treatment with methyl jasmonate can induce significant changes in the metabolite profile of tomato fruits during ripening, positively impacting the nutritional and sensory fruit quality. This treatment, associated with the blocking of ethylene receptors with 1-methylcyclopropene, proved effective in avoiding potential effects on the post-harvest life of the fruits due to the increase in ethylene synthesis caused by methyl jasmonate. Our experimental design involved four replicates, and the consistency of the results indicates that the effects have good reproducibility. However, they were conducted only on the Grape cultivar, and it would be interesting to reproduce these treatments in other cultivars to assess the influence of genotype on responses to treatment with methyl jasmonate and 1-methylcyclopropene. From the viewpoint of applicability, the presented protocol has good commercial potential, since the concentrations of methyl jasmonate and 1-methylcyclopropene used were low and, consequently, did not require large volumes of the compounds. The volatility of the compounds and simplicity of the method of exposure of the fruits make the treatments feasible for larger environments, such as commercial chambers. Although both substances are considered generally recognized as safe (GRAS), further studies about the impact on sensory quality would be important to assess consumer acceptability for the treated fruit.

Supplementary Materials: The following are available online at https://www.mdpi.com/article/10.3390/foods10040877/s1, Table S1: Carotenoids contents (μg.g^{-1} FW) in tomato pericarp (*Solanum lycopersicum* L. cv. Grape) exposed to 1-methylcyclopropene (MCP) and both 1-methylcyclopropene and methyl jasmonate (MCP+ MeJA) treatments at 04, 10, and 21 days after harvest (DAH), detected by high performance liquid chromatography (HPLC); Table S2: Tocopherols, phytol, and phytosterols in tomato pericarp (*Solanum lycopersicum* L. cv. Grape) exposed to 1-methylcyclopropene (MCP) and both 1-methylcyclopropene and methyl jasmonate (MCP+ MeJA) treatments at 04, 10, and 21 days after harvest (DAH), detected by gas chromatography-mass spectrometry (GC-MS).

Author Contributions: Investigation, Writing and original draft preparation: S.L.R.M.; Writing—review and editing: I.L.M., and E.d.C.T.; methodology, software: G.B.P.; data curation: I.L.M.; supervision: E.P. All authors have read and agreed to the published version of the manuscript.

Funding: This research was funded by the São Paulo Research Foundation (FAPESP, Grant 2013/07914-8) and Coordination for the Improvement of Higher Education Personnel (CAPES, Grant 88882.376974/2018-01).

Institutional Review Board Statement: Not applicable.

Informed Consent Statement: Not applicable.

Data Availability Statement: https://repositorio.uspdigital.usp.br/ accessed on 15 April 2021.

Acknowledgments: The authors gratefully acknowledge the financial support of the São Paulo Research Foundation (FAPESP, Grant 2013/07914-8), Coordination for the Improvement of Higher Education Personnel (CAPES, Grant 88882.376974/2018-01), and the National Council for Scientific and Technological Development (CNPq, Grant 312605/2019-6). The authors also thank T. M. Shiga, A. de Oliveira, L. F. L. Macedo, and L. H. J. da Silva for the technical assistance.

Conflicts of Interest: The authors declare no conflict of interest.

References

1. Tang, N.; Na, J.; Deng, W.; Gao, Y.; Chen, Z.; Li, Z. Metabolic and transcriptional regulatory mechanism associated with postharvest fruit ripening and senescence in cherry tomatoes. *Postharvest Biol. Technol.* **2020**, *168*, 111274. [CrossRef]
2. Tieman, D.; Zhu, G.; Resende, M.F.; Lin, T.; Nguyen, C.; Bies, D.; Rambla, J.L.; Ortiz Beltran, K.S.; Taylor, M.; Zhang, B.; et al. A chemical genetic roadmap to improved tomato flavor. *Science* **2017**, *355*, 391–394. [CrossRef] [PubMed]
3. Bramley, B.M. Regulation of carotenoid formation during tomato fruit ripening and development. *J. Exp. Bot.* **2002**, *53*, 2107–2113. [CrossRef] [PubMed]
4. Almeida, J.; Asís, R.; Molineri, V.N.; Sestari, I.; Lira, B.S.; Carrari, F.; Pereira Peres, L.E.; Rossi, M. Fruits from ripening impaired, chlorophyll degraded and jasmonate insensitive tomato mutants have altered tocopherol content and composition. *Phytochemistry* **2015**, *111*, 72–83. [CrossRef]
5. Moreau, R.A.; Nyström, L.; Whitaker, B.D.; Winkler-Moser, J.K.; Baer, D.J.; Gebauerf, S.K.; Hicks, K.B. Phytosterols and their derivatives: Structural diversity, distribution, metabolism, analysis, and health-promoting uses. *Prog. Lipid Res.* **2018**, *70*, 35–61. [CrossRef] [PubMed]
6. Kumar, A.; Singh, P.K.; Parihar, R.; Dwivedi, V.; Lakhotia, S.C.; Ganesh, S. Decreased O-linked GlcNAcylation protects from cytotoxicity mediated by huntingtin Exon1 protein fragment. *J. Biol. Chem.* **2014**, *289*, 13543–13553. [CrossRef] [PubMed]
7. Prasanna, V.; Prabha, T.N.; Tharanathan, R.N. Fruit ripening phenomena—An overview. *Crit. Rev. Food Sci. Nutr.* **2007**, *47*, 1–19. [CrossRef]
8. Farmer, E.E.; Ryan, C.A. Octadecanoid precursors of jasmonic acid activate the synthesis of wound-inducible proteinase inhibitors. *Plant Cell* **1992**, *4*, 129–134. [CrossRef]
9. Fan, X.; Mattheis, J.P.; Fellman, J.K. A role for jasmonates in climacteric fruit ripening. *Planta* **1998**, *204*, 444–449. [CrossRef]
10. Zhang, Z.; Huber, D.J.; Rao, J. Delay of tomato fruit ripening in response to 1-methylcyclopropene is influenced by internal ethylene levels. *Postharvest Biol. Technol.* **2009**, *54*, 1–8. [CrossRef]
11. Su, H.; Gubler, W.D. Effect of 1-methylcyclopropene (1-MCP) on reducing postharvest decay in tomatoes (*Solanum lycopersicum* L.). *Postharvest Biol. Technol.* **2012**, *64*, 133–137. [CrossRef]
12. Cliff, M.; Lok, S.; Lu, C.; Toivonen, P.M.A. Effect of 1-methylcyclopropene on the sensory, visual, and analytical quality of greenhouse tomatões. *Postharvest Biol. Technol.* **2009**, *53*, 11–15. [CrossRef]
13. Fabi, J.P.; Cordenunsi, B.R.; Barreto, G.P.d.M.; Mercadante, A.Z.; Lajolo, F.M.; Nascimento, J.R.O. Papaya fruit ripening: Response to ethylene and 1-methylciclopropene. *J. Agric. Food Chem.* **2007**, *55*, 6118–6123. [CrossRef] [PubMed]
14. Lisec, J.; Schauer, N.; Kopka, J.; Willmitzer, L.; Fernie, A.R. Gas chromatography mass spectrometry-based metabolite profiling in plants. *Nat. Protoc.* **2006**, *1*, 387–396. [CrossRef]
15. Kind, T.; Wohlgemuth, G.; Lee, D.Y.; Lu, Y.; Palazoglu, M.; Shahbaz, S.; Fiehn, O. FiehnLib: Mass spectral and retention index libraries for metabolomics based on quadrupole and time-of-flight gas chromatography/mass spectrometry. *Anal. Chem.* **2009**, *81*, 10038–10048. [CrossRef] [PubMed]
16. Bligh, E.G.; Dyer, W.J. A rapid method of total lipid extraction and purification. *Can. J. Biochem. Physiol.* **1959**, *37*, 911–917. [CrossRef] [PubMed]
17. Ichihara, K.I.; Fukubayashi, Y. Preparation of fatty acid methyl esters for gas-liquid chromatography. *J. Lipid Res.* **2010**, *51*, 635–640. [CrossRef] [PubMed]
18. Fiehn, O.; Kopka, J.; Dörmann, P.; Altmann, T.; Trethewey, R.N.; Willmitzer, L. Metabolite profiling for plant functional genomics. *Nat. Biotechnol.* **2000**, *18*, 1157–1161. [CrossRef]
19. Sérino, S.; Gomez, L.; Costagliola, G.; Gautier, H. HPLC assay of tomato carotenoids: Validation of a rapid microextraction technique. *J. Agric. Food Chem.* **2009**, *57*, 8753–8760. [CrossRef]

20. da Silva Souza, M.A.; Peres, L.E.; Freschi, J.R.; Purgatto, E.; Lajolo, F.M.; Hassimotto, N.M. Changes in flavonoid and carotenoid profiles alter volatile organic compounds in purple and orange cherry tomatoes obtained by allele introgression. *J. Sci. Food Agric.* **2019**, *100*, 1662–1670. [CrossRef]
21. Chong, J.; Soufan, O.; Li, C.; Caraus, I.; Li, S.; Bourque, G.; Wishart, D.S.; Xia, J. MetaboAnalyst 4,0: Towards more transparent and integrative metabolomics analysis. *Nucl. Acids Res.* **2018**, *46*, W486–W494. [CrossRef] [PubMed]
22. Guillén, F.; Castillo, S.; Zapata, P.; Martínez-Romero, D.; Serrano, M.; Valero, D. Efficacy of 1-MCP treatment in tomato fruit. 1. Duration and concentration of 1-MCP treatment to gain an effective delay of postharvest ripening. *Postharvest Biol. Technol.* **2007**, *43*, 23–27. [CrossRef]
23. Blankenship, S.M.; Dole, J.M. 1-Methylcyclopropene: A review. *Postharvest Biol. Technol.* **2003**, *28*, 1–25. [CrossRef]
24. Beauvoit, B.; Belouah, I.; Bertin, N.; Cakpo, C.B.; Colombié, S.; Dai, Z.; Gautier, H.; Génard, M.; Moing, A.; Roch, L.; et al. Putting primary metabolism into perspective to obtain better fruits. *Ann. Bot.* **2018**, *122*, 1–21. [CrossRef]
25. Durazzo, A.; Azzini, E.; Foddai, M.S.; Nobili, F.; Garaguso, I.; Raguzzini, A.; Finotti, E.; Tisselli, V.; Del Vecchio, S.; Piazza, C.; et al. Influence of different crop management practices on the nutritional properties and benefits of tomato *-Lycopersicon esculentum* cv Perfectpeel-. *Int. J. Food Sci. Technol.* **2010**, *45*, 2637–2644. [CrossRef]

 *foods*

MDPI

Article

Preparation of an Amidated Graphene Oxide/Sulfonated Poly Ether Ether Ketone (AGO/SPEEK) Modified Atmosphere Packaging for the Storage of Cherry Tomatoes

Yao Cheng [1], Hao Dong [2], Yuanyue Wu [1] and Kaijun Xiao [1,*]

[1] School of Food Science and Technology, South China University of Technology, 381, Wushan Rd., Tianhe District, Guangzhou 510641, China; yaoyao950920@163.com (Y.C.); wuyuanyue213@163.com (Y.W.)
[2] School of Food Science and Technology, Zhongkai University of Agriculture and Engineering, 24, Dongsha Street, Fangzhi Rd., Haizhu District, Guangzhou 510225, China; donghao@zhku.edu.cn
* Correspondence: fekjxiao@scut.edu.cn; Tel.: +86-020-87113848

Abstract: The shelf life of cherry tomatoes is short so that new and efficient preservation techniques or procedures are required to reduce postharvest losses. This study focused on the development of a sulfonated poly ether ether ketone (SPEEK) film incorporated with amidated graphene oxide (AGO), for the storage of cherry tomatoes in modified atmosphere packaging. The mechanical properties, gas permeability, and moisture permeability were subsequently tested. The evolution of attributes related to shelf life, such as gas composition, physicochemical properties, and sensory properties were also monitored during storage trials. AGO, as an inorganic filler, increases the thermal stability and mechanical properties of SPEEK-based films, while it reduces the water absorption, swelling rate, and moisture permeability. Importantly, all the AGO/SPEEK films exhibited enhanced gas permeability and selective permeability of CO_2/O_2 relative to the SPEEK film. Moreover, 0.9% (*w/w*) AGO/SPEEK film showed an enhanced permeability coefficient of CO_2, corresponding to an increase of 50.7%. It could further improve the selective coefficient of CO_2/O_2 to 67.1%. The results of preservation at 8 °C revealed that: 0.9% (*w/w*) AGO/SPEEK film was significantly effective at maintaining the quality and extending the shelf life of cherry tomatoes from 15 to 30 days, thereby suggesting the potential for applying AGO-incorporated SPEEK films for food packaging materials.

Keywords: amidated graphene oxide; sulfonated poly ether ether ketone; modified atmosphere film; cherry tomatoes; food packaging

Citation: Cheng, Y.; Dong, H.; Wu, Y.; Xiao, K. Preparation of an Amidated Graphene Oxide/Sulfonated Poly Ether Ether Ketone (AGO/SPEEK) Modified Atmosphere Packaging for the Storage of Cherry Tomatoes. *Foods* **2021**, *10*, 552. https://doi.org/10.3390/foods10030552

Academic Editor: Eleni Tsantili

Received: 7 December 2020
Accepted: 28 February 2021
Published: 7 March 2021

1. Introduction

Cherry tomatoes (*Lycopersicon esculentum var. cerasiforme cv.*) consist of beneficial nutrients, such as minerals, vitamins, phenolic compounds, and lycopene [1]. However, cherry tomatoes deteriorate rapidly without proper preservation, and thus their shelf life ranges from only 5 to 7 days after harvest [2]. Therefore, it is very important to reduce decay incidence and extend the postharvest life of cherry tomatoes. Currently, different treatments, such as drying, edible coatings, modified atmosphere packaging, etc., have been applied to maintain the postharvest quality of cherry tomatoes [2–4]. However, modern consumers' convenient lifestyle and their demand for minimally processed fruits and vegetables put forward higher requirements for freshness preservation [4]. Modified atmosphere packaging, as one of the most widely used techniques for fresh produce preservation, has attracted significant research interest for its time-saving convenience and ease of use. It reduces gas exchange, moisture evaporation, and respiration reaction rates and suppresses physiological disorders of fruits and vegetables [5]. Existing modified atmosphere packaging usually involves the use of polymer materials, such as polyethylene (PE), polyvinyl chloride (PVC), and polypropylene (PP) [4]. However, these films have limited gas selectivity or gas permeability, which can no longer meet the requirements for the preservation of cherry tomatoes [6]. From this perspective, it is of great practical significance to develop an atmosphere-modified packaging that matches the respiration rate of cherry tomatoes [7].

Among diverse polymers, sulfonated poly ether ether ketone (SPEEK), as an aromatic thermoplastic engineering plastic, has attracted increasing attention for the development of food packaging materials due to its excellent gas permeability [8]. For instance, He et al. [9] prepared a SPEEK/PVDF self-balancing packaging film, which significantly shortened the equilibrium time of CO_2 or O_2 concentration in the package, making the shelf life of broccoli more than double. However, there is still much room to improve the selective permeability of SPEEK. Noteworthily, these polymers can be further improved as functional packaging materials by incorporating some new compounds, such as nanoparticles [10]. Graphene oxide (GO), as a newly emerging carbon nanomaterial with a two-dimensional layered structure, has a large specific surface area and strong CO_2 adsorption capacity, and thus acts as one of the ideal materials [11,12]. At present, the methods for the surface modification of GO can be categorized as covalent and non-covalent functionalization [13]. Among them, the amidated form of GO (AGO) better improves the adsorption of CO_2 and effectively prevents the deterioration of mechanical properties of the film. However, the effects of AGO/SPEEK films on postharvest attributes and flavor characteristics in cherry tomatoes have not been reported yet.

In the present study, GO was modified by amidation and then added to SPEEK in the form of inorganic fillers to prepare a film with a high CO_2/O_2 separation ratio. The produced films were then characterized to analyze the efficacy of modification of SPEEK with AGO. Subsequently, the mechanical properties, water uptake ratio (WUR), swelling ability (SA), water vapor permeability (WVP), and gas permeability of the pure SPEEK film and AGO/SPEEK films were also examined and compared. Finally, the efficacy of the resulting package was assessed by performing storage trials of cherry tomatoes.

2. Materials and Methods

2.1. Materials and Chemicals

Cherry tomatoes were obtained from a local farm in Guangdong province, China. PP boxes (1 L) were obtained from a local store (Walmart, Guangzhou, China), with no obvious visual defects. GO was obtained from Sigma-Aldrich Corp. (Shanghai, China). PEEK (Victrex™ 450 G) was obtained from Victrex Co., Ltd. (Victrex, UK). CO_2 (99.99%), O_2 (99.99%), and N_2 (99.99%) were obtained from Air Chemical Co., Ltd. (Guangzhou, China). All other analytical chemical reagents were achieved from Aladdin Industrial Corporation (Shanghai, China) and used as received.

2.2. Synthesis of AGO

GO powder (~200 mg) was dispersed in *N,N*-dimethylformamide (DMF) by ultrasonication. The mixture was mechanically stirred at 30 °C for 1 h and then slowly dropped into ethylenediamine (EDA, 300 mL) under rapid stirring. The reaction was allowed to proceed for 24–48 h under nitrogen protection. After the reaction, the mixture was diluted in ethanol (1600 mL) and washed several times to ensure the removal of excess EDA. Finally, AGO could be collected after filtration and dried overnight in a vacuum oven at 50 °C [13–15].

2.3. Preparation of AGO/SPEEK Films

SPEEK (DS 79.8%) and AGO/SPEEK films were prepared by the method proposed by He et al. [9,12,15]. SPEEK was dissolved in DMF to form a 10% (*w/w*) mixed solution under magnetic stirring. Then, AGO (0.1%, 0.5%, and 0.9%) was added to react with SPEEK for 24 h at 60 °C in order to obtain a casting solution. The casting solution was defoamed at room temperature under vacuum. Subsequently, it was cast on a clean glass plate, and its thickness was controlled at 40 μm using an adjustable scraper. The casted glass plate was placed horizontally into a vacuum oven at 60 °C and dried for 24 h to completely remove the solvent. Finally, the dried film was peeled off from the glass plate for further tests.

2.4. Characterization

The structure and functional groups of the surface of the films were analyzed by Fourier-transform infrared spectroscopy (FTIR, 2000 GX spectrometer, Perkin Elmer, Waltham, MA, USA) at 4000 to 400 cm^{-1}. The structure of the molecules was determined using a Renishaw micro-Raman spectroscopy system with a laser wavelength of 532 nm (Raman, RENISHAW PLC, Kingswood, UK). Crystalline properties were evaluated by X-ray diffraction (XRD, D8 Advance X-ray diffractometer, Bruker, Germany). The morphology and microstructure of the films were inspected by scanning electron microscopy (SEM, SU-3500, Hitachi, Japan) at an accelerating voltage of 10 kV and by transmission electron microscopy (TEM, JEM-2100F, Tokyo, Japan). The thermal properties of the films were determined from 25 to 800 °C at a heating rate of 10 °C min^{-1} in a nitrogen atmosphere by thermogravimetric analysis (TGA, Metter-Toledo, Switzerland). The composition and chemical state of the elements in the films were determined by X-ray photoelectron spectroscopy (XPS, K-Alpha$^+$ spectrometer, Thermo-Fisher, Waltham, MA, USA).

2.5. Film Properties

2.5.1. Mechanical Properties

Mechanical properties, including elongation at break (EAB), tensile strength (TS), and elastic modulus (EM), were determined by tensile testing, using an LR 5K universal testing machine (Lloyd Instrument, Bognor Regis, UK). The film was cut into rectangular strips with dimensions of 60 mm × 20 mm. The deformation was recorded at a crosshead speed of 10 mm min^{-1} and an initial gage length of 50 mm. Averages were calculated from five replicates of each sample.

2.5.2. Swelling Ability

Film swelling was determined by following the method proposed by Banerjee et al. [16]. A film (30 mm × 30 mm) was dried at 100 °C for 2 h and immersed in distilled water at room temperature for 24 h. The WUR and area swelling ratio (ASR) of the film were calculated by the weight difference of the film before and after water absorption.

2.5.3. Water Vapor Permeability

The films were cut into pieces with size of 4 cm × 4 cm. The water vapor permeability of the film was measured following the method provided by Shima Jafarzadeh [17]. The water vapor transmission rate (WVTR) was measured from the slopes (linear) of the steady-state portion of weight loss of the cup versus time curve. Further, the WVP of the films was measured as follows:

$$\mathrm{WVP} = \mathrm{WVTR} \times \frac{\mathrm{D}}{\Delta \mathrm{p}} \quad (1)$$

where D is the thickness of the film and Δp is the partial water vapor pressure difference over the two sides of the film.

2.5.4. Gas Permeability

The theory of measuring the gas permeability of the film uses the traditional differential pressure method. The films were cut into a 10 cm × 10 cm size and were measured with a BSV-1A gas permeability tester (Lab Stone, Guangzhou, China). The single gas permeability of the film was determined by changing the gas source (O_2, CO_2, or N_2) in the room temperature. The gas permeability coefficient of the film (P) is calculated as follows:

$$\mathrm{P} = \frac{\Delta \mathrm{p}}{\Delta \mathrm{t}} \times \frac{\mathrm{V}}{\mathrm{S}} \times \frac{\mathrm{T_0}}{\mathrm{P_0} \times \mathrm{T}} \times \frac{\mathrm{D}}{(\mathrm{p_1} - \mathrm{p_2})} \quad (2)$$

The ideal gas selectivity of the film can be obtained by calculating the quotient of the permeability of the film under different gas sources.

Where $\Delta p/\Delta t$ is the arithmetic mean of changes in hourly gas pressure in the infiltration chamber; V is the volume of the infiltration chamber; S is the area of the film; T is the test temperature; $p1-p2$ is the pressure difference between the two sides of the film; T_0 and p_0 are the temperature and pressure under standard conditions, which are 273.15 K and 101,315 Pa, respectively, and D is the thickness of the film.

2.6. Pretreatment and Storage of Cherry Tomatoes

Fresh cherry tomatoes without mechanical damage were manually screened and precooled at 4 ± 1 °C and 90% relative humidity for 24 h. The samples (250 ± 10 g) were randomly packed in a 1 L packaging box (PP), respectively. The surface of the box was sealed with AGO/SPEEK films using a JSMAP-1D100F-S semi-automatic modified atmosphere packaging machine (JuGang Machinery Manufacturing Co., Ltd., Shanghai, China). The temperature, pressure, and heat-sealing time of the packaging machine were 110 ± 5 °C, 1.8 kPa, and 1 s, respectively. Samples without packaging were set as control group. The packaged samples were stored in a thermo-hygrostat at 8 ± 1 °C for 30 days. The physiological and biochemical properties of cherry tomatoes were evaluated every 5 days.

2.6.1. Internal Atmosphere Composition

Concentrations of CO_2 and O_2 in the box were monitored using an OXYBABY M+ gas analyzer (Zillion Corporation, Shanghai, China). During the measurement, a customized sealing gasket with a size of 1 cm × 1 cm was pasted on the package, and then the needle of the gas analyzer was inserted into the package. The sensor of the gas analyzer that connected to the needle automatically collects gas samples for measurement in 10 s and the actual values of O_2 and CO_2 volume ratio (%) would be displayed on the screen of the gas analyzer in real time.

2.6.2. Sensory Properties

In the sensory evaluation experiment, 10 professional sensory evaluators conducted the experiment on color, odor, texture, and other indicators of cherry tomatoes according to sensory scoring standards. A 5-point scale (0, unacceptable to 5, excellent) was used. In the sensory laboratory, under a cool white fluorescent lamp, numbered samples were randomly placed in small dishes and provided to each sensory evaluator [9].

2.6.3. Physicochemical Properties

The hardness of cherry tomatoes was measured using a GY 2 handheld hardness meter (Minks Testing Equipment Co., Ltd., Xi'an, China). Hardness values were determined as maximum force registered in the force vs. time curves. Measurements were made on five fruits per experimental unit.

Color parameters (L*, a*, b*) of cherry tomatoes were measured using a CR20 colorimeter (Konica Minolta Sensing Inc, Tokyo, Japan). For each experimental unit, 10 tomatoes were measured and the results are presented as ΔE [18].

Cherry tomatoes (125 g) were homogenized with distilled water (125 mL) for 5 min using a high shear mixing homogenizer, and the pH of the mixture was determined using an FE20 digital pH meter (Mettler Toledo, Zurich, Switzerland).

The weight of cherry tomatoes was recorded at the initial day and sampling day by using an AUY220 analytical balance (Shimadzu Corporation, Kyoto, Japan). The weight loss (%) was calculated following the method described by Araguez et al. [19].

The total soluble solid (TSS) content of cherry tomatoes was determined using an AR 2008 Abbe refractometer (Kruess, Hamburg, Germany). A drop of clear tomato juice obtained by squeezing the fruit in a muslin cloth was placed in the reading chamber, and the value of TSS was read (%) [20].

The content of vitamin C (V_C) was determined using a UV1601 spectrophotometer (Rayleigh Analytical Instruments Co., Ltd., Beijing, China) following the method reported by Hou et al. [21]. The V_C content was calculated and expressed as mg per 100 g of sample based on weight.

2.7. Statistical Analysis

Statistical analyses were performed by using the SPSS software (SPSS Inc., Chicago, IL, USA). All experiments were conducted in triplicates and all data presented as mean ± standard deviation were subjected to analysis of variance (ANOVA). Duncan's multiple range tests were used to compare the difference among mean values at a level of 0.05.

3. Results and Discussion

3.1. Analysis of AGO

SEM was performed to characterize the microscopic morphology of AGO. Figure 1a shows that AGO was stacked together to form a multilayer sheet structure, which is consistent with the reported of Ramanathan et al. [20–22]. Compared with GO, Figure 1b shows TEM images, further showing a transparent yarn-like structure with fewer wrinkles and smoother surface of AGO [15,20]. It was attributed to the partial reduction of GO by EDA and the interaction between the AGO plates, which together led to decrimping and blurring of the edges [14,15,23].

The FTIR spectra of AGO are shown in Figure 1c. The obvious disappearance of stretching peaks of O−H (3430 cm^{-1}) and C=O (1729 cm^{-1}), and the appearance of stretching peaks of N−H (3260 cm^{-1}) and C−H (3050, 2885 cm^{-1}) both indicate the partial modification of GO by EDA [23,24]. Moreover, for AGO, three new peaks appear at 1665, 1536, and 1440 cm^{-1}, which are assigned to the C=O stretching of the amide I band and the combined absorption caused by N−H bending and C–N stretching in the amide II band, respectively. It is attributed to the amidation reaction or the substitution reaction between EDA and GO, which is consistent with the XRD results (Figure 1d).

The Raman spectra of GO and AGO are shown in Figure 1e. A Raman D-band and a G-band of GO are observed at 1354 and 1600 cm^{-1}, corresponding to the structure defects on the graphene sheets and the sp^2 hybridization of the hexagonal carbon structure, respectively. Compared to the Raman spectra of GO, the D-band and G-band of the AGO shift to 1340 and 1595 cm^{-1}, respectively. In general, the relative intensity of the I_D/I_G ratio partly indicates the quality of grapheme [25]. Herein, the I_D/I_G ratio of GO (1.41) was lower than that of AGO (1.60), indicating a slight reduction of GO, which was attributed to the conversion of carbon atoms from the sp^3 to the sp^2 state [26].

The composition and chemical state of elements of AGO were further researched by XPS. Figure 1f, demonstrates that the full spectrum of AGO shows a significant decrease in O1s and an increase in C1s, accompanied by the generation of a new peak N1s, indicating the successful modification of GO. Through further analysis, it was found that the changes in the intensity of O1s component in Figure 1f were attributed to the disappearance or the decrease of the intensity of the C–OH group (285.2 eV), C–O–C group (286.8 eV), and O=C–O (288.9 eV) group in Figure 1f-1. The new peak at the binding energy of 285.9 eV of C1s component in Figure 1f-4 corresponded to the C–N group, which was proved by the three main Gaussian peaks at binding energies of 399.0, 400.0, and 401.5 eV of N1s component in Figure 1f-3, assigned to –NH_2, C–N, and –CO–NH groups, respectively. Furthermore, the O1 spectrum of AGO shown in Figure 1f-5 also indicates the decrease in the strength of the oxygen-containing groups to varying degrees, which further reveals that the amidation reaction between EDA and GO is mainly caused by amino groups and oxygen-containing groups [23,27].

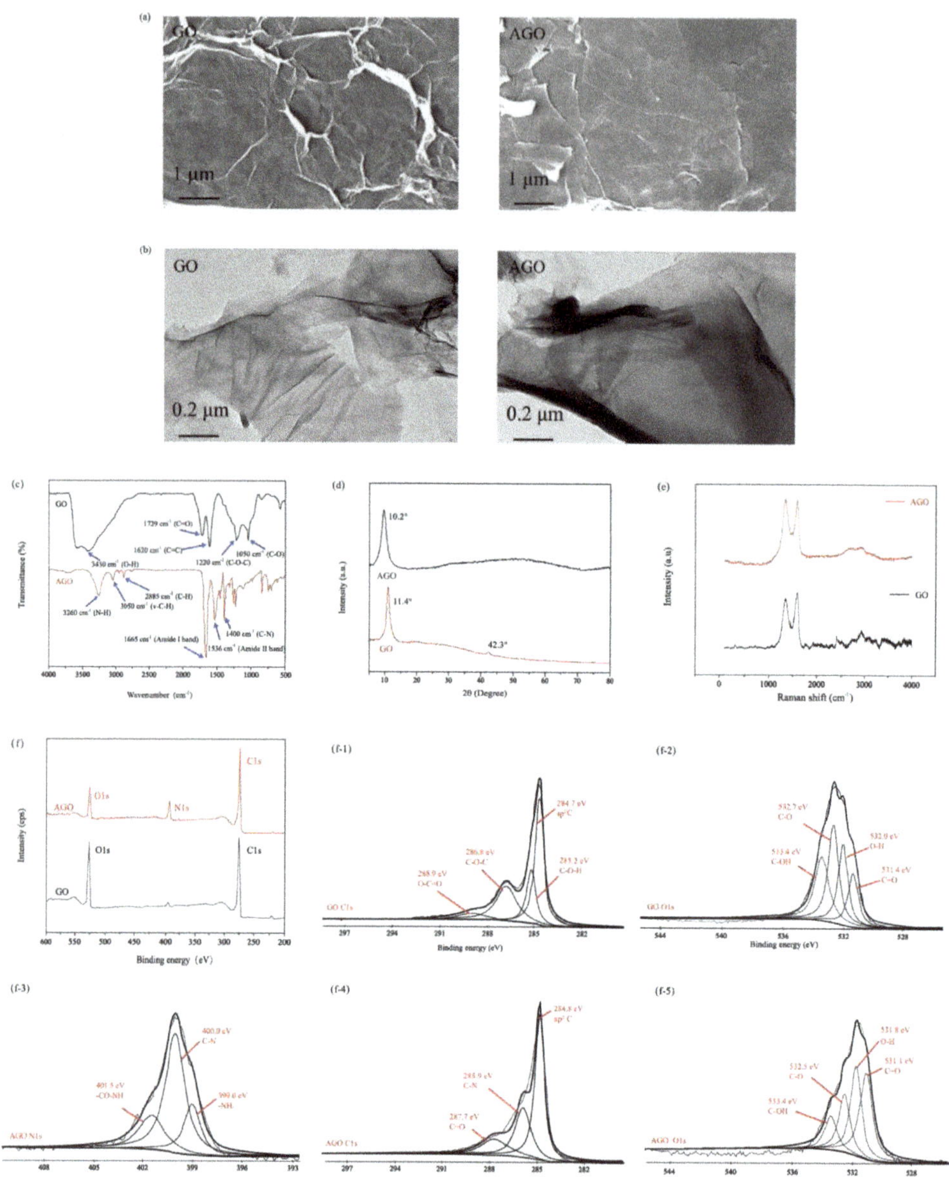

Figure 1. Characterization of amidated graphene oxide (AGO). (**a**) SEM, (**b**) TEM, (**c**) FTIR spectra, (**d**) XRD pattern, (**e**) Raman spectra, (**f**) XPS survey spectra, (**f-1**) C1s spectra of GO, (**f-2**) O1s spectra of GO, (**f-3**) N1s spectra of AGO, (**f-4**) C1s spectra of AGO, and (**f-5**) O1s spectra of AGO.

3.2. Characterization Analysis of AGO/SPEEK Films

For any blending system, an important effect that should be addressed is the miscibility among different compounds as it determines structural stability and physicochemical properties of the blend. In this study, good miscibility between SPEEK and AGO resulted in the formation of transparent and homogeneous blending solutions and films. More obvious details were obtained based on the microstructure of the material. As shown in Figure 2a,b, there was a partial depressions on the GO/SPEEK films with no sheet structure, which is similar to the result of Dai et al. [12,28,29]. The surface of AGO/SPEEK films displayed a relatively rough morphology with a more obvious layered structure compared to GO/SPEEK films, consistent with the typical morphology of AGO [12,23]. Notably, most of the AGO was evenly dispersed in the SPEEK matrix, while small-scale accumulation gradually appeared with an increase in the AGO loading, indicating that the two compounds were homogeneous [28].

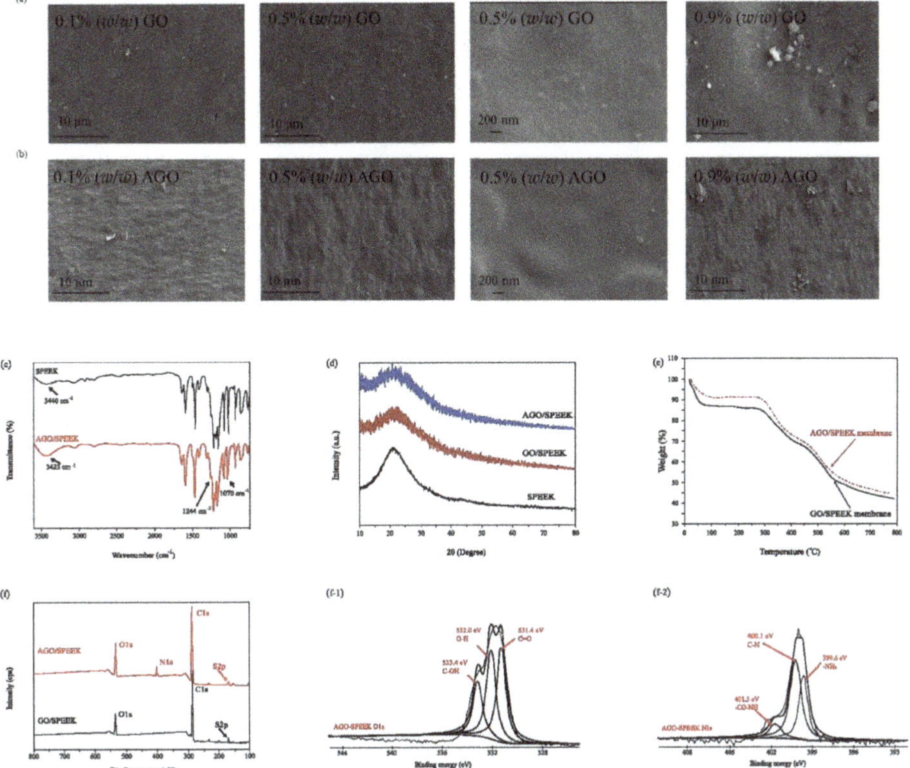

Figure 2. Characterization of films. (**a**) SEM of GO/SPEEK films, (**b**) SEM of AGO/SPEEK films, (**c**) FTIR spectra, (**d**) XRD pattern, (**e**) TGA curves, (**f**) XPS survey spectra, (**f-1**) O1s spectra and (**f-2**) N1s spectra of 0.5% (*w/w*) AGO/SPEEK films.

Numerous oxygen-containing functional groups, such as carbonyl (C=O), hydroxyl (–OH), carboxyl (–COOH), as well as amino groups, are present on both sides of AGO; therefore, the interfacial interactions between AGO and SPEEK matrices should be verified. Figure 2c shows the FTIR spectra of AGO/SPEEK films. The AGO/SPEEK film shows a characteristic broad peak corresponding to intermolecular hydrogen bonds, which is reflected in the shift of the −OH group band from 3440 to 3423 cm^{-1}. The bands at 1080 and 1250 cm^{-1} are associated with asymmetrical and symmetrical stretching vibrations of the O=S=O group, while for AGO/SPEEK film, the corresponding bands are shifted to 1070 and 1244 cm^{-1}. The composition and chemical state of elements of AGO/SPEEK films were also researched by XPS (Figure 2f), which was found to be similar to the XPS results of AGO and revealed the blending mechanism between AGO and SPEEK. Figure 2d shows XRD results, clearly indicating the formation of hydrogen bonds between the sulfonated acid groups in SPEEK and polar or amino groups on AGO. Furthermore, TGA (Figure 2e) confirmed that the interface interaction caused by hydrogen bonds between AGO and

SPEEK matrix was stronger, leading to the shift in the TG curve toward high temperature, indicating the improvement of thermal stability [12,29].

3.3. Analysis of Film Properties

3.3.1. Mechanical Properties

The mechanical properties of films represent their ability to maintain their integrity and endure external stress during the processing, transportation, handling, and storage of packaged materials [30]. As shown in Figure 3, AGO/SPEEK film presented the highest TS (Figure 3a) and EM (Figure 3b) but the lowest EAB (Figure 3c) under the maximum load of 0.9% (*w/w*). It was attributed to the existence of numerous hydrogen bonds in the AGO/SPEEK film that caused strong cohesive energy density [12,31]. Likewise, AGO/SPEEK films also presented higher TS and EM under the same load, which was ascribed to the reaction of more $-NH_2$ groups with SPEEK as indicated by the decrease of crystallinity shown in XRD pattern [29]. The EAB of AGO/SPEEK films (Figure 3c) was lower than that of the original film by varying degrees, which was imputed to the loading of AGO that restricted the movement of SPEEK polymer chains [32]. Moreover, extensive hydrogen bonds increased the interfacial adhesion, thereby enhancing the mechanical properties of the film [12,31]. Noteworthily, when the filling amount of the inorganic filler was between 0.1% and 0.9% (*w/w*), the EAB of films remained above 40%, showing that it still had sufficient mechanical strength and extensibility, which is required for food packaging applications [30].

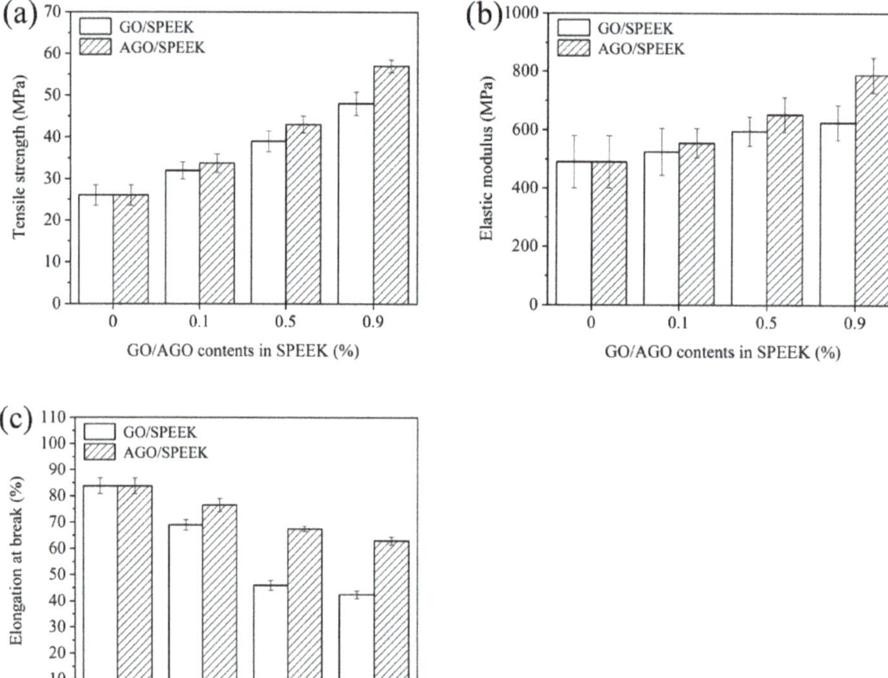

Figure 3. Mechanical properties of AGO/SPEEK films. (**a**) Tensile strength, (**b**) elastic modulus, (**c**) elongation at break.

3.3.2. Swelling Ability

The swelling ability is an important factor in composite films and represents its water absorption resistance property and type of film use. The water absorption tendency of the film affects the swelling ability to a certain extent. Consequently, the water absorption of the film shows a trend similar to that of the swelling ability. As shown in Figure 4a,b, the swelling ability of both the films shows a trend that first increases and then decreases, and this was mainly attributed to the influence of functional groups. When the added amount of the inorganic filler (GO/AGO) was less than 0.1% (*w/w*), the hydrophilic functional groups in the inorganic filler resulted in temporary increase in the swelling ability of the film. However, when the amount was more than 0.1% (*w/w*), "blocking effect" caused by the increase of functional groups played the main role, leading to a decline in swelling [15,31]. Specifically, the area swelling ratio of films with 0.5% and 0.9% (*w/w*) AGO was significantly reduced, reaching 15% and 10%, respectively. Moreover, the area swelling ratio of the AGO/SPEEK film was lower than that of the GO/SPEEK film under the same loading, which was ascribed to the less hydrophilic amide groups introduced by AGO, and it narrowed the gap of the SPEEK polymer chain. Thereby, it can be concluded that the incorporation of AGO and SPEEK matrix can improve the usability of films for high-humidity food packaging.

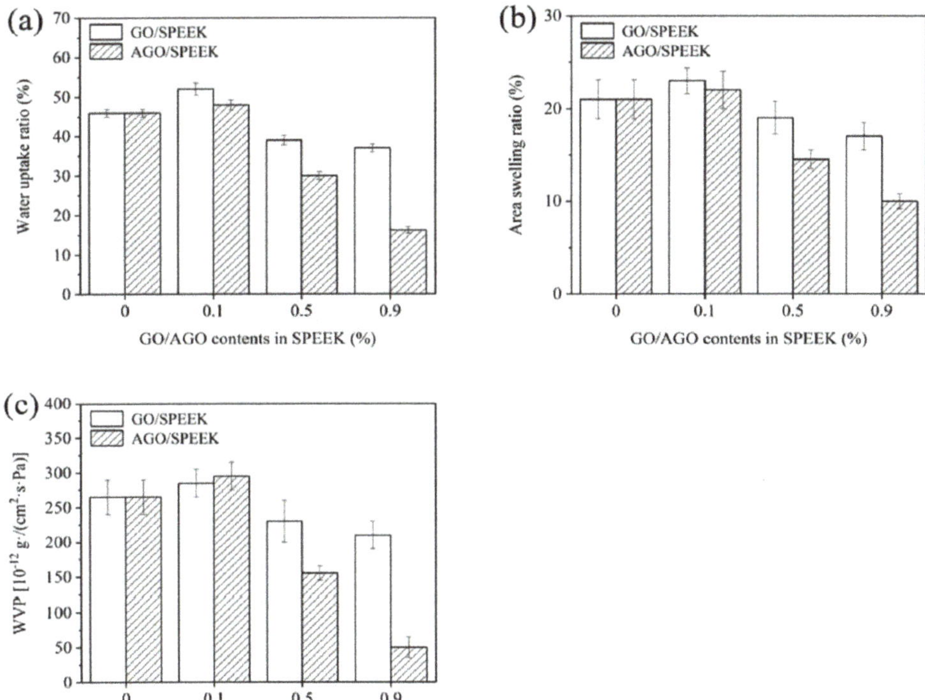

Figure 4. Water permeability of AGO/SPEEK films. (**a**) Water uptake ratio, (**b**) area swelling rate, (**c**) water vapor permeability.

3.3.3. Water Vapor Permeability

WVP is one of the most important functional properties for a food packaging film [30]. Film with low WVP has a good water vapor barrier property, which can effectively reduce moisture transfer between the surrounding atmosphere and the food environment. As shown in Figure 4c, the SPEEK film has relatively high moisture permeability, which is attributed to the presence of abundant sulfonic acid groups that gather together to form hydrophilic domains that generate hydrophilic transport channels [33]. The WVP result of AGO/SPEEK films was found to be similar to the results of WUR. One reason was that dispersed AGO constituted a barrier and hindered the migration of water molecules through the film, while the other was that AGO could extend the path of water molecules through the film, thereby delaying the time to penetrate the film [31]. However, it can be seen that the WVPs of AGO/SPEEK films were both above 50×10^{-12} g·cm/(cm^2·s·Pa), which is higher than that of PE [34]. Therefore, the AGO/SPEEK films were found to be suitable for the preservation of fruits with high moisture content.

3.3.4. Gas Permeability

Blending changes the porosity of film and thus also the gas permeability for different gases [9]. The pure gas permeability and ideal selectivity of AGO/SPEEK films are summarized in Table 1. SPEEK has superior gas permeability performance because of the presence of a polar $-SO_3H$ group chain, which interacts with different gas molecules and then improves upon the solubility coefficients with the gas particles [9,35]. Comparative analysis indicates that the gas permeability of GO/SPEEK films decreases slightly due to the consumed $-SO_3H$ groups, which is consistent with the results of Xin et al. [36]. Notably, the CO_2 permeability of AGO/SPEEK films was significantly increased, while the permeability of O_2 or N_2 did not change significantly. It was attributed to the presence of $-NH_2$ groups in AGO that have an excellent affinity for CO_2. Moreover, the reaction between CO_2 and $-NH_2$ groups intensified in the wet state, resulting in increased CO_2 permeability of AGO/SPEEK films [34]. Therefore, for both CO_2/O_2 and CO_2/N_2, ideal selectivity increased from 7.3 and 22.2 to 12.2 and 29.1, respectively, with increasing AGO composition in the blended films.

Table 1. Gas permeability of films for pure gas.

SPEEK Films	Permeability (Barrer)			Ideal Gas Selectivity (α)	
	CO_2	O_2	N_2	CO_2/O_2	CO_2/N_2
0	14.68	2.02	0.66	7.3	22.2
0.1% (*w/w*) GO	13.26	1.96	0.60	6.7	22.1
0.5% (*w/w*) GO	12.31	1.80	0.65	6.8	18.9
0.9% (*w/w*) GO	13.92	1.85	0.62	7.5	22.4
0.1% (*w/w*) AGO	16.81	2.06	0.66	8.1	25.5
0.5% (*w/w*) AGO	19.40	1.84	0.68	10.5	28.5
0.9% (*w/w*) AGO	22.12	1.82	0.76	12.2	29.1

3.4. Quality Analysis of Cherry Tomatoes

3.4.1. Internal Atmosphere Composition

The gas permeability of the film is an important factor affecting the exchange of internal and external gases. The respiration of fruits and vegetable produces CO_2 and consumes O_2 in the microenvironment of package. On account of relative gas pressure differences between the inside and outside of the package, CO_2 gets released to the atmosphere and O_2 will enter the package. Thus, films with appropriate gas permeability to O_2 and CO_2 can significantly prolong the shelf life of fruit and vegetables. As shown in Figure 5a, the CO_2 concentration inside the package maintained an equilibrium concentration of 4–7%, while the corresponding O_2 concentration was 4–6% (Figure 5b). The difference in gas concentration in the final package was not obvious, although the loading of AGO was different, which may be caused by the originally low loading of AGO. However, the higher the loading of AGO, the lower the final equilibrium concentration of CO_2 in the package. Accordingly, it was speculated herein that the modified atmosphere packaging with an AGO loading of 0.5–0.9% (*w/w*) is more suitable for preservation.

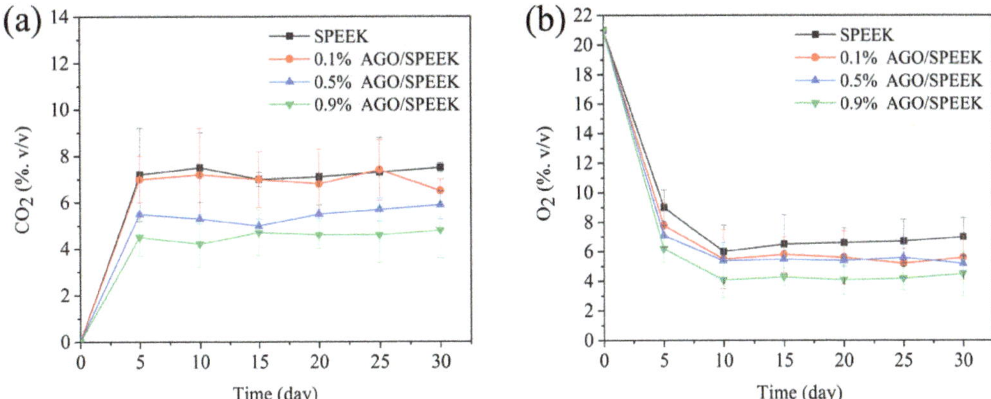

Figure 5. Internal atmosphere composition under AGO/SPEEK films packaging in the storage of cherry tomatoes. (**a**) CO_2, (**b**) O_2.

3.4.2. Sensory Properties

The sensory properties such as color, hardness, and overall acceptability of cherry tomatoes packaged with AGO/SPEEK films were determined during storage, and their results are presented in Figure 6. In general, cherry tomatoes were ranked almost similar with respect to their sensory properties; however, packaged cherry tomatoes had better peel color when the AGO content in the package was higher. Interestingly, cherry tomatoes packaged with 0.9% (*w/w*) AGO/SPEEK films maintained acceptable sensory scores even after 30 days of storage, showing the potential of AGO/SPEEK film in extending the shelf life.

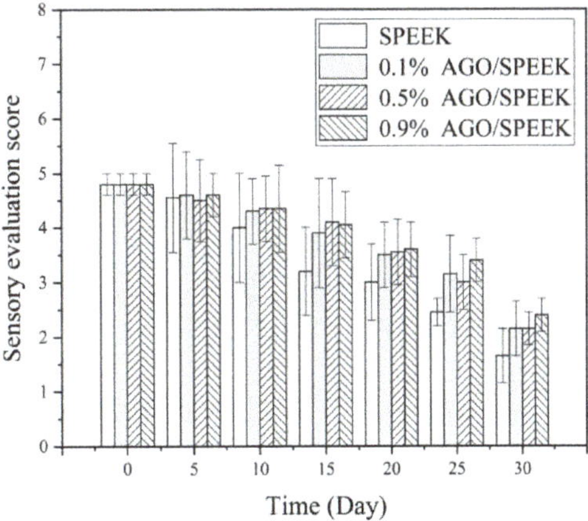

Figure 6. Sensory evaluation scores of AGO/SPEEK films in the storage of cherry tomatoes (a 5-point scale: 1 means unacceptable, 2 means barely accepted, 3 means general, 4 means good, and 5 means excellent).

3.4.3. Physicochemical Properties

Changes in color, hardness, weight loss, pH, TSS content, and Vc content of cherry tomatoes packaged with AGO/SPEEK films during storage are shown in Figure 7. Color is the main external quality criteria affecting the marketability [4]. Figure 7a demonstrates that the ΔE of cherry tomatoes in different packages shows an increasing trend, and color changes of AGO/SPEEK films packaged are relatively minimal in different periods, probably owing to the reduction of ethylene synthesis influenced by the high content of CO_2. This is in good agreement with the gas permeability results of AGO/SPEEK films discussed earlier. Specifically, high loading of AGO leads to enhanced CO_2/O_2 selective permeability of the film and it is easier to form a high CO_2 and low O_2 gas atmosphere, which thus slows down the accumulation of pigment and the browning process of tissues caused by the aging of cherry tomatoes [4,21].

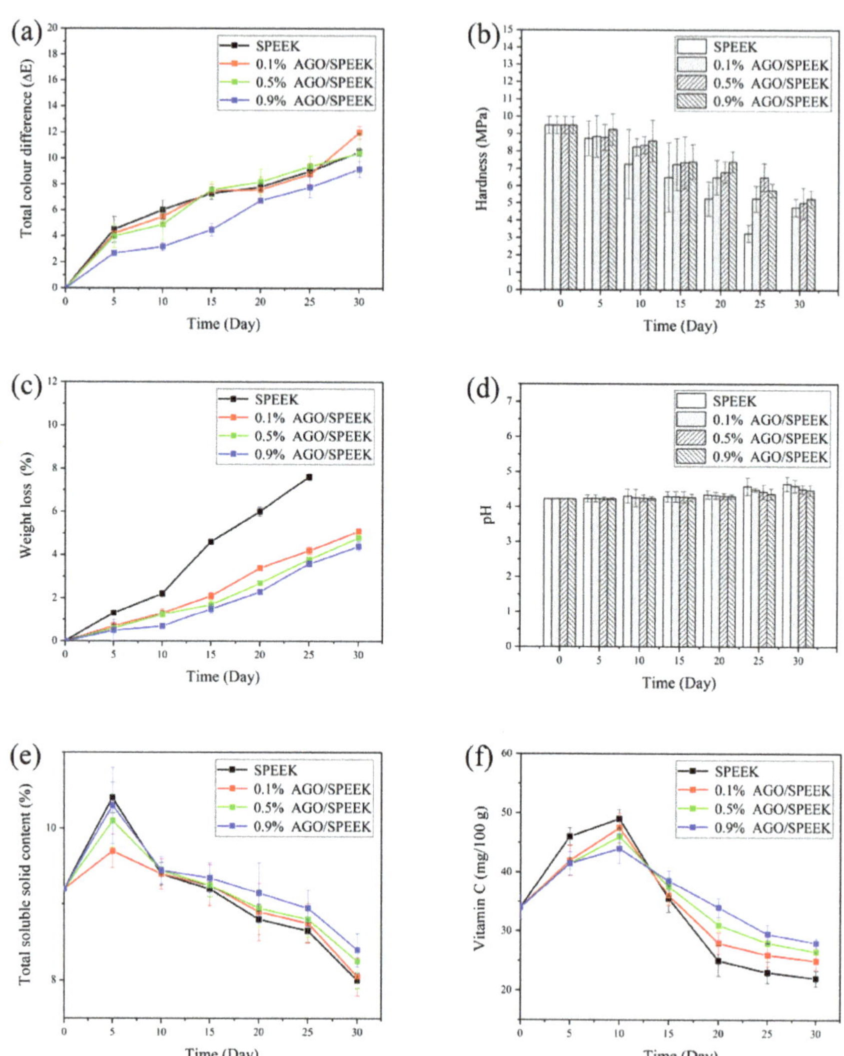

Figure 7. Physicochemical properties of AGO/SPEEK films in the storage of cherry tomatoes. (**a**) Total color difference, (**b**) hardness, (**c**) weight loss, (**d**) pH, (**e**) total soluble solid content, (**f**) vitamin C.

Furthermore, hardness of cherry tomatoes is an important factor for the estimation of maturity [37]. Figure 7b illustrates that the hardness of cherry tomatoes in all groups exhibited a decreasing trend with the extension of storage time, and the control group was more significant, which depends on the progress of the conversion of insoluble pectin to soluble pectin or even pectin acid [38]. The hardness of cherry tomatoes packaged with

0.9% (*w/w*) AGO/SPEEK film decreased by 47.4%, which is similar to the result reported by Zhang et al. [37]. Thus, the AGO/SPEEK film can efficiently slow down the softening of the cherry tomatoes.

Transpiration and respiration of the fruits during storage lead to loss of water and weight, which can cause a wilting phenomenon and shorten the postharvest life of fresh fruit. According to the reported, cherry tomatoes will cause severe wilting when the weight loss is about 11%, while at about 6% weight loss they will not wilt [39,40]. Figure 7c shows that the weight loss of cherry tomatoes was the least when the packaging was done with 0.9% (*w/w*) AGO/SPEEK film and exhibited an increasing trend in the following order: 0.5% (*w/w*), 0.1% (*w/w*), and the original film. Undoubtedly, the samples in the control group were significantly softened and cracked and exhibited the highest loss of weight of up to 8% after 30 days of storage, while that of others increased to 3–4%, which benefits from the water resistance of films inhibiting the evaporation of water. Notably, that is coherence with the results of WVP shown in Figure 4c.

Moreover, pH is an important reference for evaluating the quality of cherry tomatoes during storage because it uses organic acids as a substrate for respiration [19]. Figure 7d shows a slight increasing trend of pH from 4.2 to 4.8, which coincided with the results of hardness caused by increasing pectin acid. The pH of control samples increased more significantly mainly resulting from no packaging, which is consistent with the report by D'Aquino et al. [41].

TSS refers mainly to soluble sugars and is a very important indicator for determining consumer acceptability [42]. Figure 7e exhibits that the TSS content increased at the beginning of storage (5 days), benefiting from the accumulation of nutrients brought by maturity. However, it decreased to varying degrees in the later stage, caused by the hydrolysis of sucrose for respiration and maintaining the physiological activity of plant raw materials [42]. Noteworthily, cherry tomatoes packaged in a 0.9% (*w/w*) AGO/SPEEK film showed a slower decrease in TSS compared with others, indicating a delay in the ripening of cherry tomatoes under high concentration of CO_2 [43].

Figure 7f shows that the Vc content in cherry tomatoes increased initially and then decreased during the storage period, similar to the behavior described by Zhou et al. [30]. The increased V_C content of cherry tomatoes was attributed to the accumulation of post-ripening effect, while the decrease resulted from the high rates of respiration, oxidative deterioration, and accumulation of CO_2 [30,41]. Noteworthily, the V_C content of cherry tomatoes packaged in 0.5% (*w/w*) or 0.9% (*w/w*) AGO/SPEEK films was basically maintained at 30 mg/100 g at the end of storage, much higher than others, benefited from favorable oxygen barrier properties of AGO/SPEEK films that effectively prevented the oxidation of V_C.

4. Conclusions

AGO/SPEEK films were prepared by physical blending for modified atmosphere packaging. A series of data revealed that the loading of AGO improves the mechanical properties, gas permeability, and selective permeability of CO_2/O_2 compared to the original film, while reducing the water uptake ratio, the swelling ability, and the water vapor permeability. Preservation experiments revealed that 0.9% (*w/w*) AGO/SPEEK films delayed the maturation process and doubled the shelf life of cherry tomatoes. Overall, these findings demonstrated that AGO/SPEEK films have great application potential in the preservation of fruits with high moisture content and low respiration rate. Undeniably, many more systematic explorations are still demanded to confirm the safety and service life of the film when storing various food products, which will be pursued in the future.

Author Contributions: Conceptualization, K.X., Y.C., and H.D.; methodology, Y.C. and H.D.; formal analysis, K.X., Y.C., and H.D.; data curation, Y.C. and H.D.; writing—original draft preparation, Y.C., H.D., and Y.W. Both Y.C. and H.D. are first authors and contributed equally to the manuscript and are listed in alphabetical order. All authors have read and agreed to the published version of the manuscript.

Funding: The research was financed by the Research and Development Plan in Key Areas of Guangdong Province (2019B020222001), Guangzhou Rural Science and Technology Commissioner Project (GZKTP201932), and Guangdong Basic and Applied Basic Research Foundation (2019A1515111213).

Institutional Review Board Statement: Not applicable.

Informed Consent Statement: Not applicable.

Data Availability Statement: The study did not report any data.

Conflicts of Interest: The authors declare no conflict of interest.

References

1. Feng, L.; Zhang, M.; Adhikari, B.; Guo, Z. Application of high-pressure argon for improving postharvest quality of cherry tomato. *J. Food Process. Eng.* **2018**, *41*, e12882. [CrossRef]
2. An, K.; Li, H.; Zhao, D.; Ding, S.; Tao, H.; Wang, Z. Effect of Osmotic Dehydration with Pulsed Vacuum on Hot-Air Drying Kinetics and Quality Attributes of Cherry Tomatoes. *Dry Technol.* **2013**, *31*, 698–706. [CrossRef]
3. Li, C.; Tao, J.; Zhang, H. Peach gum polysaccharides-based edible coatings extend shelf life of cherry tomatoes. *3 Biotech* **2017**, *7*. [CrossRef] [PubMed]
4. Paulsen, E.; Barrios, S.; Lema, P. Ready-to-eat cherry tomatoes: Passive modified atmosphere packaging conditions for shelf life extension. *Food Packag. Shelf* **2019**, *22*, 100407. [CrossRef]
5. Qu, P.; Zhang, M.; Fan, K.; Guo, Z. Microporous modified atmosphere packaging to extend shelf life of fresh foods: A review. *Crit. Rev. Food Sci.* **2020**. [CrossRef] [PubMed]
6. Mangaraj, S.; Goswami, T.K. Modified atmosphere packaging—An ideal food preservation technique. *J. Food Sci. Technol.* **2009**, *46*, 399–410.
7. Park, H.B.; Kamcev, J.; Robeson, L.M.; Elimelech, M.; Freeman, B.D. Maximizing the right stuff: The trade-off between membrane permeability and selectivity. *Science* **2017**, *356*, 6343. [CrossRef]
8. Ye, J.; Cheng, Y.; Sun, L.; Ding, M.; Wu, C.; Yuan, D.; Zhao, X.; Xiang, C.; Jia, C. A green SPEEK/lignin composite membrane with high ion selectivity for vanadium redox flow battery. *J. Membr. Sci.* **2019**, *572*, 110–118. [CrossRef]
9. He, Q.; Xiao, K. Quality of broccoli (*Brassica oleracea* L. var. italica) in modified atmosphere packaging made by gas barrier-gas promoter blending materials. *Postharvest Biol. Technol.* **2018**, *144*, 63–69. [CrossRef]
10. Jie, X.; Shen, P. Research progress in polymer modified SPEEK membranes. *Battery Bimon.* **2010**, *40*, 50–53.
11. Pumera, M. Electrochemistry of graphene, graphene oxide and other graphenoids: Review. *Electrochem. Commun.* **2013**, *36*, 14–18. [CrossRef]
12. Dai, W.; Shen, Y.; Li, Z.; Yu, L.; Xi, J.; Qiu, X. SPEEK/Graphene oxide nanocomposite membranes with superior cyclability for highly efficient vanadium redox flow battery. *J. Mater. Chem. A* **2014**, *2*, 12423–12432. [CrossRef]
13. Zheng, L.; Zhen, W. Preparation and characterization of amidated graphene oxide and its effect on the performance of poly(lactic acid). *Iran Polym. J.* **2018**, *27*, 239–252. [CrossRef]
14. Xue, B.; Zhu, J.; Liu, N.; Li, Y. Facile functionalization of graphene oxide with ethylenediamine as a solid base catalyst for Knoevenagel condensation reaction. *Catal. Commun.* **2015**, *64*, 105–109. [CrossRef]
15. Feng, M.; Huang, Y.; Cheng, Y.; Liu, J.; Liu, X. Rational design of sulfonated poly(ether ether ketone) grafted graphene oxide-based composites for proton exchange membranes with enhanced performance. *Polymer* **2018**, *144*, 7–17. [CrossRef]
16. Banerjee, S.; Kar, K.K. Synergistic effect of aluminium phosphate and tungstophosphoric acid on the physicochemical properties of sulfonated poly ether ether ketone nanocomposite membrane. *J. Appl. Polym. Sci.* **2016**, *133*. [CrossRef]
17. Jafarzadeh, S.; Rhim, J.; Alias, A.K.; Ariffin, F.; Mahmud, S. Application of antimicrobial active packaging film made of semolina flour, nano zinc oxide and nano-kaolin to maintain the quality of low-moisture mozzarella cheese during low-temperature storage. *J. Sci. Food Agric.* **2019**, *99*, 2716–2725. [CrossRef]
18. Guerrero, A.; Ferrero, S.; Barahona, M.; Boito, B.; Lisbinski, E.; Maggi, F.; Sanudo, C. Effects of active edible coating based on thyme and garlic essential oils on lamb meat shelf life after long-term frozen storage. *J. Sci. Food Agric.* **2020**, *100*, 656–664. [CrossRef] [PubMed]
19. Araguez, L.; Colombo, A.; Borneo, R.; Aguirre, A. Active packaging from triticale flour films for prolonging storage life of cherry tomato. *Food Packag. Shelf* **2020**, *25*, 100520. [CrossRef]
20. Boonsiriwit, A.; Xiao, Y.; Joung, J.; Kim, M.; Singh, S.; Lee, Y.S. Alkaline halloysite nanotubes/low density polyethylene nanocomposite films with increased ethylene absorption capacity: Applications in cherry tomato packaging. *Food Packag. Shelf* **2020**, *25*, 100533. [CrossRef]
21. Hou, Y.; Gao, J.; Gu, L.; Wang, S.; Zeng, R. Effects of agaro-oligosaccharide treatment on postharvest quality of cherry tomatoes during cold storage. *J Food Process. Preserv.* **2015**, *39*, 949–955. [CrossRef]
22. Ramanathan, T.; Fisher, F.T.; Ruoff, R.S.; Brinson, L.C. Amino-Functionalized Carbon Nanotubes for Binding to Polymers and Biological Systems. *Chem. Mater.* **2005**, *17*, 1290–1295. [CrossRef]
23. Yan, J.; Chen, G.; Cao, J.; Yang, W.; Xie, B.; Yang, M. Functionalized graphene oxide with ethylenediamine and 1, 6-hexanediamine. *New Carbon Mater.* **2012**, *27*, 370–376. [CrossRef]

24. Sharma, S.; Biswal, B.K.; Kumari, D.; Bindra, P.; Kumar, S.; Stobdan, T.; Shanmugam, V. Ecofriendly Fruit Switches: Graphene Oxide-Based Wrapper for Programmed Fruit Preservative Delivery To Extend Shelf Life. *ACS Appl. Mater. Interfaces* **2018**, *10*, 18478–18488. [CrossRef] [PubMed]
25. Qiu, Y.; Wang, M.; Zhang, W.; Liu, Y.; Li, Y.V.; Pan, K. An asymmetric graphene oxide film for developing moisture actuators. *Nanoscale* **2018**, *10*, 14060–14066. [CrossRef]
26. Li, W.; Yang, J.; Wu, Z.; Wang, J.; Li, B.; Feng, S.; Deng, Y.; Zhang, F.; Zhao, D. A Versatile Kinetics-Controlled Coating Method To Construct Uniform Porous TiO_2 Shells for Multifunctional Core-Shell Structures. *J. Am. Chem. Soc.* **2012**, *134*, 11864–11867. [CrossRef]
27. Kim, N.H.; Kuila, T.; Lee, J.H. Simultaneous reduction, functionalization and stitching of graphene oxide with ethylenediamine for composites application. *J. Mater. Chem. A* **2013**, *1*, 1349–1358. [CrossRef]
28. Zheng, L.; Geng, Z.; Zhen, W. Preparation, characterization, and reaction kinetics of poly (lactic acid)/amidated graphene oxide nanocomposites based on reactive extrusion process. *J. Polym. Res.* **2019**, *26*, 78. [CrossRef]
29. Yin, Y.; Wang, H.; Cao, L.; Li, Z.; Li, Z.; Gang, M.; Wang, C.; Wu, H.; Jiang, Z.; Zhang, P. Sulfonated poly(ether ether ketone)-based hybrid membranes containing graphene oxide with acid-base pairs for direct methanol fuel cells. *Electrochim. Acta* **2016**, *203*, 178–188. [CrossRef]
30. Zhou, X.; Yang, R.; Wang, B.; Chen, K. Development and characterization of bilayer films based on pea starch/polylactic acid and use in the cherry tomatoes packaging. *Carbohydr. Polym.* **2019**, *222*, 114912. [CrossRef] [PubMed]
31. Heo, Y.; Im, H.; Kim, J. The effect of sulfonated graphene oxide on Sulfonated Poly (Ether Ether Ketone) membrane for direct methanol fuel cells. *J. Membr. Sci.* **2013**, *425*, 11–22. [CrossRef]
32. Mo, S.; Peng, L.; Yuan, C.; Zhao, C.; Tang, W.; Ma, C.; Shen, J.; Yang, W.; Yu, Y.; Min, Y.; et al. Enhanced properties of poly(vinyl alcohol) composite films with functionalized graphene. *RSC Adv.* **2015**, *5*, 97738–97745. [CrossRef]
33. Parnian, M.J.; Rowshanzamir, S.; Gashoul, F. Comprehensive investigation of physicochemical and electrochemical properties of sulfonated poly (ether ether ketone) membranes with different degrees of sulfonation for proton exchange membrane fuel cell applications. *Energy* **2017**, *125*, 614–628. [CrossRef]
34. Xin, Q.; Li, Z.; Li, C.; Wang, S.; Jiang, Z.; Wu, H.; Zhang, Y.; Yang, J.; Cao, X. Enhancing the CO2 separation performance of composite membranes by the incorporation of amino acid-functionalized graphene oxide. *J. Mater. Chem. A* **2015**, *3*, 6629–6641. [CrossRef]
35. Khan, A.L.; Li, X.; Vankelecom, I.F.J. SPEEK/Matrimid blend membranes for CO2 separation. *J. Membr. Sci.* **2011**, *380*, 55–62. [CrossRef]
36. Qiu, X.; Dong, T.; Ueda, M.; Zhang, X.; Wang, L. Sulfonated reduced graphene oxide as a conductive layer in sulfonated poly(ether ether ketone) nanocomposite membranes. *J. Membr. Sci.* **2017**, *524*, 663–672. [CrossRef]
37. Zhang, X.; Zhang, X.; Liu, X.; Du, M.; Tian, Y. Effect of polysaccharide derived from Osmunda japonica Thunb-incorporated carboxymethyl cellulose coatings on preservation of tomatoes. *J. Food Process. Preserv.* **2019**, *43*, 43. [CrossRef]
38. Zhang, L.; Chen, F.; Zhang, P.; Lai, S.; Yang, H. Influence of Rice Bran Wax Coating on the Physicochemical Properties and Pectin Nanostructure of Cherry Tomatoes. *Food Bioprocess Technol.* **2017**, *10*, 349–357. [CrossRef]
39. Mendes, K.F.; Mendes, K.F.; Guedes, S.F.; Aparecida Santos Silva, L.C.; Arthur, V. Evaluation of physicochemical characteristics in cherry tomatoes irradiated with Co-60 gamma-rays on post-harvest conservation. *Radiat. Phys. Chem.* **2020**, *177*, 109139. [CrossRef]
40. Guo, X.; Chen, B.; Wu, X.; Li, J.; Sun, Q. Utilization of cinnamaldehyde and zinc oxide nanoparticles in a carboxymethylcellulose-based composite coating to improve the postharvest quality of cherry tomatoes. *Int. J. Biol. Macromol.* **2020**, *160*, 175–182. [CrossRef] [PubMed]
41. D'Aquino, S.; Mistriotis, A.; Briassoulis, D.; Di Lorenzo, M.L.; Malinconico, M.; Palma, A. Influence of modified atmosphere packaging on postharvest quality of cherry tomatoes held at 20 °C. *Postharvest Biol. Technol.* **2016**, *115*, 103–112. [CrossRef]
42. Pobiega, K.; Przybyl, J.L.; Zubernik, J.; Gniewosz, M. Prolonging the Shelf Life of Cherry Tomatoes by Pullulan Coating with Ethanol Extract of Propolis during Refrigerated Storage. *Food Bioprocess Technol.* **2020**, *13*, 1675. [CrossRef]
43. Akbudak, B.; Akbudak, N.; Seniz, V.; Eris, A. Effect of pre-harvest harpin and modified atmosphere packaging on quality of cherry tomato cultivars "Alona" and "Cluster". *Br. Food J.* **2012**, *114*, 180–196. [CrossRef]

Article

Preharvest Application of Chitosan Improves the Postharvest Life of 'Garmrok' Kiwifruit through the Modulation of Genes Related to Ethylene Biosynthesis, Cell Wall Modification and Lignin Metabolism

H. M. Prathibhani C. Kumarihami [1], Jin Gook Kim [1,2], Yun-Hee Kim [3], Mockhee Lee [4], Young-Suk Lee [5], Yong-Bum Kwack [6] and Joonyup Kim [7,*]

1 Division of Applied Life Science, Graduate School, Gyeongsang National University, Jinju 52828, Korea; prathibhanick@gmail.com (H.M.P.C.K.); jgkim119@gnu.ac.kr (J.G.K.)
2 Institute of Agriculture and Life Science, Gyeongsang National University, Jinju 52828, Korea
3 Department of Biology Education, Gyeongsang National University, Jinju 52828, Korea; cefle@gnu.ac.kr
4 Namhae Branch, National Institute of Horticultural and Herbal Science, Rural Development Administration, Namhae 52430, Korea; mockey92@korea.kr
5 Department of Horticulture Research, Gyeongsangnam-do Agricultural Research and Extension Service, Jinju 52733, Korea; yseve77@korea.kr
6 Department of Fruit Science, Korea National College of Agriculture and Fisheries, Jeonju 54874, Korea; kwack@korea.kr
7 Department of Horticultural Science, Chungnam National University, Daejeon 34134, Korea
* Correspondence: jykim12@cnu.ac.kr; Tel.: +82-42-821-5738

Abstract: The influence of the preharvest application of chitosan on physicochemical properties and changes in gene expression of 'Garmrok' kiwifruit during postharvest cold storage (0 °C; RH 90–95%; 90 days) was investigated. Preharvest treatment of chitosan increased the fruit weight but had no significant effect on fruit size. The chitosan treatment suppressed the ethylene production and respiration rate of kiwifruit during the cold storage. The reduction of ethylene production of chitosan-treated kiwifruit was accompanied with the suppressed expression of ethylene biosynthesis genes. Moreover, preharvest application of chitosan diminished weight loss and delayed the changes in physicochemical properties that include firmness, soluble solids content, titratable acidity, total sugars, total acids, total phenols, and total lignin. As a result, the preharvest application of chitosan delayed the maturation and ripening of fruit. Expression of genes related to cell wall modification was down-regulated during the early maturation (ripening) period, while those related to gene expression for lignin metabolism were up-regulated at the later stages of ripening. These results demonstrate that the preharvest application of chitosan maintained the fruit quality and extends the postharvest life of 'Garmrok' kiwifruit, possibly through the modulation of genes related to ethylene biosynthesis, cell wall modification, and lignin metabolism.

Keywords: cell wall modification; chitosan; ethylene biosynthesis; fruit quality; lignin metabolism; postharvest quality; preharvest treatment

Citation: Kumarihami, H.M.P.C.; Kim, J.G.; Kim, Y.-H.; Lee, M.; Lee, Y.-S.; Kwack, Y.-B.; Kim, J. Preharvest Application of Chitosan Improves the Postharvest Life of 'Garmrok' Kiwifruit through the Modulation of Genes Related to Ethylene Biosynthesis, Cell Wall Modification and Lignin Metabolism. *Foods* **2021**, *10*, 373. https://doi.org/10.3390/foods10020373

Academic Editors: Eleni Tsantili and Jinhe Bai
Received: 19 January 2021
Accepted: 6 February 2021
Published: 9 February 2021

1. Introduction

Kiwifruit (*Actinidia* sp.) is widely accepted by consumers for its organoleptic and nutritional properties. It is an excellent source of vitamin C (L-ascorbic acid) [1,2], and is generally known as "Chinese gooseberry", "China's miracle fruit", and "the horticultural wonder of New Zealand" [3]. The genus *Actinidia* contains more than 60 species with a wide diversity, but only two of them (*A. deliciosa*, and *A. chinensis*) are being produced commercially [2,4,5]. Due to their high nutritional value and desirable taste, the demand for kiwifruits has been increasing. Year-round production and supply of kiwifruit to meet the demands inevitably requires proper postharvest management.

Kiwifruit is extremely perishable, with a typical climacteric ripening pattern [1,5]. It is harvested at an unripe but physiologically mature stage [5]. After harvest, the physicochemical properties of fruit decline rapidly due to the influence of internal biochemical reactions and the external environment [6,7]. The postharvest performance of kiwifruit is strongly affected by the maturity or physiological state of the fruit at harvest, in conjunction with the applied postharvest management. In particular, the postharvest storage life of kiwifruit is mostly limited by its relatively high metabolic activity and extreme sensitivity to ethylene during storage [5–8]. Accordingly, the conditions for the preservation of kiwifruit for prolonged periods are undoubtedly important. The kiwifruit industry highly relies on low temperature [5,9] in combination with controlled/modified atmospheres [7,10] to extend their postharvest life. Besides, several other preservation technologies including edible coatings [7,10,11], and treatments with chemical agents such as methyl jasmonate [12], salicylic acid [6], and 1-methyl cyclopropane (1-MCP) [12] have been trying to extend the kiwifruit postharvest life. Among these, the application of edible coatings has been reported as one of the novel technologies with great potential for extending the postharvest life of kiwifruit.

The edible coatings on fresh fruit serve as an alternative to modified atmosphere packaging, as they reduce the quality changes and quantity losses through control of the internal atmosphere of the individual fruit [13]. It has been shown that edible coatings have the potential to reduce moisture loss, rate of respiration, ethylene production, and ripening, while they maintain quality along with storability [7,11,13,14]. In postharvest management, polysaccharide-, protein-, and lipid-based solutions have been demonstrated as applicable edible coatings that prolonged the postharvest life of the whole kiwifruit or other fruits [1,7,10,11,14–16].

Recently, the natural compound chitosan (poly β-(1,4) *N*-acetyl-d-glucosamine) has been widely used as an edible coating. It is a polysaccharide derived from the deacetylation of chitin [6,17–20]. It has been shown that the use of chitosan as an edible coating enhances the quality and postharvest life of various fruits owing to its excellent film-forming, non-toxic, biocompatible, biodegradable, and antifungal properties [2,8,18,20–22]. Besides, several studies have demonstrated that the postharvest use of chitosan as an edible coating maintains quality and extends the postharvest life of kiwifruits such as green-fleshed (*A. deliciosa*) kiwifruit [3,18,23], yellow-fleshed kiwifruit (*A. chinensis*) [6,24], red kiwifruit (*A. melanandra*) [2], hardy kiwifruit (*A. arguta*) [13], and arctic kiwifruit (*A. kolomikta*) [21]. Although chitosan can be applied in either preharvest or postharvest treatments, the reports on preharvest treatments of chitosan and the effects on the postharvest control are still limited. Nonetheless, the preharvest application of chitosan is highly feasible and can be applied on the fruit around harvest time [19].

In our previous studies, we evaluated the effectiveness of the application of preharvest chitosan in combination with calcium chloride (Ca-chitosan) that resulted in the enhancement of fruit quality and postharvest life of hardy kiwifruit (*A. arguta* 'Saehan') [25] and green-fleshed kiwifruit (*A. deliciosa* 'Garmrok') [8]. In a separate study, Zhang et al. [22] reported that the preharvest spraying of chitosan composite films (chitosan, calcium, dextrin, ferulic acid, and auxiliaries) had several positive effects on the postharvest quality and control of diseases in kiwifruit (*A. deliciosa* 'Guichang'). Although these results suggest the efficacy of preharvest treatment of chitosan in maintaining the postharvest properties of kiwifruit, the physiological changes and underlying molecular events involved in maintaining fruit physicochemical properties need to be further investigated. In the current study, we show that the preharvest application of chitosan maintains the fruit quality and extends the postharvest life of 'Garmrok' kiwifruit possibly through the modulation of genes related to ethylene biosynthesis, cell wall modification, and lignin metabolism.

2. Materials and Methods

2.1. Plant Materials and Preharvest Chitosan Treatment

The study was carried out from 4 October 2019 to 3 February 2020, on kiwifruit (*A. deliciosa* 'Garmrok'). Kiwifruit was harvested from an orchard in Namhae Sub-Station, National Institute of Horticultural and Herbal Science, Rural Development Administration, Korea. 'Garmrok' kiwifruit vines were cultivated on a pergola trellis system and general kiwifruit cultivation recommendations were implemented.

The experiment comprised three treatments that included control (0), 100, and 500 mg·L^{-1} chitosan. Untreated fruits served as the control treatment. Chitosan (from shrimp shells; ≥75% deacetylation; Sigma-Aldrich, Seoul, Korea) at a concentration of 20 g·L^{-1} was dissolved in distilled water consisting of 0.1 mol·L^{-1} acetic acids by stirring to make a stock solution. The resultant stock solution was further diluted with distilled water to make working solutions at 100 and 500 mg·L^{-1} concentrations. Tween 80 (0.1%) was added to improve wettability. The chitosan concentrations used in the study were chosen based on our previous studies [8,25]. Each treatment comprised three biological replicates (trees) in a randomized complete block design in which a single tree corresponds to the experimental unit. Kiwifruit on vines was fully dipped in a cup of chitosan solution until the fruit were completely wet. The chitosan was applied at four times: 4 October 2019 (146 days after full bloom (DAFB)), 12 October 2019 (154 DAFB), 19 October 2019 (161 DAFB), and 28 October 2019 (170 DAFB). The fruit from each treatment were harvested on 2 November 2019 (175 DAFB).

Kiwifruit were culled out for uniformity of size and lacking defects. The selected fruit were packed in corrugated cardboard boxes (10 kg capacity) laid with a perforated polyethylene film liner. All experiments were performed in triplicate. Fruit were stored for 90 days at 0 °C with relative humidity (RH) of 90–95%. The fruit physicochemical attributes and relative gene expressions were evaluated at 30-day intervals.

2.2. Physicochemical Quality Attributes of Fruit

Kiwifruit weight, size, core firmness, and flesh firmness were measured for 10 biological replicates (fruit) according to our previous study [8]. A rheometer (RHEO TEX SD-700, Sun Scientific Inc., Tokyo, Japan) fitted with an 8 mm round, flat-ended probe, compressing at a depth of 3 mm and a crosshead speed of 120 mm·min^{-1} was used to measure fruit firmness. The fruit were sliced into longitudinal halves and each half was measured for flesh firmness in the central zone after the peel (~2 mm thick) was removed. The fruit were cut through the equator (2 cm radial slice) and the core tissue was measured for core firmness [8]. Firmness was represented in Newton (N). Fruit weight loss was measured using a digital balance from the beginning to the end of the storage and expressed as the percentage of weight loss relative to the initial weight.

Changes in respiration rate and ethylene production rate were measured in triplicate using a gas chromatograph. After weighing the individual fruit, each of them was placed in a 630 mL volume airtight polypropylene container (HPL851-2.1L, Locknlock, Seoul, Korea) fitted with a rubber septum, for 3 h at room temperature. Air samples of the headspace were removed from the septum with a syringe and injected into a gas chromatograph (GC-7890B; Agilent Technologies, Santa Clara, CA, USA) that is equipped with a stainless steel column (2.0 m × 3.0 mm i.d.) packed with Porapak Q (Shinwa, Kyoto, Japan) and a flame ionization detector (FID) to measure the ethylene production. The respiration rate was determined using a gas chromatograph (GC 6890, Agilent Technologies, USA) that is equipped with a stainless steel column (2.0 m × 3.0 mm i.d.) packed with Shincarbon ST (Shinwa, Kyoto, Japan) and a thermal conductivity detector (TCD). The measurements were expressed in mg·CO_2 kg^{-1} h^{-1} and μL·C_2H_4 kg^{-1} h^{-1}.

Surface of preharvest chitosan-treated and untreated 'Garmrok' kiwifruit peel was examined by scanning electron microscopy (SEM, LEO 1420VP). The micrographs were viewed at an accelerating voltage of 20 kV in high vacuum conditions. Before SEM observation, the samples were dried using a freeze dryer at −78 °C for 2 days (Ilshin

BioBase, Gyeonggi-do, Korea). The dried samples were coated with an ultrathin layer of palladium/gold layer (Pd/Au) on an ion sputtering machine (Quorum Techn, SC7620) and comparable magnifications of kiwifruit peel of coated and non-coated fruits were photographed.

The soluble solids content (SSC, %) and titratable acidity (TA, %) were measured as described in our previous study [8]. Ten biological replicates were measured for SSC, while TA was measured in three replications in which one replicate contained the extract of ten fruits. The contents of total sugar, organic acid, phenolic, and lignin were measured in three replications in which one replicate contained the extract of three fruits. Sugars (fructose, glucose, and sucrose) and organic acids (oxalic, quinic, malic, and citric) were analyzed by HPLC (Agilent 1200 Chemstation, Agilent, CA, USA), according to the method of Kim et al. [4] with slight modifications. A 250 mm × 4.6 mm i.d. Shodex Asahipak NH2P-50 4E column (Showa Denko, Tokyo, Japan) was used for sugar analysis. The organic acid analysis was done using a 250 mm × 4.6 mm i.d. Develosil RPAQUEOUS-AR-3 column (Nomura Chemical Co., Ltd., Seto, Japan). The peaks were detected at 214 nm with UV/VIS detector (1200 Variable Wavelength Detector, Agilent, CA, USA). Sugars and organic acids were detected by the retention time in the chromatograms compared to the standards. The results were expressed in g/100 g on a fresh weight basis. The total sugar content was detected in a mixture of fructose, glucose, and sucrose (Table S1). The total organic acid was detected in a mixture of oxalic, quinic, malic, and citric acids (Table S1).

The content of total phenolic was analyzed spectrophotometrically using the modified Folin–Ciocâlteu method, following Kim et al. [26]. The absorbance was recorded at 725 nm using a UV–vis spectrophotometer (Varioskan Flash Multimode Reader, Thermo Fisher Scientific, Waltham, MA, USA). The content of total phenolic was calculated using a standard curve of gallic acid and expressed as mg of gallic acid equivalent (GAE) per 100 g on a fresh weight basis.

The total lignin content of kiwifruit flesh was assessed by a two-step acid hydrolysis procedure following the methodology of the National Renewable Energy Laboratory (NREL/TP-510-42618) [27]. The primary hydrolysis step was done with 72% H_2SO_4 at 30 °C for 60 min. At the secondary hydrolysis step, the reaction mixture was diluted to 4% H_2SO_4 and autoclaved at 121 °C for 1 h. The autoclaved hydrolysis solution was vacuum filtered into the previously weighed crucibles of medium porosity (10 to 15 μm). The solid residue remaining in the filter crucibles was oven-dried at 105 °C overnight and is considered to be the acid-insoluble lignin. The content of acid-soluble lignin in the hydrolysates was also quantified. The acid-soluble lignin was calculated based on measuring absorbance at 320 nm in a UV-visible spectrophotometer (UV–vis, 7205, Jenway Co., Staffordshire, UK). The content of total lignin in the sample was determined as the sum of the acid-insoluble lignin and acid-soluble lignin, expressed as a percentage.

2.3. RNA Extraction, cDNA Synthesis, and Gene Expression Analysis by Quantitative Real-Time PCR

The total RNA of kiwifruit flesh tissue was isolated from three biological replicates of fruit at every 30 days of storage using the Ribospin™ Plant Total RNA Purification Kit (GeneAll Biotechnology Co., Ltd., Seoul, Korea) by following the manufacturer's protocol. The extracted RNA was quantified by Nanodrop spectrophotometer (Thermo Scientific™ NanoDrop 2000, Waltham, MA, USA). One microgram of total RNA was used for cDNA synthesis using the TOPscript™ RT DryMIX (*dT18 plus*) Kit (Enzynomics; Republic of Korea) according to the manufacturer's protocol.

The real-time qPCR was performed with gene-specific primers (Table S2). Each reaction was performed in triplicate. For each sample in 20 μL final volume were contained 1 μL cDNA, 1 μL specific primers, and 10 μL of 2X Real-Time PCR Master Mix (Including SFCgreen® I in the mixture) according to the manufacturer's protocol (BioFACT™, Yuseong-Gu, Daejeon, Korea).

Three genes of cell wall modification, *A. deliciosa* polygalacturonase C (*AdPGC*), *A. deliciosa* expansin 1 (*AdEXP1*), *A. deliciosa* expansin 2 (*AdEXP2*), and two genes of ethylene biosynthesis, *A. deliciosa* ACC (1-aminocyclopropane-1-carboxylic acid) synthase 2 (*AdACS2*), *A. deliciosa* ACC oxidase 2 (*AdACO2*) were selected based on our previous studies [28,29]. The lignin metabolism-related genes, *A. chinensis* phenylalanine ammonia-lyase (*AcPAL*), *A. chinensis* cinnamyl-alcohol dehydrogenase (*AcCAD*), and *A. chinensis* peroxidase 2 (*AcPOD2*) were selected according to Li et al. [12]. For cell wall modification genes and ethylene biosynthesis genes quantitative RT-PCR was performed using the Rotor-Gene Q detection system (Qiagen, Hilden, Germany) and included initial annealing of 5 min at 95 °C, followed by 50 cycles of 15 s at 95 °C, 30 s at 60 °C, and 45 s at 72 °C [28,29], followed by a melting-curve analysis at the end of each run. The cycling procedure for lignin metabolism-related genes included initial annealing of 3 min at 95 °C, followed by 44 cycles of 10 s at 95 °C, 30 s at 57 °C, and 30 s at 72 °C [12]. The kiwifruit actin gene, *A. deliciosa* actin 1 (*AdACT1*) was used as the internal control (housekeeping gene) to normalize expression differences in each sample. Quantitative RT-PCR data were analyzed using the $2^{-\Delta\Delta CT}$ method [30], and changes in gene expression caused by chitosan treatment was compared to untreated fruit (control) of each sampling date ($p \leq 0.05$).

2.4. Statistical Analysis

The data were subjected to analysis of variance (ANOVA) using statistical analysis software (SAS, version 9.4, SAS Institute Inc., Cary, NC, USA). The significance of differences between the mean values were determined by least significant difference (LSD) test ($p \leq 0.05$). Statistical analysis of qRT-PCR data was performed with Student's *t*-test ($p \leq 0.05$). A two-way ANOVA compared the independent groups (preharvest treatment and storage days at 0 °C) based on the dependent variables as shown in Supplementary Tables S3–S6. The graphics of the analyzed results were generated using SigmaPlot v12.0 (Systat Software, Inc., SigmaPlot for Windows, San Jose, CA, USA).

3. Results

3.1. Fruit Weight and Size at Harvest

The chitosan-treated fruit had the highest fruit weight of 86.27 and 91.79 g ($p \leq 0.05$) at 100 and 500 mg·L^{-1} concentrations, respectively (Table 1) when compared with the average weight of the control fruit (81.49 g). The fruit size was measured by means of fruit length, longitudinal width, and transverse width. There was no difference in fruit length and longitudinal width between the treatments, except for the transverse width (Table 1). The smaller values of transverse width of 50.28 and 51.86 mm were measured for fruit treated with 100 and 500 mg·L^{-1} chitosan, respectively, and a larger value of transverse width was observed for the control fruit (52.06 mm).

Table 1. Effects of preharvest chitosan application on fruit weight and fruit size at harvest in 'Garmrok' kiwifruit. Experimental data represent means ± standard error with n = 10. * in each column indicate significant differences between treatments on harvest day, according to the least significant difference (LSD) test at $p \leq 0.05$.

Treatment	Fruit Weight (g)	Fruit Size (mm)		
		Length	Longitudinal Width	Transverse Width
Control	81.49 ± 1.1	55.17 ± 0.7	55.85 ± 0.7	52.06 ± 0.6 *
Chitosan 100 mg·L^{-1}	86.27 ± 2.4 *	54.72 ± 1.2	54.25 ± 0.5	50.28 ± 0.4
Chitosan 500 mg·L^{-1}	91.79 ± 2.9 *	55.17 ± 1.1	55.45 ± 0.5	51.86 ± 0.6

3.2. Ethylene Production (C_2H_4) and Respiration (CO_2) Rates

The differences were not found in either C_2H_4 production or respiration rates (CO_2 production) at harvest date (Figure 1a,b). After 30 days of storage, C_2H_4 production was highest ($p \leq 0.05$) in fruit treated with 500 mg·L^{-1} chitosan while it was lowest ($p \leq 0.05$) in fruit treated with 100 mg·L^{-1} chitosan (Figure 1a). For the rest of the storage period, C_2H_4

production was lower ($p \leq 0.05$) for both 100 and 500 mg·L^{-1} chitosan-treated fruit than that for the control samples. As for the respiration, control fruit displayed the highest respiration rate during the cold storage than the preharvest chitosan-treated fruit (Figure 1b). While there was a lower CO_2 production with the higher chitosan concentration, the mean values of CO_2 production in 100 and 500 mg·L^{-1} chitosan pre-treated fruit were not statistically significant. At the end of the storage, the highest CO_2 production was found in control fruit (20.0 mg·kg^{-1}·h^{-1}), followed by 100 mg·L^{-1} (16.4 mg·kg^{-1}·h^{-1}) and 500 mg·L^{-1} (14.3 mg·kg^{-1}·h^{-1}) chitosan-treated kiwifruit. Moreover, a significant effect of the days of storage at 0 °C was observed for C_2H_4 production and respiration rates (Table S3).

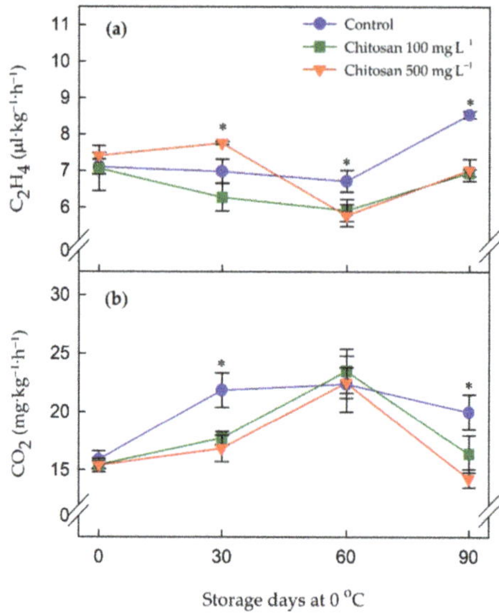

Figure 1. Effects of preharvest chitosan application on (**a**) ethylene production and (**b**) respiration rates during cold storage in 'Garmrok' kiwifruit. Vertical bars indicate SE with n = 3. * indicates significant differences between treatments at each sampling date, according to the least significant difference (LSD) test at $p \leq 0.05$.

3.3. Loss of Fruit Weight

The loss of fruit weight for all treatments displayed a similar increasing pattern as the storage period proceeded (Table 2 and Table S3). Preharvest chitosan application (100 and 500 mg·L^{-1}) deterred the weight loss of 'Garmrok' kiwifruit during the cold storage, and the highest ($p \leq 0.05$) weight loss was observed in the control fruit (Table 2). Although the weight loss appeared to be slightly higher at 500 mg·L^{-1} chitosan than 100 mg·L^{-1} chitosan, they were not statistically significant. The greatest loss of fruit weight for each treatment was observed at the end of storage (90 days), as shown as weight loss of 1.8%, 2.2%, and 2.3% for 100, 500 mg·L^{-1} chitosan and control treatment, respectively.

Table 2. Effects of preharvest chitosan application on fruit weight loss in 'Garmrok' kiwifruit during cold storage. Experimental data represent means ± standard error with n = 10. * in each column indicate significant differences between treatments at each sampling date, according to the least significant difference (LSD) test at $p \leq 0.05$.

Treatment	Storage Days at 0 °C			
	0	30	60	90
	Fruit weight loss (%)			
Control	0.0±0.0	0.9 ± 0.0	1.5 ± 0.1	2.3 ± 0.1
Chitosan 100 mg·L^{-1}	0.0±0.0	0.5 ± 0.1 *	1.0 ± 0.1 *	1.8 ± 0.1 *
Chitosan 500 mg·L^{-1}	0.0±0.0	0.6 ± 0.1 *	1.2 ± 0.1 *	2.2 ± 0.2

3.4. Changes in Soluble Solids Content, Titratable Acidity, Total Sugars, and Total Acids

As expected, soluble solids content (SSC) and total sugar content increased over the period of storage for all the treatments, which reached the maximum value at the end of storage (Table 3 and Table S4). Nonetheless, the preharvest treatment of chitosan was associated with slower increasing rates of SSC during storage, as significantly lower ($p \leq 0.05$) SSC at the end of storage were observed in preharvest chitosan-treated fruit (11.9% in 100 mg·L^{-1} and 11.6% in 500 mg·L^{-1}) compared with the control fruit (12.7%). In addition, preharvest chitosan-treated fruit displayed the lowest increase in total sugar content during the storage which resulted in a significantly lower ($p \leq 0.05$) total sugar content at the end of storage. At harvest, the ratio of fructose + glucose/sucrose, a potential indicator of invertase activity [31], was lower ($p \leq 0.05$) in 100 and 500 mg·L^{-1} chitosan-treated fruit. However, there was no difference in the fructose + glucose/sucrose ratio among treatments for the rest of the storage.

In regard to titratable acidity (TA) and total acid content, it was clear that both attributes displayed a decreasing trend over the storage time (Table 3 and Table S4). The preharvest chitosan-treated fruit had the greatest TA values ($p \leq 0.05$) at harvest. The fruit treated with 100 mg·L^{-1} chitosan displayed the positive effect on TA that preserved the greatest content ($p \leq 0.05$) during storage time compared with the control and 500 mg·L^{-1} chitosan-treated fruit. As for the total acid content, preharvest chitosan treatments resulted in higher ($p \leq 0.05$) values at harvest time than the control. Throughout the whole storage, the application of 100 mg·L^{-1} chitosan was most effective in preserving the greatest total acid content, followed by 500 mg·L^{-1} chitosan when compared with the control. In addition, the application of preharvest chitosan with 100 and 500 mg·L^{-1} caused the greatest ($p \leq 0.05$) citric acid/quinic acid ratio at harvest; however, there was no significant difference in this ratio among treatments during the rest of storage.

3.5. Changes in Total Phenolic and Lignin Contents

In general, there was an increasing tendency in the content of total phenolic as the storage period prolonged in all the treatments (Figure 2a and Table S5). The application of preharvest chitosan at both concentrations caused the greatest content of total phenolic at harvest and this trend continued during the cold storage. Although 100 mg·L^{-1} chitosan-treated fruit maintained higher total phenolic content during the cold storage it was not statistically significant with untreated fruit at 30 and 60 days of storage. However, at the end of the storage period, the untreated fruit had the lowest total phenolic content (106.4 mg/100 g), while chitosan-treated fruit had total phenolic content of 121.3 and 124.2 mg/100 g for 100 and 500 mg·L^{-1} applications, respectively.

As shown in Figure 2b, the total lignin content increased over the whole storage in all the treatments (Table S5). There were no differences in total lignin content at harvest; however, the changes ($p \leq 0.05$) in the content of total lignin between treatments became apparent from 30 days of storage. The application of preharvest chitosan at 500 mg·L^{-1} was associated with the greatest ($p \leq 0.05$) total lignin content throughout the storage time, and this was followed by the 100 mg·L^{-1} chitosan-treated fruit.

Table 3. Effects of preharvest chitosan application on soluble solids content, titratable acidity, total sugar content (HPLC analysis), total acid content (HPLC analysis), and their ratios in 'Garmrok' kiwifruit during cold storage. Experimental data represent means ± standard error with n = 10 (SSC) and n = 3 (titratable acidity, total sugar content, total acid content, and their ratios). *, in each column indicate significant differences between treatments at each sampling date, according to the least significant difference (LSD) test at $p \leq 0.05$.

Treatment	Storage Days at 0 °C			
	0	30	60	90
		Soluble solids content (SSC, %)		
Control	6.2 ± 0.2	9.6 ± 0.1	11.8 ± 0.2	12.7 ± 0.1
Chitosan 100 mg·L^{-1}	6.0 ± 0.2	9.3 ± 0.2 *	11.7 ± 0.2	11.9 ± 0.1 *
Chitosan 500 mg·L^{-1}	6.0 ± 0.2	9.9 ± 0.2	11.4 ± 0.1	11.6 ± 0.3 *
		Titratable acidity (TA, %)		
Control	1.7 ± 0.0	1.5 ± 0.0	1.2 ± 0.1	1.1 ± 0.0
Chitosan 100 mg·L^{-1}	1.9 ± 0.1 *	1.5 ± 0.0	1.3 ± 0.0 *	1.0 ± 0.0
Chitosan 500 mg·L^{-1}	1.8 ± 0.0 *	1.5 ± 0.0	1.2 ± 0.0	0.9 ± 0.0 *
		Total sugar content (g/100 g)		
Control	6.9 ± 0.2	9.7 ± 0.4	9.2 ± 0.5	9.4 ± 0.5
Chitosan 100 mg·L^{-1}	6.8 ± 0.2	9.2 ± 0.3	8.9 ± 0.2	9.2 ± 0.3
Chitosan 500 mg·L^{-1}	6.7 ± 0.2	8.4 ± 0.2 *	8.5 ± 0.4	8.1 ± 0.2 *
		Total acid content (g/100 g)		
Control	1.3 ± 0.0	1.3 ± 0.0	1.2 ± 0.0	1.3 ± 0.0
Chitosan 100 mg·L^{-1}	1.4 ± 0.0*	1.4 ± 0.0 *	1.3 ± 0.0 *	1.4 ± 0.0 *
Chitosan 500 mg·L^{-1}	1.4 ± 0.0*	1.3 ± 0.0	1.3 ± 0.0 *	1.3 ± 0.0
		Fructose + glucose/sucrose ratio		
Control	4.3 ± 0.3	2.3 ± 0.2	2.5 ± 0.0	2.6 ± 0.1
Chitosan 100 mg·L^{-1}	3.6 ± 0.2*	2.1 ± 0.1	2.3 ± 0.3	3.1 ± 0.7
Chitosan 500 mg·L^{-1}	3.4 ± 0.1*	2.1 ± 0.2	2.4 ± 0.0	2.6 ± 0.2
		Citric acid/quinic acid ratio		
Control	1.2 ± 0.0	1.2 ± 0.0	1.2 ± 0.0	1.2 ± 0.0
Chitosan 100 mg·L^{-1}	1.3 ± 0.0 *	1.2 ± 0.0	1.1 ± 0.0 *	1.1 ± 0.0
Chitosan 500 mg·L^{-1}	1.4 ± 0.0 *	1.1 ± 0.1	1.2 ± 0.0	0.9 ± 0.1

3.6. Changes in Firmness

It is clear from Figure 3 and Table S3 that the firmness of kiwifruit (flesh and core) decreased gradually during the cold storage. At harvest, the firmness of flesh (Figure 3a) was not different between treatments, while the firmness of core (Figure 3b) was greatest ($p \leq 0.05$) in the fruits treated with the preharvest chitosan. Throughout the whole storage, it was apparent that the treatment of chitosan at 100 and 500 mg·L^{-1} retained the higher flesh firmness when compared to that of control. Although the mean core firmness in preharvest chitosan-treated kiwifruits were higher than the control, they were not significantly different from the control fruit during the storage.

3.7. Expression of Ethylene Biosynthesis-, Cell Wall Modification- and Lignin Metabolism-Related Genes

Expressions of several genes associated with ethylene biosynthesis were examined (Figure 4 and Table S6). In general, expression of *AdACS2* for the kiwifruit treated with preharvest chitosan at both concentrations was maintained at a relatively lower level than that of control during the storage (Figure 4a). This was apparent especially for the later stages of storage (i.e., 60 and 90 days). Another gene responsible for the biosynthesis of ethylene, *AdACO2*, displayed a similar pattern during the cold storage (Figure 4b). Preharvest treatment of chitosan at both concentrations suppressed expression of *AdACO2*, and this pattern of lower expression was observed until 60 days of cold storage.

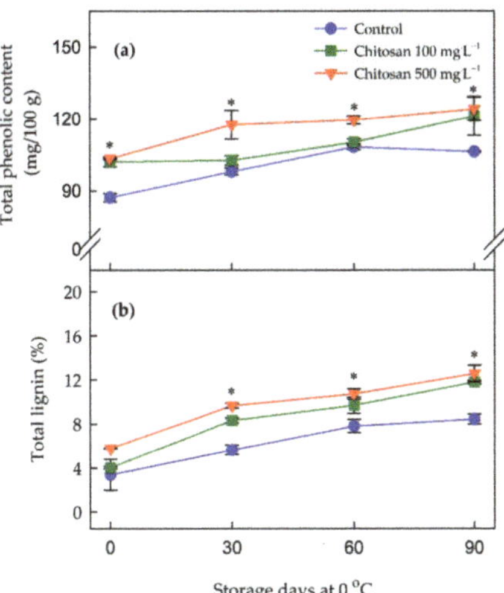

Figure 2. Effects of preharvest chitosan application on (**a**) total phenolic content and (**b**) total lignin content during cold storage in 'Garmrok' kiwifruit. Vertical bars indicate SE with n = 3. * indicates significant differences between treatments at each sampling date, according to the least significant difference (LSD) test at $p \leq 0.05$.

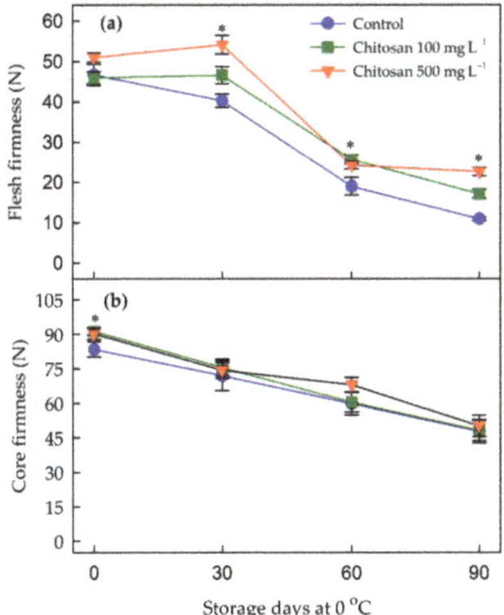

Figure 3. Effects of preharvest chitosan application on (**a**) flesh firmness and (**b**) core firmness during cold storage in 'Garmrok' kiwifruit. Vertical bars indicate SE with n = 10. * indicates significant differences between treatments at each sampling date, according to the least significant difference (LSD) test at $p \leq 0.05$.

We examined several genes related to cell wall modification and the pattern of expression is shown in Figure 5 and Table S6. The preharvest treatment of chitosan suppressed the expression of *AdPGC* (Figure 5a) and *AdEXP1* (Figure 5b) at 30 and 60 days of the cold storage. The suppressed expression of these two genes was observed primarily with lower concentration of chitosan (100 mg·L^{-1}). The effect of suppressed expression for *AdEXP2* (Figure 5c) was greater with a higher concentration of chitosan (500 mg·L^{-1}).

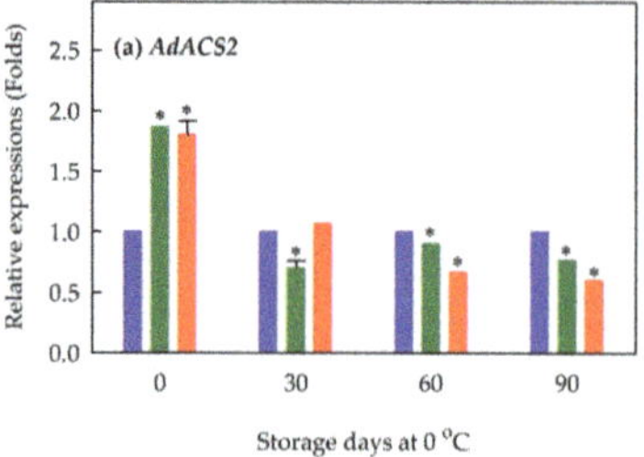

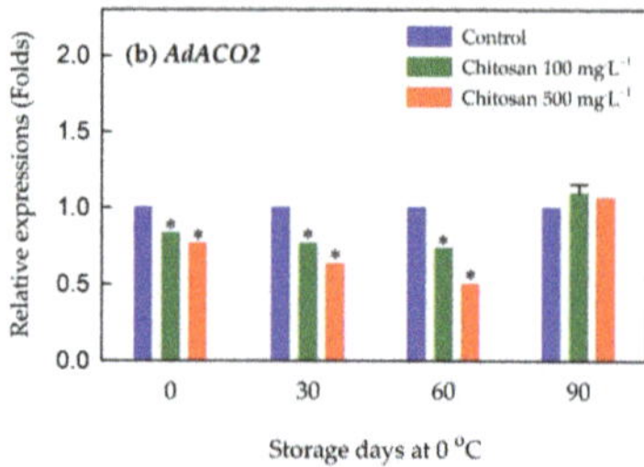

Figure 4. Effects of preharvest chitosan application on relative expression of ethylene biosynthesis-related genes (**a**) *AdACS2* and (**b**) *AdACO2* during cold storage in 'Garmrok' kiwifruit. Vertical bars indicate SE with n = 3. Actin gene was used as internal control and the quantitative RT-PCR data were analyzed using the $2^{-\Delta\Delta CT}$ method. * Statistical difference caused by chitosan treatment was compared to untreated fruit (control) of each sampling date ($p \leq 0.05$).

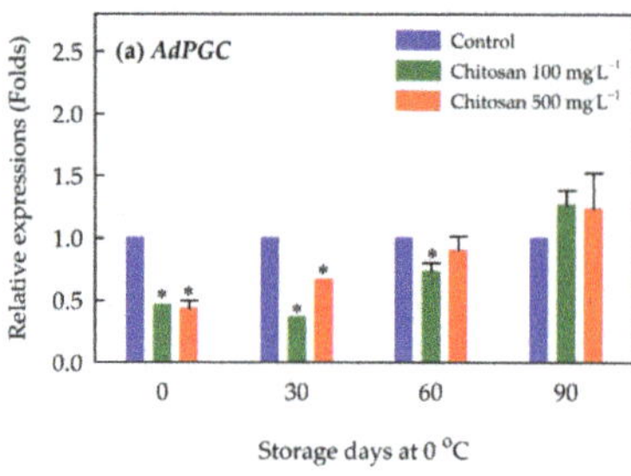

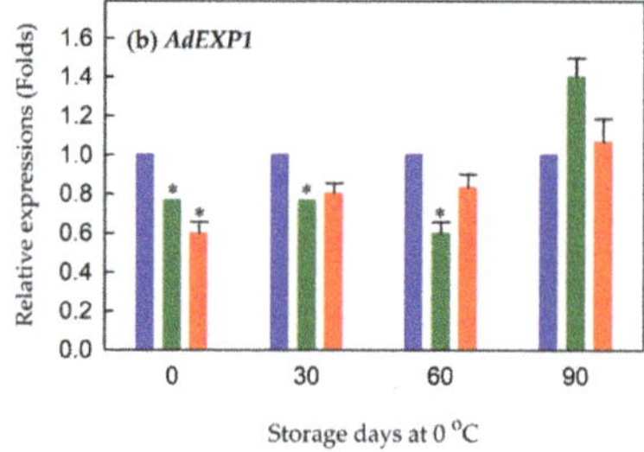

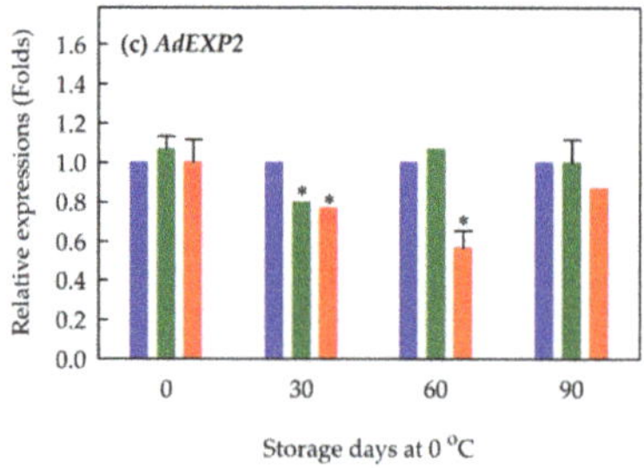

Figure 5. Effects of preharvest chitosan application on relative expression of cell wall-modification genes (**a**) *AdPGC*, (**b**) *AdEXP1*, and (**c**) *AdEXP2* during cold storage in 'Garmrok' kiwifruit. Vertical bars indicate SE with n = 3. Actin gene was used as internal control and the quantitative RT-PCR data were analyzed using the $2^{-\Delta\Delta CT}$ method. * Statistical difference caused by chitosan treatment was compared to untreated fruit (control) of each sampling date ($p \leq 0.05$).

As for genes associated with lignin metabolism, expression of *AcPAL*, a gene that is known to catalyze the first committed step in general phynylropanoid metabolism [32], was upregulated by the preharvest treatment of chitosan at both concentrations (Figure 6a and Table S6). The upregulation of *AcPAL* was most obvious at 30 days of cold storage. In addition, a gene considered to be an indicator of lignin biosynthesis [32], *AcCAD*, was upregulated at the end of cold storage (i.e., 90 days of storage) by both concentrations (Figure 6b and Table S6). A gene that is involved in the polymerization of monolignols to yield the lignin polymer [33], *AcPOD2*, was upregulated by the preharvest treatment of chitosan at both concentrations, especially toward the end of cold storage (Figure 6c and Table S6).

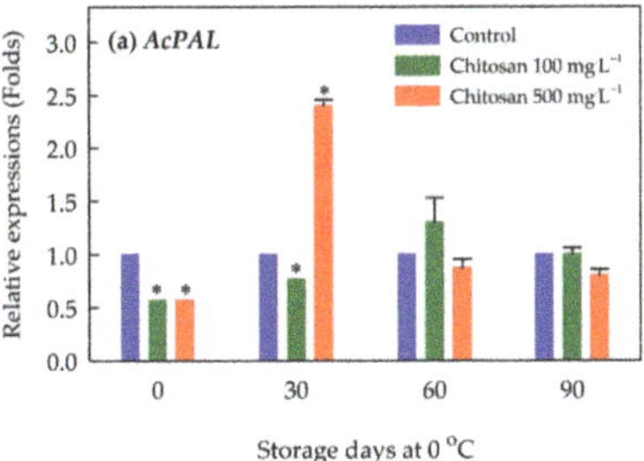

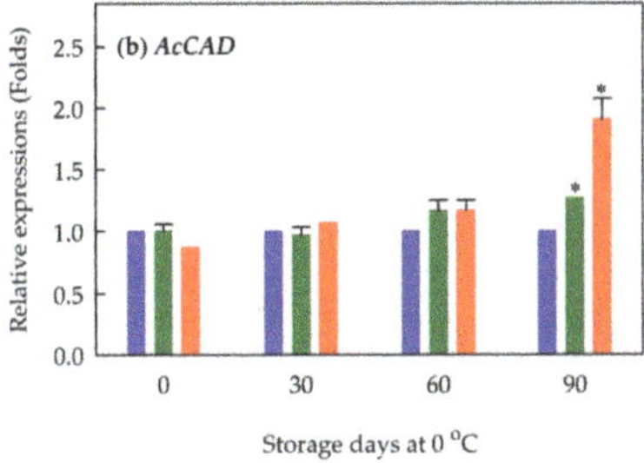

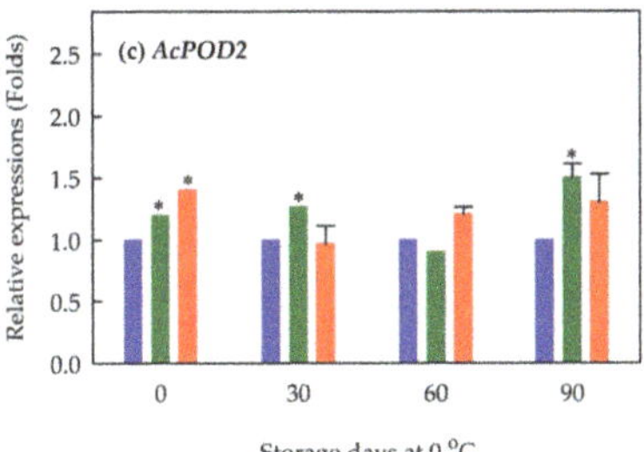

Figure 6. Effects of preharvest chitosan application on relative expression of lignin metabolism-related genes (**a**) *AcPAL*, (**b**) *AcCAD*, and (**c**) *AcPOD2* during cold storage in 'Garmrok' kiwifruit. Vertical bars indicate SE with n = 3. Actin gene was used as internal control and the quantitative RT-PCR data were analyzed using the $2^{-\Delta\Delta CT}$ method. * Statistical difference caused by chitosan-treatment was compared to untreated fruit (control) of each sampling date ($p \leq 0.05$).

4. Discussion

4.1. Preharvest Application of Chitosan Increased the Weight of 'Garmrok' Kiwifruit at Harvest

In our previous study, we showed that the preharvest treatment of calcium chitosan (100 mg·L^{-1}) increased the fruit weight with no effect on the size of kiwifruit at harvest [8]. A recent similar study demonstrated that the preharvest spray of chitosan increased the single fruit weight with the equivalent yield of kiwifruit without affecting the size of the fruit [22]. Here, we show that the application of preharvest chitosan also increased the fruit weight and delayed maturation and ripening of the kiwifruit. The influence of preharvest treatment of chitosan manifested throughout the cold storage is reflected on the physicochemical and molecular attributes in kiwifruit.

It was considered that the increased fruit weight by the treatment of chitosan is associated with the formation of a semipermeable barrier on the fruit surface that may have reduced the water loss from transpiration and respiration [8]. When we examined the surface of the 'Garmrok' kiwifruit peel under a scanning electron microscope, this semipermeable barrier-forming characteristic of chitosan was observed as a coated film on the kiwifruit peel (Figure 7). The film coating of chitosan on the fruit surface can absorbs moisture due to the hydrophilic nature of chitosan solution [34] and this could be attributed to the increased weight of kiwifruit, which in turn reduced water loss from fresh fruit. Alternatively, the increased weight of chitosan-treated kiwifruit may be attributed to the activity as a biostimulant and growth elicitor, as demonstrated in *Capsicum annuum* 'Yolo Wonder' bell pepper [35], *Solanum lycopersicon* 'Faridah' tomato [36], and *A. deliciosa* 'Guichang' kiwifruit [22]. Future investigations on the detailed mechanistic insights as to the increased weight of fruits at harvest pretreated by chitosan further awaits.

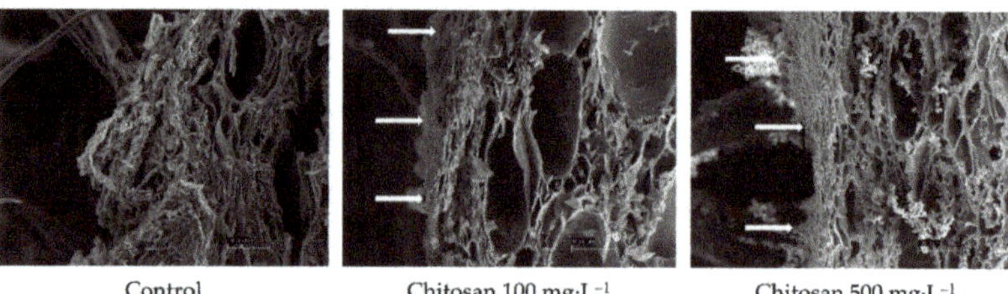

Figure 7. Scanning electron micrographs (SEMs) of preharvest chitosan-treated and untreated 'Garmrok' kiwifruit peel surfaces. The arrows indicate the chitosan film on kiwifruit peel after coating. All micrographs are at 500X and the scale bars in each image represent the lengths of 100 μm.

4.2. *Preharvest Application of Chitosan Is Associated with the Reduced Production of Ethylene and Rate of Respiration during Cold Storage*

During the cold storage, we observed a noticeable effect of preharvest chitosan application on the reduction of ethylene production and respiration rate of 'Garmrok' kiwifruit. The reduced ethylene production and respiration rate can be explained by the film-forming property of chitosan [17,19]. Chitosan has excellent selective permeability to the respiratory gases, particularly by blocking the oxygen that reduces respiration [20]. Kiwifruit is a climacteric fruit and the ripening process is largely regulated by the plant hormone ethylene. Ethylene production in climacteric fruit is mainly governed by the activity of two enzymes, i.e., 1-aminocyclopropane-1-carboxylic acid (ACC) synthase (ACS) and ACC oxidase (ACO) [37,38]. The conversion of ACC to ethylene by ACO is O_2-dependent [37,38]. The enzyme activity of ACO is suppressed by low O_2 and high CO_2, while high CO_2 concentrations prevent the auto-induction of ACS [37–39]. The effect of low O_2 and high CO_2 concentrations during storage on ethylene production and ripening for the shelf-life has been well reported in *A. deliciosa* 'Hayward' kiwifruit [39,40] and also in *Pyrus communis* 'Conference' pear [41].

The application of preharvest chitosan that acted as a barrier of oxygen might have affected the low production of ethylene in the kiwifruit by suppressing the key genes in the ethylene biosynthesis. Recently, He et al. [42] also observed that chitosan oligosaccharides (COS) suppressed the expression of genes involved in the ethylene biosynthesis (*FaACS* and *FaACO*), which reduced the softness and increased the shelf-life of strawberry fruit (*Fragaria ananassa* 'Qingxiang'). In addition, the resultant reduced production of ethylene and respiration rates might have positively influenced the retention of various quality attributes that prolonged the postharvest life of 'Garmrok' kiwifruit.

4.3. Preharvest Application of Chitosan Reduced the Weight Loss of 'Garmrok' Kiwifruit during Cold Storage

The proper status of water in fresh fruit before and during postharvest handling plays a critical role in maintaining the quality of the crop [36]. In general, weight loss in fresh fruits is caused by water loss associated with the metabolic activities occurring during respiration and transpiration processes [19]. In this study, we show that the application of preharvest chitosan reduces the weight loss of 'Garmrok' kiwifruit during cold storage. Previous studies have demonstrated that several features of chitosan treatment are linked to the reduction of weight loss in various fruits [13,17,19,36,43–46]. These include chitosan coatings not just as a barrier of transpiration, but also as a protectant and a wound sealer of fruit skin from mechanical injuries that delayed dehydration.

4.4. Preharvest Application of Chitosan Possesses Maturity- and Ripening-Delaying Effects on 'Garmrok' Kiwifruit

Soluble solids content (SSC), titratable acidity (TA), and total sugar and acid contents are important chemical attributes associated with the edible quality of ripe kiwifruit and are used as maturity- and ripening-related indices in quality measurements [5]. The increase in SSC during fruit ripening could be attributed to the breakdown of carbohydrates into simple sugars [5,19,44]. The reduction in TA during fruit ripening, on the other hand, indicates the degradation of organic acids, which are used as respiratory substrates [5,17,44]. It has been shown that low respiration and ethylene production repress the hydrolysis of carbohydrates, which results in a low SSC, total sugar contents, and delayed ripening [36]. In another report, the ratio of fructose + glucose/sucrose at harvest, a possible indicator of invertase activity [31], was shown to be associated with maturity- and ripening-delaying effects by preharvest application of chitosan, which supports our current results and also our previous study [8].

Our results revealed the slower increasing rate of SSC with lower values of SSC at the end of cold storage in 'Garmrok' kiwifruit when treated with preharvest chitosan. In addition, the kiwifruit treated with preharvest chitosan displayed a slower increase in total sugar content followed by the lowest total sugar content at the end of cold storage. These changes may have to do with the reduced rates of respiration and ethylene production, which in turn slowed the metabolic activity in preharvest chitosan-treated 'Garmrok' kiwifruit. Furthermore, the slow degradation of organic acids observed in our study may have resulted from the reduced respiration rate, which slowed the reduction in TA and total acid content. Our findings are consistent with the results of those observed in other fruits by preharvest treatment with chitosan [17,19,44–46].

4.5. Preharvest Application of Chitosan Might Improve the Antioxidant Activity of 'Garmrok' Kiwifruit by Positively Affecting Phenolic Metabolism

Phenolic compounds, the major secondary metabolites in plants, are responsible for the flavor and color of fruits with antioxidant potential [22,26,36,42]. The (+)-catechin, chlorogenic acid, rutin, (−)-epicatechin, quercetin, and tannin are the major phenolics identified in kiwifruit [26] and are responsible for the antioxidant capacity of fruit while being involved in the antioxidant metabolism [22,26]. Over the 90 days of storage period at 0 °C, preharvest application of chitosan increased the total phenolic content in 'Garmrok' kiwifruit, as similarly shown in other fruits treated with preharvest chitosan [22,34,36,42,45,47,48]. The low level of total phenolic content observed in the untreated kiwifruit might be due to the rapid breakdown of cell structure compared with the chitosan-treated kiwifruits. The fact that kiwifruit treated with preharvest chitosan that retained higher total content of phenolic compounds may be associated with the reduced activity of polyphenol oxidase (PPO) that oxidizes the phenolic compounds. The increased content of total phenolic compounds could possibly have obtained the antioxidant capacity of kiwifruit during the cold storage [36,42], which not only elicited immune responses but also improved the quality of 'Garmrok' kiwifruit during cold storage that delayed the ripening.

4.6. Preharvest Application of Chitosan Maintained the Firmness of 'Garmrok' Kiwifruit during Cold Storage

Loss of fruit firmness is a deleterious quality attribute that determines the postharvest life, fruit quality, commercial value, and consumer acceptability. In our study, preharvest application of chitosan exerted a beneficial effect on kiwifruit firmness during the cold storage. These results are in agreement with those obtained in our previous study on *A. deliciosa* 'Garmrok' kiwifruit [8] and Zhang et al. [22] on *A. deliciosa* 'Guichang' kiwifruit. It was considered that maintenance of firmness by chitosan coatings results from the formation of a semipermeable film on fruit surface that can act as a protective barrier to reduce respiration rate, thereby slowing the metabolic activity and textural changes [17,19,22,44,46,49]. As kiwifruit is a typical climacteric fruit, the reduced production of respiration and ethylene affected by chitosan treatment appear to be linked to the kiwifruit softening. Whether ethylene directly regulates expression of cell wall-modifying enzymes in kiwifruit, as observed in other climacteric, fruit remains to be investigated.

In addition, fruit firmness is closely associated with changes in the biochemical properties of the cell wall. Polygalacturonase (PG) is a key enzyme involved in cell wall modifications during fruit ripening, with its important functions in degradation of soluble pectin and depolymerization of solubilized pectin [50–52]. In kiwifruit, there are three polygalacturonase genes, namely, *PGA*, *PGB*, and *PGC*. Of these, *PGC* is a predominant gene expressed throughout the softening of kiwifruit [50,51,53]. The suppression of the *AdPGC* gene in preharvest chitosan-treated 'Garmrok' kiwifruit during the early phase of storage (i.e., 30 and 60 days) may be associated with the delayed kiwifruit softening and retained firmness. Expansins (EXPs) are non-enzymatic proteins that have been considered to play a role in loosening and extension of the cell wall (the cellulose-xyloglucan hydrogen bonds) [50,52,54]. Yang et al. [54] reported that two expansin genes *AdEXP1* and *AdEXP2* are actively involved in kiwifruit (*A. deliciosa* 'Bruno') ripening and softening. In our study, we observed a negative effect on expression of expansin genes (*AdEXP1* and *AdEXP2*) during the postharvest storage of 'Garmrok' kiwifruit. Thus, preharvest chitosan treatment was effective in delaying fruit softening, maintaining kiwifruit firmness, and extend their storage life by possibly inhibiting the activities of EXPs.

Further, a higher flesh firmness of 'Garmrok' kiwifruit treated with preharvest chitosan can be ascribed to the increase in lignin content. Lignin is one of the most abundant polyphenolic polymers in higher plants that mostly functions as a structural material which enhances the strength and rigidity of plant cells [12,55]. The increase of total lignin content in the chitosan-treated 'Garmrok' kiwifruit during the cold storage indicates that preharvest chitosan could induce lignin biosynthesis, as shown by He et al. [42] and Saavedra et al. [48]. Lignin biosynthesis involves the activities of three major enzymes (PAL, CAD, and POD), as well as the coordinated expression of these enzyme genes [12,32,33,55,56]. In our study, the higher expression of *AcPAL* during the early period of storage may have influenced the lignin accumulation in the preharvest chitosan-treated kiwifruit. In addition, the increased level of expression in *AcCAD* and *AcPOD2* at the later stage of storage is associated with the increased content of lignin in the preharvest chitosan-treated kiwifruit. Collectively, expression of lignin metabolism-related genes affected by preharvest chitosan treatment may have enhanced lignin metabolism that conferred greater firmness in 'Garmrok' kiwifruits.

5. Conclusions

Our results demonstrate that *A. deliciosa* 'Garmrok' kiwifruit can be stored at 0 °C for more than two months when preharvest chitosan was applied to kiwifruit. Preharvest application of chitosan reduced weight loss, maintained firmness by modulating the expression of ethylene biosynthesis, cell wall modification, and lignin metabolism-related genes, possibly evoked antioxidant activity by phenolic metabolism, and displayed a maturity-

and the ripening-delaying effect that improved the overall quality and extended the storage life of kiwifruit. This work also revealed that the treatment with 500 mg·L^{-1} chitosan was more effective in maintaining overall physicochemical attributes of 'Garmrok' kiwifruit during postharvest storage. The results open a promising strategy for maintaining the postharvest properties of kiwifruit. Detailed mechanistic insights that underlie physicochemical attributes need to be further investigated. In addition, the possible effect on sensory attributes of chitosan-coated kiwifruit and lignin accumulation for the consumer's acceptance should be taken into consideration in future studies.

Supplementary Materials: The following are available online at https://www.mdpi.com/2304-8158/10/2/373/s1, Table S1: Effects of preharvest chitosan application on sugars (fructose, glucose, and sucrose) and organic acids (oxalic, quinic, malic, and citric) (HPLC analysis) in 'Garmrok' kiwifruit during cold storage. Experimental data represent means ± standard error with n = 3. *, in each column indicate significant differences between treatments at each sampling date, according to the least significant difference (LSD) test at $p \leq 0.05$, Table S2: The primer sequences and gene information used in this study, Table S3: Ethylene production, respiration, fruit weight loss, flesh firmness and core firmness of 'Garmrok' kiwifruit as affected by chitosan-treatment and days of storage at 0 °C, Table S4: Soluble solids content (SSC), titratable acidity (TA), total sugar content (HPLC analysis), total acid content (HPLC analysis), and their ratios of 'Garmrok' kiwifruit as affected by chitosan-treatment and days of storage at 0 °C, Table S5: Total phenolic content and total lignin content of 'Garmrok' kiwifruit as affected by chitosan-treatment and days of storage at 0 °C, Table S6: Relative expression of ethylene biosynthesis-related genes (*AdACS2* and *AdACO2*), cell wall-modification genes (*AdPGC*, *AdEXP1*, and *AdEXP2*), and lignin metabolism-related genes (*AcPAL*, *AcCAD*, and *AcPOD2*) of 'Garmrok' kiwifruit as affected by chitosan-treatment and days of storage at 0 °C.

Author Contributions: Conceptualization, J.G.K.; methodology, Y.-H.K.; formal analysis, H.M.P.C.K.; investigation, Y.-S.L.; resources, M.L. and Y.-B.K.; writing—original draft preparation, H.M.P.C.K.; writing—review and editing, J.K.; supervision, J.G.K. and J.K.; funding acquisition, J.K. All authors have read and agreed to the published version of the manuscript.

Funding: This work was supported by research fund of Chungnam National University (J.K).

Acknowledgments: We acknowledge Seon-In Yeom (Gyeongsang National University, Republic of Korea) for providing access to their laboratories for conducting gene analysis studies and real-time qPCR studies. We thank Yang Jae Kyung and Ji Young Jung (Gyeongsang National University, Republic of Korea) for allowing conducting lignin analysis experiments at their laboratory. Also, we thank Seon-Gi Jeong (Gyeongsangnam-do Agricultural Research and Extension Services, Republic of Korea) for the help on the scanning electron microscope.

Conflicts of Interest: The authors declare no conflict of interest.

References

1. Benítez, S.; Achaerandio, I.; Sepulcre, F.; Pujolà, M. Aloe vera based edible coatings improve the quality of minimally processed "Hayward" kiwifruit. *Postharvest Biol. Technol.* **2013**, *81*, 29–36. [CrossRef]
2. Kaya, M.; Česoniene, L.; Daubaras, R.; Leskauskaite, D.; Zabulione, D. Chitosan coating of red kiwifruit (*Actinidia melanandra*) for extending of the shelf life. *Int. J. Biol. Macromol.* **2016**, *85*, 355–360. [CrossRef]
3. Vivek, K.; Subbarao, K.V. Effect of edible chitosan coating on combined ultrasound and NaOCl treated kiwifruits during refrigerated storage. *Int. Food. Res. J.* **2018**, *25*, 101–108.
4. Kim, J.G.; Beppu, K.; Fukuda, T.; Kataoka, I. Evaluation of fruit characteristics of Shima sarunashi (*Actinidia rufa*) indigenous to warm regions in Japan. *J. Jpn. Soc. Hortic. Sci.* **2009**, *78*, 394–401. [CrossRef]
5. Cha, G.H.; Kumarihami, H.M.P.C.; Kim, H.L.; Kwack, Y.B.; Kim, J.G. Storage temperature influences fruit ripening and changes in organic acids of kiwifruit treated with exogenous ethylene. *Hortic. Sci. Technol.* **2019**, *37*, 618–629. [CrossRef]
6. Huang, Z.; Li, J.; Zhang, J.; Gao, Y.; Hui, G. Physicochemical properties enhancement of Chinese kiwifruit (*Actinidia chinensis* Planch) via chitosan coating enriched with salicylic acid treatment. *J. Food. Meas. Charact.* **2017**, *11*, 184–191. [CrossRef]
7. Xu, S.; Chen, X.; Sun, D.W. Preservation of kiwifruit coated with an edible film at ambient temperature. *J. Food. Eng.* **2001**, *50*, 211–216. [CrossRef]
8. Kumarihami, H.M.P.C.; Cha, G.H.; Kim, J.G.; Kim, H.L.; Lee, M.; Kwack, Y.B.; Cho, J.G.; Kim, J. Effect of preharvest Ca-Chitosan application on postharvest quality of 'Garmrok' kiwifruit during cold storage. *Hortic. Sci. Technol.* **2020**, *38*, 239–248. [CrossRef]

9. Shin, M.H.; Kwack, Y.B.; Kim, Y.H.; Kim, J.G. Storage temperature affects the ripening characteristics of 'Garmrok', 'Hayward', 'Goldone', and 'Jecy Gold' kiwifruit treated with exogenous ethylene. *Hortic. Sci. Technol.* **2018**, *36*, 730–740. [CrossRef]
10. Mastromatteo, M.; Mastromatteo, M.; Conte, A.; Del Nobile, M.A. Combined effect of active coating and MAP to prolong the shelf-life of minimally processed kiwifruit (*Actinidia deliciosa* cv. Hayward). *Food Res. Int.* **2011**, *44*, 1224–1230. [CrossRef]
11. Xu, S.; Da Xu, L.; Chen, X. Determining optimum edible films for kiwifruits using an analytical hierarchy process. *Comput. Oper. Res.* **2003**, *30*, 877–886. [CrossRef]
12. Li, H.; Suo, J.; Han, Y.; Liang, C.; Jin, M.; Zhang, Z.; Rao, J. The effect of 1-methylcyclopropene, methyl jasmonate and methyl salicylate on lignin accumulation and gene expression in postharvest 'Xuxiang' kiwifruit during cold storage. *Postharvest Biol. Technol.* **2017**, *124*, 107–118. [CrossRef]
13. Fisk, C.L.; Silver, A.M.; Strik, B.C.; Zhao, Y. Postharvest quality of hardy kiwifruit (*Actinidia arguta* 'Ananasnaya') associated with packaging and storage conditions. *Postharvest Biol. Technol.* **2008**, *47*, 338–345. [CrossRef]
14. Benítez, S.; Achaerandio, I.; Pujolà, M.; Sepulcre, F. Aloe vera as an alternative to traditional edible coatings used in fresh-cut fruits: A case of study with kiwifruit slices. *LWT Food Sci. Technol.* **2015**, *61*, 184–193. [CrossRef]
15. Allegra, A.; Inglese, P.; Sortino, G.; Settanni, L.; Todaro, A.; Liguori, G. The influence of *Opuntia ficus-indica* mucilage edible coating on the quality of "Hayward" kiwifruit slices. *Postharvest Biol. Technol.* **2016**, *120*, 45–51. [CrossRef]
16. Diab, T.; Biliaderis, C.G.; Gerasopoulos, D.; Sfakiotakis, E. Physicochemical properties and application of pullulan edible films and coatings in fruit preservation. *J. Sci. Food Agric.* **2001**, *81*, 988–1000. [CrossRef]
17. El-Badawy, H.E.M. Effect of chitosan and calcium chloride spraying on fruits quality of Florida Prince. *Res. J. Agric. Biol. Sci.* **2012**, *8*, 272–281.
18. Fortunati, E.; Giovanale, G.; Luzi, F.; Mazzaglia, A.; Kenny, J.M.; Torre, L.; Balestra, G.M. Effective postharvest preservation of kiwifruit and romaine lettuce with a chitosan hydrochloride coating. *Coatings* **2017**, *7*, 196. [CrossRef]
19. Gayed, A.A.N.A.; Shaarawi, S.A.M.A.; Elkhishen, M.A.; Elsherbini, N.R.M. Pre-harvest application of calcium chloride and chitosan on fruit quality and storability of 'Early Swelling' peach during cold storage. *Ciênc. Agrotecnol.* **2017**, *41*, 220–231. [CrossRef]
20. Elsabee, M.Z.; Abdou, E.S. Chitosan based edible films and coatings: A review. *Mater. Sci. Eng. C* **2013**, *33*, 1819–1841. [CrossRef]
21. Drevinskas, T.; Naujokaitytė, G.; Maruška, A.; Kaya, M.; Sargin, I.; Daubaras, R.; Česonienė, L. Effect of molecular weight of chitosan on the shelf life and other quality parameters of three different cultivars of *Actinidia kolomikta* (kiwifruit). *Carbohydr. Polym.* **2017**, *173*, 269–275. [CrossRef] [PubMed]
22. Zhang, C.; Long, Y.H.; Wang, Q.P.; Li, J.H.; Wu, X.M.; Li, M. The effect of preharvest 28.6% chitosan composite film sprays for controlling the soft rot on kiwifruit. *Hortic. Sci.* **2019**, *46*, 180–194. [CrossRef]
23. Du, J.; Gemma, H.; Iwahori, S. Effects of chitosan coating on the storage of peach, Japanese pear, and kiwifruit. *J. Jpn. Soc. Hortic. Sci.* **1997**, *66*, 15–22. [CrossRef]
24. Zheng, F.; Zheng, W.; Li, L.; Pan, S.; Liu, M.; Zhang, W.; Liu, H.; Zhu, C. Chitosan controls postharvest decay and elicits defense response in kiwifruit. *Food Bioprocess Technol.* **2017**, *10*, 1937–1945. [CrossRef]
25. Kim, J.G.; Park, Y.; Shin, M.H.; Muneer, S.; Lerud, R.; Michelson, C.; Kang, D., II; Min, J.H.; Kumarihami, H.M.P.C. Application of NIR-Spectroscopy to predict the harvesting maturity, fruit ripening and storage ability of Ca-chitosan treated baby kiwifruit. *J. Stored Prod. Postharvest Res.* **2018**, *9*, 44–53. [CrossRef]
26. Kim, J.G.; Beppu, K.; Kataoka, I. Varietal differences in phenolic content and astringency in skin and flesh of hardy kiwifruit resources in Japan. *Sci. Hortic.* **2009**, *120*, 551–554. [CrossRef]
27. Sluiter, A.; Hames, B.; Ruiz, R.; Scarlata, C.; Sluiter, J.; Templeton, D.; Crocker, D. NREL/TP-510-42618 analytical procedure—Determination of structural carbohydrates and lignin in Biomass. *Lab. Anal. Proced.* **2012**, *17*, 1–15. Available online: http://www.nrel.gov/docs/gen/fy13/42618.pdf (accessed on 15 December 2019).
28. Shin, M. Fruit Ripening Characteristics of Kiwifurit by Exogenous Ethylene Tretment and Storage Temperature. Ph.D. Thesis, Gyeongsang National University, Jinju, Korea, 2018.
29. Shin, M.H.; Kwack, Y.B.; Kim, Y.H.; Kim, J.G. Fruit ripening and related gene expression in 'Goldone' and 'Jecy Gold' kiwifruit by exogenous ethylene application. *Hortic. Sci. Technol.* **2019**, *37*, 54–64. [CrossRef]
30. Livak, K.J.; Schmittgen, T.D. Analysis of relative gene expression data using real-time quantitative PCR and the $2^{-\Delta\Delta CT}$ method. *Methods* **2001**, *25*, 402–408. [CrossRef]
31. Macrae, E.A.; Bowen, J.H.; Stec, M.G.H. Maturation of kiwifruit (*Actinidia deliciosa* cv Hayward) from two orchards: Differences in composition of the tissue zones. *J. Sci. Food Agric.* **1989**, *47*, 401–416. [CrossRef]
32. Whetten, R.; Sederoff, R. Lignin biosynthesis. *Plant Cell* **1995**, *7*, 1001–1013. [CrossRef] [PubMed]
33. Xie, M.; Zhang, J.; Tschaplinski, T.J.; Tuskan, G.A.; Chen, J.G.; Muchero, W. Regulation of lignin biosynthesis and its role in growth-defense tradeoffs. *Front. Plant. Sci.* **2018**, *9*, 1–9. [CrossRef] [PubMed]
34. Meng, X.; Li, B.; Liu, J.; Tian, S. Physiological responses and quality attributes of table grape fruit to chitosan preharvest spray and postharvest coating during storage. *Food Chem.* **2008**, *106*, 501–508. [CrossRef]
35. Mahmood, N.; Abbasi, N.A.; Hafiz, I.A.; Ali, I.; Zakia, S. Effect of biostimulants on growth, yield and quality of bell pepper cv. Yolo Wonder. *Pak. J. Agric. Sci.* **2017**, *54*, 311–317. [CrossRef]

36. Almunqedi, B.; Kassem, H.; Almunqedhi, B.M.; Kassem, H.A.; Al-Harbi, A.E.A.R. Effect of preharvest chitosan and/or salicylic acid spray on quality and shelf-life of tomato fruits. *Artic. J. Sci. Eng. Res.* **2017**, *4*, 114–122. Available online: https://www.researchgate.net/publication/322138527 (accessed on 25 January 2020).
37. Kubo, Y. Ethylene, oxygen, carbon dioxide, and temperature in postharvest physiology. In *Abiotic Stress Biology in Horticultural Plants*; Kanayama, Y., Kochetov, A., Eds.; Springer: Berlin/Heidelberg, Germany, 2015; pp. 14–45. [CrossRef]
38. Paul, V.; Pandey, R. Role of internal atmosphere on fruit ripening and storability—A review. *J. Food Sci. Technol.* **2014**, *51*, 1223–1250. [CrossRef]
39. Rothan, C.; Nicolas, J. High CO_2 levels reduce ethylene production in kiwifruit. *Physiol. Plant.* **1994**, *92*, 1–8. [CrossRef]
40. Antunes, M.D.C.; Sfakiotakis, E.M. Ethylene biosynthesis and ripening behaviour of "Hayward" kiwifruit subjected to some controlled atmospheres. *Postharvest Biol. Technol.* **2002**, *26*, 167–179. [CrossRef]
41. De Wild, H.P.J.; Otma, E.C.; Peppelenbos, H.W. Carbon dioxide action on ethylene biosynthesis of preclimacteric and climacteric pear fruit. *J. Exp. Bot.* **2003**, *54*, 1537–1544. [CrossRef]
42. He, Y.; Bose, S.K.; Wang, W.; Jia, X.; Lu, H.; Yin, H. Pre-harvest treatment of chitosan oligosaccharides improved strawberry fruit quality. *Int. J. Mol. Sci.* **2018**, *19*, 2194. [CrossRef]
43. El-wahab, S.M.A.; El-aziz, A.A.F.A.; El-hafeez, A.E.A.A.; Emam, I.A.I. Effect of pre-harvest treatments and different cold storage temperatures on fruit quality of " Wonderful " pomegranate. *Middle East. J. Agric.* **2017**, *6*, 1057–1077.
44. Mohamed, M.A.A.; El-khalek, A.F.A.; Elmehrat, H.G. Preharvest applications of potassium silicate, chitosan and calcium chloride to improve mango cv. 'Zibda' fruit quality and storability. *Egypt J. Hortic.* **2017**, *44*, 17–32. [CrossRef]
45. Mazaro, S.M.; Deschamps, C.; De Mio, L.L.M.; Biasi, L.A.; De Gouvea, A.; Sautter, C.K. Post harvest behavior of strawberry fruits after preharvest treatment with chitosan and acibenzolar-s-methyl. *Rev. Bras. Frutic.* **2008**, *30*, 185–190. [CrossRef]
46. Tezotto-Uliana, J.V.; Fargoni, G.P.; Geerdink, G.M.; Kluge, R.A. Chitosan applications pre- or postharvest prolong raspberry shelf-life quality. *Postharvest Biol. Technol.* **2014**, *91*, 72–77. [CrossRef]
47. Meng, X.; Tian, S. Effects of preharvest application of antagonistic yeast combined with chitosan on decay and quality of harvested table grape fruit. *J. Sci. Food. Agric.* **2009**, *89*, 1838–1842. [CrossRef]
48. Saavedra, G.M.; Figueroa, N.E.; Poblete, L.A.; Cherian, S.; Figueroa, C.R. Effects of preharvest applications of methyl jasmonate and chitosan on postharvest decay, quality and chemical attributes of *Fragaria chiloensis* fruit. *Food Chem.* **2016**, *190*, 448–453. [CrossRef]
49. Reddy, M.V.B.; Belkacemi, K.; Corcuff, R.; Castaigne, F.; Arul, J. Effect of pre-harvest chitosan sprays on post-harvest infection by *Botrytis cinerea* quality of strawberry fruit. *Postharvest Biol. Technol.* **2000**, *20*, 39–51. [CrossRef]
50. Fullerton, C.G.; Prakash, R.; Ninan, A.S.; Atkinson, R.G.; Schaffer, R.J.; Hallett, I.C.; Schröder, R. Fruit from two kiwifruit genotypes with contrasting softening rates show differences in the xyloglucan and pectin domains of the cell wall. *Front. Plant Sci.* **2020**, *11*, 1–19. [CrossRef]
51. Schröder, R.; Atkinson, R.G. Kiwifruit cell walls: Towards an understanding of softening? *N. Z. J. For. Sci.* **2006**, *36*, 112–129.
52. Payasi, A.; Mishra, N.N.; Chaves, A.L.S.; Singh, R. Biochemistry of fruit softening: An overview. *Physiol. Mol. Biol. Plants* **2009**, *15*, 103–113. [CrossRef]
53. Wang, Z.Y.; MacRae, E.A.; Wright, M.A.; Bolitho, K.M.; Ross, G.S.; Atkinson, R.G. Polygalacturonase gene expression in kiwifruit: Relationship to fruit softening and ethylene production. *Plant Mol. Biol.* **2000**, *42*, 317–328. [CrossRef]
54. Yang, S.; Xu, C.; Zhang, B.; Li, X.; Chen, K. Involvement of both subgroups A and B of expansin genes in kiwifruit fruit ripening. *HortScience* **2007**, *42*, 315–319. [CrossRef]
55. Liu, Q.; Luo, L.; Zheng, L. Lignins: Biosynthesis and biological functions in plants. *Int. J. Mol. Sci.* **2018**, *19*, 335. [CrossRef]
56. Cai, C.; Xu, C.J.; Li, X.; Ferguson, I.; Chen, K.S. Accumulation of lignin in relation to change in activities of lignification enzymes in loquat fruit flesh after harvest. *Postharvest Biol. Technol.* **2006**, *40*, 163–169. [CrossRef]

Article

Optical Absorption and Scattering Properties at 900–1650 nm and Their Relationships with Soluble Solid Content and Soluble Sugars in Apple Flesh during Storage

Li Fang [1], Kangli Wei [1], Li Feng [1], Kang Tu [1], Jing Peng [1], Jiahong Wang [2] and Leiqing Pan [1,*]

[1] College of Food Science and Technology, Nanjing Agricultural University, Nanjing 210095, China; 2019108040@njau.edu.cn (L.F.); 2016108043@njau.edu.cn (K.W.); fengli@njau.edu.cn (L.F.); Kangtu@njau.edu.cn (K.T.); jpeng@njau.edu.cn (J.P.)

[2] College of Light Industry and Food Engineering, Nanjing Forestry University, Nanjing 210037, China; njfuwjh@126.com

* Correspondence: pan_leiqing@njau.edu.cn; Tel.: +86-25-8439-9016

Received: 10 November 2020; Accepted: 15 December 2020; Published: 17 December 2020

Abstract: Soluble solid content (SSC) is regarded as the most significant internal quality associated with the taste and maturity in fruits. Evaluating the relationship between the optical properties and soluble sugars facilitates exploration of the mechanism of optical techniques in SSC assessment. In this research, absorption coefficient (μ_a) and reduced scattering coefficient (μ'_s) of Fuji apple during storage were obtained using automatic integrating sphere (AIS) at 905–1650 nm. Relationships between μ_a, μ'_s and SSC, and soluble sugars contents, were investigated. The result showed that SSC, the content of total soluble sugars (TSS), fructose, glucose and sucrose were all decreasing after storage, and the same trend appeared in the change of μ_a and μ'_s. In the whole wavelength range, both μ_a and μ'_s were positively related to SSC and soluble sugars contents. The correlations between μ_a and SSC, and soluble sugars contents, showed increasing tendencies with increasing wavelengths, while for μ'_s, correlations had the opposite trend. The strongest correlations between μ_a and SSC, and soluble sugars contents, were observed in the correlation of μ_a with sucrose, with an r of 0.934. Furthermore, a partial least square (PLS) model for sucrose based on μ_a was built with the coefficient of determination of prediction (R_p^2) and the root mean square error of prediction (RMSEP) of 0.851 and 1.047, respectively. The overall results demonstrate that optical properties at the range of 905–1650 nm could be used to evaluate SSC of apples and this may due to the strong correlation between sucrose content and μ_a.

Keywords: apple flesh; absorption; scattering; soluble sugars; 905–1650 nm

---

1. Introduction

Apple fruit is highly valued by consumers for its crispy texture and sweet and sour taste. Soluble sugars in apple fruit largely affect sweetness, and they are the main components of soluble solids [1]. Compared with obtaining the content of soluble sugars, it is faster and more economical to measure soluble solids content (SSC). SSC is measured via the fruit juice. Therefore, SSC is used as a significant property for assessing the sweetness of sugars in apples, and the non-destructive evaluation of SSC by optical techniques have been investigated for many years [2–6], inclusive of near-infrared spectroscopy (NIR) and hyperspectral imaging (HSI).

Light transfer in tissues contains absorption or scattering, and the process of interaction mainly depends on the different chemical components and structures of tissues. By collecting light that is reflected back from or transmitted through the tissue based on NIR and HSI techniques, the chemical as well as structural characteristics of tissues can be acquired [7]. The prediction models for SSC can be built based on chemometrics methods by using the captured spectra and the standard measurements, and some satisfactory results have been obtained [8–12]. The calibration models can then be used to predict new samples after establishing and validating. However, it leads to limitation in illustrating the chemical and structural properties of interest from the meaningful information presented in obtained spectra due to the overarching behaviors of absorption and scattering in tissues collected by NIR and HSI techniques which cannot separate these two properties [13]. Besides, the accuracy of NIR and HSI techniques is limited, because the techniques are based on the Lambert-Beer law, which often discards the scattering effect. Furthermore, the obtained properties of reflectance or transmittance are dependent on the instrument types and light source/detecting probe setup [14]. There are, in general, inherent shortcomings with NIR and HSI techniques, which present great challenge in internal quality assessment of agri-food products.

The optical parameters measurement resulted in separate and quantized information of the absorption and scattering properties by calculating the absorption coefficient (μ_a) and reduced scattering coefficient (μ'_s) [15,16]. The current measurement techniques are developed to measure absorption and scattering properties, mainly including integrating sphere (IS) [17,18], spatially-resolved (SR) [19], time-resolved (TR) [20] and frequency-domain (FD) [21]. In addition, the techniques are based on the light transfer theory, which involves inverse adding double (IAD), Monte Carlo (MC), and diffuse approximation (DA), and the optical properties can be obtained by IS with IAD [22,23]. Such methods have opened up the opportunity to attribute internal quality parameters, both chemical composition and structure information, to optical properties of tissues for investigating the one to one relationship between them [24–27]. He et al. [28] applied an automatic integrating sphere (AIS) system combined with IAD strategies to acquiring the μ_a and μ'_s spectra of pear tissues in 400–1150 nm. Using a single integrating sphere, Wang et al. [29] estimated the absorption and scattering properties of healthy and diseased onion skin and flesh over the range of 550–880 nm and 950–1650 nm.

Many studies in Vis-NIR (400–1100 nm) have been conducted. The findings showed that μ_a around 525 and 675 nm were affected by anthocyanin and chlorophyll. In addition, μ_a at 630 nm is attributed to total galacturonic acid (GA) in residue insoluble pectin (RIP), with an r of 0.64 [30]. μ_a and μ'_s at 675 nm are noticed for area and equivalent diameter of apple tissue cells (r = 0.581–0.941) [31]. Nowadays, there are few studies on optical properties in longer wavelength ranges (above 1100 nm). However, the absorption property features above 1100 nm contain more important and obvious bonds, C-H, O-H, C-C, than the features observed in the Vis-NIR range [24,32]. Thus, enhanced sensitivity of optical properties related to tissue components can be obtained compared to the wavelength range above 1100 nm [33,34]. The increase in SSC causes an increase in the signal of absorption at 1198 nm, while water cored tissue caused a characteristic decline in light scattering coefficients [35]. However, the relationship between them has not been quantified, and the relationships between absorption and scattering properties and the compositions in SSC are also not clear yet.

Therefore, this research aimed to comprehensively clarify the mechanism of detecting SSC by optical technology and provide information for accurately developing improved SSC detectors of Fuji apple during storage by optical technology. In the work, the μ_a and μ'_s of apple fleshes in longer wavelength range NIR (905–1650 nm) were calculated based on the AIS system with the IAD method, and internal quality (SSC and the contents of soluble sugars) was measured. Finally, relationships between μ_a, μ'_s and SSC, the contents of total soluble sugar (TSS), fructose, glucose and sucrose were investigated, then the potential for predicting the SSC was evaluated.

2. Materials and Methods

2.1. Sample Collection and Storage

The Fuji apples were provided by an orchard in Xuzhou, Jiangsu Province, China. In total, 60 apple samples were stored at 0 ± 1 °C with relative humidity (RH 95%) in normal atmosphere

(NA) for 150 days. The measurement took place during 0–150 days, at 30-day intervals and in total 10 individual apples were picked and measured randomly after placing at room temperature for 5 h on each occasion. Another 60 samples were stored at 25 °C with RH 95% in NA for 50 days, and 10 apples were picked from 0 day at about 10-day intervals up to 50 days.

2.2. Optical Properties

The automatic integrating sphere system with IAD was employed to measure the μ_a and μ'_s. The diffuse reflectance and diffuse transmittance of the samples, reference and dark noise were needed to calculate μ_a and μ'_s, and the spectral signals (905–1650 nm) were collected by a spectrometer (SW2520, Isuzu Optics, Taiwan, China). Different measurement configurations were presented in Figure 1, while specific components of instruments, and measurements' detailed descriptions, are presented by Wei et al. [36] and Prahl [37].

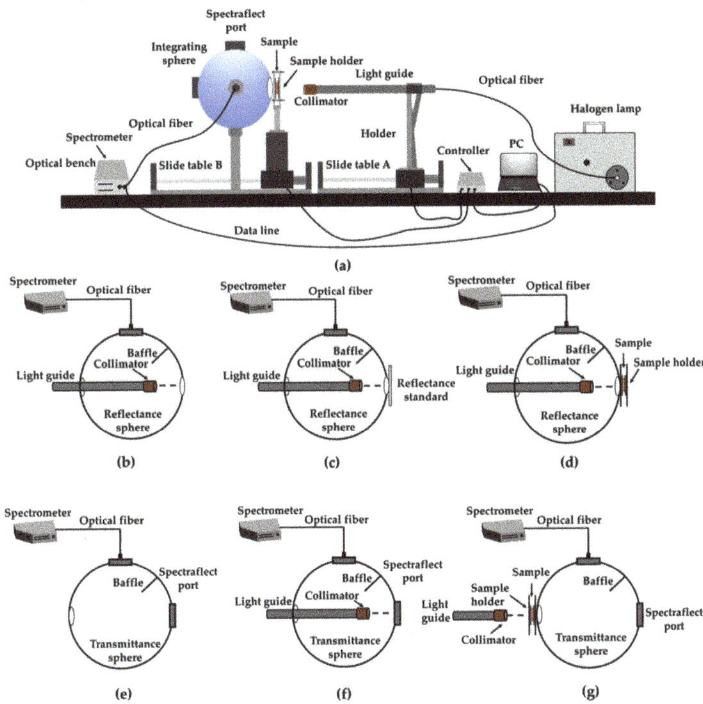

Figure 1. Illustration of the setup of automatic single integrating sphere (**a**), and the schematics for measurements on the reflectance of dark noise R_d (**b**), reference R_r (**c**) and sample R_s (**d**), the transmittance of dark noise T_d (**e**), reference T_r (**f**) and sample T_s (**g**).

The accuracy of the estimated μ_a and μ'_s was validated with both pure water and 1% Intralipid® (Sigma Aldrich, St. Louis, MO, USA). The main ingredients were soybean oil, glycerin and egg yolk phospholipids. In addition, the particle size distribution in the solution was relatively uniform and stable, also very close to the main scattering particles (chylomicrons) in biological tissues. Thus, it was taken as standard scattering resolution. The scattering resolution was established by diluting 20% stock solution with water. The reference μ_a and μ'_s values were described by Van et al. [38] and Deng et al. [39]. The performance of this methodology was validated and the results were reported by Wei et al. [36].

According to the Monte Carlo simulation of Ma et al. [35], the penetration depth of light at 1198 nm is 3.3 mm in the whole tissue of Fuji apple. To ensure the penetration effect of light and the sample volume needed for chemical detection, two 2.5 mm thick flesh slices were cut from each sample apple in the opposite side. The samples were then cut into slices with 2.5 mm thickness × 30 mm width × 35 mm length and the thickness of each slice was measured by using digital calipers. The sample slices were then sandwiched with two pieces of quartz glasses to measure the spectral signals. The specific description of measurements can be found in Wei et al. [36] and Ma et al. [40].

2.3. SSC and Soluble Sugars Contents Measurements

The juice of the fresh slices was obtained to assess the SSC by a hand-held refractometer (PAL-1, ATAGO, Tokyo, Japan) after spectral measurement. The measurement range, resolution and accuracy of the instrument were 0.0–53.0% °Brix, 0.1% °Brix and ±0.2% °Brix, respectively. In addition, fructose, glucose, and sucrose contents were quantified by high performance liquid chromatography (HPLC) as described further in Wei et al. [36]. For each sample, two grams of apple flesh homogenate was taken in a centrifuge tube with 30 mL of distilled water in a water bath at 80 °C and centrifuged at 12,000 r/min for 20 min. Supernatants were collected and diluted five-fold., and the solution was filtered through a 0.45 μm filter. Sugars were separated by a Zorbax carbohydrate 70-A column (Agilent, Santa Clara, CA, USA) using a solvent of acetonitrile:water (75:25, *v/v*). The flow rate was 0.8 mL/min at 40 °C. Different concentrations of the sugar standard solution were made for glucose, fructose and sucrose, then the content of the three sugars in the samples was established by comparing each peak retention time and peak area with those of the standard.

2.4. Statistical Analysis

All the quantitative comparations between the different soluble sugars of the examined apple flesh were subjected to Variance analysis, which were tested for $p < 0.05$ using the SPSS Statistics 20 package (IBM). The correlation between the contents of soluble sugars and soluble solids was investigated using the Pearson method of analysis with significant $p < 0.05$ in IBM SPSS 20. The partial least square regression (PLS) models of SSC, TSS and soluble sugars contents were built based on μ_a and μ'_s separately using MATLAB R2010b® in order to choose the most important soluble sugar parameter that determined optical properties of apple flesh. Finally, the model performance was evaluated by the determination coefficient of prediction ($R_p{}^2$) with the root mean square error of prediction (*RMSEP*), and the determination coefficient of calibration ($R_c{}^2$) with the root mean square error of calibration (*RMSEC*).

3. Results and Discussion

3.1. Validation Results of Optical Properties

The system in Figure 2a shows the curves of μ_a of water obtained by AIS and μ_a reported by Deng et al. [39]. The results are in agreement with previous reports, which find that there are three absorption peaks at 1485, 1169 and 980 nm. In addition, the relative error of μ_a values at 905–1600 nm ranged from 0.03% to 10.25%. Figure 2b presents the curve of μ'_s values of 1% Intralipid based on AIS. A comparison of the measured μ'_s and reference value based on Mie theory revealed that both profiles showed a downward trend with the increase in wavelengths [38], and the mean relative error was 9.20%. However, the relative error of μ_a and μ'_s after 1600 nm clearly increases. This is due to a low signal-to-noise ratio of the spectrometer, or more detected interference signals. Therefore, μ_a and μ'_s at 905–1600 nm were taken as the effective data in this study.

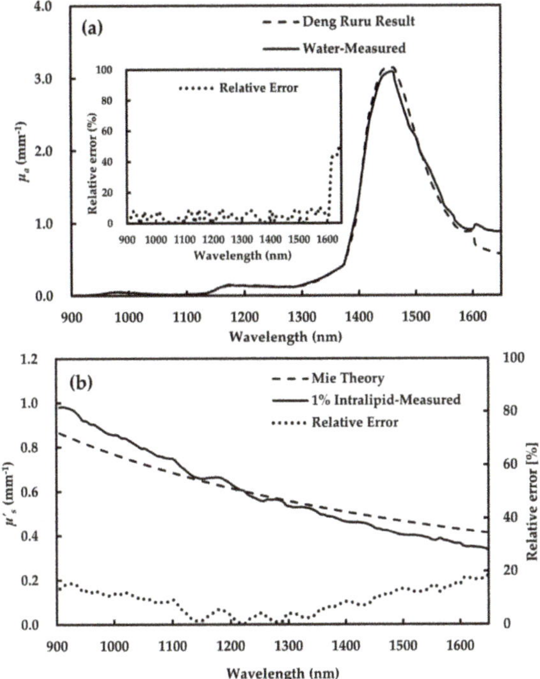

Figure 2. Absorption (μ_a) estimation on water (**a**) and scattering (μ'_s) estimation on 1% Intralipid (**b**).

3.2. Spectral Analysis

3.2.1. Absorption Coefficient

The results obtained from the mean μ_a spectra of apple samples storage at 25 and 0 °C are set out in Figure 3a,b. The graphs show that there has been a similar spectral trend, peaking at 980, 1169 and 1485 nm, which corresponds to the vibration absorption of O-H, C-H and C-C [41]. In addition, the absorption at 1485 nm was significantly stronger than that at 980 and 1169 nm, with values (1.52–1.88 mm^{-1}) being 10 times higher than that of the others. In addition, the μ_a value peaked at 1485 nm when the apple was stored at 25 °C after 10 days. There was no significant difference between day 0 and day 20, and μ_a decreased significantly on the 30th, 40th and 50th days. At 0 °C, μ_a at 1485 nm showed a similar trend with that at 25 °C, with the largest value on the 30th day during storage.

The μ_a values of apple flesh stored at 25 and 0 °C were within 0.04 and 1.88, and within 0.05 and 1.67 mm^{-1}, respectively. Generally, the μ_a values in this study were comparable to those reported in other research. In the study by Ma et al. [35], the μ_a values of apple samples were 0.00–0.50 mm^{-1}, measured by HIS at 1000–1800 nm. The research of Saeys et al. reported the μ_a values of Royal Gala (0.00–2.20 mm^{-1}) based on IS at 350–2200 nm [33]. An important issue that emerged from the data is the dramatic difference in absorption values of apple flesh measured by different techniques. The μ_a obtained using IS is much higher than that with HIS. This may be due to the different sample tissue state, different varieties of apples, different wavelength ranges as well as different measured method in the absorption property [42].

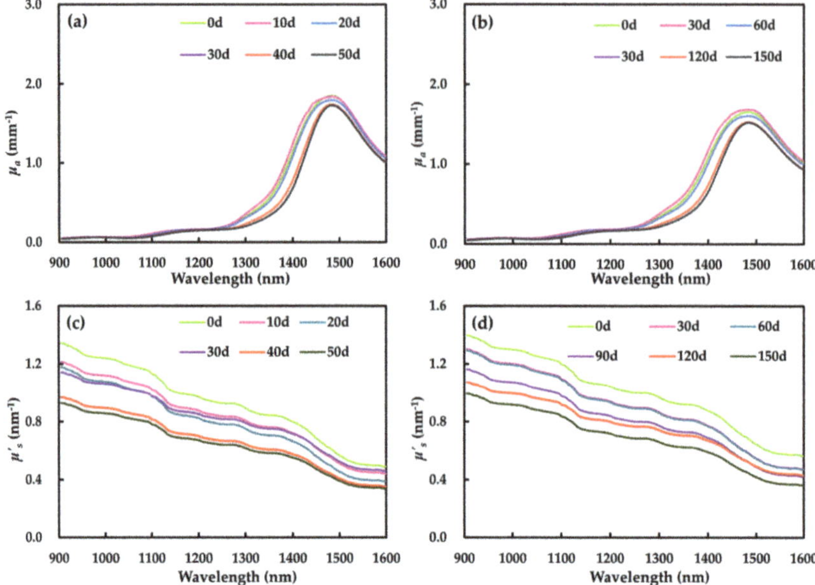

Figure 3. Average optical properties of Fuji apple flesh for six storage times: absorption (μ_a) and scattering (μ'_s) at 25 °C (**a**,**c**); absorption (μ_a) and scattering (μ'_s) at 0 °C (**b**,**d**). d, day.

3.2.2. Reduced Scattering Coefficient

Compared with absorption coefficient, the reduced scattering coefficient displays different changing patterns in the spectral region of 905–1600 nm (Figure 3c,d). The μ'_s spectral was flatter and steadily decreased with increasing wavelength. In addition, the μ'_s spectral changed consistently as storage time progressed, which was in accordance with studies of He et al. [28] and Cen et al. [31]. The scattering coefficient at 905–1600 nm was in 0.34–1.40 mm^{-1}, which was smaller than μ'_s of Royal Gala (1.00–1.50 mm^{-1}) [33]. This may be related to different apple cultivars [43]. Within the short wave near infrared region (905–1100 nm), the scattering property was still stronger than the absorption property. μ'_s was more than seven times the value of μ_a, ranging from 0.80 to 1.40 mm^{-1} at 25 °C, and from 0.04 to 0.11 mm^{-1} at 0 °C. This result was also reported in Wei et al. [36]. In the range of medium wave near infrared (1100–1600 nm), the scattering property gradually declined, and the absorption became the dominant optical property as the wavelength increased [33].

3.3. SSC and Soluble Sugars Contents

3.3.1. Changes of Contents of Soluble Solid and Soluble Sugars

At 25 and 0 °C, SSC showed the trend of increasing before decreasing, and separately reached the maximum value on the 10th and 30th day (Figure 4). As shown in Figure 5, the TSS, fructose and glucose contents of flesh increased significantly due to starch hydrolysis, while the sucrose content remained at a high level, with no significant change in the early stage of storage at 25 and 0 °C [44]. This was also consistent with the phenomenon that the fruit will turn sweet during the storage after harvest. After storage, SSC declined significantly ($p < 0.05$), from 13.85% to 11.26% at 25 °C and from 13.97% to 12.26% at 0 °C.

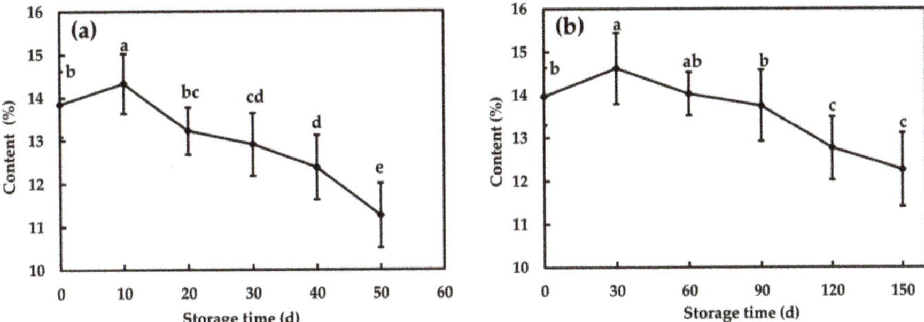

Figure 4. Soluble solid content (SSC) of Fuji apple flesh for six storage times at 25 °C (**a**) and 0 °C (**b**). Data at six storage time with different letters are significantly different with $p < 0.05$. The vertical bars denote standard deviations. d, day.

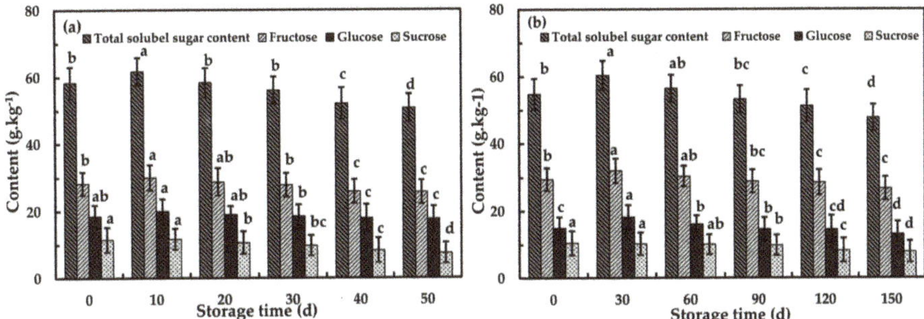

Figure 5. Soluble sugars contents of Fuji apple flesh for six storage times at 25 °C (**a**) and 0 °C (**b**). Columns with different letters for the same cultivar are significantly different with $p < 0.05$. Standard error is indicated. d, day.

Increasing with storage time, the contents of TSS, fructose, glucose and sucrose gradually declined, and reached the lowest value at the end of storage. At 25 °C, the contents of TSS, fructose and glucose reached the highest value on the 10th day. After 50 days storage, the contents of TSS, fructose and glucose were reduced to 50.72, 25.67 and 17.62 g kg^{-1}, respectively. In addition, the content of sucrose decreased from 11.52 to 7.44 g kg^{-1}. At 0 °C, the TSS, fructose and glucose contents reached the highest value at a slower rate than at 25 °C, and reached the peak on the 30th day, with levels of 60.34, 31.95 and 18.14 g kg^{-1}, respectively. On the 150th day, the TSS, fructose and glucose contents were the lowest, 47.62, 26.66 and 13.09 g kg^{-1}, respectively. The content of sucrose did not change significantly during 60-day storage and reached the minimum value of 7.86 g kg^{-1} on the 150th day.

3.3.2. Relationships among SSC and Soluble Sugar Contents

Relationships among the contents of TSS, fructose, glucose, sucrose and SSC were measured on a one to one basis, and SSC presented a positive association profile with TSS, fructose, glucose and sucrose during storage at 25 and 0 °C ($p < 0.05$) in Tables 1 and 2. It is clear that among TSS and the three soluble sugars, sucrose had the strongest association with the changes of SSC with high corresponding r values of 0.973 at 25 °C and 0.963 at 0 °C. Other soluble sugars had lower contributions to the correlations with SSC, shown as a second relation with TSS and fructose, and finally glucose. The results were consistent with the report by Wei et al. [36]. It should be noted that significant positive correlations were found between TSS with three types of soluble sugars ($r > 0.847$), and fructose contributed the most to the correlation with TSS, with r values of 0.974 at 25 °C and 0.975 at 0 °C. This could be expected as the content of fructose comprised the largest fraction of the TSS content.

Table 1. Pearson's linear correlation coefficients among SSC and soluble sugars for 50 days at 25 °C.

	SSC	TSS	Fructose	Glucose	Sucrose
SSC	1	0.956 **	0.921 **	0.868 *	0.973 **
TSS	\	1	0.974 **	0.942 **	0.921 **
Fructose	\	\	1	0.957 **	0.943 **
Glucose	\	\	\	1	0.841 *
Sucrose	\	\	\	\	1

* $p < 0.05$, ** $p < 0.01$, '\': without value, $n = 60$ individual apples; soluble solid content (SSC), total soluble sugars (TSS).

Table 2. Pearson's linear correlation coefficients among SS and soluble sugars for 150 days at 0 °C.

	SSC	TSS	Fructose	Glucose	Sucrose
SSC	1	0.963 **	0.933 **	0.870 *	0.950 **
TSS	\	1	0.975 **	0.967 **	0.847 *
Fructose	\	\	1	0.982 **	0.795
Glucose	\	\	\	1	0.685
Sucrose	\	\	\	\	1

* $p < 0.05$, ** $p < 0.01$, '\': without value, $n = 60$ individual apples.

3.4. Optical Properties–SSC–Soluble Sugars Contents Relations

3.4.1. Absorption Coefficient–SSC–Soluble Sugars Contents Relations

The r values between the absorption coefficient of apple flesh at 905–1600 nm and SSC, TSS content and soluble sugars contents at 25 and 0 °C are shown in Figure 6a,b, respectively. The μ_a was positively associated with SSC, TSS content and soluble sugars contents in the whole range, exhibiting high r values of 0.873–0.960 at 25 °C and 0.863–0.963 at 0 °C. A previous study reported a similar result, specifically that SSC is positively correlated with μ_a at 1198 nm [35]. At 905–1600 nm, the r profile between μ_a and the SSC, TSS content and soluble sugars contents had similar patterns, with almost increasing tendencies with increasing wavelengths. Therefore, from Figure 6a,b, we could observe that SSC and soluble sugars presented stronger correlation in the medium-wave near infrared region than in the short-wave near infrared region. Compared to the report of Wei et al. [36], the average r values μ_a and the SSC and TSS content and soluble sugars contents at 905–1050 nm were different. This may be due to different samples used in the two studies and different rates of accumulation or decomposition of soluble sugars. However, due to increasing response to the content information of chemical components, the correlation between absorption properties and soluble sugars increases with the increasing of the wavelength in the whole visible-medium wave near infrared region in general [45].

Overall, SSC ($r = 0.933$) and sucrose ($r = 0.934$) had the highest average correlation coefficients with μ_a, followed by TSS ($r = 0.928$), and finally fructose ($r = 0.915$) and glucose ($r = 0.899$) at 25 °C. The phenomena may be connected to the strong correlation between SSC and sucrose content. At 0 °C, r values of μ_a and the SSC, TSS content and soluble sugars contents ranged from 0.890 to 0.930. The result also showed that μ_a peaked at 980, 1169 and 1485 nm, and the correlation at the three peaks increased successively, with r ranges of 0.881–0.922, 0.894–0.938 and 0.913–0.961, which was in good agreement with the successively stronger spectral signal. To further investigate the correlations with SSC and sucrose content, Table 3 plotted the direct linear correlations of μ_a at 980, 1169 and 1485 nm with SSC and sucrose content (25 °C). The coefficients of determination (R^2) of the equations were between 0.848 and 0.921. Results from the equations indicate that SSC and sucrose content gave good linear correlations with μ_a, with an R^2 of 0.915 and 0.921, respectively, which confirms the results shown in Figure 6a.

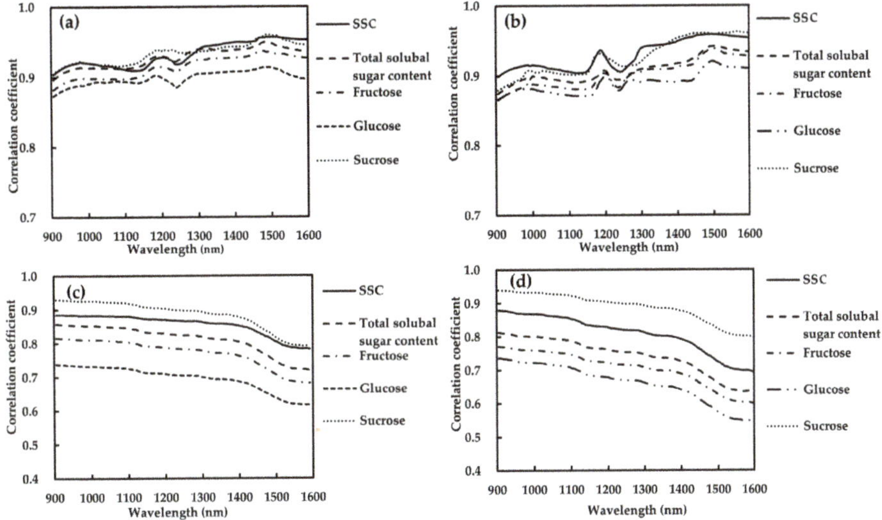

Figure 6. Correlations coefficient of μ_a and μ'_s with content of soluble solid (SSC), total soluble sugars (TSS) and three types of soluble sugars of apples at 25 °C (**a**,**c**) and 0 °C (**b**,**d**).

Table 3. Linear fitting equations of soluble solid contents, sucrose contents and the absorption coefficients at 980, 1189 and 1498 nm for apple flesh.

Parameters (Independent Variable x)	μ_a (Dependent Variable y)	Linear Fitting Equation	R^2
SSC	980 nm	$y = 0.0021x + 0.0383$	0.848
	1189 nm	$y = 0.0053x + 0.1039$	0.859
	1498 nm	$y = 0.0581x + 1.0107$	0.915
Sucrose	980 nm	$y = 0.0022x + 0.0361$	0.850
	1189 nm	$y = 0.005x + 0.1064$	0.880
	1498 nm	$y = 0.036x + 1.4102$	0.921

3.4.2. Scattering Coefficient–SSC–Soluble Sugars Contents Relations

The values of r between μ_a and either SSC or contents of soluble sugars, when compared with the respective r between μ'_s and the same variables, exhibited different patterns; μ'_s decreased gradually with the increase in wavelength (Figure 6c,d) which may be related to the decreased values of μ'_s over the whole investigated wavelength range. It was also observed that soluble solid and soluble sugars were positively correlated with scattering properties. At 25 °C, the average r values of SSC, TSS fructose, glucose and sucrose with μ'_s were 0.855, 0.811, 0.771, 0.695 and 0.884, respectively. At 0 °C, the average values of r were 0.808, 0.742, 0.704, 0.659 and 0.887, respectively. Moreover, SSC, TSS, fructose and sucrose have higher correlation with μ'_s than glucose. Among the three types of soluble sugars, μ'_s was most related to sucrose (r = 0.789–0.939), followed by fructose (r = 0.599–0.816) and glucose (r = 0.544–0.738). The same result was found in the correlation of SSC and μ_a with the three soluble sugars. The findings proved that the sucrose contributed the most to the strong relationship between optical properties and SSC.

According to the results of Sections 3.4.1 and 3.4.2, the correlation between soluble sugars and μ_a was stronger than that between soluble sugars and μ'_s in the whole visible-medium wave near infrared region. Besides, with the increase in wavelength, the correlation between soluble sugars and μ_a increased, while the correlation between soluble sugars and μ'_s decreased. It could be inferred that μ_a was more associated with chemical components, while μ'_s was less correlated with them [7,30,36,46].

Furthermore, previous study presented that scattering property was influenced by microstructure. Rowe et al. [26] measured the optical spectra of apples over the range of 400–1050 nm using IS and obtained an average r of–0.68 between the μ'_s and firmness. Liu et al. showed that μ'_s of kiwifruit had stronger correlation with firmness (r = 0.74–0.87) than with SSC (r = 0.49–0.81) and moisture content (r = 0.59–0.80) [7]. These studies indicated that the absorption and scattering properties had potential to determine the internal quality of apple.

3.5. SSC and Soluble Sugars Contents Predictions

Table 4 generalized the statistics of PLS model results of calibration and prediction using μ_a and μ'_s at 905–1600 nm for SSC and the contents of soluble sugars of Fuji apple. The R_c^2 of the modeling set was higher than the R_p^2 of the prediction set, thus these models were reliable. Among the three prediction models of soluble sugars based on μ_a and μ'_s, the best model was for sucrose content ($R_p^2 > 0.7365$, $RMSEP < 1.458$ g/kg). In addition, the models relied on μ_a, and the μ'_s for fructose and glucose content were slightly worse, the R_p^2 values of which were in the range of 0.635–0.781 and 0.593–0.725, and the *RMSEP* values were 1.570–2.216 g/kg and 1.344–1.906 g/kg. A previous study reported similar results in that the prediction results on optical properties for sucrose were better than fructose and glucose [36].

Table 4. Model results of soluble solid contents and soluble sugar contents for apples.

Parameters	Optical Property	Calibration		Prediction	
		R_c^2	*RMSEC*	R_p^2	*RMSEP*
SSC	μ_a	0.851	0.314	0.833	0.329
	μ'_s	0.739	0.693	0.726	0.722
Total soluble sugar	μ_a	0.837	2.055	0.830	2.062
	μ'_s	0.710	3.657	0.701	3.699
Fructose	μ_a	0.782	1.567	0.781	1.570
	μ'_s	0.658	1.935	0.635	2.216
Glucose	μ_a	0.769	1.338	0.752	1.344
	μ'_s	0.611	1.821	0.593	1.906
Sucrose	μ_a	0.860	1.032	0.851	1.047
	μ'_s	0.744	1.437	0.736	1.458

The root mean square error of calibration (*RMSEC*); the coefficient of determination of prediction (R_p^2); the root mean square error of prediction (RMSEP).

In addition, the R_p^2 values of the five prediction models based on μ_a were in the range of 0.752–0.851, and the *RMSEP* values of the SSC and soluble sugar models were 0.329% and 1.047–2.062 g/kg, respectively. Li et al. [47] assessed the SSC of apple based on near infrared spectroscopy, with the results of R_p^2 and RMSEP in the ranges of 0.576–0.854 and 0.547–0.673%, and the prediction accuracy of SSC was equivalent to the present study. In addition, according to the model results (Table 4), the prediction results based on μ_a were better than the model results based on μ'_s, which further proved that SSC and soluble sugars were more correlated with μ_a than μ'_s. The findings obtained from this study made a contribution to the fact the light absorption links to the chemical compositions of materials, and that scattering of photons is a physical procedure that greatly depends on the wave characteristics of photons and the structures of mediums.

In general, the prediction models for sucrose based on μ_a in this study performed well ($R_p^2 = 0.851$, $RMSEP = 1.047$). The models for sucrose in 400–1050 nm built with the AIS system, according to the report of Wei et al. [36], were second to the models of our study. According to their study, the models' R_p^2 and *RMSEP* values were 0.804 and 1.099, respectively. The result might be caused by the presence of the more obvious bonds C-H, O-H, C-C with a longer wavelength, which indicated that the absorption properties of the mid-wave near infrared region are more suitable for the quantitative detection of soluble solids and soluble sugars.

4. Conclusions

The optical properties of Fuji apple during storage in two different storage conditions (25 °C, 50 days; 0 °C, 150 days) were evaluated by the AIS system at 905–1600 nm, and the exact relationships of absorption and scattering properties with SSC and soluble sugars contents were explored. The changes of μ_a, μ'_s, SSC and soluble sugars contents showed a similar pattern at the two experimented temperatures during storage. Three prominent absorption peaks at 980, 1169 and 1485 nm were noticed for the spectra of μ_a. In addition, μ_a and μ'_s were positively associated with SSC and soluble sugars contents (r = 0.659–0.936), and their values are greatly relied on the wavelength. Thus, it could be easily understood that the decreases in SSC and soluble sugar content were accompanied by declines in μ_a and μ'_s. Additionally, the μ_a had stronger correlation with these soluble components than with μ'_s. Another important conclusion was that both SSC and μ_a were more strongly correlated with sucrose content, with r values of 0.933, 0.934 (at 25 °C) and 0.930, 0.927 (at 0 °C). Moreover, partial least square regression (PLS) models for SSC and soluble sugars contents were developed, with the best performance of $R_p^2 = 0.851$ and $RMSEP = 1.047$, which appeared in the model of sucrose based on μ_a. Furthermore, due to better measuring bonds of C-H, O-H, C-C at the longer wavelength range, the prediction model at 905–1600 nm performed better than the models developed in the 400–1050 nm range in the report of Wei et al. [36]. Thus, these results made the mechanism for detecting SSC of apples based on optical techniques clear. The strong association between the absorption properties and the sucrose content provided potential and theoretical support in assessing SSC based on optical techniques.

However, this study was carried out on a relatively limited number of fruit and only on apples from one cultivar. Further work needs to involve the investigation of exact relationships between absorption and scattering properties and soluble sugars contents of other varieties of apple flesh, as well as the correlation of optical properties with other components which contribute to quality, such as to firmness.

Author Contributions: Conceptualization, K.T. and L.P.; methodology and data curation, K.W. and L.F. (Li Fang); formal analysis, J.P., L.F. (Li Feng) and J.W.; investigation, K.W.; writing—original draft preparation, L.F. (Li Fang); writing—review and editing, L.P. and K.T.; visualization, L.F. (Li Feng) and J.W.; supervision, K.T. and J.P.; project administration, K.T. and L.P.; funding acquisition, L.P. All authors have read and agreed to the published version of the manuscript.

Funding: This research was funded by Chinese National Foundation of Nature and Science (31671926), Fundamental Research Funds for the Central Universities (KYLH202003) and the Priority Academic Program Development of Jiangsu Higher Education Institutions (PAPD).

Conflicts of Interest: The authors declare no conflict of interest.

References

1. Li, M.J.; Feng, F.J.; Cheng, L.L. Expression patterns of genes involved in sugar metabolism and accumulation during apple fruit development. *PLoS ONE* **2012**, *7*, 33055. [CrossRef] [PubMed]
2. Wei, X.; He, J.; Zheng, S.; Ye, D. Modeling for SSC and firmness detection of persimmon based on NIR hyperspectral imaging by sample partitioning and variables selection. *Infrared Phys. Technol.* **2019**, 103099. [CrossRef]
3. Li, J.; Zhang, H.; Zhan, B.; Wang, Z.; Jiang, Y. Determination of SSC in pears by establishing the multi-cultivar models based on visible-NIR spectroscopy. *Infrared Phys. Technol.* **2019**, *102*, 103006. [CrossRef]
4. Fan, S.; Wang, Q.; Tian, X.; Yang, G.; Xia, Y.; Li, J.; Huang, W. Non-destructive evaluation of soluble solids content of apples using a developed portable Vis/NIR device. *Biosyst. Eng.* **2020**, *198*, 138–148. [CrossRef]
5. Guo, Z.; Wang, M.M.; Agyekum, A.A.; Wu, J.J.; Chen, Q.S.; Zuo, M. Quantitative detection of apple watercore and soluble solids content by near infrared transmittance spectroscopy. *J. Food Eng.* **2020**, *279*, 109955. [CrossRef]
6. Sun, Y.; Wei, K.L.; Liu, Q.; Pan, L.; Tu, K. Classification and discrimination of different fungal diseases of three infection levels on peaches using hyperspectral reflectance imaging analysis. *Sensors* **2018**, *14*, 1295. [CrossRef]

7. Liu, Q.; Wei, K.L.; Xiao, H.; Tu, S.C.; Sun, K.; Sun, Y.; Pan, L.Q.; Tu, K. Near-infrared hyperspectral imaging rapidly detects the decay of postharvest strawberry based on water-soluble sugar analysis. *Food Anal. Methods* **2019**, *12*, 936–946. [CrossRef]
8. Valero, C.; Ruizaltisent, M.; Cubeddu, R.; Pifferi, A.; Taroni, P.; Torricelli, A.; Valentini, G.; Johnson, D.S.; Dover, C.J. Detection of internal quality in kiwi with time-domain diffuse reflectance spectroscopy. *Appl. Eng. Agric.* **2004**, *20*, 223–232. [CrossRef]
9. Oliveira-Folador, G.; Bicudo, M.D.O.; de Andrade, E.F.; Renard, C.M.-G.C.; Bureau, S.; de Castilhos, F. Quality traits prediction of the passion fruit pulp using NIR and MIR spectroscopy. *LWT Food Sci. Technol.* **2018**, *95*, 172–178. [CrossRef]
10. Adebayo, S.E.; Hashim, N.; Abdan, K.; Hanafi, M.; Mollazade, K. Prediction of quality attributes and ripeness classification of bananas using optical properties. *Hortic. Sci.* **2016**, *212*, 171–182. [CrossRef]
11. Mo, C.Y.; Kim, M.S.; Kim, G.; Lim, J.; Delwiche, S.R.; Chao, K.L.; Li, H.; Cho, B.-K. Spatial assessment of soluble solid contents on apple slices using hyperspectral imaging. *Biosyst. Eng.* **2017**, *159*, 10–21. [CrossRef]
12. Pan, L.Q.; Lu, R.F.; Zhu, Q.B.; McGrath, J.M.; Tu, K. Measurement of moisture, soluble solids, sucrose content and mechanical properties in sugar beet using portable visible and near-infrared spectroscopy. *Postharvest Biol. Technol.* **2015**, *102*, 42–50. [CrossRef]
13. Xie, D.D.; Guo, W.C. Measurement and Calculation Methods on Absorption and Scattering Properties of Turbid Food in Vis/NIR Range. *Food Bioprocess Technol.* **2020**, *13*, 229–244. [CrossRef]
14. Lu, R.F.; Van, B.R.; Saeys, W.; Li, C.; Cen, H. Measurement of optical properties of fruits and vegetables: A review. *Postharvest Biol. Technol.* **2020**, *159*, 111003. [CrossRef]
15. Hu, D.; Fu, X.P.; Wang, A.C.; Ying, Y.B. Measurement methods for optical absorption and scattering properties of fruits and vegetables. *Trans. ASABE* **2015**, *58*, 1387–1401. [CrossRef]
16. Tuchin, V. *Tissue Optics: Light Scattering Methods and Instruments for Medical Diagnosis*; SPIE Press: Bellingham, WA, USA, 2007.
17. Pickering, J.W. Double-integrating-sphere system for measuring the optical properties of tissue. *Appl. Opt.* **1993**, *32*, 399–410. [CrossRef]
18. López Maestresalas, A.; Aernouts, B.; Van Beers, R.; Arazuri, S.; Jarén, C.; De Baerdemaeker, J.; Saeys, W. Bulk optical properties of potato flesh in the 500–1900 nm range. *Food Bioprocess Technol.* **2015**, *9*, 463–470. [CrossRef]
19. Nguyen, N.D.T.; Erkinbaev, C.; Tsuta, M.; De Baerdemaeker, J.; Nicolaï, B.; Saeys, W. Spatially resolved diffuse reflectance in the visible and near-infrared wavelength range for non-destructive quality assessment of 'Braeburn' apples. *Postharvest Biol. Technol.* **2014**, *91*, 39–48. [CrossRef]
20. Vanoli, M.; Rizzolo, A.; Grassi, M.; Farina, A.; Pifferi, A.; Spinelli, L.; Torricelli, A. Time-resolved reflectance spectroscopy nondestructively reveals structural changes in 'Pink Lady®' apples during storage. *Procedia Food Sci.* **2011**, *1*, 81–89. [CrossRef]
21. He, X.M.; Fu, X.P.; Li, T.W.; Rao, X.Q. Spatial frequency domain imaging for detecting bruises of pears. *J. Food Meas. Charact.* **2018**, *12*, 1266–1273. [CrossRef]
22. Prahl, S.A.; Martin, J.C.; Germet, V.; Welch, A.J. Determining the optical properties of turbid media by using the adding-doubling method. *Appl. Opt.* **1993**, *32*, 559–568. [CrossRef] [PubMed]
23. Wang, L.H.; Jacques, S.L.; Zheng, L.Q. Conv-convolution for responses to a finite diameter photon beam incident on multi-layered tissues. *Comput. Meth. Prog. Biol.* **1997**, *54*, 141–150. [CrossRef]
24. Zhang, M.Y.; Li, C.Y.; Yang, F.Z. Optical properties of blueberry flesh and skin and Monte Carlo multi-layered simulation of light interaction with fruit tissues. *Postharvest Biol. Technol.* **2019**, *150*, 28–41. [CrossRef]
25. Huang, Y.P.; Lu, R.F.; Chen, K. Assessment of Tomato Soluble Solids Content and pH by Spatially-Resolved and Conventional Vis/NIR Spectroscopy. *J. Food Eng.* **2018**, *236*, 19–28. [CrossRef]
26. Rowe, P.I.; Künnemeyer, R.; McGlone, A.; Talele, S.; Martinsen, P.; Seelye, R. Relationship between tissue firmness and optical properties of 'Royal Gala' apples from 400 to 1050 nm. *Postharvest Biol. Technol.* **2014**, *94*, 89–96. [CrossRef]

27. Zhang, S.; Wu, X.H.; Zhang, S.H.; Cheng, Q.L.; Tan, Z.J. An effective method to inspect and classify the bruising degree of apples based on the optical properties. *Postharvest Biol. Technol.* **2017**, *127*, 44–52. [CrossRef]
28. He, X.M.; Fu, X.P.; Rao, X.Q.; Fang, Z.H. Assessing firmness and SSC of pears based on absorption and scattering properties using an automatic integrating sphere system from 400 to 1150 nm. *Postharvest Biol. Technol.* **2016**, *121*, 62–70. [CrossRef]
29. Wang, W.; Li, C.; Gitaitis, R. Optical properties of healthy and diseased onion tissues in the visible and near-infrared spectral region. *Trans. ASABE* **2014**, *57*, 1771–1782. [CrossRef]
30. Vanoli, M.; Zerbini, P.; Spinelli, L.; Torricelli, A.; Rizzolo, A. Polyuronide content and correlation to optical properties measured by time-resolved reflectance spectroscopy in 'Jonagored' apples stored in normal and controlled atmosphere. *Food Chem.* **2009**, *115*, 1450–1457. [CrossRef]
31. Cen, H.Y.; Lu, R.F.; Mendoza, F.A.; Beaudry, R.M. Relationship of the optical absorption and scattering properties with mechanical and structural properties of apple tissue. *Postharvest Biol. Technol.* **2013**, *85*, 30–38. [CrossRef]
32. Hjalmarsson, P.; Thennadil, S.N. Spatially resolved in vivo measurement system for estimating the optical properties of tissue in the wavelength range 1000–1700 nm. In *Proceedings of SPIE the International Society for Optical Engineering*; Schweitzer, D., Fitzmaurice, M., Eds.; SPIE: Bellingham, DC, USA, 2007; Volume 6628, p. 662805.
33. Saeys, W.; Velazco-Roa, M.A.; Thennadil, S.N.; Ramon, H.; Nicolaï, B.M. Optical properties of apple skin and flesh in the wavelength range from 350 to 2200 nm. *Appl. Opt.* **2008**, *47*, 908–919. [CrossRef] [PubMed]
34. Wilson, R.H.; Nadeau, K.P.; Jaworski, F.B.; Tromberg, B.J.; Durkin, A.J. Review of short-wave infrared spectroscopy and imaging methods for biological tissue characterization. *J. Biomed. Opt.* **2015**, *20*, 30901. [CrossRef] [PubMed]
35. Ma, T.; Li, X.Z.; Inagaki, T.; Yang, H.; Tsuchikawa, S. Noncontact evaluation of soluble solids content in apples by near-infrared hyperspectral imaging. *J. Food Eng.* **2018**, *224*, 53–61. [CrossRef]
36. Wei, K.L.; Ma, C.; Sun, K.; Liu, Q.; Zhao, N.; Sun, Y.; Tu, K.; Pan, L. Relationship between optical properties and soluble sugar contents of apple flesh during storage. *Postharvest Biol. Technol.* **2020**, *159*, 111021. [CrossRef]
37. Prahl, S.A. Everything I Think You Should Know about Inverse Adding-Doubling. Available online: https://omlc.ogi.edu/software/iad (accessed on 12 March 2019).
38. Van Staveren, H.J.; Moes, C.J.; Van, M.J. Light scattering in Intralipid-10% in wavelength range of 400–1100 nm. *Appl. Opt.* **1991**, *30*, 4507–4514. [CrossRef]
39. Deng, R.R.; He, Y.Q.; Qin, Y.; Chen, Q.D. Measuring pure water absorption coefficient in the near-infrared spectrum (900–2500 nm). *J. Remote Sens.* **2012**, *16*, 192–206. [CrossRef]
40. Ma, C.; Feng, L.; Pan, L.; Wei, K.; Liu, Q.; Tu, K.; Zhao, L.; Peng, J. Relationships between optical properties of peach flesh with firmness and tissue structure during storage. *Postharvest Biol. Technol.* **2020**, *163*, 111134. [CrossRef]
41. Van, B.R.; Aernouts, B.; Watte, R.; Schenk, A.; Nicolaï, B.; Saeys, W. Effect of maturation on the bulk optical properties of apple skin and cortex in the 500–1850 nm wavelength range. *J. Food Eng.* **2017**, *214*, 79–89. [CrossRef]
42. Fang, Z.H.; Fu, X.P.; He, X.M. Investigation of absorption and scattering characteristics of kiwifruit tissue using a single integrating sphere system. *J. Zhejiang Univ. Sci. B* **2016**, *17*, 484–492. [CrossRef]
43. Cubeddu, R.; D'Andrea, C.; Pifferi, A.; Torricelli, A.; Valentini, G.; Johnson, D. Time-resolved reflectance spectroscopy applied to the nondestructive monitoring of the internal optical properties in apples. *Appl. Spectrosc.* **2001**, *55*, 1368–1374. [CrossRef]
44. Blažek, J.; Hlušičková, I.; Varga, A. Changes in quality characteristics of Golden Delicious apples under different storage conditions and correlations between them. *Hortic. Sci.* **2018**, *30*, 81–89. [CrossRef]
45. Davirai, M.; Helene, H.; Philip, R.; Eyéghè-Bickong, H.A.; Vivier, M.A. A rapid qualitative and quantitative evaluation of grape berries at various stages of development using Fourier-transform infrared spectroscopy and multivariate data analysis. *Food Chem.* **2016**, *190*, 253–262. [CrossRef]
46. Tijskens, L.M.M.; Zerbini, P.E.; Schouten, R.E.; Vanoli, M.; Jacob, S.; Grassi, M. Assessing harvest maturity in nectarines. *Postharvest Biol. Technol.* **2007**, *45*, 204–213. [CrossRef]

47. Li, X.; Huang, J.; Xiong, Y.; Tan, X.; Zhang, B. Determination of soluble solid content in multi-origin 'Fuji' apples by using FT-NIR spectroscopy and an origin discriminant strategy. *Comput. Electron. Agric.* **2018**, *155*, 23–31. [CrossRef]

Review

Nanoemulsions as Edible Coatings: A Potential Strategy for Fresh Fruits and Vegetables Preservation

Josemar Gonçalves de Oliveira Filho [1], Marcela Miranda [1], Marcos David Ferreira [2,*] and Anne Plotto [3,*]

1 School of Pharmaceutical Sciences, São Paulo State University (UNESP), Rodovia Araraquara—Jaú Km 1, Araraquara 14800-903, SP, Brazil; josemar.gooliver@gmail.com (J.G.d.O.F.); mmiranda.bio@gmail.com (M.M.)
2 Embrapa Instrumentação, Rua XV de Novembro, 1452, São Carlos 13560-970, SP, Brazil
3 ARS Horticultural Research Laboratory, United States Department of Agriculture, 2001 South Rock Road, Fort Pierce, FL 34945, USA
* Correspondence: marcos.david@embrapa.br (M.D.F.); anne.plotto@usda.gov (A.P.)

Abstract: Fresh fruits and vegetables are perishable commodities requiring technologies to extend their postharvest shelf life. Edible coatings have been used as a strategy to preserve fresh fruits and vegetables in addition to cold storage and/or controlled atmosphere. In recent years, nanotechnology has emerged as a new strategy for improving coating properties. Coatings based on plant-source nanoemulsions in general have a better water barrier, and better mechanical, optical, and microstructural properties in comparison with coatings based on conventional emulsions. When antimicrobial and antioxidant compounds are incorporated into the coatings, nanocoatings enable the gradual and controlled release of those compounds over the food storage period better than conventional emulsions, hence increasing their bioactivity, extending shelf life, and improving nutritional produce quality. The main goal of this review is to update the available information on the use of nanoemulsions as coatings for preserving fresh fruits and vegetables, pointing to a prospective view and future applications.

Keywords: nanotechnology; wax coating; natural antimicrobials; essential oils; nanocoatings; post-harvest; bioactive compounds; quality; preservation methods; nanomaterials

Citation: de Oliveira Filho, J.G.; Miranda, M.; Ferreira, M.D.; Plotto, A. Nanoemulsions as Edible Coatings: A Potential Strategy for Fresh Fruits and Vegetables Preservation. *Foods* **2021**, *10*, 2438. https://doi.org/10.3390/foods10102438

Academic Editor: Like Mao

Received: 23 August 2021
Accepted: 6 October 2021
Published: 14 October 2021

1. Introduction

Fruits and vegetables are important sources of minerals, vitamins, and fibers, which are essential for human's well-being, and their consumption has been associated with several beneficial effects on human health. The demand for those benefits has considerably increased over the years due to consumer preference for natural products and changes in lifestyle [1]. In this sense, fruits and vegetables are an important component of the human diet.

After they are harvested, fruits and vegetables continue the respiration process, consuming O_2 and releasing CO_2 and water. Consequently, lipids, proteins, organic acids, and carbohydrates are metabolized and energy replacement is compromised, as the vegetable or fruit is separated from the mother plant [2]. Over time, quality characteristics such as color, flavor, weight, nutritional value, and bioactive compounds continue to deteriorate as a result of senescence [3]. The water released during the respiration process plays an important role in the postharvest quality of fresh fruits and vegetables and can result in loss of nutritional value, soft texture, sagging, wrinkling, and withering [4].

Although waxes were used to preserve citrus fruit in ancient China, it was not until the twentieth century that edible coatings based on emulsions were developed to preserve the quality of fresh fruits and vegetables [5]. These emulsions are typically formulated from oils (vegetable- or animal-derived), waxes (paraffin, carnauba wax, candelilla, or beeswax), and resins (shellac or wood rosin). Furthermore, polymer-based coating solutions can have additional functionality when formulated with plant essential oils having antimicrobial activity [6]. Due to their hydrophobic nature, oils and waxes have proven to be an efficient technology for fruits and vegetables preservation post-harvest, as they are able to minimize water loss and gas exchange and improve and/or preserve the physicochemical properties, such as color, firmness, fresh appearance, and microbial protection [7–10].

Recently, nanotechnology was introduced as a new tool for making coatings based on emulsions with improved properties and functionalities. Coatings are made of macro- or microemulsions (conventional) or nanoemulsions, for which the latter can be considered a conventional emulsion with very small particles. Droplets in nanoemulsions are on a nanoscale (particle radius less than 100 nm) dispersed in an aqueous solution [11]. This changes the physical properties of the coating by further reducing moisture migration, gas exchange, oxidative reactions, and suppressing pathogenic growth (microorganisms), product deterioration and enhancing control of physiological disorders [12]. In addition, coatings based on nanoemulsions have shown to be promising vehicles for several active compounds, such as oil-soluble vitamins, antimicrobials, flavors, and nutraceuticals, which may further contribute to maintenance of food product quality attributes [13].

Figure 1 shows a survey of published scientific manuscripts on nanoemulsions as edible coatings for fruits and vegetables. The number of studies on the topic has increased considerably over the past few years, demonstrating the scientific community's increased interest in the topic. However, studies concerning in vivo biological efficiencies are limited [14] and applications on fruits and vegetables are even fewer. Thus, more research is essential to determine this technology's potential for future application on a commercial scale. In this context, the objective of this review is to update the available information on the use of nanoemulsions as coatings for preserving fresh fruits and vegetables.

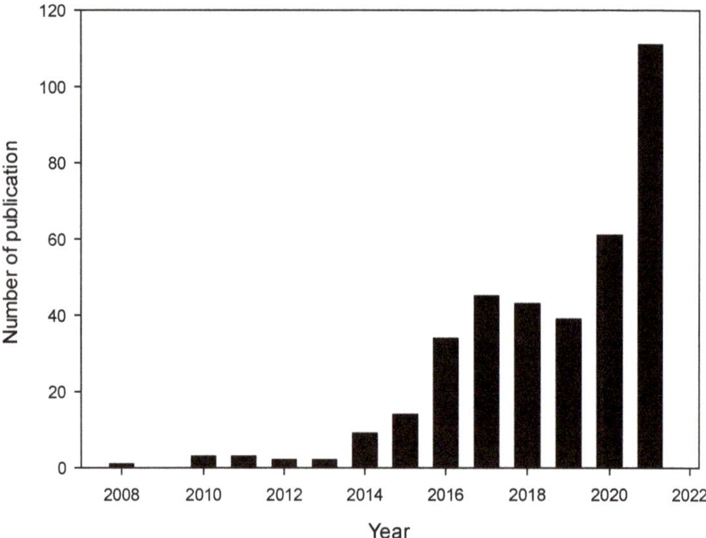

Figure 1. The distribution of publications related to 'nanoemulsion as edible coating for fruits and vegetables' (2005–2021): ScienceDirect databases. Data for 2021 is as of September 2021.

2. Edible Coatings—An Overview

The first reports of the use of coatings on fruits appeared in the 12th century in China, where wax was applied to citrus (lemons and oranges) to reduce mass loss and preserve the fruit [15]. However, it was only in 1922 that the commercial scale application of waxes began in order to increase postharvest conservation of fruits and vegetables, thus reducing postharvest losses [16].

Currently, edible coatings are used as a strategy to increase the shelf life and postharvest quality of many fresh fruits and vegetables during storage [17,18]. Edible coatings are defined as thin layers applied on the fruit surface, forming clear films produced from food-grade materials and adding to, or as a substitute for, the waxes naturally present on the fruit surface. As these films become part of the food and are consumed as such (for fruits where the peel is consumed), the materials used in their composition must be GRAS (Generally Recognized as Safe), that is, be non-toxic and safe for food [19].

Edible coatings are formulated from various biopolymers such as polysaccharide, lipid, and protein compounds, or by combining materials resulting in improved properties (Table 1). They act as an obstacle to water vapor, gases, and solutes [20] as shown in Figure 2.

Table 1. Summary of diverse structural materials frequently used for edible coating.

Material	Main Matrices	Positive Points	Negative Points	References
Polysaccharide	Starch, chitosan, alginate, cellulose, and its derivatives, and pectin	Good gas and mechanical barrier properties	Poor moisture barrier due to hydrophilic nature	[21,22]
Lipid	Animal, vegetable waxes and resins, vegetable oil, and fatty acids	Good moisture barrier properties with a shiny appearance	Poor mechanical and gas barrier properties	[18,23,24]
Protein	Gelatin, casein, whey protein, zein, soy protein, myofibrillar protein, and quinoa protein	Good gas barrier properties without anaerobic conditions	Brittle and susceptible to cracking	[25]
Composite	Combination of polysaccharide and/or protein with lipids	Good moisture and gas barrier properties	Formation of non-homogeneous emulsion	[26–29]

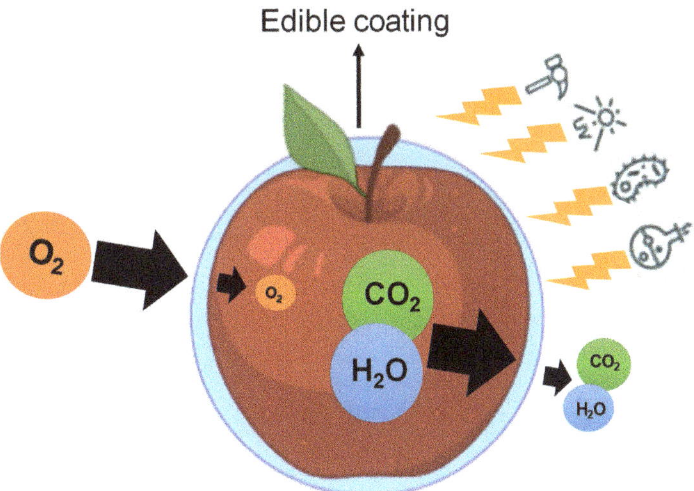

Figure 2. Main functions of edible coatings on fruits and vegetables.

The mechanism of action for coatings on fruit is similar to packaging with a modified atmosphere; the coating produces a physical barrier that modifies gas exchange between the interior of the fruit and the surrounding atmosphere, increasing the concentration of CO_2 and decreasing O_2 [30]. This environment can effectively decrease respiration rate, conserve stored energy, delay microbial growth, and therefore, extend the useful life of the fruit [31]. The coating efficiency depends on the coating thickness formed on the fruit surface, since there is a negative correlation between thickness and coating permeability [32]. Another important point is related to low permeability coatings, based on resins such as shellac, for example, which can restrict gas exchange almost entirely, leading to the accumulation of CO_2 within the fruit, and the production of compounds resulting from the fermentation process that can cause off-flavor, such as acetaldehyde and ethanol, thus affecting fruit quality [18,33].

In addition to maintaining quality and postharvest conservation of fruits and vegetables, the coating materials can also act as carriers of compounds such as food coloring, flavoring, antimicrobials, antioxidants, antagonistic microorganisms, among others [34,35]. In this sense, several natural bioactive compounds have been incorporated into edible coating materials such as essential oils [36–38], plant extracts [39,40], vitamins [34], antagonistic microorganisms [41,42], and antibrowning or firming agents in fresh cut fruit. [43,44].

3. Methods to Apply Edible Coatings

The effectiveness of coatings in preserving fresh fruits and vegetables is influenced by the application method, which will be chosen according to the nature of the food to be coated, the surface attributes, the rheological properties of the solution, and the main purpose of the coating [45]. The adhesion of coatings to food surfaces is essential for performance of their intended function [16,46]. Wettability is used to quantify the interfacial interaction that occurs between the food surface and the coating. This variable must be taken into account when assessing the performance of the coating solution on the food surface [31].

Dipping (Figure 3a), spraying (Figure 3b), and hand coating (Figure 3c) techniques are the most common methods for applying edible coatings to fresh fruits and vegetables. Other techniques such as fluidized bed and foaming are also available; however, these techniques are rarely used on commercial and laboratory scales [45].

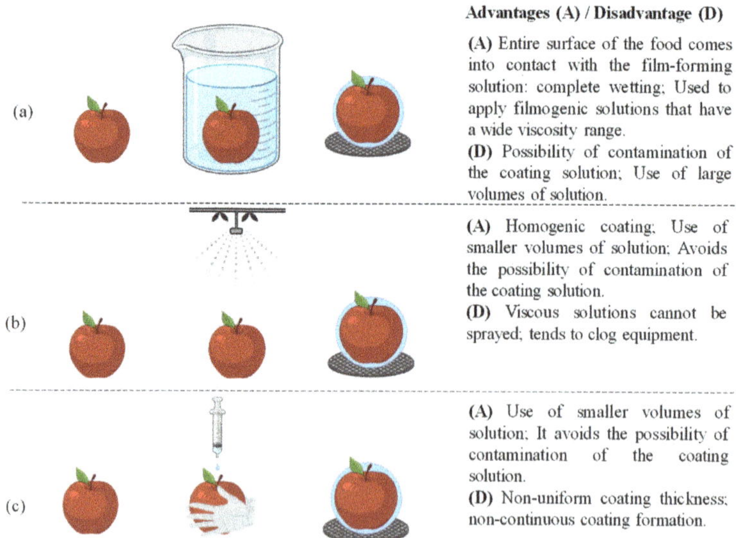

Figure 3. Dipping (**a**), spraying (**b**), and (**c**) hand coating techniques to apply edible coatings.

On a laboratory scale, immersion is one of the main methods used for coating fruits due to its simplicity, without dependence on equipment, and uniformity of film obtained. In this method, the entire surface of the food is submerged in the film-forming solution at a constant speed, allowing full surface coverage, ensuring complete surface wetting [47]. After application, the excess solution is drained to eliminate the overload of film-forming solution on the fruit surface [48]. Finally, the food is dried with the excess solvent and liquid being evaporated to leave the film in contact with the food surface. Drying can take place at room temperature or using a heated air tunnel after draining the solution. This technique allows the application of coating solutions with a wide viscosity range [46]. A negative point of this technique is the possibility of cross-contamination from fruit to fruit during the immersion process due to the accumulation of residues and microbial organisms [45].

To avoid this problem, products that will be coated must be properly cleaned and sanitized, and the coating solution replaced frequently [15]. According to Raghav et al. [16], in general, fruits and vegetables are immersed for 5–30 s in the coating solution.

In turn, the spraying technique, most popular in packing houses, provides a homogeneous and attractive coating. In addition, it avoids the possibility of contaminating the coating solution [49]. This process increases the liquid surface through the formation of drops and distributes them over the food surface [45]. During spray application, the fruit or vegetable is placed on a plate or rotating rollers at a coordinated speed, under dispersing nozzles activated manually or automatically. This procedure is repeated until the desirable coating thickness is achieved. A drawback of this technique is that viscous solutions cannot be sprayed as they clog the equipment [50].

Another method to apply a filmogenic solution is by gloved hands to the fruit surface. Fruits can be coated by spreading a uniform amount of coating solution by hand while wearing latex gloves. It is appropriate on a laboratory scale to avoid solution contaminations and to minimize waste of experimental coating solutions during screenings. However, a negative aspect consists of the non-homogeneous film thickness formed on the entire fruit surface [18,35].

4. Nanomaterials in Edible Coatings

In recent years, nanotechnology has been used as an important tool to increase the storage period for food products. The application of nanoscale particles provides different and improved properties compared to particles with larger size. Related to foods, nanotechnology has a wide spectrum of uses in films and coatings due to the improved features they impart [51].

Figure 4 shows the advances in the development of nanosystems incorporated with food-grade ingredients, which makes it feasible to explore functional modifications in food coating materials that include nanoemulsions, polymeric nanoparticles, nanostructured lipid transporters, nanotubes, nanocrystals, nanofibers, and others [52]. Nanosystems, when incorporated into matrices based on hydrocolloids (proteins or carbohydrates), give rise to nanocomposites, which are the combination of two or more materials, one of which is on a nanoscale, in order to improve coating properties [52,53].

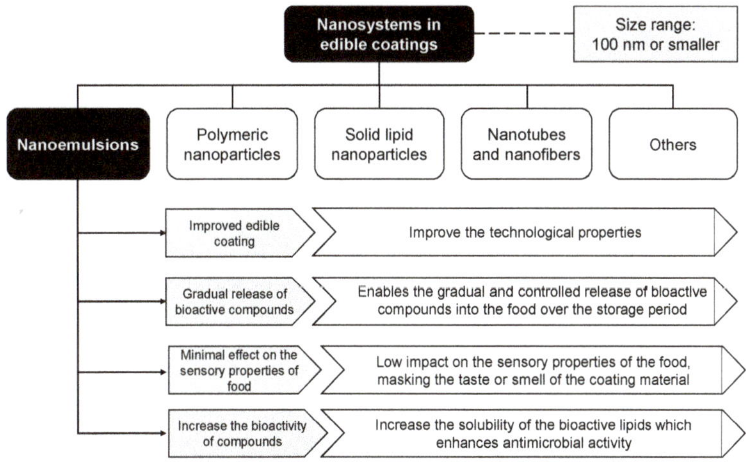

Figure 4. Nanomaterials in edible coatings.

The main changes due to use of nanosystems in nanocomposite coatings refer to the water barrier, optical and microstructural mechanical properties, and the antimicrobial and antioxidant effects. Nanoparticles in coatings potentiate these activities when antimicrobial or antioxidant compounds are incorporated in the coating, by enabling their gradual and controlled release over the period of fruit storage, sometimes under different storage conditions, hence improving bioavailability of these compounds over time [52,54]. The improvements in these properties are important to guarantee food quality maintenance as well as to reduce the development of decay microorganisms (bacteria, filamentous fungi, and yeasts) and action of free radicals that deteriorate food and reduce shelf life [55]. Another advantage of adding active agents to nanosystems is that a smaller proportion of these substances is necessary to obtain good activity; therefore, the use of these compounds in low concentrations does not negatively affect food sensory properties [12].

5. Fundamentals of Nanoemulsions

Emulsions are generally made of two immiscible liquids, commonly oil and water, forming a relatively stable mixture. Generally, emulsions are systems that contain a dispersed and continuous phase and can be classified according to the three-dimensional organization of the oil and water phases. Oil-droplets dispersed within an aqueous phase is named *oil-in-water* (O/W) emulsion, whereas water-droplets dispersed in the oil phase is classified as *water-in-oil* (W/O) emulsion, and they are the most common emulsions [14,56,57]. Figure 5 shows the schematically structures of O/W (Figure 5A) and W/O (Figure 5B) emulsions, emphasizing the micelle structures dispersed in the continuous phase.

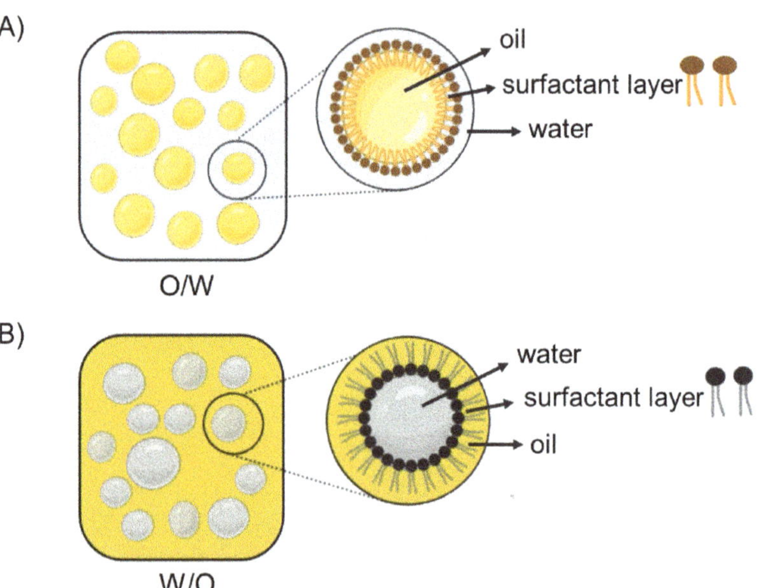

Figure 5. Schematic representation of (**A**) *oil-in-water* (O/W) and (**B**) *water-in-oil* (W/O) emulsions, representing micelle structure dispersed in continuous phase for each system.

Emulsions are classified into three main classes according to thermodynamic stability, stable mechanisms, and physical properties: macroemulsion or conventional emulsion, nanoemulsion, and microemulsion. Conventional and nanoemulsions are thermodynamically unstable, while the microemulsion is stable. The droplet mean radius for conventional emulsions are bigger, which distinguishes them from nanoemulsions with a radius of less than 100 nm [11,57,58]. The droplet size in nanoemulsions is a key-point that influences their capability to improve the bioavailability of added hydrophobic substances, such as carotenoids [58], and increase antimicrobial essential oil properties [59] or oil compounds [60]. The nanoemulsion classes will be further discussed in this article, with the focus of nanoemulsions as edible nanocoatings.

5.1. Nanoemulsions and Production Methods

The small size of particles in nanoemulsions allows potential advantages over conventional emulsions, such as greater stability concerning particle aggregation and gravitational separation, in addition to high optical transparency, modification of the physical properties of the coating, and increased bioavailability of bioactive-loaded lipid droplets [57]. Free nanoemulsion-based delivery systems increased the bioaccessibility of vitamins (D) and carotenoids (β-carotene and curcumin) [58,61]; however, studies have demonstrated that bioactive-loaded nanoemulsions prepared with a biopolymer mixture can be trapped in the matrices and decrease bioaccessibility.

Nanoemulsions need energy for their formation, which is provided by mechanical equipment or physical and chemical properties of the system. Procedures using mechanical energy are called high energy methods and use microfluidizers, high-pressure homogenizers, and ultrasonic homogenizers. The methods that employ the system's physical and chemical properties are categorized as low energy, such as spontaneous emulsification, phase inversion temperature, and emulsion inversion methods [54,57].

When high-energy methods are employed, the surfactants help break oil-droplets inside the homogenizer by decreasing interfacial tension, thus promoting smaller droplets and preventing droplet aggregation. A high shear rate is necessary to break the droplet to form nano-droplets, and is generally achieved by high-pressure homogenizers, as the use of high energy generates forces that can break the droplets in the dispersed phase [56,57]. Those methods are well established in the food industry and can be adapted for nanoemulsion production. On the other hand, for low energy methods, surfactants promote small droplet spontaneous formation due to their ability to generate extremely low interfacial tensions under specific conditions. Therefore, the surfactants utilized are extremely important because the emulsion pH stability, ionic strength, heating, cooling, and storage are mainly determined by the amphiphilic molecule chosen [56,57].

The amphiphilic material, such as surfactants, phospholipids, proteins, and polysaccharides, reduces the interfacial tension and maintains droplet stability. Emulsions (*O/W* or *W/O*) (Figure 6A,B) are the most stable systems; however in unusual regimes, multiple emulsions such as *W/O/W* and *O/W/O* (Figure 6C,D) may be formed and are usually extremely unstable to coalescence [14,54,56]. Most fruits and vegetables contain a high-water volume; therefore, among emulsions the *O/W* type (Figure 6A) is the most explored for food systems due to the possibility of loading the oil-droplets with lipophilic key-compounds surrounded by water [14,54].

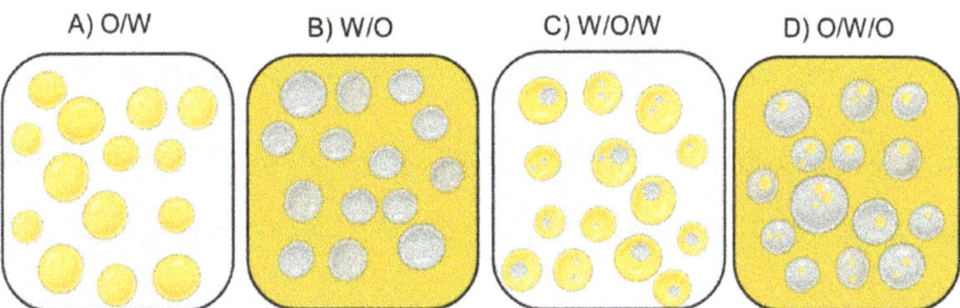

Figure 6. Representation of most common emulsion (**A**) *oil-in-water* (O/W) and (**B**) *water-in-oil* (W/O), and multiple emulsions (**C**) *water-in-oil-in-water* (W/O/W) and (**D**) *oil-in-water-in-oil* (O/W/O).

5.2. Surfactants

Surfactants can be classified according to their electrical characteristics as ionic, non-ionic, and zwitterionic. Most foods surfactants are ionic, such as esterified monoglycerides, which are mainly negatively charged and can form nanoemulsions using low or high energy. Non-ionic surfactants also can be used for both methods and have low toxicity and irritability, including compounds such as Tween® (condensate of sorbitol fatty acid esters and ethylene oxide) and Span® (a family of fatty acids sorbitan). On the other hand, zwitterionic surfactants contain two or more ionizable groups with opposite charges, and consequently, they can have a negative, positive, or neutral charge depending on the pH solution. For example, this group includes lecithin, a phospholipid widely used in foods [57,62].

One of the main aspects of an emulsion formulation is the choice of surfactant. The hydrophilic–lipophilic Balance (HLB) system was developed, which represents the balance of the size and strength of the polar and non-polar groups [62]. It demonstrates molecule properties as amphiphilic compounds using a numerical scale, assigning higher HLB values as the substance is more hydrophilic [62]. However, the HLB system only considers the properties of the surfactant itself. For this reason, the hydrophilic–lipophilic deviation (HLD) system is another approach to the behavior exhibited by surfactant–oil–water and usually more suitable in formulations [57,63]. In addition, proteins, polymers with amphiphilic properties, and combinations of polymers and surfactants can act as emulsifiers [64].

Studies have demonstrated the importance of modulating nanoemulsions composition and structure to achieve higher digestion and absorption in the gastrointestinal tract and to efficiently deliver compounds such as vitamins and nutraceuticals [54,58,65,66]. Therefore, the choice of emulsifier is of extreme importance since it can improve carotenoids bio-accessibility, for example. In a study performed by Yao et al. (2019) [65], the authors demonstrated the relationship between carotenoids bio-accessibility from spinach and co-ingesting with excipient nanoemulsions: nanoemulsions containing different ratios of medium or long-chain triglycerides in the oil phase composition decreased β-carotene bioaccessibility when the ratio of medium-chain triglycerides was increased. The findings were credited to the formed micelle's ability to hold the carotenoids in their hydrophobic domains.

6. Plant-Based Nanoemulsions as Edible Coatings on Fruits and Vegetables Postharvest

6.1. Coatings Based on Essential Oil Nanoemulsions

One of the main features can be in the form of antimicrobial nanoemulsions, for example, nanoemulsions based on plant essential oils, which are associated with biopolymers such as alginate, chitosan, and starch, among others. It has been shown that when essential oils are encapsulated in nanoemulsions, they have less impact on the sensory properties of the food, masking the taste or smell of the core material (coating), yet providing better biological activity of essential oils due to the increase in the surface area [67]. In this way, it is possible to use low doses of bioactive material, increasing the transport of active ingredients through biological membranes, thus intensifying the bioavailability of bioactive compounds, in addition to less interaction with other components of the food matrix. Other advantages are the low mass transport of compounds into and out of the coating, less impact on optical, barrier, and microstructural properties and greater coating stability [68,69].

Essential oils have received special attention as active ingredients applicable in food coatings, due to their potent antimicrobial and antioxidant activities [70]. Essential oils are volatile aromatic substances of low molecular weight (for example, phenolic compounds, such as monoterpenes, flavonoids, and phenolic acids) produced by plants (for example, cinnamon, thyme, lavender, ginger, palmarosa, lemongrass, mint, citrus fruits, and fennel) or their isolated components (for example, eugenol, geraniol, menthol, limonene, carvacrol, and linalool) that can reduce microbial growth in food, and have been studied as natural antimicrobials in food for decades [71]. However, their volatile nature, low water solubility, and strong aroma limit their applications in foods. In this sense, using nanotechnological approaches is a promising strategy to enable the application of essential oils as natural antimicrobials in foods, overcoming their limitations and increasing their antimicrobial activity [52].

Table 2 presents the main types of nanoemulsions as edible coatings classified by matrix type and their impact on fruit and vegetable shelf life.

Table 2. Main types nanoemulsions as edible coatings and their impact on fruit and vegetable shelf life.

Matrix	Bioactive Substance or Lipid Compound	Production Technique	Functionality	Fruit or Vegetable	Reference
Modified chitosan	Lemon, mandarin, oregano, or clove essential oils	High-pressure homogenization (HPH)	Increase the antimicrobial activity of the essential oil and improve the homogeneity and stability of the emulsion	Arugula leaf (*Eruca sativa*)	[72]
Chitosan	Carvacrol, bergamot, mandarin, and lemon essential oils	High-pressure homogenization	Increase the antimicrobial activity of essential oils	Green beans (*Phaseolus vulgaris*)	[73]
Sodium alginate	Basil essential oil	Ultrasound	Increase the antimicrobial activity of essential oil	Okra (*Abelmoschus esculentus*)	[74]
Pullulan	Cinnamon essential oil	Ultrasound	Improve the distribution of oil in the matrix and increase its antimicrobial activity	Strawberry (*Fragaria × ananassa*)	[75]
Carnauba wax	Lemongrass essential oil	Dynamic high pressure	Increase the antimicrobial activity of the essential oil and improve the homogeneity and stability of the emulsion	Plums (*Prunus salicina*)	[76]
Carnauba wax	Lemongrass essential oil	High shear probe and high-pressure dynamic processing (DHP)	Increase the antimicrobial activity of essential oil	Grape berry (*Vitis labruscana* Bailey)	[77]
Candelilla wax	Extract of tarbush	High-speed stirrer	Improved the wettability of the nanocoating on the Fuji apple surface	Fuji apple (*Malus domestica 'Fuji*)	[78]
Quinoa protein/chitosan	Thymol	1200 rpm agitation	Increase the antimicrobial activity of the active compound and improve dispersion in the matrix	Strawberry (*Fragaria × ananassa*)	[79]
Sodium alginate	Lemongrass essential oil	Microfluidization	Improve the stability of the emulsion and increase the antimicrobial activity of the essential oil	Fresh-cut Fuji apples (*Malus domestica 'Fuji*)	[59]
Hydroxypropyl methylcellulose	Carnauba wax nano-emulsion	High-pressure homogenization (HPH) and mechanical stirring	Reduce gas permeability and moisture loss	'Redtainung' Papaya (*Carica papaya*)	[28]
Sodium alginate	Citral	Ultrasound	Improve the dispersion of the active compound in the matrix and increase its antimicrobial activity	Fresh cut pineapples (*Ananas comosus*)	[60]
Carnauba wax	Oleic acid and Carnauba wax	High-pressure homogenization (HPH)	Improve optical properties, and emulsion stability	'Nova' mandarins (*Citrus reticulata*) and 'Unique' tangors (*C. reticulata C. sinensis*)	[18]
Chitosan	Cellulose nanocrystal and oleic acid	Ultra turrax homogenizer	Increase coating stability at high humidity, adhesion on fruit surface and delayed ripening of pears	Bartlett pears (*Pyrus communis*)	[80]

Edible coatings based on nanoemulsions of essential oils have been studied as an alternative to prolong fresh fruit and vegetable shelf life. For example, a coating based on the nanoemulsion of lemon essential oil and chitosan increased the shelf life of arugula leaves by 7 days compared to a coating of chitosan or lemon oil alone [72]. Likewise, coatings based on modified chitosan and carvacrol nanoemulsions completely inhibited the growth of *Escherichia coli* on fresh green beans during the 11-day period under refrigeration [73]. Gundewadi et al. [74] also reported that the nanoemulsification of basil essential oil in an alginate coating was more effective than its respective microemulsion and presented better coating stability. In addition, when applied to okra fruits, nanoemulsion was more efficient in preserving texture, color, and sensory characteristics compared to control fruits. The essential oil of nanoemulsified basil showed greater antifungal activity against fungal pathogens than microemulsions. Chu et al. [75] developed a pullulan coating with a cinnamon essential oil nanoemulsion for strawberry storage. The nanoemulsion-based coating was more effective than other coatings in reducing loss of mass, firmness, total soluble solids, acidity, and controlling the growth of fungi and bacteria during fruit storage.

In another study, Prakash, Baskaran, and Vadivel [60], evaluated the effect of a coating based on sodium alginate and citral nanoemulsion on the quality of fresh cut pineapples. Coatings based on nanoemulsions were effective at reducing microbial growth during storage. In addition, at a concentration of 0.2% of citral nanoemulsion, the coating reduced the presence of *Salmonella enterica* and *Listeria monocytogenes* after artificial inoculation [60]. The coating based on nanoemulsions of lemongrass essential oil, Tween® 80 and alginate was more effective at preserving the characteristics of minimally processed Fuji apples than their respective conventional emulsions. The nanoemulsion coating inhibited the growth of artificially inoculated *E. coli* on fruits faster than conventional emulsions [59].

6.2. Coatings Based on Plant-Based Wax Nanoemulsions

Commercial coatings based on approved waxes must meet state/national fruit and vegetable additive regulations and be considered safe for consumption. However, to improve the characteristics of wax-based coatings, they are combined with synthetic chemicals to prevent microbiological deterioration and to ensure homogeneous stability of the coating during product storage. Commercial coatings are typically formulated using oxidized polyethylene wax (a by-product of the petroleum industry), carnauba wax (from the leaves of the carnauba palm, *Copernicia cerifera*), candelilla wax (from the candelilla shrub, *Euphorbia cerifera*), and shellac (from the insect bug *Kerria lacca*) as matrices, combined with water and other agents such as oleic acid, morpholine, ammonia, polydimethylsiloxane antifoam, and others [81].

The compounds combined with waxes used as emulsifying, moisturizing, and antimicrobial agents in commercial coatings, are mostly synthetic chemical products and could be a concern for human health [82]. As an example, morpholine is a base acting as a counterion to facilitate fatty acids emulsification in waxes. In the presence of nitrite/nitrate, morpholine can form N-nitrosomorpholine, a potent mutagen and carcinogen [83,84]. N-nitrosomorpholine was not found on coated fruit surface, but the possibility of its formation in the gut from reaction of morpholine with dietary nitrates was considered; it was found at concentrations less than the safe dose of 4.3 ng/kg body weight/day, not enough to raise concerns [85]. Ammonia could be used as a replacement for morpholine [86], but its highly volatile and irritant nature makes it less easy to use than morpholine.

Consumers are increasingly concerned about the safety and quality of food, driving the demand for so-called "environmentally friendly coatings", that is, coatings based on natural products of plant origin that do not present any harm to the consumer's health if consumed. The use of waxes and compounds of animal origin has been limited by vegan and vegetarian consumers, consumers who are allergic to animal products (such as chitosan) and religious beliefs that do not encourage the consumption of animals [87]. Therefore, the demand for plant-based wax-based coatings is an important market for fresh fruits and vegetables, and nanotechnology is a promising tool to meet this demand by improving the properties of these coatings, especially wax-based ones, reducing the need of synthetic additives.

Nanotechnology has been successfully used to produce plant-based waxes nanoemulsions, such as carnauba wax [18] and candelilla wax [78] without the addition of morpholine. Wax-based nanoemulsions can have improved barrier properties due to the small size of the droplets, promoting greater homogeneity compared to conventional emulsions, greater transparency, improved physico-chemical properties (optical, mechanical, and barrier) and greater stability in comparison with conventional emulsions [18,54,78]. In addition, these nanoemulsions can be used for the development of nanocomposite coatings, in combination with hydrocolloid components (polysaccharides and proteins) in order to improve the water barrier properties of these compounds and minimize the impact of the incorporation of lipid compounds in the matrix hydrocolloids [29].

Lipid nanoemulsions made from plant-based waxes have shown greater effectiveness as edible coatings than conventional emulsions on fresh fruits and vegetables preservation (Table 2). The carnauba wax nanoemulsion coating showed less water loss, conferred gloss, and caused less ethanol production than shellac in coated 'Nova' mandarins (*Citrus reticulata*) and 'Unique' tangors (*C. sinensis*) [18]. In addition, the coating based on carnauba wax nanoemulsion exhibited less changes in the fruit internal atmosphere and volatile profile, and consequently, better flavor compared to the conventional carnauba wax emulsion and commercial shellac [18].

Lipid nanoemulsions produced from waxes, such as carnauba or candelilla, have been shown to be suitable vehicles for carrying bioactive compounds, such as plant extracts and essential oils [78,88]. They can improve the physical stability of the active substances, and improve the bioactivity of these compounds, and due to the prolonged and slow diffusion, they reduce the impact of these substances on the sensory properties of fruits and vegetables [88]. De Léon-Zapata et al. [78] developed candelilla wax nanoemulsions added with tarbush extract and evaluated its effect on the preservation of Fuji apples. The combination of extract and nanocoating reduced the size of the droplets and improved the zeta potential and optical properties of the coating. When applied to Fuji apples, the nanocoating effectively reduced physico-chemical and microbiological changes and delayed fruit senescence in comparison with the control treatment.

In another study, a nanoemulsion of carnauba wax combined with lemongrass essential oil nanoemulsion was applied to plums [76]. The coatings were able to inhibit the growth of *S. typhimurium* (*S. enterica*) and *E. coli* O157: H7 inoculated plums during storage, and did not significantly affect their taste and appearance (brightness). In addition, nanoemulsion coatings were effective at reducing weight loss, ethylene production and respiration rate. Fruit coated with nanoemulsions showed greater firmness and increase in phenolic compounds content during storage in comparison with uncoated fruits [76]. A similar result was observed in another study carried out by these authors with grape berries. The coating based on carnauba wax and lemongrass essential oil nanoemulsion inhibited the growth of *S. typhimurium* and *E. coli* O157: H7 inoculated fruit. Lemongrass in nanoemulsions did not affect berry taste and improved their brightness. Coatings based on nanoemulsions were also able to reduce weight loss and maintain firmness, phenolic compounds, and antioxidant activity in berries. The coatings demonstrated the potential to reduce microbiological contamination of grape berries by foodborne pathogens and prolong their shelf life. [77].

7. Trends in Materials Based on Nanoemulsions with Potential for Application in the Preservation of Fruits and Vegetables

New coating materials based on nanoemulsions with potential for application in fruits and vegetables have been developed in the last two years with the aim of contributing even more to the preservation of these products. One way to develop these functionalized materials is to combine composites with different properties to develop a functionalized coating. For example, de Oliveira Filho et al. [89] developed a functionalized coating combining arrowroot starch (biopolymeric matrix), carnauba wax nanoemulsion (to improve the water barrier properties of the coating), cellulose nanocrystals (to improve mechanical properties and stabilize the emulsion), and essential oils (to confer antimicrobial activity). The combination of compounds resulted in a coating material with excellent water barrier, mechanical, thermal, optical, microstructural, and antimicrobial properties against fungi that attack fruits during post-harvest.

Another increasingly explored trend in the development of new coatings based on nanoemulsions with better stabilities is the use of solid particles to form Pikering nanoemulsions, that is, nanoemulsions stabilized with solid particles such as cellulose nanocrystals [90], starch nanocrystals [91], γ-Al2O3 nanoparticles [92], cyclodextrin [93], among others.

Pickering nanoemulsions have excellent stability due to irreversible adsorption that occurs between solid particles at the oil–water interface due to the high adsorption energy [94]. Another characteristic of these nanoemulsions is the ability to release active ingredients encapsulated under specific conditions, such as pH and temperature [93]. Almasi, Azizi, and Amjadi [95] compared two coating materials based on pectin, one with marjoram essential oil encapsulated in a whey protein/inulin stabilized Pickering nanoemulsion, and the other with marjoram essential oil nanoemulsified with Tween 80. Coatings based on pectin with Pickering nanoemulsions presented mechanical and water barrier properties superior to those based on standard nanoemulsion. In another study, López-Monterrubio et al. [96] developed highly efficient β-carotene nanoemulsions stabilized by a complex formed by hydrolyzed whey protein and pectin. The nanoemulsions showed good stability during the 30-day storage period with low formation of clumps.

Deng et al. [80] developed coatings based on chitosan and Pickering nanoemulsion of oleic acid stabilized with cellulose nanocrystals and evaluated their effects on the postharvest conservation of green D'Anjou and Bartlett pears (*Pyrus communis* L.). The coating formulated with 5% cellulose nanocrystals showed strong adhesion to the fruit surface, showing greater gas barrier property compared to the commercial Semperfresh™ product, and presented a more homogeneous matrix, being effective in delaying ripening and increased the shelf life of pears during storage.

Although the above new materials have been little studied in food systems, the results described in the literature are very encouraging.

8. Potential Toxicity, Limitations, and Regulatory Aspects of Nanoemulsions

Nanoemulsions, due to the nanometric size of the droplets, may partially remain intact during digestion, representing potential safety risks related to the compounds used for their production (such as surfactants). They can be of concern in metabolic or hormonal dysregulation due to their rapid absorption compared to conventional emulsions, their ability to increase the bioavailability of bioactive agents to a toxic level, and the possibility of increased absorption by epithelial cells which can cause changes in the functionality of the gastrointestinal tract [97]. However, as they have a high surface area, nanoemulsions can also be quickly digested by enzymes from the gastrointestinal tract, reducing the possible toxic effect that can occur due to their accumulation in organ cells [98].

In vitro studies were performed using cell cultures, usually models of normal cells such as fibroblasts, to investigate potential toxicity of nanoemulsions. Kaur et al. [99] reported that nanoemulsions based on tocopheryl polyethylene glycol succinate (TPGS), lemon oil, Tween-80, and water did not show toxicity in Hep G2 cells. In another study, Marchese et al. [100] observed that bergamot essential oil nanoemulsions showed cytotoxic activity against Caco 2 cells at high concentrations. A limitation of these studies is the fact that authors have not previously exposed the nanoemulsions in simulated conditions of the gastrointestinal tract before contact with the cells.

Knowledge about the potential toxicity of nanoemulsions in vivo is still limited and should be investigated [97]. The effect of nanoemulsions based on antimicrobial compounds, such as essential oils, on the gastrointestinal tract is also poorly reported in the literature. This effect must be carefully studied, as antimicrobial compounds can influence the intestinal microbiota or epithelial cells of the gastrointestinal tract.

In a recent study, Hort et al. [101] evaluated the toxicity of Miglyol and egg lecithin nanoemulsions using an in vivo model (male Wistar rats). The nanoemulsions were orally administered to rats for 21 days at lipid concentrations of 200, 400, or 800 mg/kg of body weight. The results of biochemical, hematological, oxidative stress, and genotoxicity parameters showed that nanoemulsions could be considered safe for oral administration, but high doses by the parenteral route could cause toxic effects.

The few studies suggest that nanoemulsions formulated with GRAS ingredients do not exhibit strong cytotoxic effects. The nanometer size of the droplets suggests that they are rapidly transformed into monoglycerides and free fatty acids in the small intestine, which are normal digestion products and should not have toxic effects [57].

As for regulatory aspects, essential oils and other antimicrobial agents are mainly regulated by the European Food Safety Authority (EFSA) in Europe and the Food and Drug Administration (FDA) in the United States [102]. However, for nanoemulsions there is no international authority that makes this regulation. The FDA addresses the regulation of nanotechnology products as guidance for industries. The European Council and Parliament have regulated food nanotechnology as new food products or food ingredients [103].

9. Conclusions and Future Perspectives

The use of substances obtained from plant-based natural sources has emerged as a trend in the fresh fruit and vegetable market for coating applications. The application of these compounds on a nanoscale has advantages allowing a wider use in relation to particles on larger scales. Recent studies indicate that nanoemulsions play an important role in the development of a new generation of coatings with improved properties for the preservation of fresh fruits and vegetables. This emerging technology makes it possible to improve the physical stability and performance of active substances within an edible coating, bringing the possibility of increasing the quality and/or nutritional value of fruits and vegetables. Although the evidence published to date suggests that nanoemulsions applied as edible coatings can extend the life of different fruits and vegetables, there are other important aspects to explore before considering them on a commercial scale in future trends, such as the bioavailability of bioactive compounds incorporated in the nanoemulsions, potential toxicity and digestibility, for example. Most of the tested nanoemulsion coatings have antimicrobial properties; however, it can also be possible to produce and apply edible coatings with health-promoting substances.

Author Contributions: Writing—original draft preparation, J.G.d.O.F. and M.M.; writing—review and editing, M.D.F. and A.P.; visualization, J.G.d.O.F. and M.M.; supervision, M.D.F.; project administration, M.D.F.; funding acquisition, M.D.F. All authors have read and agreed to the published version of the manuscript.

Funding: This research was funded by FAPESP (process 2016/23419-5,2018/10657-0, and 2018/24612-9), CAPES (001), CNPq (process 407956/2016-6; fellowship 310728/2019-3), Empresa Brasileira de Pesquisa Agropecuária (Embrapa), Rede Agronano, and MCTI-SisNano from Brazil.

Acknowledgments: Authors thank Elizabeth Baldwin for reviewing the manuscript and constructive suggestions.

Conflicts of Interest: The authors declare no conflict of interest.

Disclaimer: Mention of a trademark or proprietary product is for identification only and does not imply a guarantee or warranty of the product by the U.S. Department of Agriculture. The U.S. Department of Agriculture prohibits discrimination in all its programs and activities on the basis of race, color, national origin, gender, religion, age, disability, political beliefs, sexual orientation, and marital or family status.

References

1. Dhandevi, P.; Jeewon, R. Fruit and vegetable intake: Benefits and progress of nutrition education interventions-narrative review article. *Iran. J. Public Health* **2015**, *44*, 1309–1321.
2. Brizzolara, S.; Manganaris, G.A.; Fotopoulos, V.; Watkins, C.B.; Tonutti, P. Primary metabolism in fresh fruits during storage. *Front. Plant Sci.* **2020**, *11*, 80. [CrossRef] [PubMed]
3. Anwar, R.; Mattoo, A.K.; Handa, A.K. Ripening and senescence of fleshy fruits. In *Postharvest Biology and Nanotechnology*; Paliyath, G., Subramanian, J., Lim, L.-T., Subramanian, K.S., Handa, A.K., Mattoo, A.K., Eds.; John Wiley & Sons, Inc.: Hoboken, NJ, USA, 2019; pp. 15–51.
4. Nunes, C.N.; Emond, J.-P. Relationship between weight loss and visual quality of fruits and vegetables. *Proc. Fla. State Hort. Soc.* **2007**, *120*, 235–245.

5. Baldwin, E.A. Edible coatings for fresh fruits and vegetables: Past, present, and future. *Edible Coat. Film. Improv. Food Qual.* **1994**, *1*, 25.
6. Acevedo-Fani, A.; Soliva-Fortuny, R.; Martín-Belloso, O. Nanoemulsions as edible coatings. *Curr. Opin. Food Sci.* **2017**, *15*, 43–49. [CrossRef]
7. Bai, J.; Baldwin, E.A.; Hagenmaier, R.H. Alternatives to shellac coatings provide comparable gloss, internal gas modification, and quality for 'Delicious' apple fruit. *HortScience* **2002**, *37*, 559–563. [CrossRef]
8. Navarro-Tarazaga, M.-L.; Perez-Gago, M.-B.; Goodner, K.; Plotto, A. A new composite coating containing HPMC, beeswax, and Shellac for 'Valencia' oranges and 'Marisol' tangerines. *Proc. Fla. State Hort. Soc.* **2007**, *120*, 228–234.
9. Valencia-Chamorro, S.A.; Pérez-Gago, M.B.; del Río, M.Á.; Palou, L. Effect of antifungal hydroxypropyl methylcellulose (HPMC)–lipid edible composite coatings on postharvest decay development and quality attributes of cold-stored 'Valencia' oranges. *Postharvest Biol. Technol.* **2009**, *54*, 72–79. [CrossRef]
10. Nawab, A.; Alam, F.; Hasnain, A. Mango kernel starch as a novel edible coating for enhancing shelf-life of tomato (Solanum lycopersicum) fruit. *Int. J. Biol. Macromol.* **2017**, *103*, 581–586. [CrossRef]
11. McClements, D.J. Nanoemulsions versus microemulsions: Terminology, differences, and similarities. *Soft Matter* **2012**, *8*, 1719–1729. [CrossRef]
12. Hasan, S.K.; Ferrentino, G.; Scampicchio, M. Nanoemulsion as advanced edible coatings to preserve the quality of fresh-cut fruits and vegetables: A review. *Int. J. Food Sci. Technol.* **2020**, *55*, 1–10. [CrossRef]
13. Salvia-Trujillo, L.; Soliva-Fortuny, R.; Rojas-Graü, M.A.; McClements, D.J.; Martín-Belloso, O. Edible nanoemulsions as carriers of active ingredients: A review. *Annu. Rev. Food Sci. Technol.* **2017**, *8*, 439–466. [CrossRef]
14. Jin, W.; Xu, W.; Liang, H.; Li, Y.; Liu, S.; Li, B. Nanoemulsions for food: Properties, production, characterization, and applications. In *Emulsions*; Grumezescu, A.M., Ed.; Elsevier: Amsterdam, The Netherlands, 2016; pp. 1–36.
15. Andrade, R.D.; Skurtys, O.; Osorio, F.A. Atomizing spray systems for application of edible coatings. *Compr. Rev. Food Sci. Food Saf.* **2012**, *11*, 323–337. [CrossRef]
16. Raghav, P.K.; Agarwal, N.; Saini, M. Edible coating of fruits and vegetables: A review. *Int. J. Sci. Res. Mod. Educ.* **2016**, *I*, 188–204.
17. Ghoora, M.D.; Srividya, N. Effect of packaging and coating technique on postharvest quality and shelf life of *Raphanus sativus* L. and *Hibiscus sabdariffa* L. microgreens. *Foods* **2020**, *9*, 653. [CrossRef]
18. Miranda, M.; Sun, X.; Ference, C.; Plotto, A.; Bai, J.; Wood, D.; Assis, O.B.G.; Ferreira, M.D.; Baldwin, E. Nano-and Micro-Carnauba Wax Emulsions versus Shellac Protective Coatings on Postharvest Citrus Quality. *J. Am. Soc. Hortic. Sci.* **2020**, *1*, 1–10. [CrossRef]
19. Burdock, G.A.; Carabin, I.G. Generally recognized as safe (GRAS): History and description. *Toxicol. Lett.* **2004**, *150*, 3–18. [CrossRef]
20. Riva, S.C.; Opara, U.O.; Fawole, O.A. Recent developments on postharvest application of edible coatings on stone fruit: A review. *Sci. Hortic.* **2020**, *262*, 109074. [CrossRef]
21. Thakur, R.; Pristijono, P.; Scarlett, C.J.; Bowyer, M.; Singh, S.; Vuong, Q.V. Starch-based films: Major factors affecting their properties. *Int. J. Biol. Macromol.* **2019**, *132*, 1079–1089. [CrossRef]
22. Mohamed, S.A.; El-Sakhawy, M.; El-Sakhawy, M.A.-M. Polysaccharides, protein and lipid-based natural edible films in food packaging: A review. *Carbohydr. Polym.* **2020**, *238*, 116178. [CrossRef]
23. Dehghani, S.; Hosseini, S.V.; Regenstein, J.M. Edible films and coatings in seafood preservation: A review. *Food Chem.* **2018**, *240*, 505–513. [CrossRef]
24. Eddin, A.S.; Ibrahim, S.A.; Tahergorabi, R. Egg quality and safety with an overview of edible coating application for egg preservation. *Food Chem.* **2019**, *296*, 29–39. [CrossRef] [PubMed]
25. Arnon-Rips, H.; Poverenov, E. Improving food products' quality and storability by using Layer by Layer edible coatings. *Trends Food Sci. Technol.* **2018**, *75*, 81–92. [CrossRef]
26. Galus, S.; Kadzińska, J. Food applications of emulsion-based edible films and coatings. *Trends Food Sci. Technol.* **2015**, *45*, 273–283. [CrossRef]
27. Dhumal, C.V.; Sarkar, P. Composite edible films and coatings from food-grade biopolymers. *J. Food Sci. Technol.* **2018**, *55*, 4369–4383. [CrossRef] [PubMed]
28. Miranda, M.; Gozalbo, A.M.; Sun, X.; Plotto, A.; Bai, J.; de Assis, O.; Ferreira, M.; Baldwin, E. Effect of mono and bilayer of carnauba wax based nano-emulsion and HPMC coatings on popst-harvest quality of 'redtainung' papaya. In Proceedings of the Embrapa Instrumentação-Artigo em anais de congresso (ALICE), São Carlos, Brazil, 3–5 December 2019.
29. De Oliveira Filho, J.G.; Bezerra, C.C.d.O.N.; Albiero, B.R.; Oldoni, F.C.A.; Miranda, M.; Egea, M.B.; de Azeredo, H.M.C.; Ferreira, M.D. New approach in the development of edible films: The use of carnauba wax micro-or nanoemulsions in arrowroot starch-based films. *Food Packag. Shelf Life* **2020**, *26*, 100589. [CrossRef]
30. Bodbodak, S.; Moshfeghifar, M. Advances in modified atmosphere packaging of fruits and vegetables. In *Eco-Friendly Technology for Postharvest Produce Quality*; Siddiqui, M.W., Ed.; Elsevier: Amsterdam, The Netherlands, 2016; pp. 127–183.
31. Nor, S.M.; Ding, P. Trends and advances in edible biopolymer coating for tropical fruit: A review. *Food Res. Int.* **2020**, *134*, 109208. [CrossRef]
32. Porat, R.; Fallik, E. Production of off-flavours in fruit and vegetables under fermentative conditions. In *Fruit and Vegetable Flavour*; Brückner, B., Wyllie, S.G., Eds.; Elsevier: Amsterdam, The Netherlands, 2008; pp. 150–164.
33. Baldwin, E.A.; Nisperos-Carriedo, M.; Shaw, P.E.; Burns, J.K. Effect of coatings and prolonged storage conditions on fresh orange flavor volatiles, degrees brix, and ascorbic acid levels. *J. Agric. Food Chem.* **1995**, *43*, 1321–1331. [CrossRef]
34. Quirós-Sauceda, A.E.; Ayala-Zavala, J.F.; Olivas, G.I.; González-Aguilar, G.A. Edible coatings as encapsulating matrices for bioactive compounds: A review. *J. Food Sci. Technol.* **2014**, *51*, 1674–1685. [CrossRef] [PubMed]
35. Sun, X.; Baldwin, E.; Ritenour, M.; Plotto, A.; Bai, J. Evaluation of natural colorants and their application on citrus fruit as alternatives to Citrus Red No. 2. *HortScience* **2015**, *50*, 1353–1357. [CrossRef]

36. Marin, A.; Sun, X.; Miranda, M.; Ference, C.; Baldwin, E.A.; Bai, J.; Ritenour, M.A.; Zhang, J.; Plotto, A. Optimizing Essential Oil Applications to Prevent Postharvest Decay in Strawberries. *Proc. Fla. State Hort. Soc* **2019**, *132*, 185–188.
37. Basaglia, R.R.; Pizato, S.; Santiago, N.G.; de Almeida, M.M.M.; Pinedo, R.A.; Cortez-Vega, W.R. Effect of edible chitosan and cinnamon essential oil coatings on the shelf life of minimally processed pineapple (Smooth cayenne). *Food Biosci.* **2021**, *41*, 100966. [CrossRef]
38. De Oliveira Filho, J.G.; da Cruz Silva, G.; de Aguiar, A.C.; Cipriano, L.; de Azeredo, H.M.C.; Junior, S.B.; Ferreira, M.D. Chemical composition and antifungal activity of essential oils and their combinations against Botrytis cinerea in strawberries. *J. Food Meas. Charact.* **2021**, *15*, 1815–1825. [CrossRef]
39. Tesfay, S.Z.; Magwaza, L.S.; Mbili, N.; Mditshwa, A. Carboxyl methylcellulose (CMC) containing moringa plant extracts as new postharvest organic edible coating for Avocado (Persea americana Mill.) fruit. *Sci. Hortic.* **2017**, *226*, 201–207. [CrossRef]
40. de Jesús Salas-Méndez, E.; Vicente, A.; Pinheiro, A.C.; Ballesteros, L.F.; Silva, P.; Rodríguez-García, R.; Hernández-Castillo, F.D.; Díaz-Jiménez, M.d.L.V.; Flores-López, M.L.; Villarreal-Quintanilla, J.Á. Application of edible nanolaminate coatings with antimicrobial extract of *Flourensia cernua* to extend the shelf-life of tomato (*Solanum lycopersicum* L.) fruit. *Postharvest Biol. Technol.* **2019**, *150*, 19–27. [CrossRef]
41. Marín, A.; Plotto, A.; Atarés, L.; Chiralt, A. Lactic acid bacteria incorporated into edible coatings to control fungal growth and maintain postharvest quality of grapes. *HortScience* **2019**, *54*, 337–343. [CrossRef]
42. Kharchoufi, S.; Parafati, L.; Licciardello, F.; Muratore, G.; Hamdi, M.; Cirvilleri, G.; Restuccia, C. Edible coatings incorporating pomegranate peel extract and biocontrol yeast to reduce *Penicillium digitatum* postharvest decay of oranges. *Food Microbiol.* **2018**, *74*, 107–112. [CrossRef]
43. Ghidelli, C.; Pérez-Gago, M. Recent advances in modified atmosphere packaging and edible coatings to maintain quality of fresh-cut fruits and vegetables. *Crit. Rev. Food Sci. Nutr.* **2018**, *58*, 662–679. [CrossRef] [PubMed]
44. Marín, A.; Baldwin, E.A.; Bai, J.; Wood, D.; Ference, C.; Sun, X.; Brecht, J.K.; Plotto, A. Edible coatings as carriers of antibrowning compounds to maintain appealing appearance of fresh-cut mango. *HortTechnology* **2021**, *31*, 27–35. [CrossRef]
45. Suhag, R.; Kumar, N.; Petkoska, A.T.; Upadhyay, A. Film formation and deposition methods of edible coating on food products: A review. *Food Res. Int.* **2020**, *136*, 109582. [CrossRef]
46. Senturk Parreidt, T.; Schmid, M.; Müller, K. Effect of dipping and vacuum impregnation coating techniques with alginate based coating on physical quality parameters of cantaloupe melon. *J. Food Sci.* **2018**, *83*, 929–936. [CrossRef] [PubMed]
47. Valdés, A.; Ramos, M.; Beltrán, A.; Jiménez, A.; Garrigós, M.C. State of the art of antimicrobial edible coatings for food packaging applications. *Coatings* **2017**, *7*, 56. [CrossRef]
48. Tahir, H.E.; Xiaobo, Z.; Mahunu, G.K.; Arslan, M.; Abdalhai, M.; Zhihua, L. Recent developments in gum edible coating applications for fruits and vegetables preservation: A review. *Carbohydr. Polym.* **2019**, *224*, 115141. [CrossRef] [PubMed]
49. Dhanapal, A.; Sasikala, P.; Rajamani, L.; Kavitha, V.; Yazhini, G.; Banu, M.S. Edible films from polysaccharides. *Food Sci. Qual. Manag.* **2012**, *3*, 9.
50. Atieno, L.; Owino, W.; Ateka, E.M.; Ambuko, J. Influence of coating application methods on the postharvest quality of cassava. *Int. J. Food Sci.* **2019**, *2019*, 2148914. [CrossRef] [PubMed]
51. Nile, S.H.; Baskar, V.; Selvaraj, D.; Nile, A.; Xiao, J.; Kai, G. Nanotechnologies in food science: Applications, recent trends, and future perspectives. *Nano-Micro Lett.* **2020**, *12*, 1–34. [CrossRef] [PubMed]
52. Zambrano-Zaragoza, M.L.; González-Reza, R.; Mendoza-Muñoz, N.; Miranda-Linares, V.; Bernal-Couoh, T.F.; Mendoza-Elvira, S.; Quintanar-Guerrero, D. Nanosystems in edible coatings: A novel strategy for food preservation. *Int. J. Mol. Sci.* **2018**, *19*, 705. [CrossRef]
53. Liu, R.; Liu, D.; Liu, Y.; Song, Y.; Wu, T.; Zhang, M. Using soy protein SiOx nanocomposite film coating to extend the shelf life of apple fruit. *Int. J. Food Sci. Technol.* **2017**, *52*, 2018–2030. [CrossRef]
54. McClements, D.J. Advances in edible nanoemulsions: Digestion, bioavailability, and potential toxicity. *Prog. Lipid Res.* **2020**, 101081.
55. Kumar, S.; Mukherjee, A.; Dutta, J. Chitosan based nanocomposite films and coatings: Emerging antimicrobial food packaging alternatives. *Trends Food Sci. Technol.* **2020**, *97*, 196–209. [CrossRef]
56. McClements, D.J. Edible nanoemulsions: Fabrication, properties, and functional performance. *Soft Matter* **2011**, *7*, 2297–2316. [CrossRef]
57. McClements, D.J.; Rao, J. Food-grade nanoemulsions: Formulation, fabrication, properties, performance, biological fate, and potential toxicity. *Crit. Rev. Food Sci. Nutr.* **2011**, *51*, 285–330. [CrossRef]
58. Zhang, R.; Zhang, Z.; McClements, D.J. Nanoemulsions: An emerging platform for increasing the efficacy of nutraceuticals in foods. *Colloids Surf. B Biointerfaces* **2020**, *194*, 111202. [CrossRef]
59. Salvia-Trujillo, L.; Rojas-Graü, M.A.; Soliva-Fortuny, R.; Martín-Belloso, O. Use of antimicrobial nanoemulsions as edible coatings: Impact on safety and quality attributes of fresh-cut Fuji apples. *Postharvest Biol. Technol.* **2015**, *105*, 8–16. [CrossRef]
60. Prakash, A.; Baskaran, R.; Vadivel, V. Citral nanoemulsion incorporated edible coating to extend the shelf life of fresh cut pineapples. *LWT* **2020**, *118*, 108851. [CrossRef]
61. Zhang, Z.; Zhang, R.; McClements, D.J. Encapsulation of β-carotene in alginate-based hydrogel beads: Impact on physicochemical stability and bioaccessibility. *Food Hydrocoll.* **2016**, *61*, 1–10. [CrossRef]
62. McClements, D.J. *Food Emulsions: Principles, Practices, and Techniques*; CRC Press: Boca Raton, FL, USA, 2015.
63. Queste, S.; Salager, J.; Strey, R.; Aubry, J. The EACN scale for oil classification revisited thanks to fish diagrams. *J. Colloid Interface Sci.* **2007**, *312*, 98–107. [CrossRef] [PubMed]
64. Blomberg, E.; Claesson, P.M. Surface forces and emulsion stability: Chapter 7. In *Food Emulsions*; Friberg, S.E., Larsson, K., Sjöblom, J., Eds.; CRC Press: New York, NY, USA, 2004; pp. 254–297.

65. Yao, K.; McClements, D.J.; Xiang, J.; Zhang, Z.; Cao, Y.; Xiao, H.; Liu, X. Improvement of carotenoid bioaccessibility from spinach by co-ingesting with excipient nanoemulsions: Impact of the oil phase composition. *Food Funct.* **2019**, *10*, 5302–5311. [CrossRef]
66. Salvia-Trujillo, L.; Martín-Belloso, O.; McClements, D.J. Excipient nanoemulsions for improving oral bioavailability of bioactives. *Nanomaterials* **2016**, *6*, 17. [CrossRef] [PubMed]
67. Tripathi, A.D.; Sharma, R.; Agarwal, A.; Haleem, D.R. Nanoemulsions based edible coatings with potential food applications. *Int. J. Biobased Plast.* **2021**, *3*, 112–125. [CrossRef]
68. Acevedo-Fani, A.; Soliva-Fortuny, R.; Martín-Belloso, O. Nanostructured emulsions and nanolaminates for delivery of active ingredients: Improving food safety and functionality. *Trends Food Sci. Technol.* **2017**, *60*, 12–22. [CrossRef]
69. Pilon, L.; Spricigo, P.C.; Miranda, M.; de Moura, M.R.; Assis, O.B.G.; Mattoso, L.H.C.; Ferreira, M.D. Chitosan nanoparticle coatings reduce microbial growth on fresh-cut apples while not affecting quality attributes. *Int. J. Food Sci. Technol.* **2015**, *50*, 440–448. [CrossRef]
70. Atarés, L.; Chiralt, A. Essential oils as additives in biodegradable films and coatings for active food packaging. *Trends Food Sci. Technol.* **2016**, *48*, 51–62. [CrossRef]
71. Maisanaba, S.; Llana-Ruiz-Cabello, M.; Gutiérrez-Praena, D.; Pichardo, S.; Puerto, M.; Prieto, A.; Jos, A.; Cameán, A. New advances in active packaging incorporated with essential oils or their main components for food preservation. *Food Rev. Int.* **2017**, *33*, 447–515. [CrossRef]
72. Sessa, M.; Ferrari, G.; Donsì, F. Novel edible coating containing essential oil nanoemulsions to prolong the shelf life of vegetable products. *Chem. Eng. Trans.* **2015**, *43*, 55–60.
73. Severino, R.; Ferrari, G.; Vu, K.D.; Donsì, F.; Salmieri, S.; Lacroix, M. Antimicrobial effects of modified chitosan based coating containing nanoemulsion of essential oils, modified atmosphere packaging and gamma irradiation against Escherichia coli O157: H7 and Salmonella Typhimurium on green beans. *Food Control* **2015**, *50*, 215–222. [CrossRef]
74. Gundewadi, G.; Rudra, S.G.; Sarkar, D.J.; Singh, D. Nanoemulsion based alginate organic coating for shelf life extension of okra. *Food Packag. Shelf Life* **2018**, *18*, 1–12. [CrossRef]
75. Chu, Y.; Gao, C.; Liu, X.; Zhang, N.; Xu, T.; Feng, X.; Yang, Y.; Shen, X.; Tang, X. Improvement of storage quality of strawberries by pullulan coatings incorporated with cinnamon essential oil nanoemulsion. *LWT* **2020**, *122*, 109054. [CrossRef]
76. Kim, I.H.; Lee, H.; Kim, J.E.; Song, K.B.; Lee, Y.S.; Chung, D.S.; Min, S.C. Plum coatings of lemongrass oil-incorporating carnauba wax-based nanoemulsion. *J. Food Sci.* **2013**, *78*, E1551–E1559. [CrossRef]
77. Kim, I.-H.; Oh, Y.A.; Lee, H.; Song, K.B.; Min, S.C. Grape berry coatings of lemongrass oil-incorporating nanoemulsion. *LWT—Food Sci. Technol.* **2014**, *58*, 1–10. [CrossRef]
78. De León-Zapata, M.A.; Pastrana-Castro, L.; Barbosa-Pereira, L.; Rua-Rodríguez, M.L.; Saucedo, S.; Ventura-Sobrevilla, J.M.; Salinas-Jasso, T.A.; Rodríguez-Herrera, R.; Aguilar, C.N. Nanocoating with extract of tarbush to retard Fuji apples senescence. *Postharvest Biol. Technol.* **2017**, *134*, 67–75. [CrossRef]
79. Robledo, N.; López, L.; Bunger, A.; Tapia, C.; Abugoch, L. Effects of antimicrobial edible coating of thymol nanoemulsion/quinoa protein/chitosan on the safety, sensorial properties, and quality of refrigerated strawberries (*Fragaria* × *ananassa*) under commercial storage environment. *Food Bioprocess Technol.* **2018**, *11*, 1566–1574. [CrossRef]
80. Deng, Z.; Jung, J.; Simonsen, J.; Wang, Y.; Zhao, Y. Cellulose nanocrystal reinforced chitosan coatings for improving the storability of postharvest pears under both ambient and cold storages. *J. Food Sci.* **2017**, *82*, 453–462. [CrossRef]
81. Kumar, R. Health effects of morpholine based coating for fruits and vegetables. *Int. J. Med. Res. Health Sci.* **2016**, *5*, 32–38.
82. Kumar, R.; Kapur, S. Morpholine: A glazing agent for fruits and vegetables coating/waxing. *Int. J. Sci. Technol. Eng.* **2016**, *2*, 694–697.
83. Ohnishi, T. Studies on mutagenicity of the food additive morpholine (fatty acid salt). *Nippon Eiseigaku Zasshi Jpn. J. Hyg.* **1984**, *39*, 729–748. [CrossRef]
84. Mirvish, S.; Salmasi, S.; Cohen, S.; Patil, K.; Mahboubi, E. Liver and forestomach tumors and other forestomach lesions in rats treated with morpholine and sodium nitrite, with and without sodium ascorbate. *J. Natl. Cancer Inst.* **1983**, *71*, 81–85.
85. Health Canada. ARCHIVED—A Summary of the Health Hazard Assessment of Morpholine in Wax Coatings of Apples. Available online: https://www.canada.ca/en/health-canada/services/food-nutrition/food-safety/information-product/summary-health-hazard-assessment-morpholine-coatings-apples.html (accessed on 29 September 2021).
86. Hagenmaier, R.D.; Baker, R.A. Edible coatings from morpholine-free wax microemulsions. *J. Agric. Food Chem.* **1997**, *45*, 349–352. [CrossRef]
87. Ncama, K.; Magwaza, L.; Mditshwa, A.; Tesfay, S. Plant-based edible coatings for managing postharvest quality of fresh horticultural produce: A review. *Food Packag. Shelf Life* **2018**, *16*, 157–167. [CrossRef]
88. Donsì, F.; Annunziata, M.; Sessa, M.; Ferrari, G. Nanoencapsulation of essential oils to enhance their antimicrobial activity in foods. *LWT-Food Sci. Technol.* **2011**, *44*, 1908–1914. [CrossRef]
89. De Oliveira Filho, J.; Albiero, B.; Cipriano, L.; Bezerra, C.d.O.; Oldoni, F.; Egea, M.; de Azeredo, H.; Ferreira, M. Arrowroot starch-based films incorporated with a carnauba wax nanoemulsion, cellulose nanocrystals, and essential oils: A new functional material for food packaging applications. *Cellulose* **2021**, *28*, 6499–6511. [CrossRef]
90. Qi, W.; Li, T.; Zhang, Z.; Wu, T. Preparation and characterization of oleogel-in-water pickering emulsions stabilized by cellulose nanocrystals. *Food Hydrocoll.* **2021**, *110*, 106206. [CrossRef]
91. Chutia, H.; Mahanta, C. Properties of starch nanoparticle obtained by ultrasonication and high pressure homogenization for developing carotenoids-enriched powder and Pickering nanoemulsion. *Innov. Food Sci. Emerg. Technol.* **2021**, *74*, 102822. [CrossRef]

92. Nandy, M.; Lahiri, B.; Philip, J. Inter-droplet force between magnetically polarizable Pickering oil-in-water nanoemulsions stabilized with γ-Al2O3 nanoparticles: Role of electrostatic and electric dipolar interactions. *J. Colloid Interface Sci.* **2021**, *607*, 1671–1686. [CrossRef]
93. Xiao, Z.; Liu, Y.; Niu, Y.; Kou, X. Cyclodextrin supermolecules as excellent stabilizers for Pickering nanoemulsions. *Colloids Surf. A Physicochem. Eng. Asp.* **2020**, *588*, 124367. [CrossRef]
94. Bao, Y.; Zhang, Y.; Liu, P.; Ma, J.; Zhang, W.; Liu, C.; Simion, D. Novel fabrication of stable Pickering emulsion and latex by hollow silica nanoparticles. *J. Colloid Interface Sci.* **2019**, *553*, 83–90. [CrossRef] [PubMed]
95. Almasi, H.; Azizi, S.; Amjadi, S. Development and characterization of pectin films activated by nanoemulsion and Pickering emulsion stabilized marjoram (Origanum majorana L.) essential oil. *Food Hydrocoll.* **2020**, *99*, 105338. [CrossRef]
96. López-Monterrubio, D.; Lobato-Calleros, C.; Vernon-Carter, E.; Alvarez-Ramirez, J. Influence of β-carotene concentration on the physicochemical properties, degradation and antioxidant activity of nanoemulsions stabilized by whey protein hydrolyzate-pectin soluble complexes. *LWT* **2021**, *143*, 111148. [CrossRef]
97. Wani, T.; Masoodi, F.; Jafari, S.M.; McClements, D. Safety of nanoemulsions and their regulatory status. In *Nanoemulsions*; Jafari, S.M., McClements, D.J., Eds.; Elsevier: Amsterdam, The Netherlands, 2018; pp. 613–628.
98. McClements, D.; Xiao, H. Potential biological fate of ingested nanoemulsions: Influence of particle characteristics. *Food Funct.* **2012**, *3*, 202–220. [CrossRef]
99. Kaur, K.; Kumar, R.; Goel, S.; Uppal, S.; Bhatia, A.; Mehta, S. Physiochemical and cytotoxicity study of TPGS stabilized nanoemulsion designed by ultrasonication method. *Ultrason. Sonochem.* **2017**, *34*, 173–182. [CrossRef]
100. Marchese, E.; D'onofrio, N.; Balestrieri, M.; Castaldo, D.; Ferrari, G.; Donsì, F. Bergamot essential oil nanoemulsions: Antimicrobial and cytotoxic activity. *Z. Nat. C* **2020**, *75*, 279–290. [CrossRef] [PubMed]
101. Hort, M.; Alves, B.d.S.; Ramires Junior, O.; Falkembach, M.; Araújo, G.d.M.; Fernandes, C.; Tavella, R.; Bidone, J.; Dora, C.; da Silva Júnior, F. In vivo toxicity evaluation of nanoemulsions for drug delivery. *Drug Chem. Toxicol.* **2021**, *44*, 585–594. [CrossRef]
102. Donsì, F.; Ferrari, G. Essential oil nanoemulsions as antimicrobial agents in food. *J. Biotechnol.* **2016**, *233*, 106–120. [CrossRef] [PubMed]
103. Amenta, V.; Aschberger, K.; Arena, M.; Bouwmeester, H.; Moniz, F.; Brandhoff, P.; Gottardo, S.; Marvin, H.; Mech, A.; Pesudo, L. Regulatory aspects of nanotechnology in the agri/feed/food sector in EU and non-EU countries. *Regul. Toxicol. Pharmacol.* **2015**, *73*, 463–476. [CrossRef] [PubMed]

MDPI
St. Alban-Anlage 66
4052 Basel
Switzerland
Tel. +41 61 683 77 34
Fax +41 61 302 89 18
www.mdpi.com

Foods Editorial Office
E-mail: foods@mdpi.com
www.mdpi.com/journal/foods

www.ingramcontent.com/pod-product-compliance
Lightning Source LLC
LaVergne TN
LVHW070141170726
843515LV00007B/1843

* 9 7 8 3 0 3 6 5 3 5 3 5 7 *